T0092622

Springer Series in Geomechanics and Geoengineering

Editors

Prof. Wei Wu
Institut für Geotechnik
Universität für Bodenkultur
Feistmantelstraße 4
1180 Vienna
Austria
E-mail: wei.wu@boku.ac.at

Prof. Ronaldo I. Borja
Department of Civil and Environmental Engineering
Stanford University
Stanford, CA 94305-4020
USA
E-mail: borja@stanford.edu

For further volumes:
http://www.springer.com/series/8069

Jacek Tejchman and Jerzy Bobiński

Continuous and Discontinuous Modelling of Fracture in Concrete Using FEM

Authors
Jacek Tejchman
Faculty of Civil and
Environmental Engineering
Gdansk University of Technology
Poland

Jerzy Bobiński
Faculty of Civil and
Environmental Engineering
Gdansk University of Technology
Poland

ISSN 1866-8755
e-ISSN 1866-8763
ISBN 978-3-642-28462-5
e-ISBN 978-3-642-28463-2
DOI 10.1007/978-3-642-28463-2
Springer Heidelberg New York Dordrecht London

Library of Congress Control Number: 2012933079

Printed on acid-free paper

Springer is part of Springer Science+Business Media (www.springer.com)

Contents

Chapter 1
Introduction

Abstract. In this chapter, a process of concrete fracture is briefly described. Methods and aspects of numerical modelling of concrete fracture using the finite element method are discussed. The outline of the book is given.

Fracture process is a fundamental phenomenon in quasi-brittle materials like concrete (Bažant and Planas 1998). It is a major reason of their damage under mechanical loading contributing to a significant degradation of the material strength which may lead to a total loss of load-bearing capacity. During a fracture process, micro-cracks first arise which change gradually into dominant macroscopic discrete cracks up to rupture. Thus, a fracture process is subdivided in general into 2 main stages: appearance of narrow regions of intense deformation (including micro-cracks) and occurrence of macro-cracks. Within continuum mechanics, strain localization can be numerically captured by a continuous approach and discrete macro-cracks can be modelled by a discontinuous one. Usually, to describe the fracture behaviour of concrete, one type approach is used.

Localization of deformation can occur as tensile zones, shear zones or mixed tensile-shear zones. Localized zones have a certain volume being not negligible as compared to the specimen size. Thus, an understanding of the mechanism of the formation of localized zones is of a crucial importance since they influence the bearing capacity of the specimen and act as a precursor of macro-cracks. The phase of localization of deformation has to be modelled in a physically consistent and mathematically correct manner. Classical FE-analyses within a continuum mechanics are not able to describe properly both the thickness of localized zones and distance between them since they do not include a characteristic length of micro-structure (Tejchman 1989, 2008). Thus, they suffer from a pathological mesh-dependency on the fineness of the spatial discretisation (its size and orientation), because differential equations of motion change their type (from elliptic to hyperbolic during quasi-static calculations) and the rate boundary value problem becomes mathematically ill-posed (Mühlhaus 1986, de Borst et al. 1992, 1993). Deformations tend to localize in a zero thickness zone in an analytical analysis and in one element wide region in FE calculations (the strain-softening domain is limited to a point and no energy dissipation takes place). As a result,

J. Tejchman, J. Bobiński: Continuous & Discontinuous Modelling of Fracture, SSGG, pp. 1–4.
springerlink.com

computed load-displacement curves are severely mesh-dependent (in particular, in a post-peak regime). Thus, classical constitutive laws require an extension by a characteristic length of micro-structure to accurately capture the formation of strain localization, to obtain an objective solution upon mesh refinement, to take into account microscopic inhomogeneities triggering strain localization (e.g. aggregate size, cement grain size) and to describe an energetic (deterministic) size effect. In turn, macro-cracks can be captured as a jump in a continuum field by means of discontinuous methods including e.g. cohesive elements (interfaces) defined along finite element edges (Ortiz and Pandolfi 1999, Gálvez et al. 2002, Zhou and Molinari 2004) or strong discontinuities using elemental or nodal enrichments wherein cracks can arbitrarily propagate through finite elements (Belytschko et al. 1988, Jirásek 2000, Belytschko et al. 2001, Simone and Sluys 2004, Oliver et al. 2006), which offers more flexibility for the crack path than interface elements. In un-cracked material, a linear elastic constitutive law is usually assumed under tensile loading. To activate a crack, a criterion for crack opening/growth is introduced. Continuous and discontinuous approaches can be used both at macro-level (concrete is described as one-phase material) and at micro-level (concrete is described as a three-phase material composed of aggregate, cement and interfacial transitional zones).

The main intention of the book is to analyze a quasi-static fracture process in the form of localization of deformation in plain concrete and reinforced concrete at macro- and meso-level by means of enhanced continuum constitutive models formulated within continuum mechanics. The analyses were performed for monotonic and cyclic loading using a finite element method. In the case of monotonic loading, three different popular continuum models: isotropic elasto-plastic, isotropic damage and smeared crack one were used. For cyclic loading, four various coupled elasto-plastic-damage formulations were applied. Constitutive models were equipped with a characteristic length of micro-structure with the aid of a non-local or a second-gradient theory. So they could describe both: the formation of localized zones with a certain thickness and spacing and a related deterministic size effect (Pamin 1994, Bažant and Jirásek 2002). FE-contours of localized zones converged to a finite size upon mesh refinement and boundary value problems became mathematically well-posed at the onset of strain localization. In addition, mesh-objective FE results of fracture in the form of macro-cracks in plain concrete under quasi-static loading using a discontinuous continuum approach with cohesive (interface) elements and XFEM were presented.

2D finite element analyses were performed with plain concrete elements under monotonic loading using continuous approaches (uniaxial compression, uniaxial tension, tensile bending and mixed shear-extension) and discontinuous approaches (uniaxial tension, tensile bending and mixed shear-extension). Plain concrete elements under cyclic loading were simulated using coupled elasto-plastic-damage formulations. In the case of reinforced concrete specimens, 2D and 3D FE calculations were carried out with bars, slender and short beams, columns, corbels and tanks under monotonic loading within enhanced both elasto-plasticity and damage mechanics. Tensile and shear failure mechanisms were studied. Attention was paid to the width and spacing of localized zones. A stochastic and deterministic size effect was carefully analyzed in plain concrete by taking into account strain localization and random spatially correlated tensile strength. They were compared with existing size effect laws. Meso-scale FE calculations of strain localization were also carried out with plain concrete (described as a three-phase material) under monotonic loading during uniaxial tension and bending. Numerical results with respect to the load-displacement diagram and geometry of localized zones were compared with corresponding both laboratory and large-scale tests from the scientific literature and some own tests.

The book includes 10 Chapters. After a short introduction in Chapter 1, Chapter 2 summarizes the most important properties of concrete and reinforced concrete. In Chapter 3, enhanced continuum constitutive models used for FE analyses are described. Discontinuous approaches with cohesive zones and XFEM are outlined in Chapter 4. Results of two-dimensional FE modelling of strain localization in concrete elements for monotonic quasi-static loading (uniaxial tension, uniaxial compression, bending and mixed shear-tension mode) and cyclic quasi-static loading (bending) are shown in Chapters 5 and 6, respectively. In Chapter 7, 2D and 3D results of FE modelling of strain localization in reinforced concrete elements and structures (bars, slender and short beams, columns, corbels and tanks) for monotonic quasi-static loading are demonstrated. Results of a deterministic and statistical size effect in plain concrete are included in Chapter 8. Chapter 9 shows results of mesoscopic modelling of strain localization in plain concrete during unaxial tension and three-point bending. Finally, general conclusions from the research and future research directions are enclosed (Chapter 10).

References

Bažant, Z., Planas, J.: Fracture and size effect in concrete and other quasi-brittle materials. CRC Press LLC (1998)

Bažant, Z., Jirásek, M.: Nonlocal integral formulations of plasticity and damage: survey of progress. Journal of Engineering Mechanics 128(11), 1119–1149 (2002)

Belytschko, T., Fish, J., Englemann, B.E.: A finite element method with embedded localization zones. Computer Methods in Applied Mechanics 70(1), 59–89 (1988)

Belytschko, T., Moes, N., Usui, S., Parimi, C.: Arbitrary discontinuities in finite elements. International Journal for Numerical Methods in Engineering 50(4), 993–1013 (2001)

de Borst, R., Mühlhaus, H.-B., Pamin, J., Sluys, L.J.: Computational modelling of localization of deformation. In: Owen, D.R.J., Onate, H., Hinton, E. (eds.) Proceedings of the 3rd International Conference on Computational Plasticity, Swansea, pp. 483–508. Pineridge Press (1992)

de Borst, R., Sluys, L.J., Mühlhaus, H.-B., Pamin, J.: Fundamental issues in finite element analyses of localization of deformation. Engineering Computations 10(2), 99–121 (1993)

Gálvez, J.C., Červenka, J., Cendón, D.A., Saouma, V.: A discrete crack approach to normal/shear cracking of concrete. Cement and Concrete Research 32(10), 1567–1585 (2002)

Jirásek, M.: Comparative study on finite element with embedded discontinuities. Computer Methods in Applied Mechanics 188(1-3), 307–330 (2000)

Mühlhaus, H.-B.: Scherfugenanalyse bei granularem Material im Rahmen der Cosserat-Theorie. Ingen. Archiv. 56(5), 389–399 (1986)

Oliver, J., Huespe, A.E., Sanchez, P.J.: A comparative study on finite elements for capturing strong discontinuities: E-FEM vs X-FEM. Computer Methods in Applied Mechanics and Engineering 195(37-40), 4732–4752 (2006)

Ortiz, M., Pandolfi, A.: Finite-deformation irreversible cohesive elements for three-dimensional crack-propagation analysis. International Journal for Numerical Methods in Engineering 44(9), 1267–1282 (1999)

Pamin, J.: Gradient-dependent plasticity in numerical simulation of localization phenomena. PhD Thesis, University of Delft (1994)

Simone, A., Sluys, L.J.: The use of displacement discontinuities in a rate-dependent medium. Computer Methods in Applied Mechanics and Engineering 193(27-29), 3015–3033 (2004)

Tejchman, J.: Scherzonenbildung und Verspannugseffekte in Granulaten unter Berücksichtigung von Korndrehungen. Veröffentlichung des Instituts für Boden- und Felsmechanik der Universität Karlsruhe 117, 1–236 (1989)

Tejchman, J.: FE modeling of shear localization in granular bodies with micro-polar hypoplasticity. In: Wu, W., Borja, R. (eds.) Springer Series in Geomechanics and Geoengineering. Springer, Heidelberg (2008)

Zhou, F., Molinari, J.F.: Dynamic crack propagation with cohesive elements: a methodology to address mesh dependency. International Journal for Numerical Methods in Engineering 59(1), 1–24 (2004)

Chapter 2
Concrete and Reinforced Concrete Behaviour

Abstract. This chapter describes briefly the most important mechanical properties of concrete, reinforcement and reinforced concrete elements in a static and dynamic regime. In addition, bond-slip between reinforcement and concrete is discussed. Attention is laid on a size effect in concrete and reinforced concrete elements.

Concrete is still the most widely used construction material in terms of volume since it has the lowest ratio between cost and strength as compared to other available materials. It is a composite phase-material consisting mainly of aggregate, cement matrix and voids containing water or air (Fig. 2.1a). As a consequence, the concrete structure is strongly heterogeneous (Fig. 2.1b). Concrete properties depend strongly on the cement and aggregate quality and the ratio between cement and water. Classical concretes are divided into 3 groups depending upon the volumetric weight γ: heavy concretes γ=28-50 kN/m^3, normal concretes γ=20-28 kN/m^3 and light concretes γ=12-20 kN/m^3. Plain concrete is a brittle or quasi-brittle material, i.e. its bearing capacity strongly and rapidly falls down during compression and tension, and localized zones and macro-cracks are created. Its behaviour is mainly non-linear (a linear behaviour is limited to a very small range of deformation). Concrete has two undesirable properties, namely: low tensile strength and large brittleness (low energy absorption capacity) that cause collapse to occur shortly after the formation of the first crack. Therefore, the application of concrete subjected to impact, earth-quaking and fatigue loading is strongly limited. To improve these two negative properties, reinforcement in the form of bars and stirrups is mainly used.

Uniaxial Compression of Concrete

Typical curves for concrete during uniaxial compression are shown in Fig. 2.2. The material elastically behaves up to 30% of its compressive strength. Above this point, the behaviour starts to be non-linear. In the vicinity of the peak on the stress-strain curve, damage begins and the curve falls down until complete failure is reached. The cracks are parallel to the loading direction. During compression, first the volume decreases linearly then non-linearly reaching its minimum at the point M of Fig. 2.2. Next, the material is subjected to dilatancy due to formation and propagation of cracks. The Poisson's ratio remains constant (0.15-0.22) up to $0.8f_c$ (f_c – uniaxial compressive strength). At failure, it may even exceed 0.5 (Fig. 2.3). The compressive

J. Tejchman, J. Bobiński: Continuous & Discontinuous Modelling of Fracture, SSGG, pp. 5–47.
springerlink.com

strength depends on the specimen size and shape (Fig. 2.4). It decreases with increasing slenderness ratio *h/b* (or *h/d*) (*h* – specimen height, *b* – specimen width, *d* – specimen diameter). It is larger for plates than for prisms and cylinders.

The higher is the compressive strength of concrete, the larger is the material brittleness (ratio between the energy consumed after and before the stress-strain peak) (Fig. 2.5).

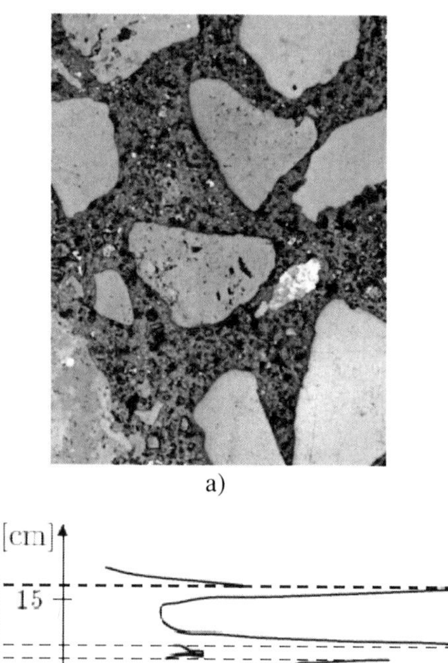

a)

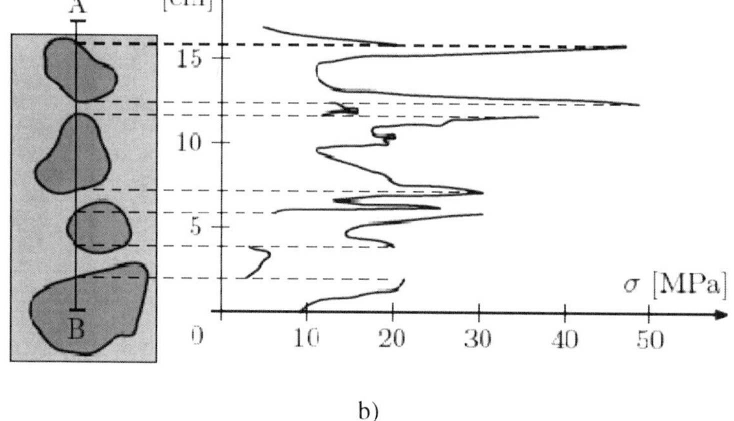

b)

Fig. 2.1 Concrete: a) non-uniform phase-structure, b) stress distribution in concrete subjected to compression by Dantu (1958) (Godycki-Ćwirko 1982, Klisiński and Mróz 1988)

Uniaxial Tension of Concrete

A typical stress-displacement diagram is shown in Fig. 2.6. The tensile strength of concrete is about 10 times lower that its compressive strength. A linear behaviour of concrete takes place up to 60% of the tensile strength f_t. A micro-crack is perpendicular to the loading direction. The material undergoes dilatancy only.

Influence of Strain Rate on Concrete

The concrete strength increases when strain rate increases due to confinement by inertial lateral restraint, shorter time for cracks to be created and viscosity of free water (Rossi 1991, Zheng and Li 2004) (Figs. 2.7 and 2.8). The material brittleness decreases with increasing strain rate. During impact loading, the compressive strength also increases; however, the material behaviour after the peak can be very different (Figs. 2.9 and 2.10). The strength of concrete after drying is not sensitive to loading rate.

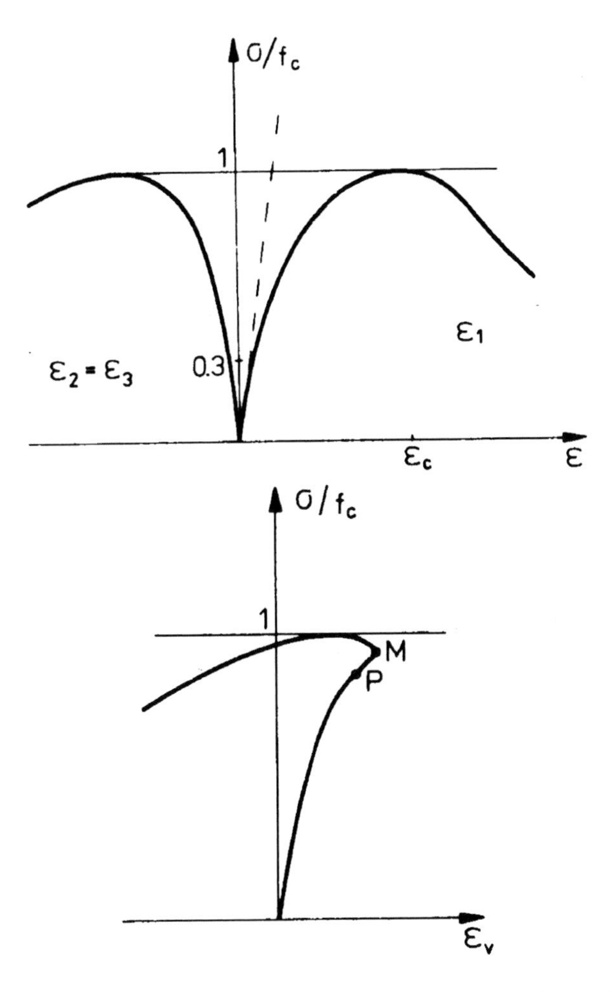

Fig. 2.2 Typical stress-strain diagram and volume changes during uniaxial vertical compression for concrete (σ - compressive normal stress, ε_1 – vertical normal strain, $\varepsilon_{2,3}$ – horizontal normal strains, ε_v – volumetric strain, f_c – uniaxial compressive strength) (Klisiński and Mróz 1988)

Biaxial Tests of Concrete
The stress-strain curves are shown in Fig. 2.11 and limit curves are depicted in Fig. 2.12. In addition, Fig. 2.13 presents volume changes. The largest increase of compressive strength during biaxial compression (by about 25%) is achieved for the ratio of the principal stresses equal to 0.6. If this ratio is equal to 1, the strength increase is about 16%.

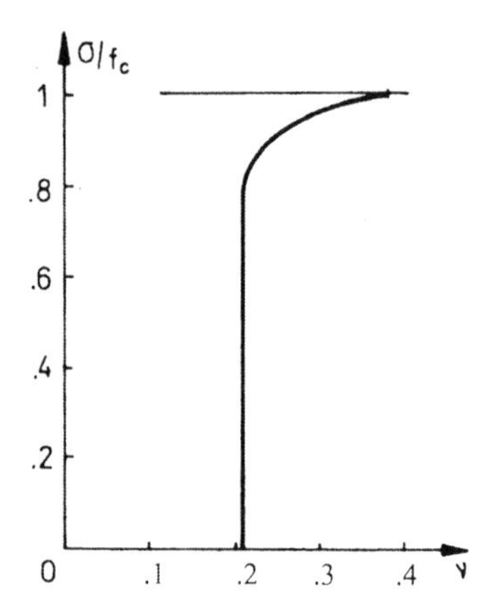

Fig. 2.3 Typical change of Poisson's ratio ν with normalized compressive stress σ/f_c during uniaxial compression (Klisiński and Mróz 1988)

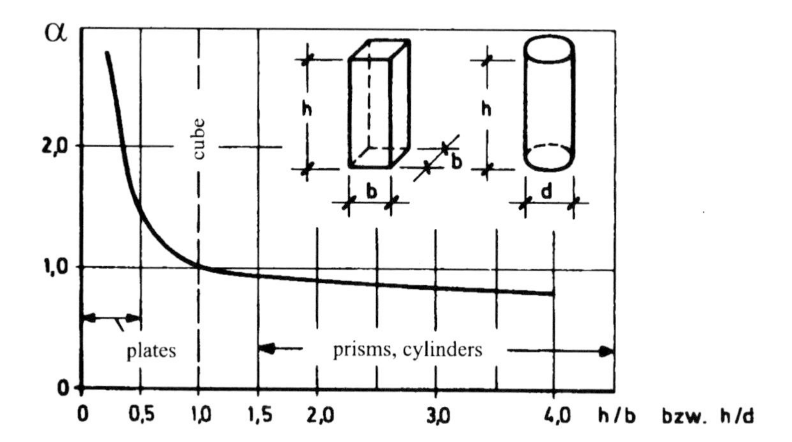

Fig. 2.4 Relationship between compressive strength of concrete prisms and compressive strength of concrete cubes α against slenderness ratio of concrete specimen h/d or h/b (Leonhard 1973)

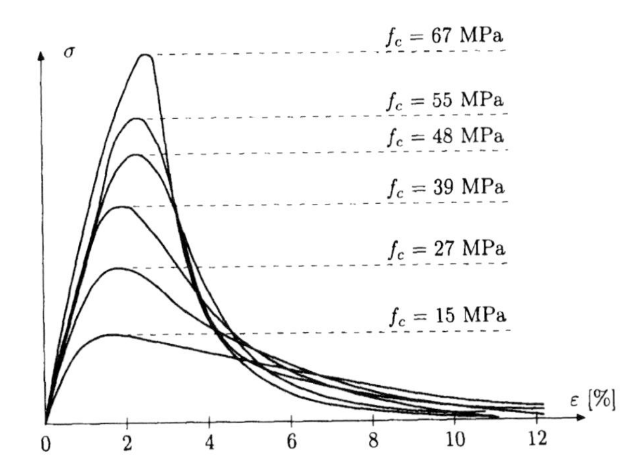

Fig. 2.5 Effect of uniaxial compression strength f_c on stress-strain relationship σ-ε (Klisiński and Mróz 1988)

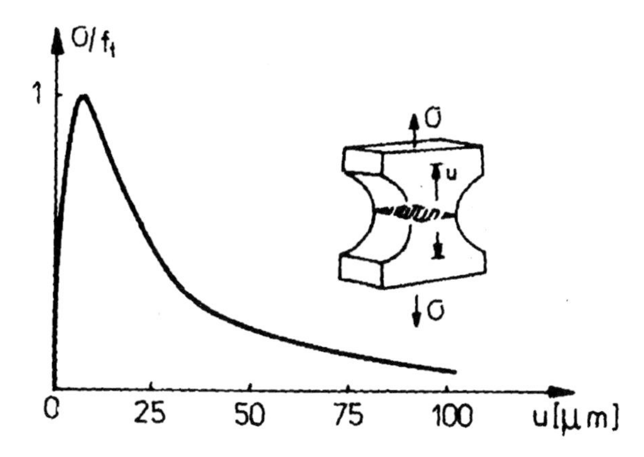

Fig. 2.6 Typical normalized stress-displacement curve during uniaxial extension for concrete (σ - vertical normal stress, u – vertical crack displacement, f_t - uniaxial tensile strength) (Klisiński and Mróz 1988)

Triaxial Tests of Concrete

Figure 2.14 demonstrates the results of monotonic usual triaxial experiments with concrete. The cylindrical specimens were initially loaded under hydrostatic confining pressure till the required value was reached. After that, horizontal confining pressure was kept constant and the specimen was subjected to increasing (Fig. 2.14) or decreasing vertical loading (Fig. 2.15). Concrete strength evidently increases with increasing confining pressure. Fig. 2.16 shows the material behaviour during hydrostatic loading, where fast material hardening is noticeable. Fig. 2.17 shows a limit curve in plane of the octahedral and hydrostatic stress for different confining pressures, which is parabolic. In turn, Fig. 2.18 depicts the results of cyclic true triaxial tests with rectangular prismatic specimens loaded in three orthogonal directions performed by Scavuzzo et al. (1983).

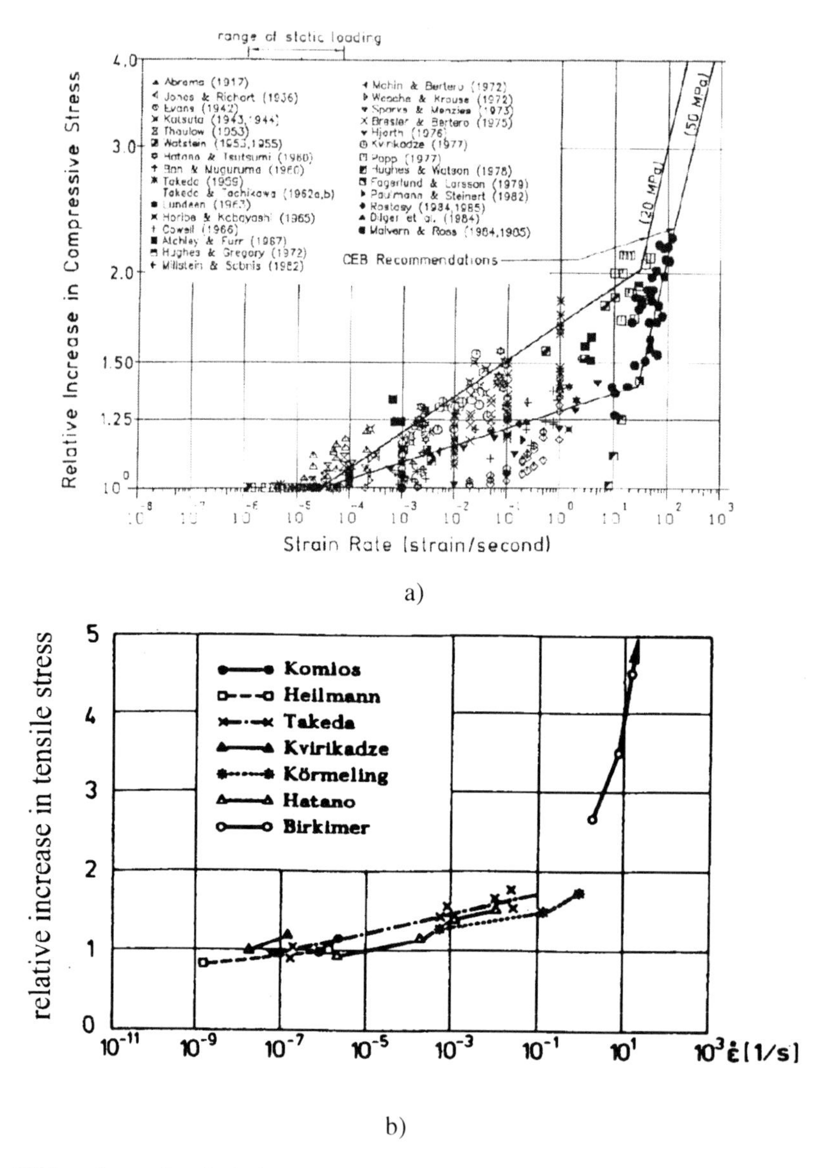

Fig. 2.7 Log-log scale of relative compressive strength increase versus strain rate (a) and relative tensile strength increase versus strain rate (b) (Bischoff and Perry 1991)

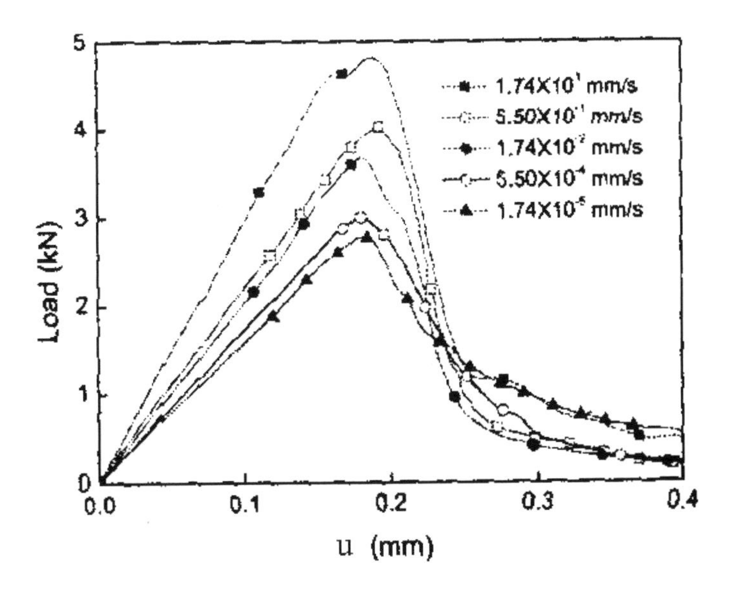

Fig. 2.8 Effect of loading velocity on concrete behaviour: load-displacement diagram during three-point bending (Zhang et al. 2009).

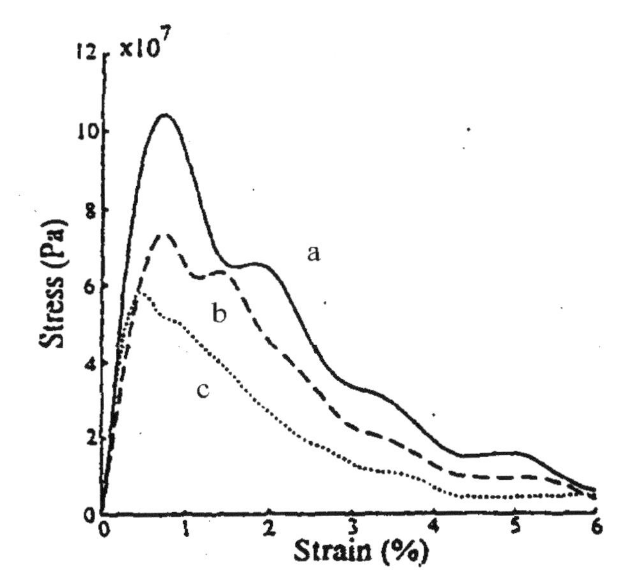

Fig. 2.9 Effect of strain rate on evolution of compressive normal stress versus normal strain: a) $d\varepsilon/dt$=700 s^{-1}, b) $d\varepsilon/dt$=500 s^{-1}, c) $d\varepsilon/dt$=300 s^{-1} (Gary 1990).

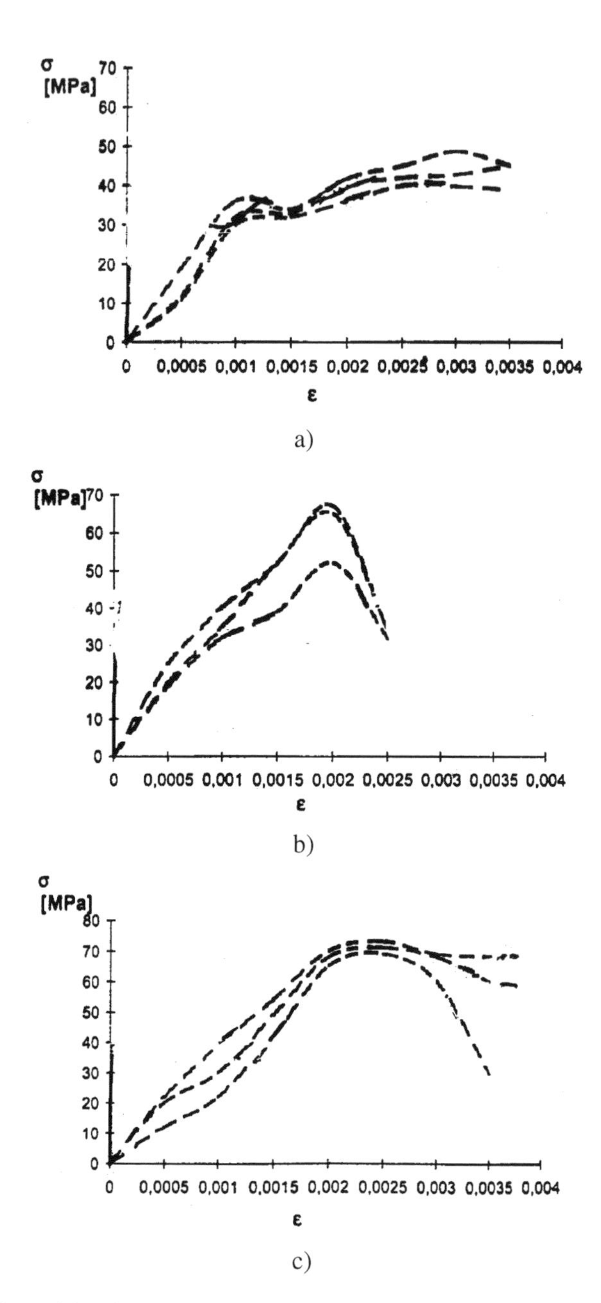

Fig. 2.10 Effect of loading velocity on stress-strain curve during impact loading (a-c): a) concrete 30 MPa, impactor mass 31.6 kg, impactor velocity 8.2 m/s, $d\varepsilon/dt$=9.0 1/s, b) concrete 50 MPa, impactor mass 31.6 kg, impactor velocity 5.3 m/s, $d\varepsilon/dt$=5.2 1/s, c) concrete 50 MPa, impactor mass 78.3 kg, impactor velocity 5.3 m/s, $d\varepsilon/dt$=5.6 1/s (Bischoff and Perry 1995)

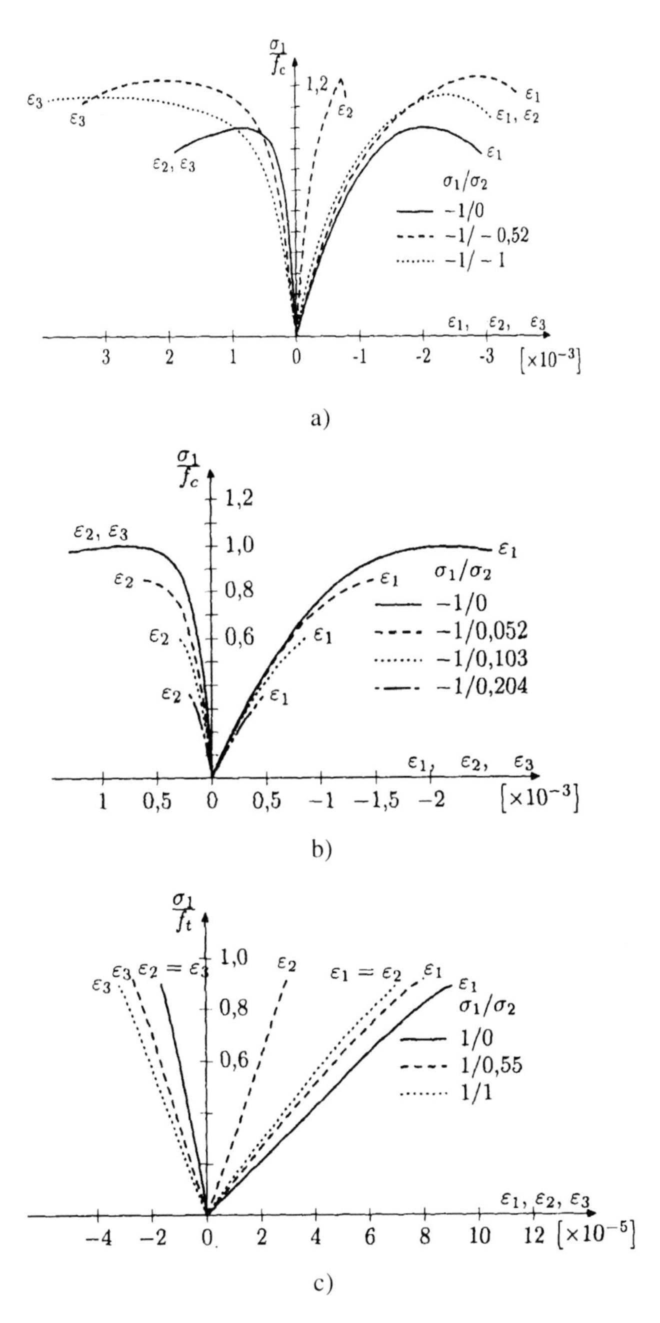

Fig. 2.11 Stress-strain curves from experiments by Kupfer et al. (1969): a) biaxial compression, b) axial compression and axial tension, c) biaxial extension (Klisiński and Mróz 1988)

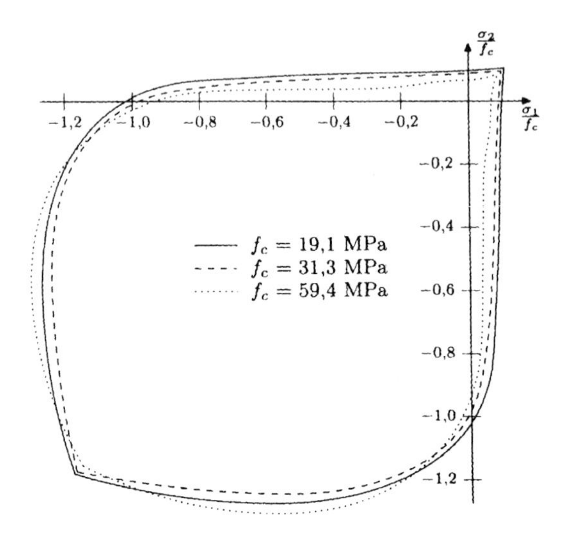

Fig. 2.12 Maximum stresses in plane stress state in biaxial tests by Kupfer et al. (1969) (Klisiński and Mróz 1988)

An approximate shape of a failure surface for concrete in the space of principal stresses based on experiments is shown in Fig. 2.19. The failure surface of concrete is symmetric against the hydrostatic line $\sigma_1=\sigma_2=\sigma_3$. The shape of the surface in a principal stress space is paraboloidal (Figs. 2.17 and 2.19). In deviatoric planes, the surface shape is approximately circular (during compression) and approximately elliptic (during tension); thus it changes from a curvilinear triangle with smoothly rounded corners to nearly circular with increasing pressure.

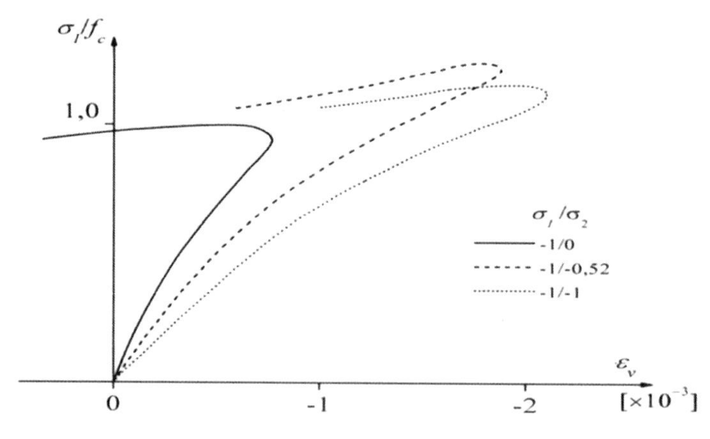

Fig. 2.13 Volume changes during biaxial tests by Kupfer et al. (1969) (Klisiński and Mróz 1988)

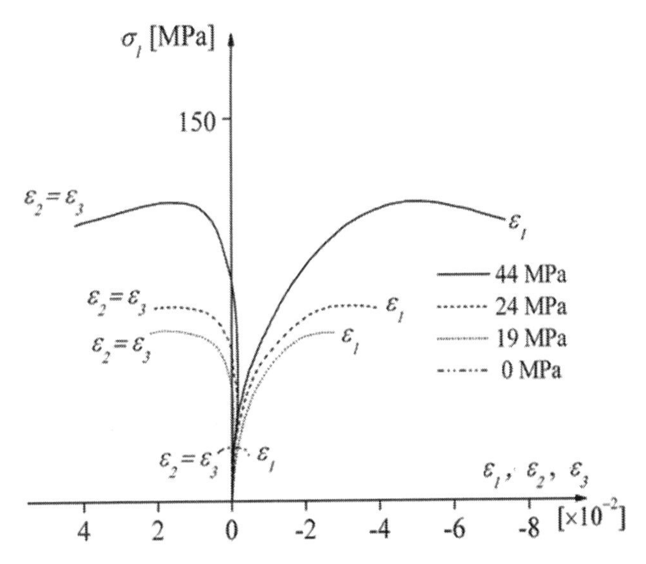

Fig. 2.14 Stress-strain curve during usual triaxial compression for different confining pressures and increasing vertical load by Kostovos and Newman (1978) and Kostovos (1980)

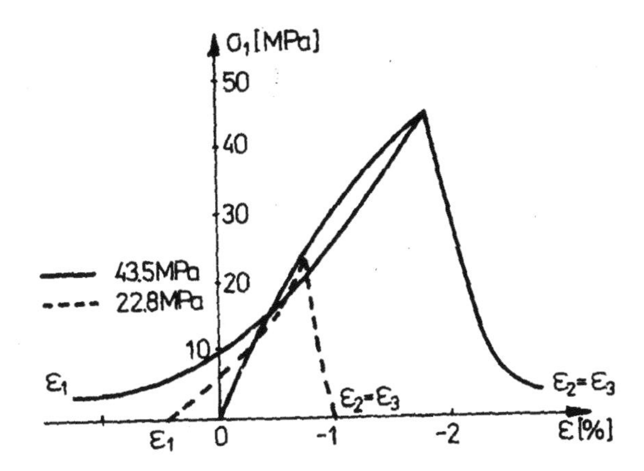

Fig. 2.15 Stress-strain curve during usual triaxial compression for different confining pressures and decreasing vertical load by Kostovos and Newman (1978) and Kotsovos (1980)

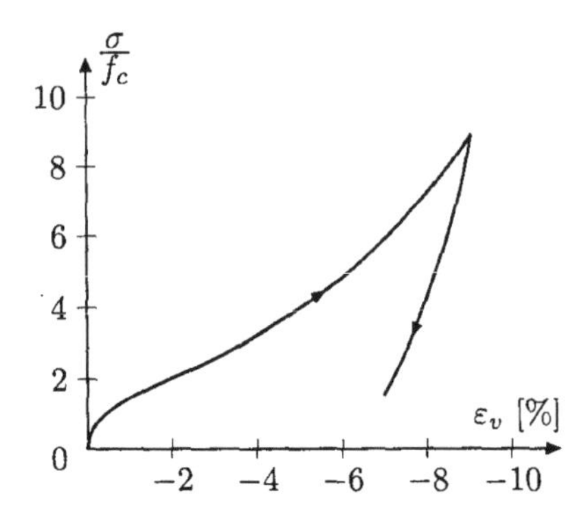

Fig. 2.16 Stress-volume strain during hydrostatic compression (Klisiński and Mróz 1998)

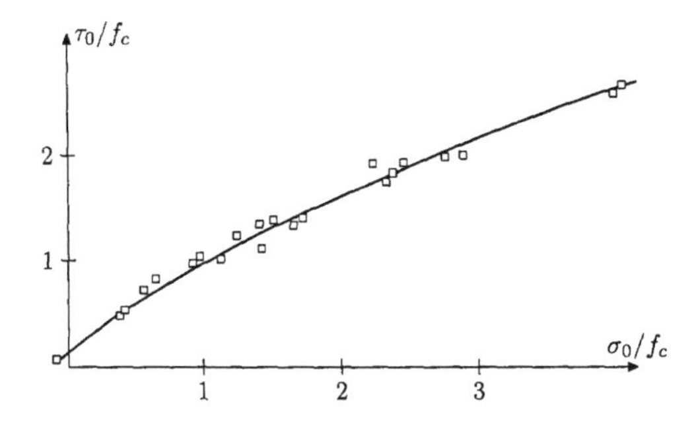

Fig. 2.17 Failure surface in plane τ_o-σ_o from usual triaxial experiments by Kotsovos (1980) (τ_o - octahedral stress, σ_o – hydrostatic stress, f_c - compressive strength) (Klisiński and Mróz 1988)

Cyclic Behaviour of Concrete

During quasi-static cyclic loading, concrete shows pronounced stiffness degradation due to fracture during both compression, tension and bending (Fig. 2.20). A hysteresis loop occurs during each cycle whose shape depends upon loading type.

Reinforcement of Concrete

The reinforcement co-operates with concrete during loading by carrying principal tensile and compressive stresses. A stress-strain diagram for different classes of reinforcement steel during uniaxial tension is shown in Fig. 2.21. A plastic region decreases with increasing tensile strength. During cyclic loading, compressive strength slightly decreases and tensile strength slightly increases (the so-called Bauschinger's effect) (Fig. 2.22).

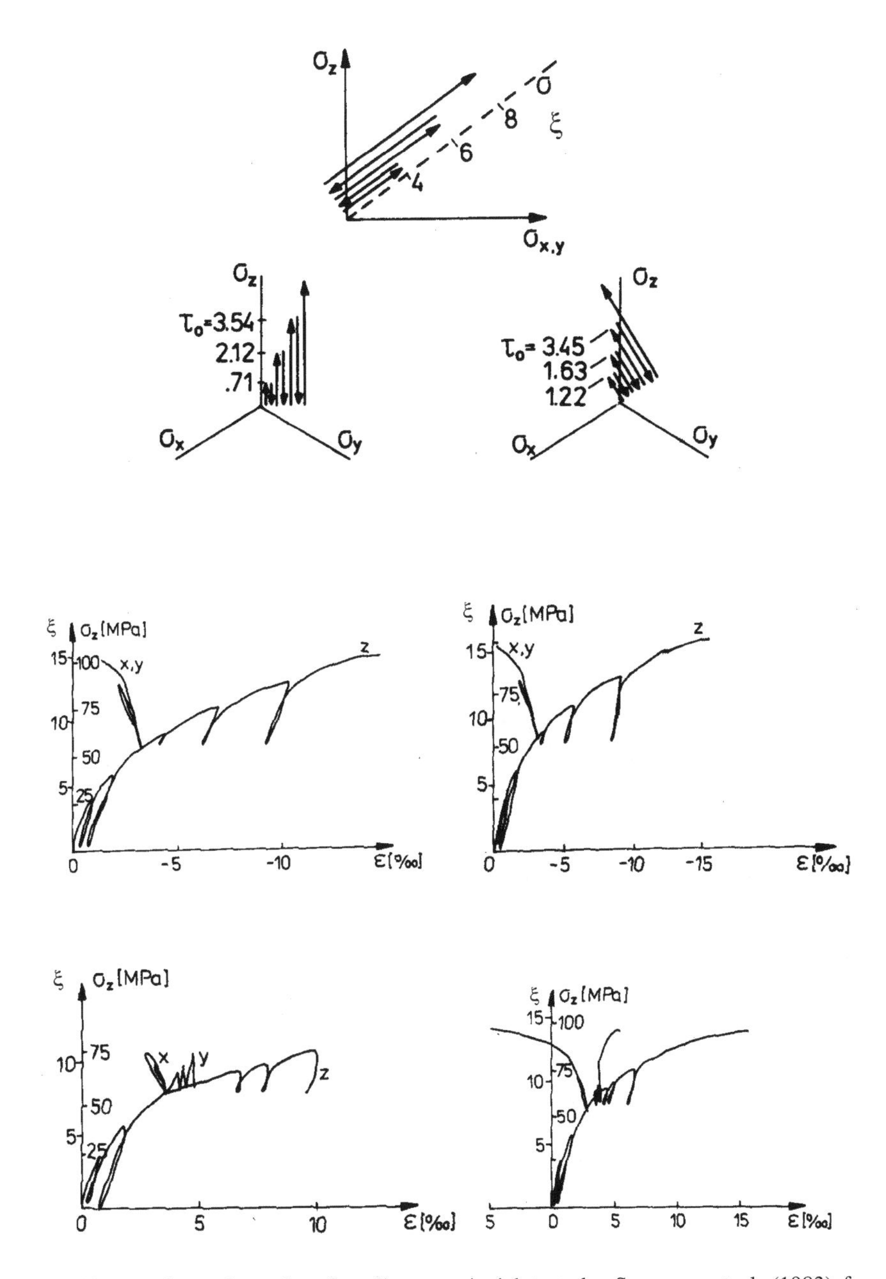

Fig. 2.18 Experimental results of cyclic true triaxial tests by Scavuzzo et al. (1983) for different stress paths (Klisiński and Mróz 1988)

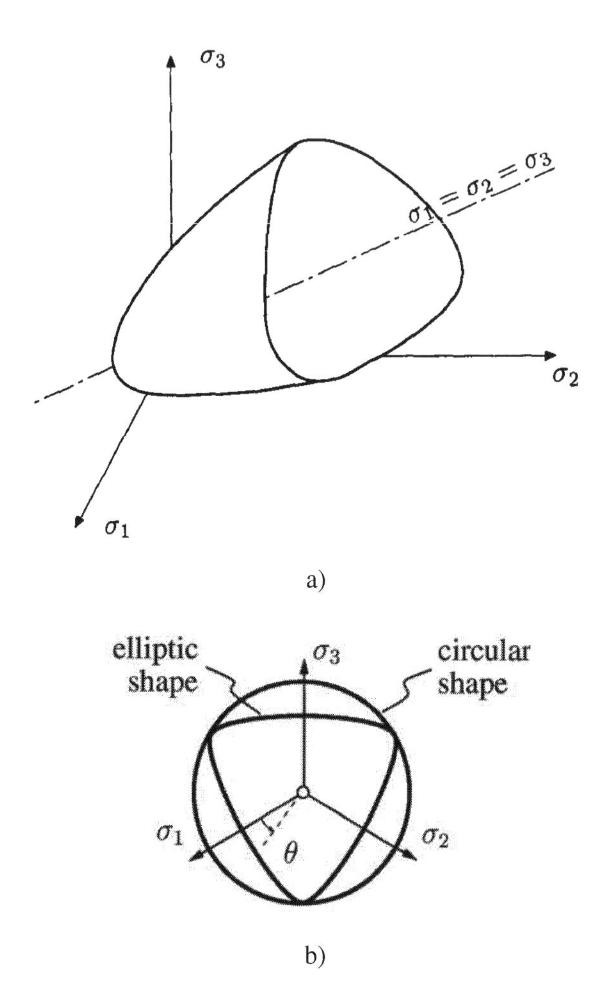

Fig. 2.19 Failure surface for concrete in space of principal stresses σ_i (a) and in deviatoric plane (θ - Lode angle) (Klisiński and Mróz 1988, Pivonka et al. 2003)

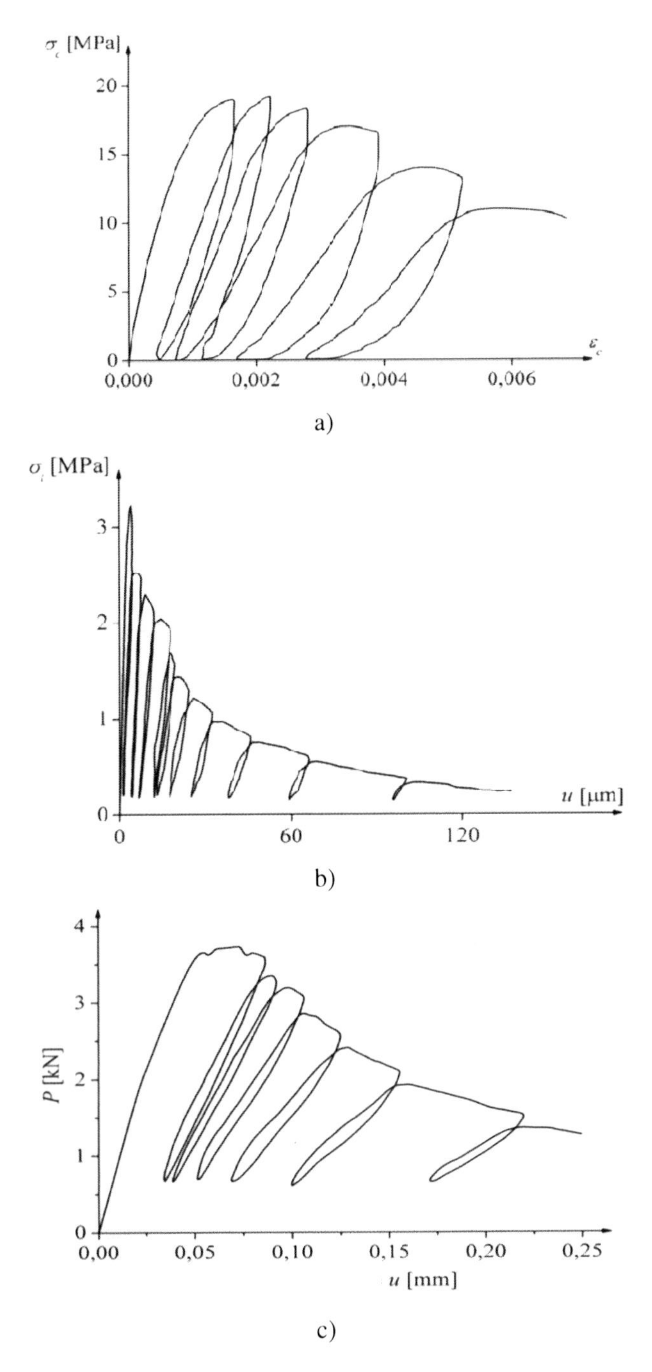

Fig. 2.20 Concrete behaviour under cyclic loading: a) stress-strain curve under uniaxial compression (Karsan and Jirsa 1969), b) stress-displacement curve under uniaxial tension (Reinhardt et al. 1986), c) force-deflection curve under four-point bending (Hordijk1991)

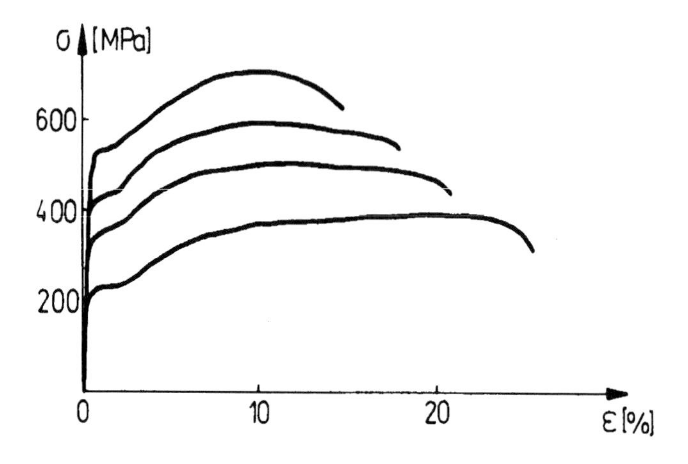

Fig. 2.21 Stress-strain diagram for reinforcement steel during uniaxial tension (Klisiński and Mróz 1988)

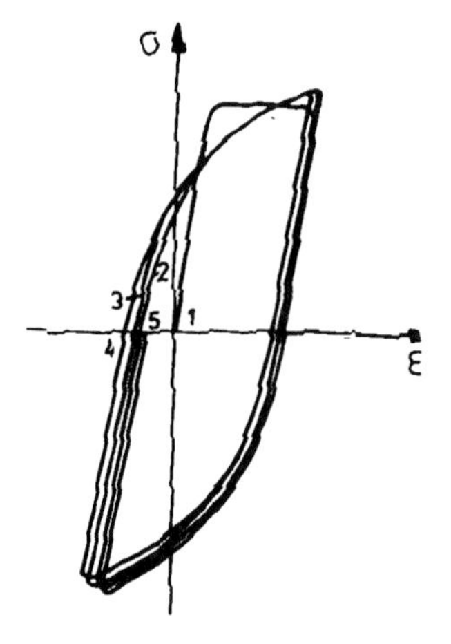

Fig. 2.22 Stress-strain diagram for reinforcement steel during cyclic loading (Klisiński and Mróz 1988)

Size Effect in Concrete Elements

Comprehensive investigations of the effect of the element size and boundary conditions along lateral surfaces during uniaxial compression were performed by van Vliet and van Mier (1995, 1996) (Fig. 2.23-2.28). The experimental results show that the both strength and ductility considerably increase with decreasing element size when large friction exists along horizontal lateral surfaces. If friction is small, the strength increase is non-significant. Fracture pattern strongly depends on boundary conditions. For large friction, cracks occur at horizontal edges only (the mid-region remains uncracked). In the case of small friction, a macro-crack is created in the form of shear zones whose number increases with a reduction of the element size.

A pronounced size effect was observed during multi-axial compression in hollow-cylinder tests using different pressures (Elkadi and van Mier 2006) (Fig. 2.29). The compressive strength reduced with increasing element size.

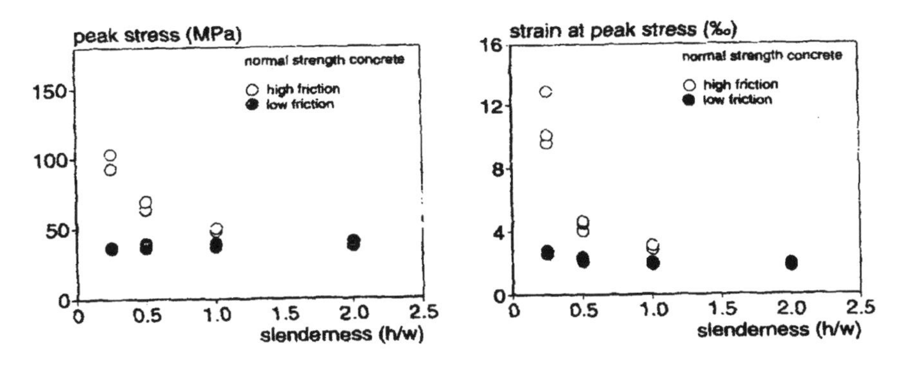

Fig. 2.23 Peak stresses and corresponding strains of the normal strength concrete during uniaxial compression specimens depending upon friction along horizontal edges (van Vliet and van Mier 1996)

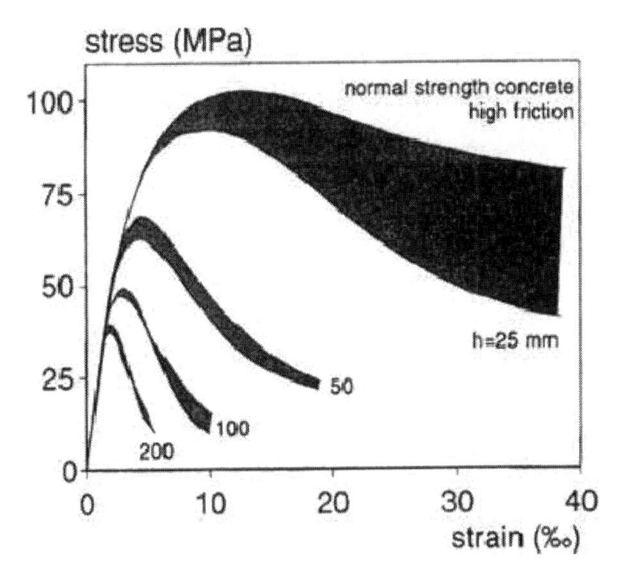

Fig. 2.24 Compressive stress-strain curves of the normal strength concrete specimens with different heights *h* and high friction boundary conditions along horizontal edges (van Vliet and van Mier 1996)

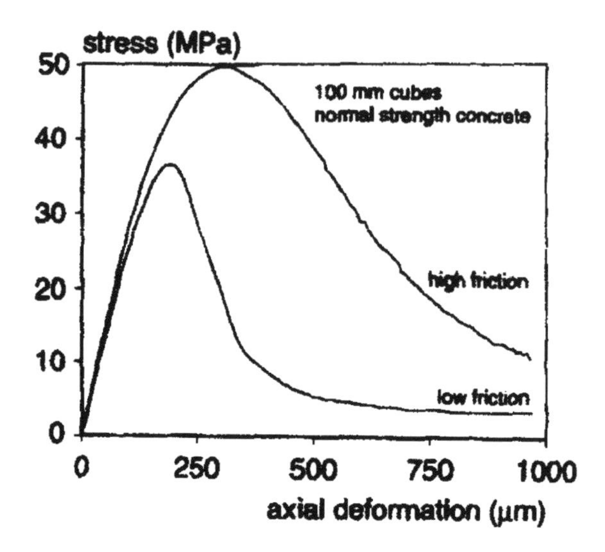

Fig. 2.25 Compressive curves for 100 mm cube specimens (normal strength concrete) loaded under high and low friction boundary conditions along horizontal edges (van Vliet and van Mier 1996)

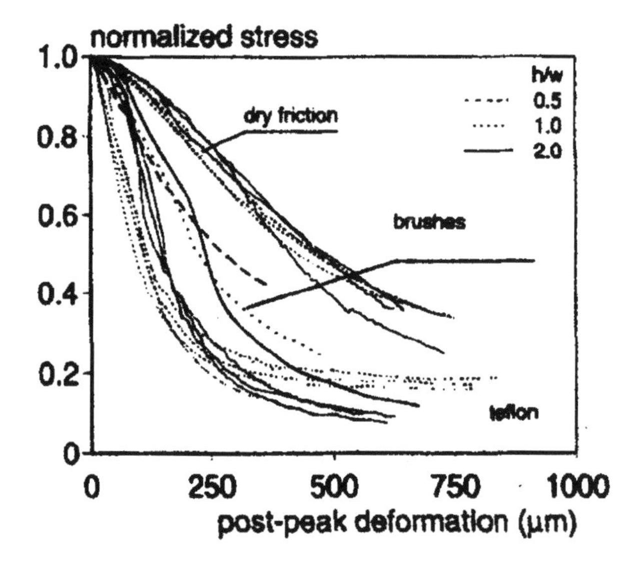

Fig. 2.26 Post-peak deformations of normal strength concrete specimens with different boundary conditions along horizontal edges (van Vliet and van Mier 1996)

A strong deterministic size effect occurred in concrete during uniaxial extension (van Vliet and van Mier 2000) (Figs. 2.30-2.32) and three-point bending (Le Bellego et al. 2003) (Figs. 2.33-2.35). In addition, a strong stochastic size effect was observed in experiments with concrete beams of the same height and different span at the same load (Koide et al. 1998, 2000) (Fig. 2.36), i.e. the maximum bending moment clearly decreased with increasing bending span.

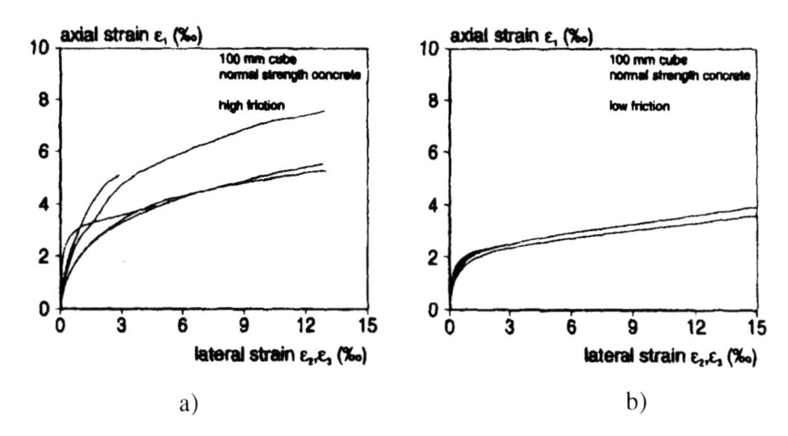

Fig. 2.27 Axial strain against lateral strain for normal strength concrete cubes with: a) high friction and b) low friction along horizontal edges (van Vliet and van Mier 1996)

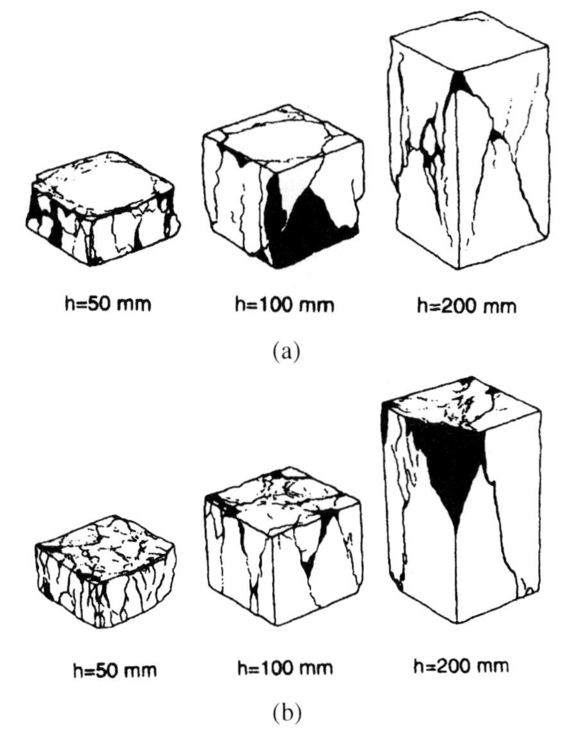

Fig. 2.28 Crack patterns of normal strength concrete specimens with high (A) and low (B) friction boundary conditions for different specimen heights h (van Vliet and van Mier 1996)

Size Effect in Reinforced Concrete Elements

A size effect occurs also in reinforced concrete elements when the failure by yielding of longitudinal steel bars is excluded in advance (brittle failure takes then place in a compressive concrete zone).

Figure 2.37 shows a size effect in reinforced concrete beams of the same slenderness without shear reinforcement in experiments by Leonhardt and Walther (1962). The size of beams was in a proportion 1:2:3:4. The reinforcement included always 2 steel bars. The reinforcement ratio was 1.35%. The bearing capacity of beams decreased with increasing beam size.

Comprehensive investigations on a size effect were carried out by Walraven and Lehwalter (1994) for different reinforced concrete beams. First, slender reinforced concrete beams loaded in shear without stirrups were investigated (Fig. 2.38). The beams were made of normal gravel concrete and lightweight concrete. The experiments were carried out with 3 different beams with the same thickness of b=200 mm: h=150 mm, l=2300 mm (small size beam '1'), h=450 mm, l=4100 mm (medium size beam '2') and h=750 mm, l=6400 mm (large size beam '3'). The reinforcement ratio was 0.79-0.83%: 1×ϕ_s8 and 2×ϕ_s10 (beam '1'), 1×ϕ_s20 and 2×ϕ_s14 (beam '2') and 3×ϕ_s22 (beam '3').

Second, short reinforced concrete beams loaded in shear without and with shear reinforcement were investigated (Fig. 2.39). The beam length L varied between 680 mm and 2250 mm and the height h was between 200 mm and 1000 mm (the beam width b was always 250 mm). In the tests, the span-to-depth ratio was always 1. The reinforcement ratio of the specimens was 1.1% (the failure by yielding of longitudinal steel bars was again excluded in advance). The shear reinforcement ratio was 0%, 0.15% and 0.30%, respectively.

The experiments show that the shear beam bearing capacity of slender and short reinforced concrete beams decreases with increasing beam size independently of stirrups (Figs. 2.40-2.42).

Size effect tests on reinforced concrete beams were recently carried out by Belgin and Sener (2008). The beams were similar in one-, two- and three-dimensions, which means that the beam width, cover thickness, bar diameter and depth of reinforcement were all proportional to the beam span and length. In all of the specimens, the uniform bending moment was obtained between the loads. Beam lengths and widths were constant 4.6 m and 0.11 m, respectively, for the one-dimensional similarity (group I, Fig. 2.43a). Beam lengths of L=1.15 m, 2.3 m, and 4.6 m, beam widths of b=0.055 m, 0.11 m and 0.22 m, beam heights of h=0.075 m, 0.15 m, 0.30 m were used for the three-dimensional similarity (group II, Fig. 2.43b). For the two-dimensional similarities, two types of tests were carried out. For group III beam heights of h=0.15 m, 0.30 m and 0.60 m were used and a beam width was constant b=0.11 m (Fig. 2.44a). For group IV beam widths of b=0.055 m, 0.11 m and 0.22 m, beam heights of h=0.075 m, 0.15 m and 0.30 m were used and beam length was constant L=2.30 m (Fig. 2.44b). The reinforcement ratio was 3% in order to induce brittle failure.

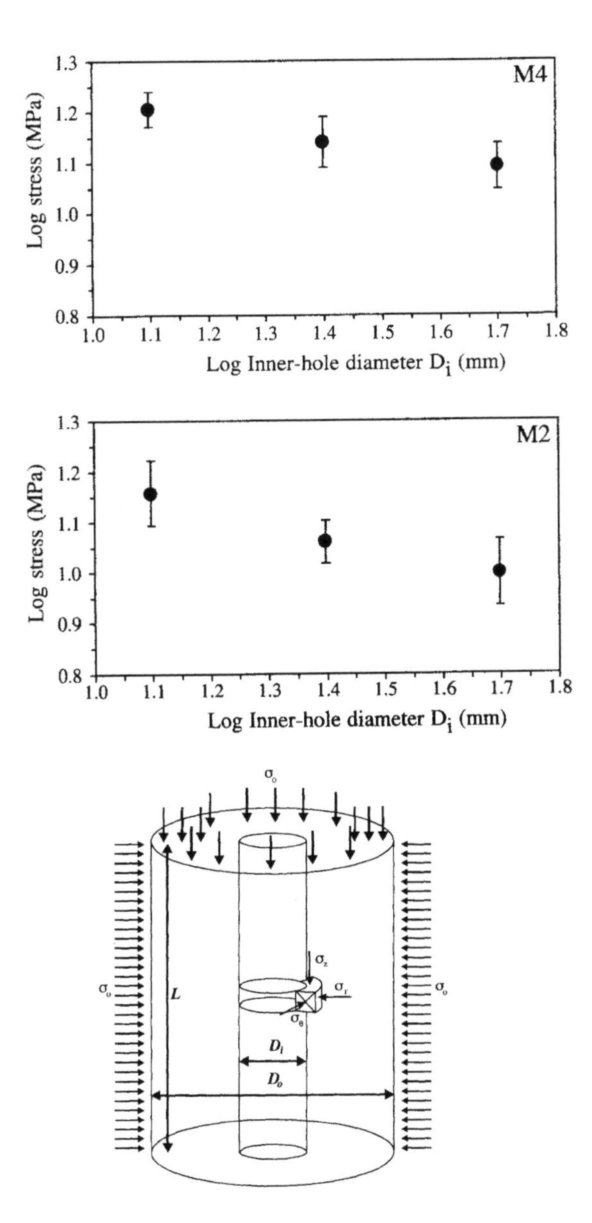

Fig. 2.29 Hollow-cylinder experiments: mean values and standard deviations for log σ versus log D_i for two different concretes (σ_o - outer stress, D_i - inner-hole diameter) (Elkadi and van Mier 2006)

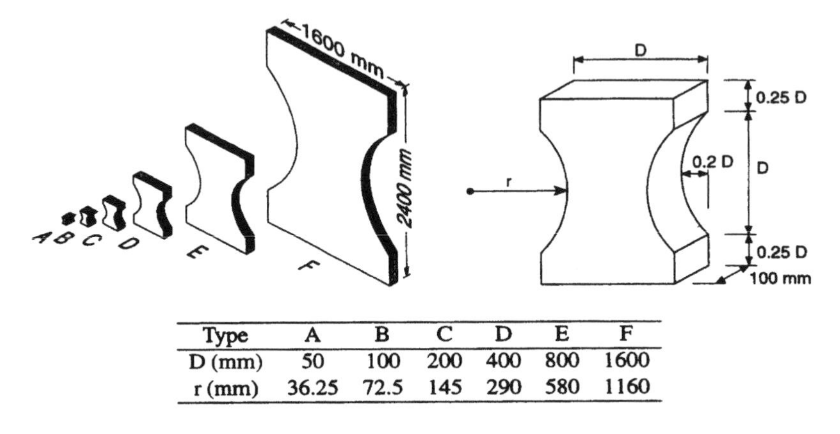

Type	A	B	C	D	E	F
D (mm)	50	100	200	400	800	1600
r (mm)	36.25	72.5	145	290	580	1160

Fig. 2.30 Specimens for size effect tests during uniaxial tension (van Vliet and van Mier 2000)

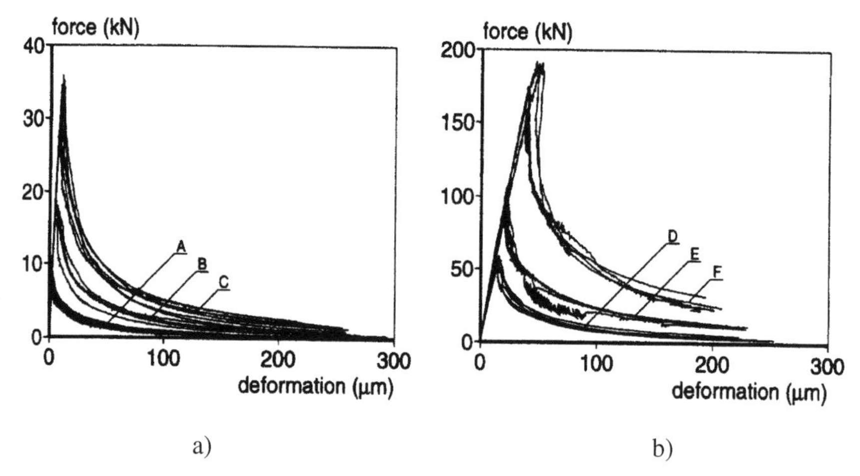

Fig. 2.31 Experimental force-deformation curves during uniaxial tension for specimens A, B, C (a) and D, E, F (b) of Fig. 2.30 (van Vliet and van Mier 2000)

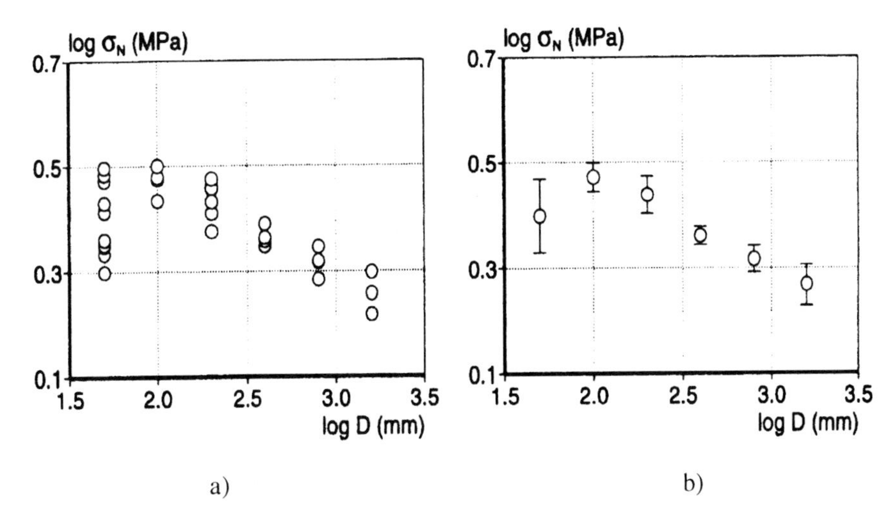

Fig. 2.32 Individual values (a) and mean values with standard deviations (b) of nominal concrete strength logσ_N versus specimen size D during uniaxial tension (van Vliet and van Mier 2000)

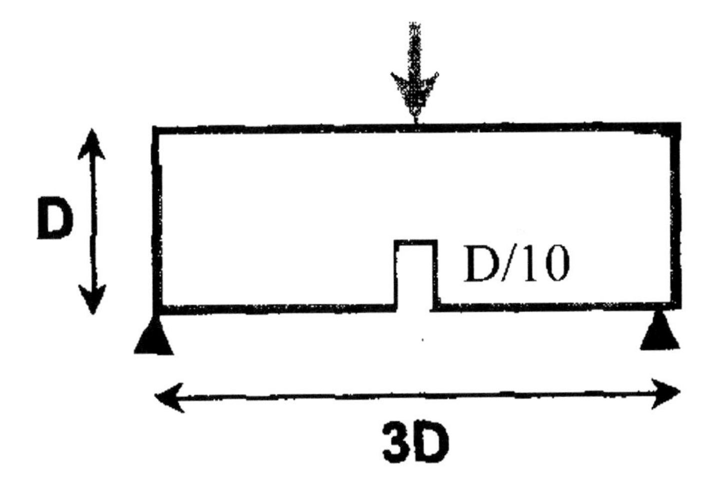

Fig. 2.33 Geometry of three-point bending size effect tests with 3 different notched beams: D_1=80 mm, D_2=160 mm, D_3=320 mm (Le Bellego et al. 2003) (D – beam height)

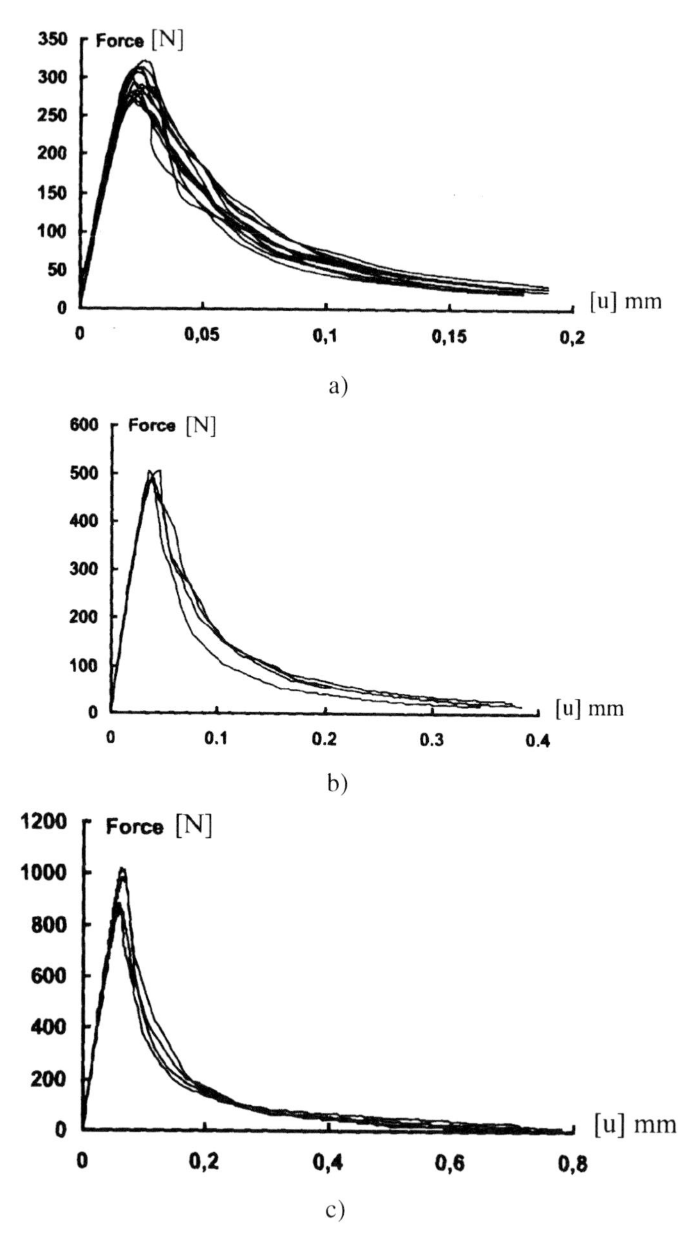

Fig. 2.34 Three point bend testing of notched beams: experimental results of vertical force versus deflection u for small- (a), medium- (b) and large-size beam (c) (Le Bellego et al. 2003)

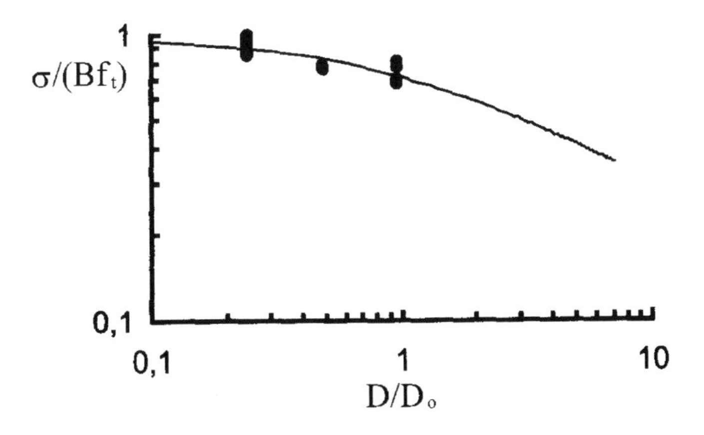

Fig. 2.35 Experimental results of size effect in three-point bending tests (points – experimental results, line – size effect law by Bažant (Bažant and Planas 1998), D – specimen size, D_o – characteristic size, B – parameter, f_t – tensile strength) (Le Bellego et al. 2003)

The beams under the constant bending moment failed by crushing of a compression zone around the peak load. Some local cracks were seen prior to crushing. The experimental results clearly confirmed the existence of a significant size effect on the nominal bending strength in beam accompanied by an increase of failure brittleness with the beam size. Figure 2.45 shows the size effect plots for the combined test results for all four groups. In the case of the two-dimensional similarity (group III and IV) tests, the maximum strengths were significantly higher than those for the one and three-dimensional similarity (group I–II) tests. For the bending tests of two-dimensional similarity (group III), the slope was e.g. 3.6 times as high as that for the three-dimensional similarity (group II) tests. Smaller beams showed a ductile behaviour compared to larger beams. The load-deflection diagrams were almost straight lines up to the peak load, after which a steeper descending branch was observed for large beams. The load-deflection diagrams for larger beams were stiffer than those for smaller beams, confirming an increase of brittleness in response to an increasing size. The post-peak behaviour was completely governed by the behaviour of concrete in a compression zone.

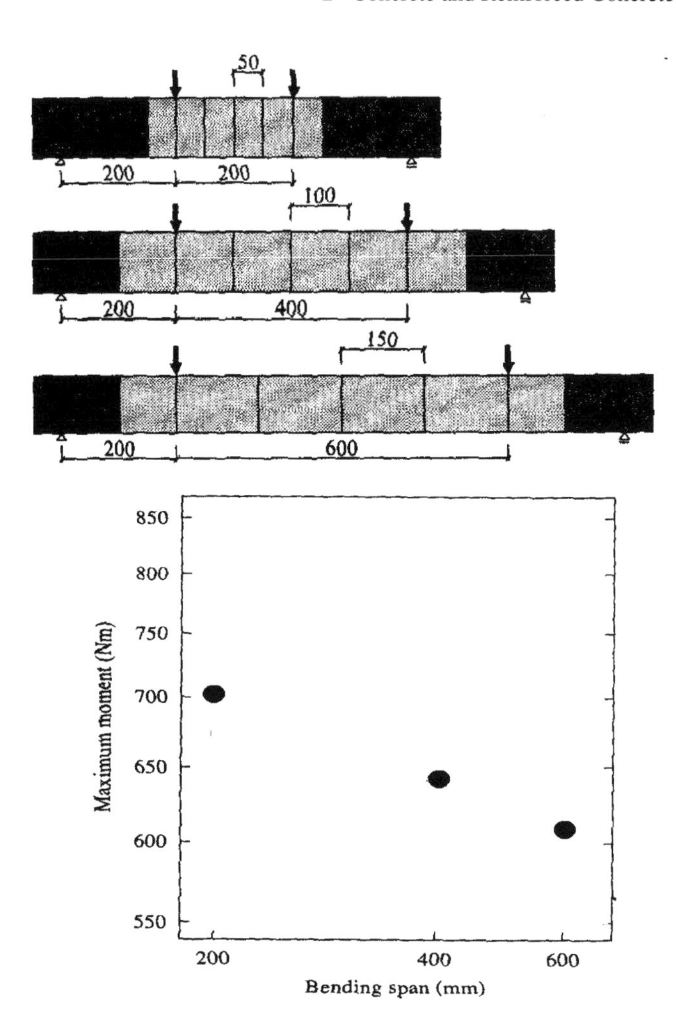

Fig. 2.36 Experimental tests with concrete beams of bending span 200 mm, 400 mm and 600 mm and measured stochastic size effect expressed by maximum bending moment against bending span (Koide et al. 1998, 2000)

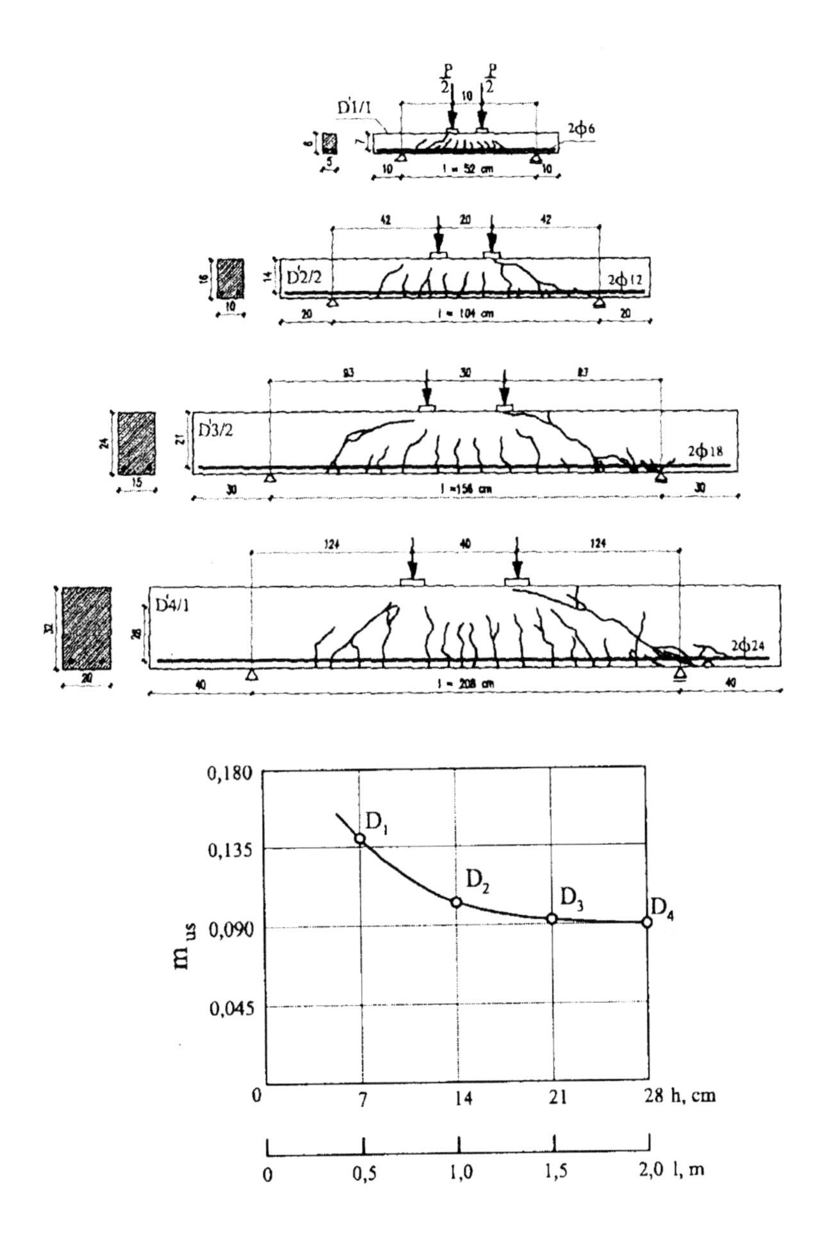

Fig. 2.37 Crack pattern and normalized failure moment against beam sizes *h* and *l* (Leonhardt and Walther 1962)

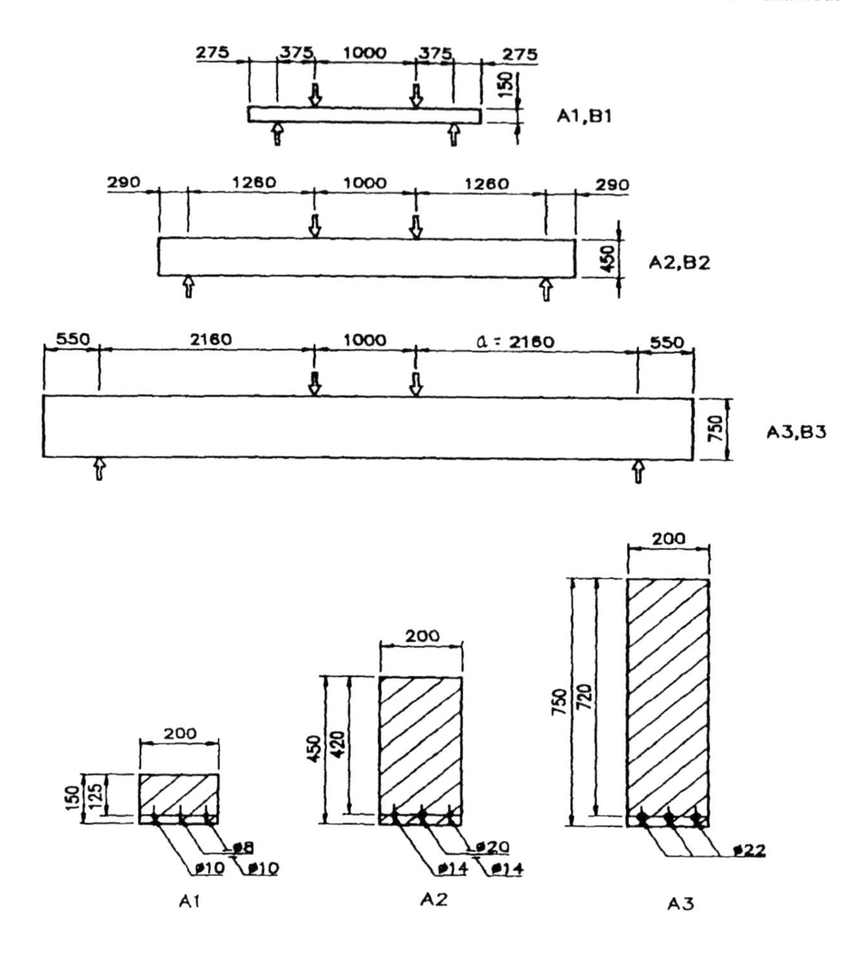

Fig. 2.38 Slender reinforced concrete beams without shear reinforcement (Walraven and Lehwalter 1994)

Bond-slip between Concrete and Reinforcement

A good bond between concrete and reinforcement is necessary for a reinforced concrete element to carry loads. A bond phenomenon was investigated by several researchers (e.g. Dörr 1980, Malvar 1992, Bolander et al. 1992, Azizinamini et al. 1993, Darwin and Graham 1993, Uijl and Bigaj 1996, Haskett et al. 2008). The experiments show that the bond stress depends on the bar roughness, bar diameter, bar location, bar ending, concrete class, bar anchorage length, direction of concrete mixing and failure mechanism (splitting or pulling out) (Idda 1999).

Fracture in concrete with a ribbed bar during a pull-out test is demonstrated in Fig. 2.46. Primary and secondary cracks occur due to the lack of bond and crushing. The effect of the bar roughness, bar location and bar ending on the evolution of the bond stress is shown in Figs. 2.47-2.49. The bond stress increases with increasing bar roughness. It is the largest if the bar is pulled out parallel to the direction of concrete mixing (Fig. 2.48). The effect of the bar ending is also of importance (Fig. 2.49).

Figure 2.50 shows a non-linear distribution of shear stresses in the contact zone between concrete and reinforcement. With increasing pull-out force, the maximum shear stress point moves into the specimen interior.

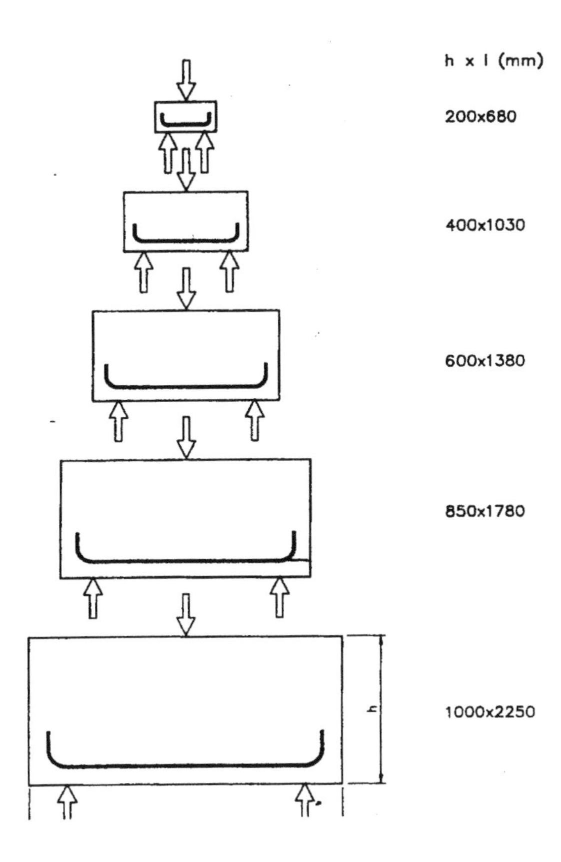

Fig. 2.39 Short reinforced concrete beams (Walraven and Lehwalter 1994)

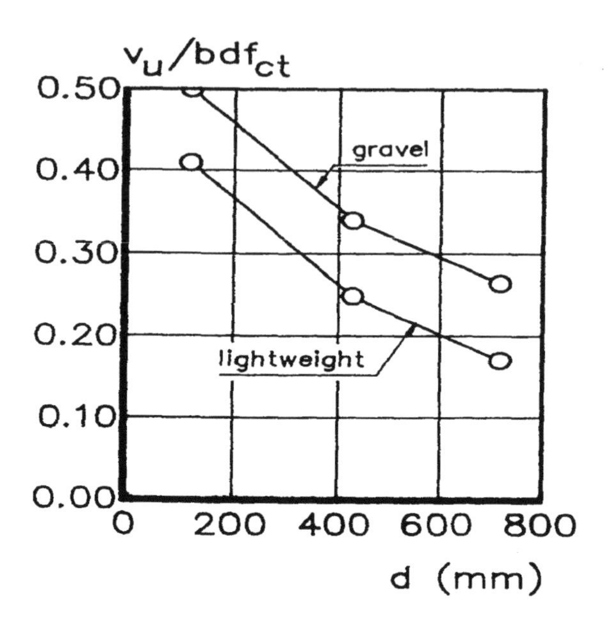

Fig. 2.40 Size effect for slender reinforced concrete beams of gravel and lightweight concrete without shear reinforcement: nominal shear strength versus effective height d (V_u - ultimate vertical force, b - beam width, f_{ct} - tensile strength) (Walraven and Lehwalter 1994)

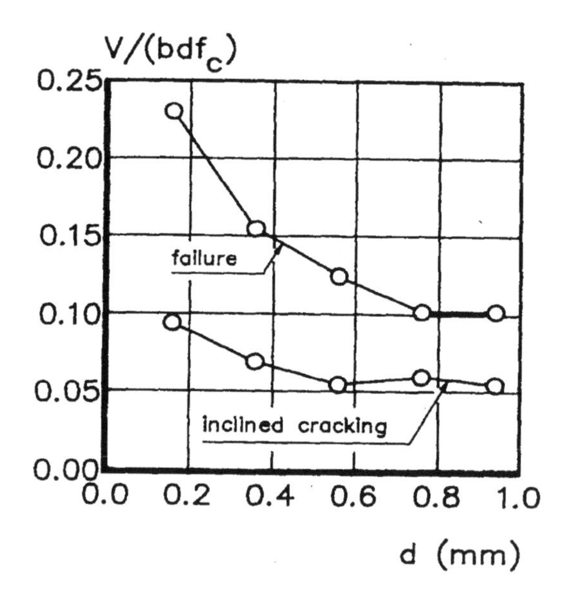

Fig. 2.41 Size effect for short reinforced concrete beams of normal concrete without shear reinforcement: nominal shear strength versus effective height d during cracking and failure (V_u - ultimate vertical force, b - beam width, f_c - compressive strength) (Walraven and Lehwalter 1994)

The bond mechanism is described in Fig. 2.51. It is accompanied by primary crack formation, bending of concrete corbels, crushing of concrete corbels connected to dilatancy and slip of separate surfaces connected to contractancy.

In turn, the experimental evolutions of the pull-out force versus slip are presented in Figs. 2.52 and 2.53. The evolution curve includes a hardening and softening phase. The pull-out force depends upon the bar diameter.

Finally, Figs. 2.54-2.57 show the effect of the reinforcement ratio, bar roughness and reinforcement cover thickness on the force-displacement curve during bending.

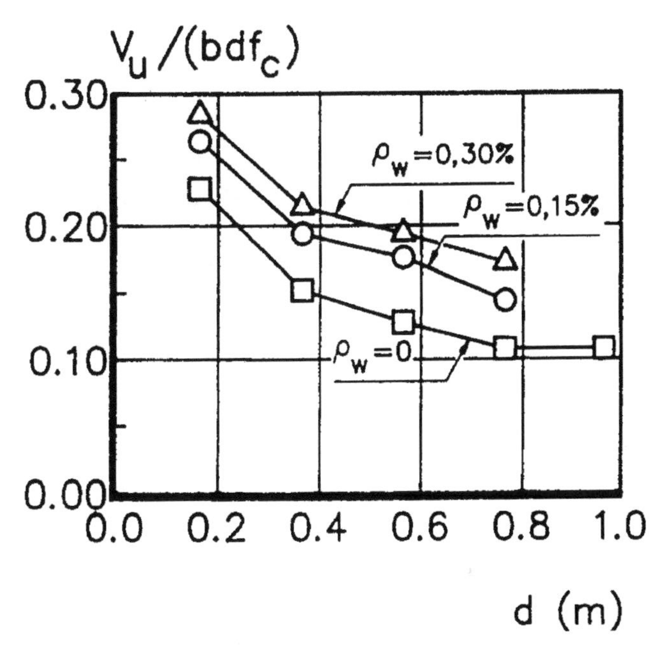

Fig. 2.42 Size effect for short reinforced concrete beams of normal concrete with shear reinforcement ratio ρ_w: nominal shear strength versus effective height d during cracking and failure (V_u - ultimate vertical force, b - beam width, f_c - compressive strength) (Walraven and Lehwalter 1994)

The results indicate that the peak and near post-peak behaviour of reinforced concrete elements are controlled by three factors: reinforcement ratio, bond-slip properties and reinforcement cover. Material softening decreases with increasing reinforcement ratio (for very low reinforcement ratios) and increases with increasing reinforcement ratio (for very high reinforcement ratios). It also decreases with increasing bar roughness. If the reinforcement cover is large enough, the specimen load exhibits a peak before the fracture zone reaches the reinforcement, then after some load decrease, the growing fracture zone reaches the reinforcement and is arrested, thus engendering hardening followed by a second peak and further softening.

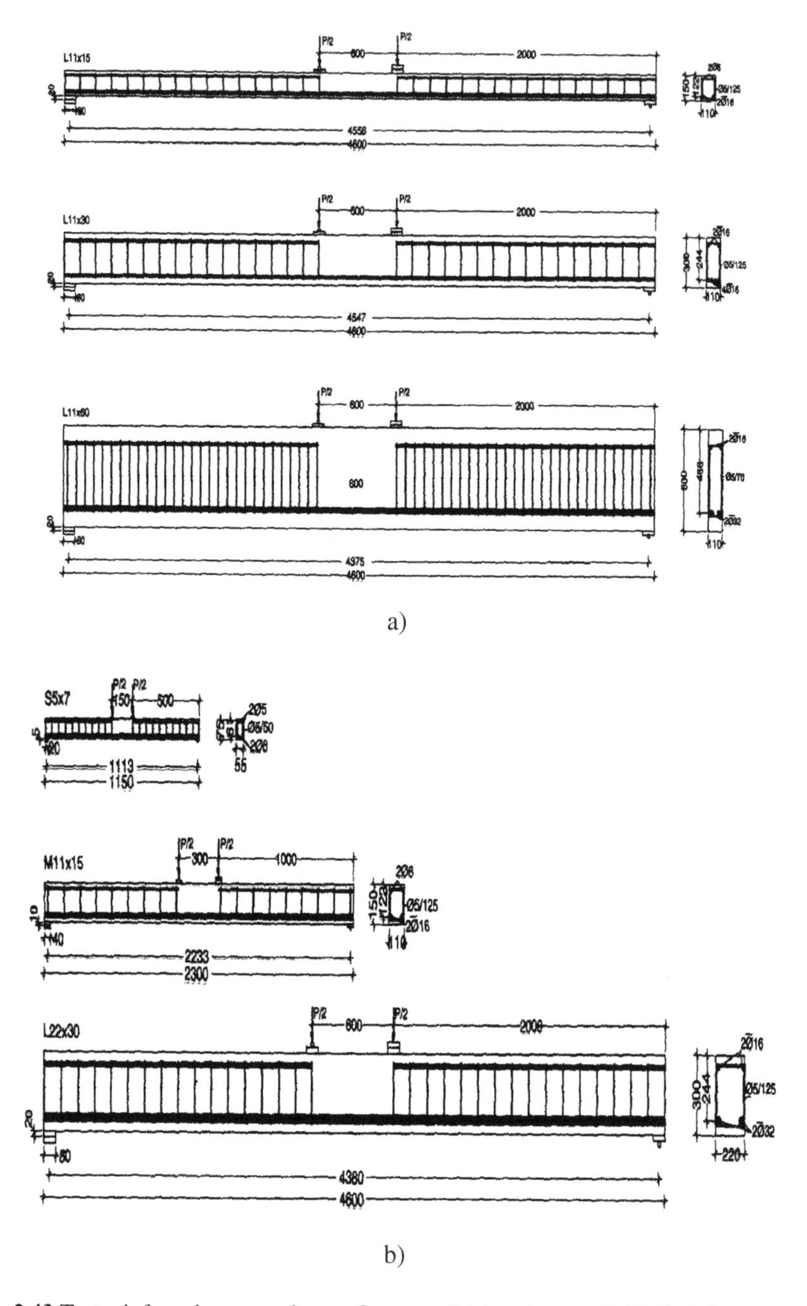

Fig. 2.43 Test reinforced concrete beams for group I (a) and group II (b) (Belgin and Sener 2008)

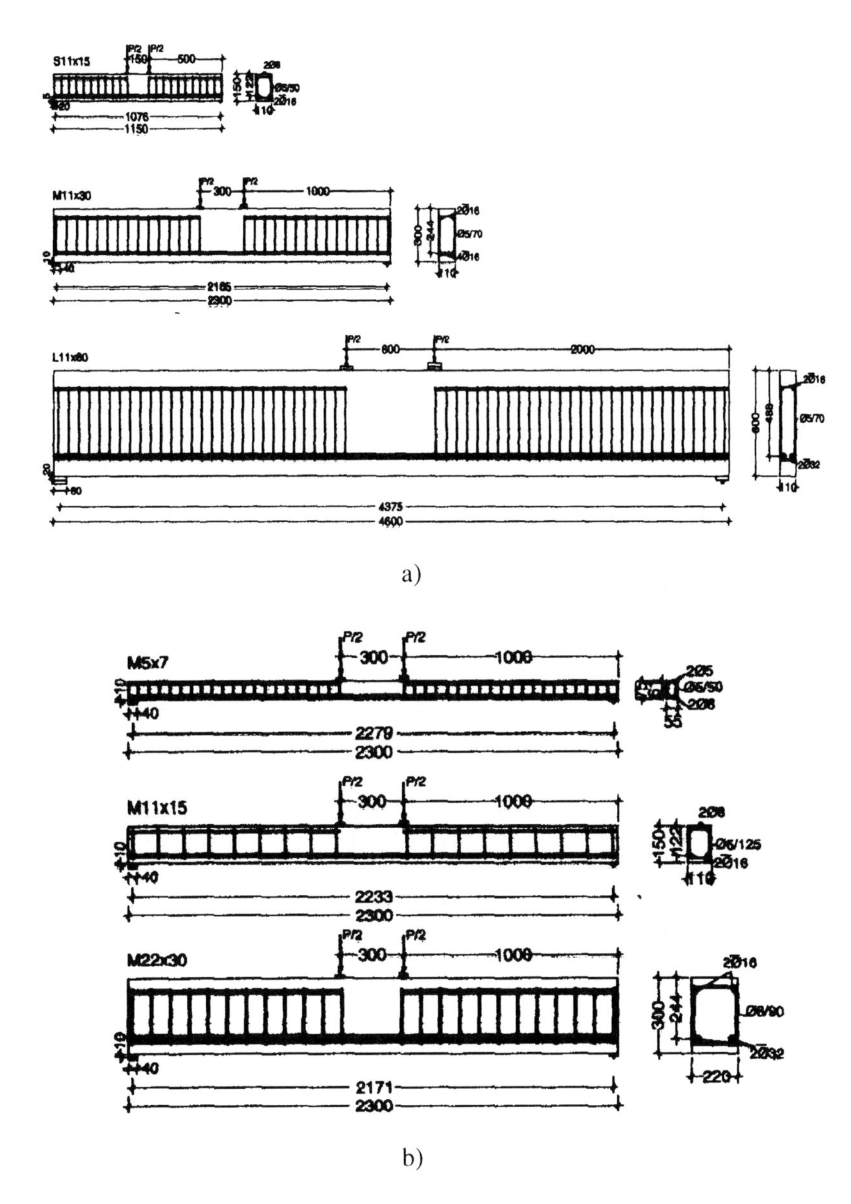

Fig. 2.44 Test reinforced concrete beams for group III (a) and group IV (b) (Belgin and Sener 2008)

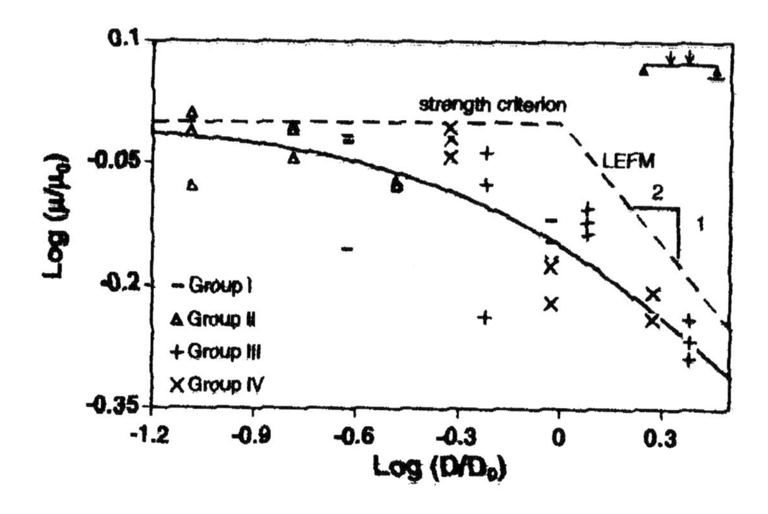

Fig. 2.45 Test results on size effect for all groups (LEFM - linear elastic fracture mechanics) (Belgin and Sener 2008)

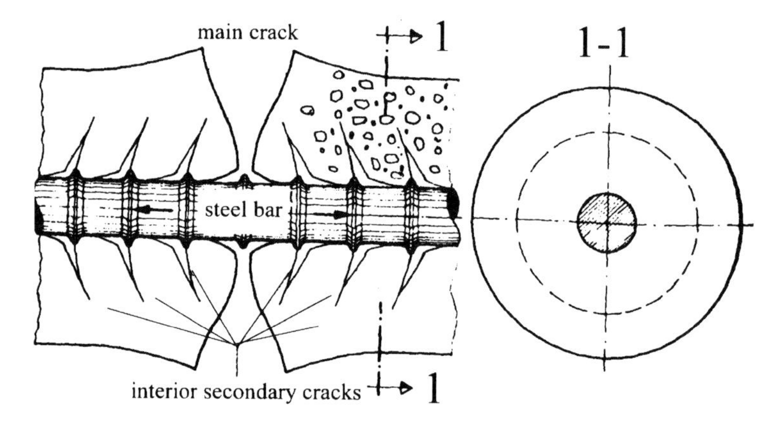

Fig. 2.46 Crack formation during pull-out test of ribbed bars (Leonhardt 1973)

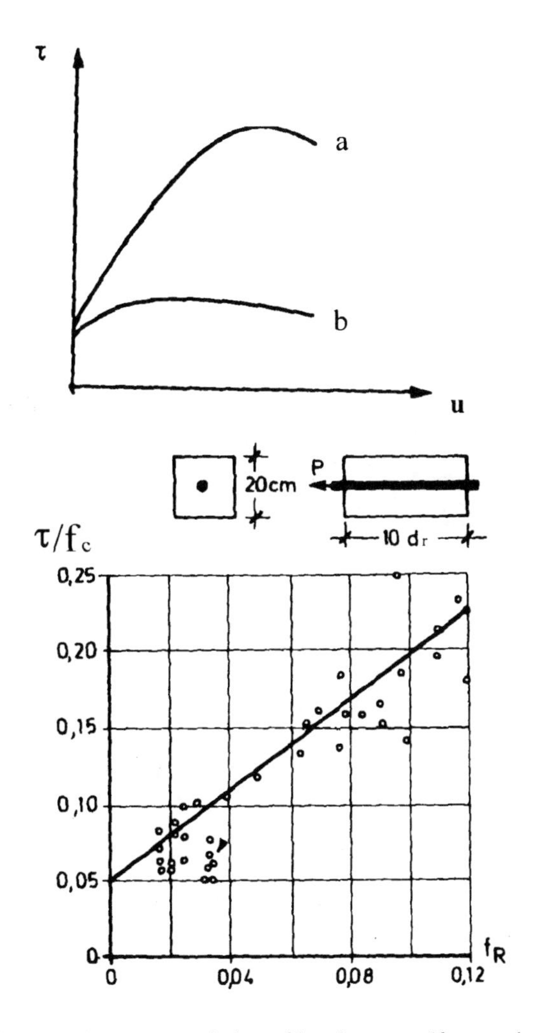

Fig. 2.47 Effect of bar roughness on evolution of bond stress τ (f_R - roughness parameter): a) very rough bars, b) smooth bars (Leonhardt 1973)

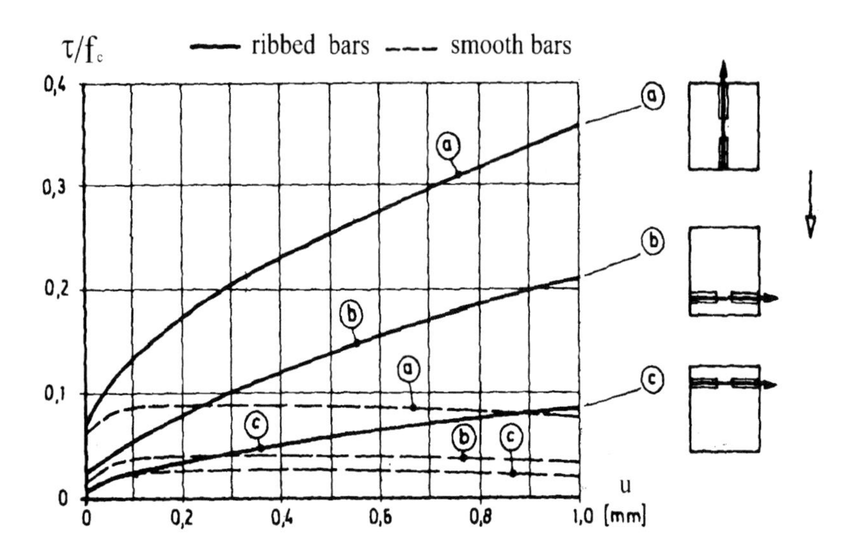

Fig. 2.48 Effect of bar location and its roughness on normalized bond stress τ/f_c (Leonhardt 1973)

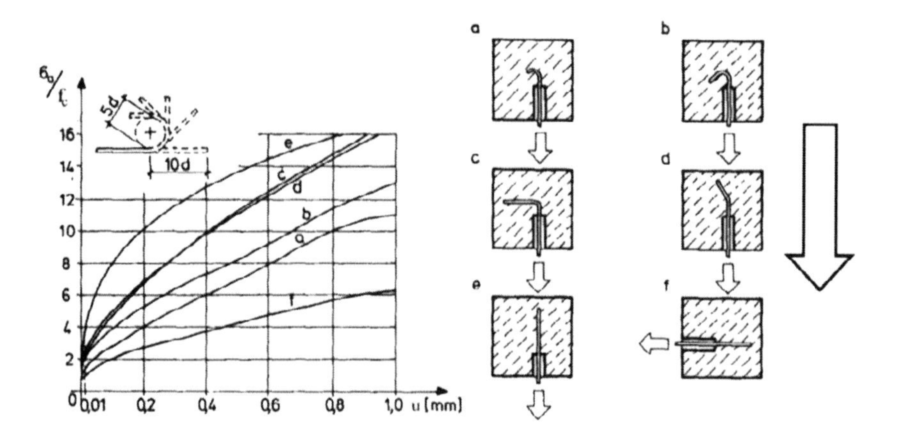

Fig. 2.49 Effect of bar ending on stresses in reinforcement (⇒ mixing direction) (Leonhardt 1973)

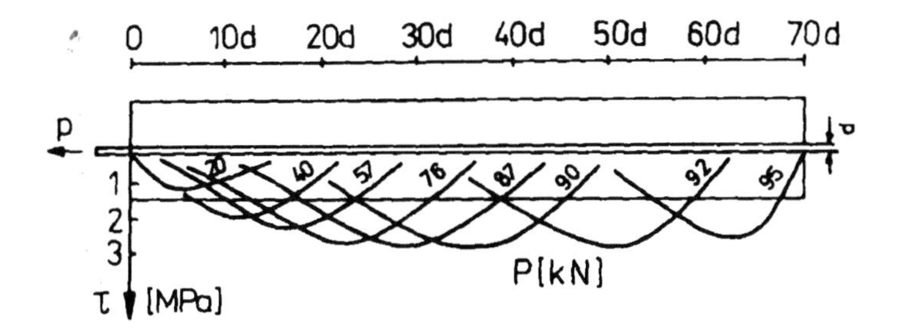

Fig. 2.50 Distribution of shear stresses τ in contact zone between concrete and reinforcement for increasing pull-out force P (Klisiński and Mróz 1988)

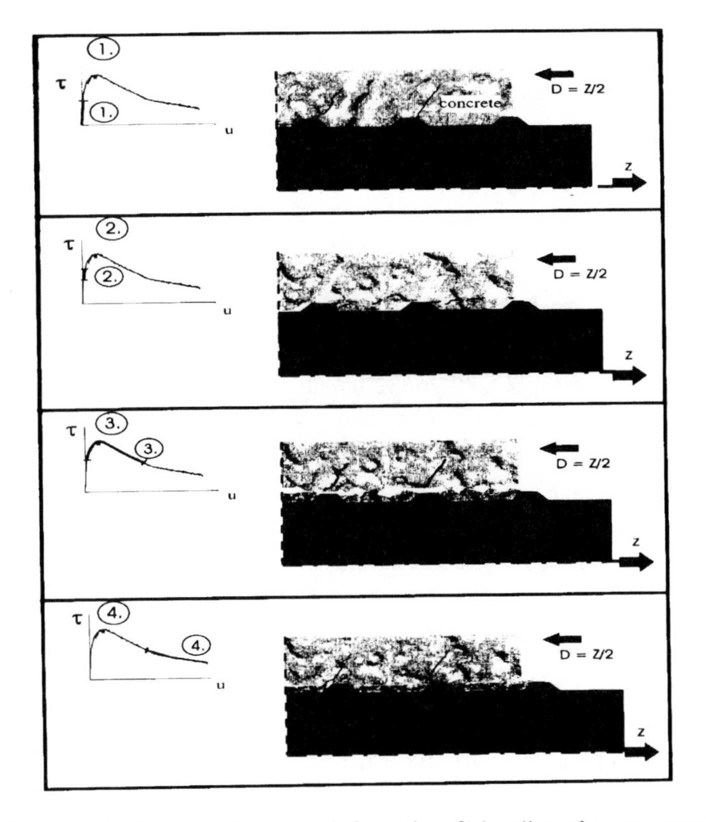

Fig. 2.51 Bond mechanism: 1. primary crack formation, 2) bending of concrete corbels, 3) crushing of concrete corbels connected to dilatancy, 4) slip of separate surfaces connected to contractancy (Idda 1999)

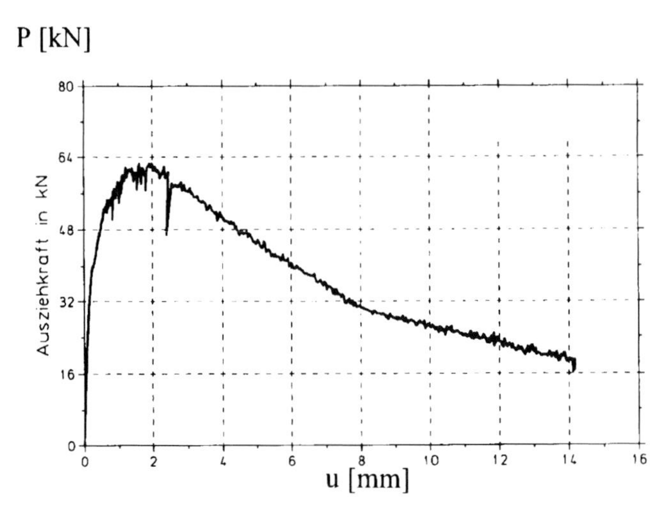

Fig. 2.52 Experimental pull-out force against slip displacement (bar diameter d_r=16 mm) (Idda 1999)

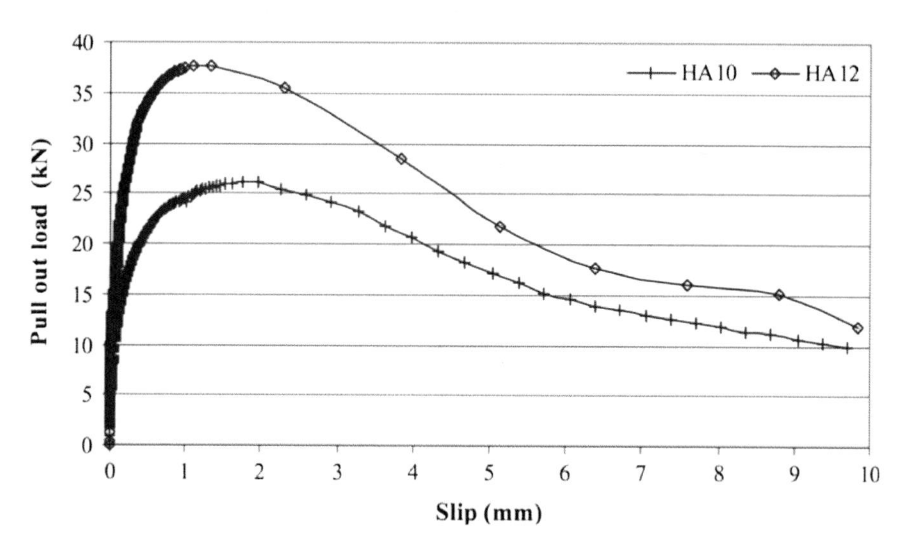

Fig. 2.53 Pull-out force versus slip for two bars with different diameter: 10 mm (HA10) and 12 mm (HA12) (Dahou et al. 2009)

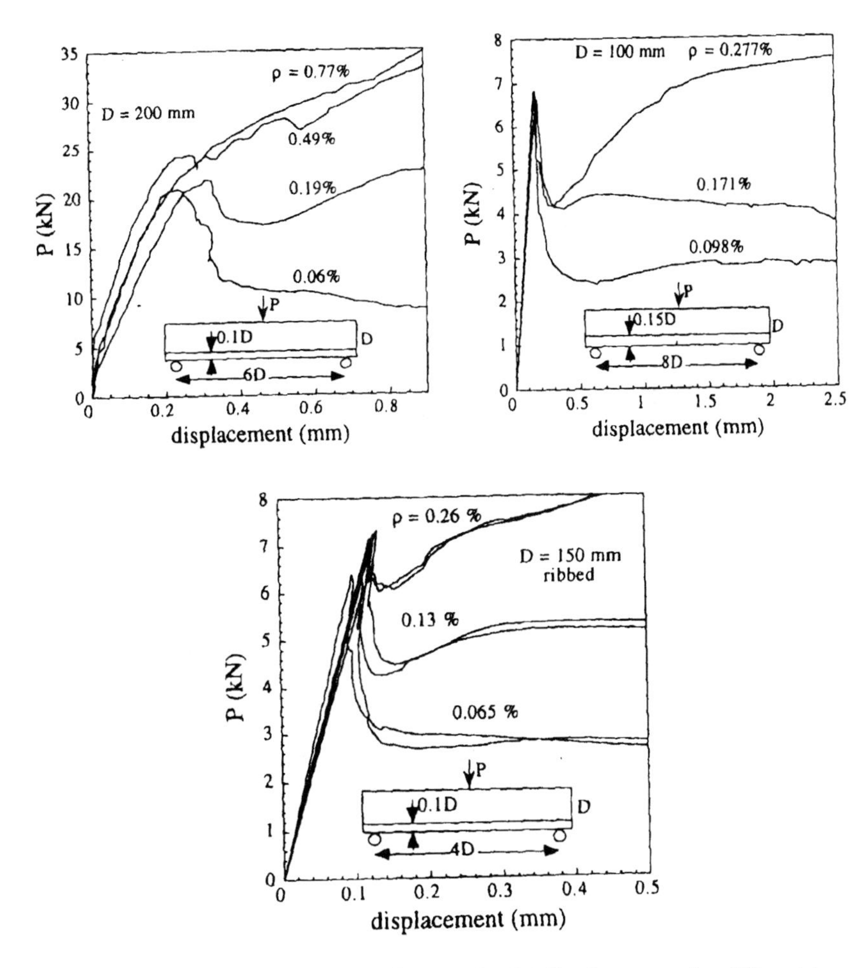

Fig. 2.54 Influence of reinforcement ratio ρ on load-deflection curve for different beam effective heights and lengths (Bažant and Planas 1998)

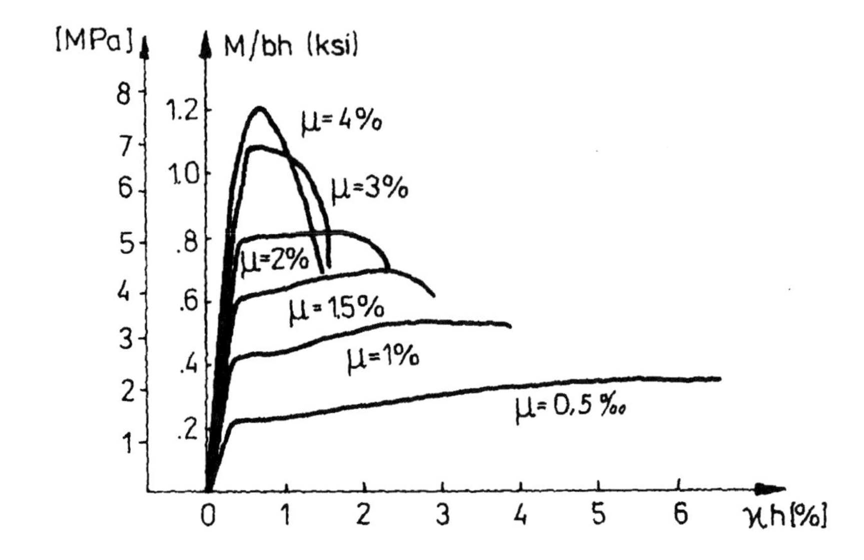

Fig. 2.55 Influence of reinforcement ratio μ on relationship between bending moment and curvature (Klisiński and Mróz 1988)

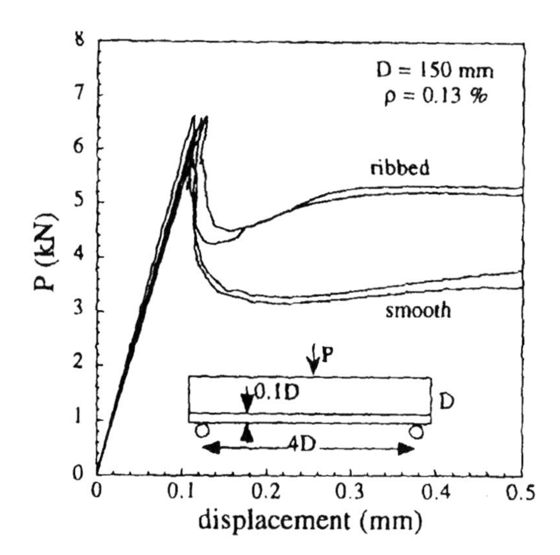

Fig. 2.56 Influence of bar roughness on load-displacement curve in reinforced concrete beams (Bažant and Planas 1998)

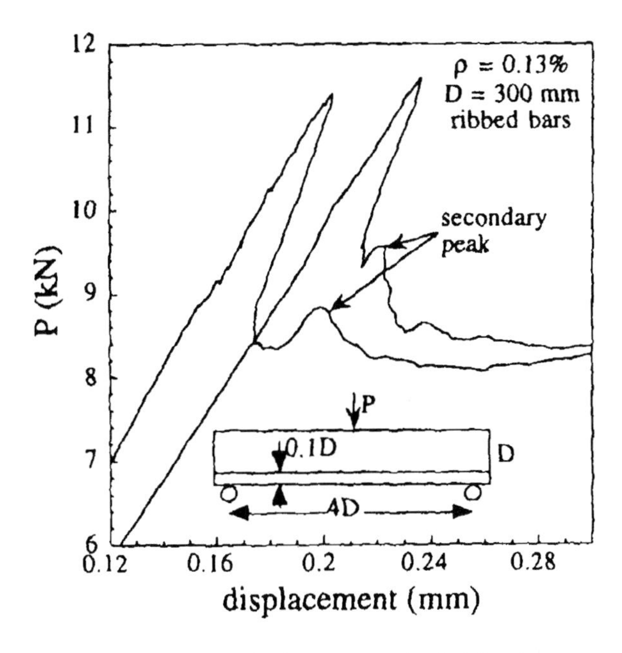

Fig. 2.57 Influence of a relatively thick cover on load-deflection curve in reinforced concrete beams (Bažant and Planas 1998)

References

Azizinamini, A., Stark, M., Roller, J.J., Ghosh, S.K.: Bond performance of reinforcing bars embedded in high-strength concrete. ACI Structural Journal (American Concrete Institute) 90(5), 554–561 (1993)

Bažant, Z., Planas, J.: Fracture and size effect in concrete and other quasi-brittle materials. CRC Press LLC (1998)

Belgin, C.M., Sener, S.: Size effect on failure of overreinforced concrete beams. Engineering Fracture Mechanics 75(8), 2308–2319 (2008)

Bischoff, P.H., Perry, S.H.: Compressive behaviour of concrete at high strain rates. Materials and Structures 24(6), 425–450 (1991)

Bischoff, P., Perry, S.H.: Impact behaviour of plane concrete loaded in uniaxial compression. Journal of Engineering Mechancis ASCE 121(6), 685–693 (1995)

Bolander, J.J., Satake, M., Hikosaka, H.: Bond degradation near developing cracks in reinforced concrete structures. Memoirs of the Faculty of Engineering 52(4), 379–395 (1992)

Dahou, Z., Mehdi, Z., Castel, A., Ghomarid, F.: Artificial neural network model for steel-concrete bond prediction. Engineering Structures 31(8), 1724–1733 (2009)

Dantu, P.: Etudes des contraintes dans les milieux heterogenes. Application au beton. Annales des l'Institut de Batiment et des Travaux Publics 11(121), 55 (1958)

Darwin, D., Graham, E.K.: Effect of deformation height and spacing on bond strength of reinforcing bars. ACI Materials Journal 90(6), 646–657 (1993)

Dörr, K.: Ein Beitag zur Berechnung von Stahlbetonscheiben unter Berücksichtigung des Verbundverhaltens. Phd Thesis, Darmstadt University (1980)
Elkadi, A.S., van Mier, J.G.M.: Experimental investigation of size effect in concrete fracture under multiaxial compression. International Journal of Fracture 140(1-4), 55–71 (2006)
Gary, G.: Essais a grande vitesse sur beton. Problemes specifiques. Scientifique Rapport GRECO edite par J.M. Reynouard, France (1990)
Godycki-Ćwirko, T.: Concrete Mechanics, Arkady (1982) (in polish)
Haskett, M., Oehlers, D.J., Mohamed Ali, M.S.: Local and global bond characteristics of steel reinforcing bars. Engineering Structures 30(2), 376–383 (2008)
Hordijk, D.A.: Local approach to fatigue of concrete, PhD Thesis, Delft University of Technology (1991)
Idda, K.: Verbundverhalten von Betonrippenstählen bei Querzug. PhD Thesis, University of Karlsruhe (1999)
Karsan, D., Jirsa, J.O.: Behaviour of concrete under compressive loadings. Journal of the Structural Division (ASCE) 95(12), 2543–2563 (1969)
Klisiński, M., Mróz, Z.: Description of inelastic deformation and damage in concrete (Opis niesprężystych deformacji i uszkodzenia betonu, in polish), Rozprawy nr.193, Poznań University of Technology, Poznań, Poland (1988)
Koide, H., Akita, H., Tomon, N.: Size effect on flexural resistance due to bending span of concrete beams. In: Mihashi, H., Rokugo, K. (eds.) Fracture Mechanics of Concrete Structures, pp. 2121–2130. Aedificatio Publishers, Freiburg (1998)
Koide, H., Akita, H., Tomon, N.: Probability model of flexural resistance on different lengths of concrete beams. In: Melchers, E., Stewart, M.G. (eds.) Application of Statistics and Probability, Balkema, vol. 2, pp. 1053–1057 (2000)
Kotsovos, M.D., Newman, J.B.: Generalized stress-strain relations for concrete. Journal of Engineering Mechanics ASCE 104(4), 845–856 (1978)
Kotsovos, M.D.: A generalized constitutive model of concrete based on fundamental material properties. Thesis, London (1980)
Kupfer, H., Hilsdorf, H.K., Rusch, H.: Behaviour of concrete under biaxial stresses. ACI Journal 65(8), 656–666 (1969)
Le Bellego, C., Dube, J.F., Pijaudier-Cabot, G., Gerard, B.: Calibration of nonlocal damage model from size effect tests. European Journal of Mechanics - A/Solids 22(1), 33–46 (2003)
Leonhardt, F., Walter, R.: Schubversuche an einfeldigen Stahlbetonbalken mit und ohne Schubbewehrung. D.A.f.Stb. 151 (1962)
Leonhardt, F.: Grundlagen zur Bemessung im Stahlbetonbau. Springer, Heidelberg (1973)
Malvar, J.: Bond of reinforcement under controlled confinement. ACI Materials Journal 189(6), 593–601 (1992)
Pivonka, P., Lackner, R., Mang, H.A.: Shapes of loading surfaces of concrete models and their influence on the peak load and failure mode in structural analyses. International Journal of Engineering Science 41(13-14), 1649–1665 (2003)
Reinhardt, H.W., Cornelissen, H.A.W., Hordijk, D.A.: Tensile tests and failure analysis of concrete. Journal of Structural Engineering ASCE 112(11), 2462–2477 (1986)
Rossi, P.: Influence of cracking in the presence of free water on the mechanical behaviour of concrete. Magazine of Concrete Research 43(154), 53–57 (1991)
Scavuzzo, R., Stankowski, T., Gerstle, K.H., Ko, H.Y.: Stress-strain curves for concrete under multiaxial load histories. Report of University of Boulder (1983)

den Uijl, J.A., Bigaj, A.: A bond model for ribbed bars based on concrete confinement. Heron 41(3), 201–226 (1996)

Walraven, J., Lehwalter, N.: Size effects in short beams loaded in shear. ACI Structural Journal 91(5), 585–593 (1994)

van Vliet, M.R.A., van Mier, J.G.M.: Softening behaviour of concrete under uniaxial compression. In: Wittmann, F.H. (ed.) Fracture Mechanics of Concrete Structures, Proc. FRAMCUS-2, pp. 383–396. Aedlflcatlo Publishers, Freiburg (1995)

van Vliet, M.R.A., van Mier, J.G.M.: Experimental investigation of concrete fracture under uniaxial compression. Mechanics of Cohesive-Frictional Materials 1(1), 115–127 (1996)

van Vliet, M.R.A., van Mier, J.G.M.: Experimental investigation of size effect in concrete and sandstone under uniaxial tension. Engineering Fracture Mechanics 65(2-3), 165–188 (2000)

Zhang, X.X., Ruiz, G., Yu, R.C., Tarifa, M.: Fracture behaviour of high-strength concrete at a wide range of loading rates. International Journal of Impact Engineering 36(10-11), 1204–1209 (2009)

Zheng, D., Li, Q.: An explanation for rate effect of concrete strength based on fracture toughness including free water viscosity. Engineering Fracture Mechanics 71(16-17), 2319–2327 (2004)

Chapter 3
Continuous Approach to Concrete

Abstract. This Chapter presents continuous models to describe concrete behaviour in a quasi-static regime during monotonic and cyclic loading. In the case of monotonic loading, isotropic elasto-plastic, isotropic damage and smeared crack model, and in the case of cyclic loading elasto-plastic-damage models are described. An integral-type non-local and a second gradient approaches to model strain localization are introduced. In addition, bond-slip laws are presented.

The concrete behaviour can be modelled with different continuum models, e.g.: within non-linear elasticity (Palaniswamy and Shah 1974), linear fracture mechanics (Bažant and Cedolin 1979, Hilleborg 1985), endochronic theory (Bažant and Bhat 1976, Bažant and Shieh 1978), micro-plane theory (Bažant and Ožbolt 1990, Jirásek 1999), plasticity (Willam and Warnke 1975, Ottosen 1977, Hsieh et al. 1982, Pietruszczak et al. 1988, Pramono and Willam 1989, Etse and Willam 1994, Menétrey and Willam 1995, Winnicki et al. 2001, Lade and Jakobsen 2002, Majewski et al. 2008), damage (Dragon and Mróz 1979, Peerlings et al. 1998, Chen 1999, Ragueneau et al. 2000, Marzec et al. 2007) and discrete ones using e.g.: interface elements with cohesive fracture constitutive laws (Carol et al. 2001, Caballero et al. 2006, 2007), a lattice approach (Herrmann et al. 1989, Vervuurt et al. 1994, Schlangen and Garboczi 1997, Cusatis et al. 2003, Bolander and Sukumar 2005, Kozicki and Tejchman 2007) and a discrete element method (DEM) (Sakaguchi and Mühlhaus 1997, Donze at al. 1999, D'Addetta et al. 2002, Hentz et al. 2004).

We used different popular non-linear continuous constitutive models to simulate the concrete behaviour under monotonic loading.

3.1 Local Models for Monotonic Loading

3.1.1 Isotropic Elasto-Plastic Model

Failure for elasto-plastic materials with isotropic hardening/softening is described by a condition

$$f(\sigma_{ij}, \kappa) = 0 \qquad (3.1)$$

J. Tejchman, J. Bobiński: Continuous & Discontinuous Modelling of Fracture, SSGG, pp. 49–93.
springerlink.com

with σ_{ij} - stress tensor and κ - hardening/softening parameter (in general there may be several hardening/softening parameters). If $f<0$, the material behaves elastically. If $f\geq 0$, the material behaves plastically. The stresses have to remain on the failure surface (consistency condition)

$$\dot{f}(\sigma_{ij},\kappa)=0=\frac{\partial f}{\partial \sigma_{ij}}:\dot{\sigma}_{ij}+\frac{\partial f}{\partial \kappa}:\dot{\kappa}. \tag{3.2}$$

Very often Equation 3.1 can be simplified by

$$f(\sigma_{ij},\kappa)=F(\sigma_{ij})-\sigma_y(\kappa)=0, \tag{3.3}$$

where F is the function of stress tensor invariants and σ_y is the yield stress. The strain increment is equal to the sum of elastic and plastic strain increments

$$d\varepsilon_{ij}=d\varepsilon_{ij}^e+d\varepsilon_{ij}^p. \tag{3.4}$$

The stress increment is related to the increment of elastic strain

$$d\sigma_{ij}=C_{ijkl}^e d\varepsilon_{kl}^e, \tag{3.5}$$

where $C^e{}_{ijkl}$ is the elastic stiffness tensor

$$C_{ijkl}^e=\lambda\delta_{ij}\delta_{kl}+\mu(\delta_{ik}\delta_{jl}+\delta_{il}\delta_{jk}). \tag{3.6}$$

λ and μ are the Lame'a constants that are connected to the modulus of elasticity E and Poisson's ratio ν

$$\lambda=\frac{\nu E}{(1+\nu)(1-2\nu)}, \qquad \mu=\frac{E}{2(1+\nu)} \tag{3.7}$$

and δ_{ij} is a Kronecker delta. The increment of plastic strain is determined with the flow rule

$$d\varepsilon_{ij}^p=d\lambda\frac{\partial g(\sigma_{ij})}{\partial \sigma_{ij}} \tag{3.8}$$

with g as the potential function and $d\lambda$ as the positive factor of proportionality. If $f=g$, the flow rule is associated. The condition of loading and unloading is equal to

$$d\lambda\geq 0, \qquad f(\sigma_{ij},\kappa)\leq 0, \qquad d\lambda f(\sigma_{ij},\kappa)=0 \tag{3.9}$$

During plastic deformation, a stress state remains on the boundary of the elastic/plastic region

$$f(\sigma_{ij}+d\sigma_{ij},\kappa+d\kappa)=0. \tag{3.10}$$

Equation 3.10 may be rewritten in a rate form as (similarly to Eq. 3.2)

$$\frac{df}{d\sigma_{ij}}d\sigma_{ij}+\frac{df}{d\kappa}d\kappa=0. \tag{3.11}$$

Equations 3.10 and 3.11 are known as consistency conditions and allow to determine the magnitude of the plastic strain increment.

The elasto-plastic stiffness matrix $C^{ep}{}_{ijkl}$ is calculated as

$$C^{ep}_{ijkl}=C^{e}_{ijkl}-\xi\frac{C^{e}_{ijmn}(\frac{\partial f}{\partial\sigma_{mn}})(\frac{\partial g}{\partial\sigma_{pq}})^{T}C^{e}_{pqkl}}{(\frac{\partial f}{\partial\sigma_{ij}})^{T}C^{e}_{ijkl}(\frac{\partial g}{\partial\sigma_{kl}})+H}, \tag{3.12}$$

where

$$H=-(\frac{\partial f}{\partial\kappa}). \tag{3.13}$$

The proportionality factor $d\lambda$ is equal to

$$d\lambda=\frac{(\frac{\partial f}{\partial\sigma_{ij}})^{T}C^{e}_{ijkl}d\varepsilon_{kl}}{(\frac{\partial f}{\partial\sigma_{ij}})^{T}C^{e}_{ijkl}(\frac{\partial g}{\partial\sigma_{kl}})\ +H}. \tag{3.14}$$

The parameter ξ=1, if f=0 and κ>0, otherwise ξ=0. The stiffness matrix $C^{ep}{}_{ijkl}$ may be non-symmetric due to $f \neq g$. The stress increment can be calculated from

$$d\sigma_{ij}=C^{e}_{ijkl}(d\varepsilon_{kl}-d\lambda\frac{\partial f}{\partial\sigma_{ij}}). \tag{3.15}$$

Usually

$$d\lambda=\eta d\kappa. \tag{3.16}$$

when η is a constant dependent upon the model.

The constitutive models use the different stress and stress tensor invariants

$$I_1=\sigma_{11}+\sigma_{22}+\sigma_{33}, \tag{3.17}$$

$$J_2=\frac{1}{2}s_{ij}s_{ij}=\frac{1}{6}[(\sigma_{11}-\sigma_{22})^2+(\sigma_{22}-\sigma_{33})^2+(\sigma_{33}-\sigma_{11})^2]+\sigma_{12}{}^2+\sigma_{23}{}^2+\sigma_{31}{}^2, \tag{3.18}$$

$$J_3=\frac{1}{3}s_{ij}s_{jk}s_{ki}, \tag{3.19}$$

$$I_1^{\varepsilon} = \varepsilon_{11} + \varepsilon_{22} + \varepsilon_{33}, \tag{3.20}$$

$$J_2^{\varepsilon} = \frac{1}{2} e_{ij} e_{ij} = \frac{1}{6}[(\varepsilon_{11} - \varepsilon_{22})^2 + (\varepsilon_{22} - \varepsilon_{33})^2 + (\varepsilon_{33} - \varepsilon_{11})^2] + \varepsilon_{12}{}^2 + \varepsilon_{23}{}^2 + \varepsilon_{31}{}^2, \tag{3.21}$$

$$J_3^{\varepsilon} = \frac{1}{3} e_{ij} e_{jk} e_{ki}, \tag{3.22}$$

where I_1 - first stress tensor invariant, J_2 – second deviatoric stress tensor invariant, J_3 – third deviatoric stress tensor invariant, I_1^{ε} – the first strain tensor invariant, J_2^{ε} – second deviatoric strain tensor invariant and J_3^{ε} – third deviatoric strain tensor invariant. In turn, J_1 (first deviatoric stress tensor invariant) and J_1^{ε} (first deviatoric strain tensor invariant) are always

$$J_1 = s_{11} + s_{22} + s_{33} = 0, \tag{3.23}$$

$$J_1^{\varepsilon} = e_{11} + e_{22} + e_{33} = 0, \tag{3.24}$$

The stress deviator s_{ij} and strain deviator e_{ij} are calculated as

$$s_{ij} = \sigma_{ij} - \frac{I_1}{3} \delta_{ij}, \tag{3.25}$$

$$e_{ij} = \varepsilon_{ij} - \frac{I_1^{\varepsilon}}{3} \delta_{ij}. \tag{3.26}$$

To describe the behaviour of concrete, a simplified elasto-plastic model was assumed. In the compression regime, a shear yield surface based on a linear Drucker-Prager criterion and isotropic hardening and softening was used (Bobiński 2006, Marzec et al. 2007, Majewski et al. 2008) (Fig. 3.1)

$$f_1 = q + p \tan \varphi - (1 - \frac{1}{3} \tan \varphi) \sigma_c(\kappa_1). \tag{3.27}$$

where q - Mises equivalent deviatioric stress, p – mean stress and φ – internal friction angle. The material hardening/softening was defined by the uniaxial compression stress $\sigma_c(\kappa_1)$, wherein κ_1 is the hardening/softening parameter corresponding to the plastic vertical normal strain during uniaxial compression. The friction angle φ was assumed as (ABAQUS 1998)

$$\tan \varphi = \frac{3(1 - r_{bc}^{\sigma})}{1 - 2 r_{bc}^{\sigma}}, \tag{3.28}$$

wherein r_{bc}^{σ} denotes the ratio between uniaxial compression strength and biaxial compression strength (r_{bc}^{σ}=1.2). The invariants q and p were defined as

$$q = \sqrt{\frac{3}{2} s_{ij} s_{ij}}, \qquad p = \frac{1}{3} \sigma_{kk}, \tag{3.29}$$

The flow potential was assumed as

$$g_1 = q + p\tan\psi \,, \tag{3.30}$$

where ψ is the dilatancy angle ($\psi \neq \varphi$). The increments of plastic strains $d\varepsilon_{ij}^{p}$ were calculated as

$$d\varepsilon_{ij}^{p} = \frac{d\kappa_1}{1-\frac{1}{3}\tan\psi}\frac{\partial g_1}{\partial \sigma_{ij}} = \frac{d\kappa_1}{1-\frac{1}{3}\tan\psi}\left(\frac{\partial q}{\partial \sigma_{ij}} + \tan\psi\frac{\partial p}{\partial \sigma_{ij}}\right). \tag{3.31}$$

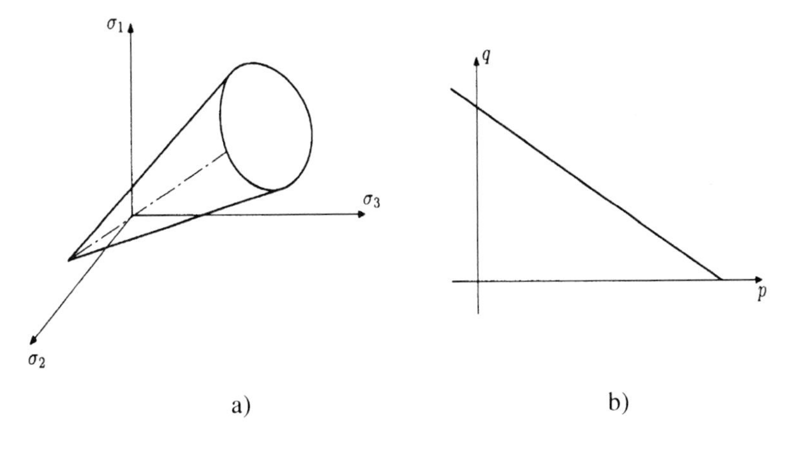

Fig. 3.1 Drucker-Prager criterion in the space of principal stresses (a) and on the plane *q-p* (b)

In turn, in the tensile regime, a Rankine criterion was used with the yield function f_2 using isotropic hardening and softening defined as (Bobiński 2006, Marzec et al. 2007, Majewski et al. 2008) (Fig. 3.2)

$$f_2 = \max\{\sigma_1, \sigma_2, \sigma_3\} - \sigma_t(\kappa_2) \,, \tag{3.32}$$

where σ_i – principal stresses, σ_t – tensile yield stress and κ_2 – softening parameter (equal to the maximum principal plastic strain ε_1^p). The associated flow rule was assumed.

The edge and vertex in the Rankine yield function were taken into account by the interpolation of 2-3 plastic multipliers according to the Koiter's rule (Pramono 1988). The same procedure was adopted in the case of combined tension (Rankine criterion) and compression (Drucker-Prager criterion).

This simple isotropic elasto-plastic model for concrete (Eqs. 3.27-3.32) requires two elastic constants: modulus of elasticity E and Poisson's ratio υ, two plastic constants: internal friction angle φ and dilatancy angle ψ, one compressive yield stress function $\sigma_c = f(\kappa_1)$ with softening and one tensile yield stress function

$\sigma_t = f(\kappa_2)$ with softening. The disadvantages of the model are the following: the shape of the failure surface in a principal stress space is linear (not paraboloidal as in reality). Thus, it is certainly not suitable in a compression regime if a large range of stress is concerned. In deviatoric planes, the shape is circular (during compression) and triangular (during tension); thus it does not gradually change from a curvilinear triangle with smoothly rounded corners to nearly circular with increasing pressure. The strength is similar for triaxial compression and extension, and the stiffness degradation due to strain localization and non-linear volume changes during loading are not taken into account.

3.1.2 Isotropic Damage Model

Continuum damage models initiated by the pioneering work of Katchanov (1986) describe a progressive loss of the material integrity due to the propagation and coalescence of micro-cracks and micro-voids. Continuous damage models (Simo and Ju 1987, Lemaitre and Chaboche 1990) are constitutive relations in which the mechanical effect of cracking and void growth is introduced with internal state variables which act on the degradation of the elastic stiffness of the material. They

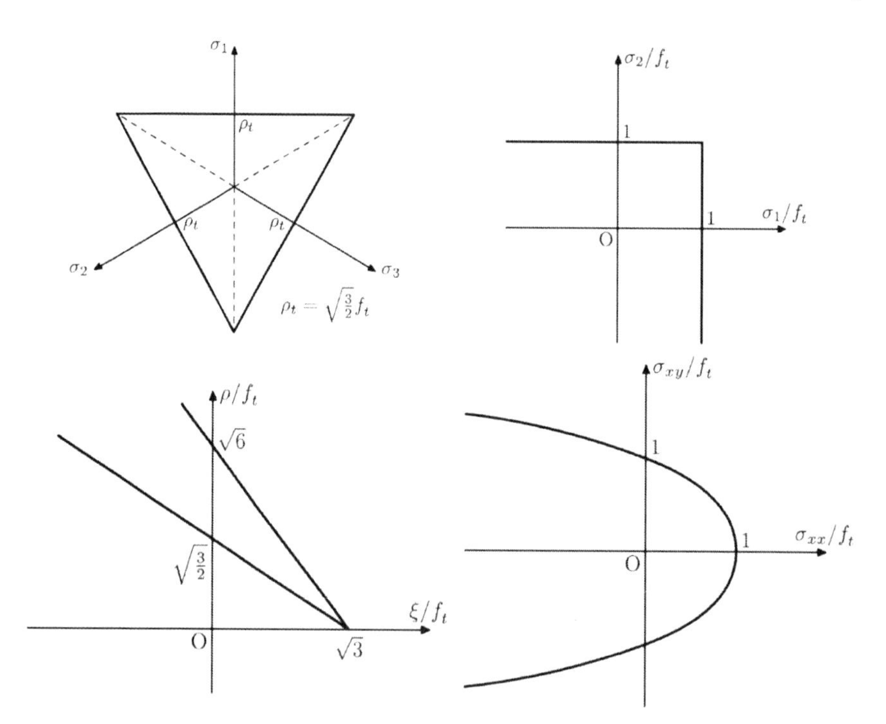

Fig. 3.2 Rankine criterion: π-plane, tensile and compressive meridian planes, σ_1-σ_2 plane, σ_{xx}-σ_{xy} plane (σ_i - principal stresses, σ_{xx} - normal stress, σ_{xy} - shear stress f_t - tensile strength, ξ - hydrostatic axis, ρ - deviatoric axis, ρ_t - deviatoric length)

can be relatively simple - isotropic (Pijaudier-Cabot 1995, Peerlings et al. 1998, Geers, et al. 1998, Huerta et al. 2003, Jirásek 2004a) or more complex - anistropic (Zhou et al. 2002, Krajcinovic and Fonseka 1981, Kuhl and Ramm 2000). The damage variable defined as the ratio between the damage area and the overall material area can be chosen as a scalar, several scalars, a second order tensor, a fourth order tensor and an eight order tensor.

A simple isotropic damage continuum model describes the material degradation with the aid of only a single scalar damage parameter D growing monotonically from zero (undamaged material) to one (completely damaged material). The stress-strain relationship is represented by

$$\sigma_{ij} = (1 - D)C^{e}_{ijkl}\varepsilon_{kl}\,, \tag{3.33}$$

where C^{e}_{ijkl} is the linear elastic stiffness matrix (including modulus of elasticity E and Poisson's ratio υ) and ε_{kl} is the strain tensor. Thus, the damage parameter D acts as a stiffness reduction factor (the Poisson's ratio is not affected by damage) between 0 and 1. The growth of damage is controlled by a damage threshold parameter κ which is defined as the maximum equivalent strain measure $\tilde{\varepsilon}$ reached during the load history up to time t. The loading function of damage is

$$f(\tilde{\varepsilon}, \kappa) = \tilde{\varepsilon} - \max\{\kappa, \kappa_0\}\,, \tag{3.34}$$

where κ_0 denotes the initial value of κ when damage begins. If the loading function f is negative, damage does not develop. During monotonic loading, the parameter κ grows (it coincides with $\tilde{\varepsilon}$) and during unloading and reloading it remains constant. To define the equivalent strain measure $\tilde{\varepsilon}$, different criteria can be used. In the book, we applied 4 different equivalent strain measures $\tilde{\varepsilon}$. First, a Rankine failure type criterion by Jirásek and Marfia (2005) was adopted

$$\tilde{\varepsilon} = \frac{\max\left\{\sigma_i^{eff}\right\}}{E}\,, \tag{3.35}$$

where σ_i^{eff} are the principal values of the effective stress

$$\sigma_i^{eff} = \sigma^{e}_{ijkl}\varepsilon_{kl}\,. \tag{3.36}$$

Second, a modified Rankine failure type criterion was applied

$$\tilde{\varepsilon} = \frac{\sigma_1^{eff} - c\left\langle -\sigma_2^{eff}\right\rangle}{E} \tag{3.37}$$

with $\sigma_1^{eff} > \sigma_2^{eff}$ and a non-negative coefficient c. This formulation is equivalent to Eq. 3.35 in a tension-tension regime, but it behaves in a different way in a mixed tension-compression regime (the coefficient c reflects the influence of the principal compressive stress). With the coefficient c=0, Eq. 3.35 is recovered.

Third, we considered a modified von Mises definition of the equivalent strain measure $\tilde{\varepsilon}$ in terms of strains (de Vree et al. 1995, Peerlings et al. 1998)

$$\tilde{\varepsilon} = \frac{k-1}{2k(1-2\nu)} I_1 + \frac{1}{2k}\sqrt{\left(\frac{k-1}{1-2\nu} I_1^{\varepsilon}\right)^2 + \frac{12k}{(1+\nu)^2} J_2^{\varepsilon}}\,. \tag{3.38}$$

The parameter k in Eq. 3.38 denotes the ratio between compressive and tensile strength of the material. A two-dimensional representation of Eq. 3.38 is given in Fig. 3.3 for k=10.

Finally, a equivalent strain measure $\tilde{\varepsilon}$ following Häuβler-Combe and Pröchtel (2005), based on the failure criterion by Hsieh-Ting-Chen (Hsieh et. al 1982), was assumed

$$\tilde{\varepsilon} = \frac{1}{2}\left(c_2\sqrt{J_2^{\varepsilon}} + c_3\varepsilon_1 + c_4 I_1^{\varepsilon} + \sqrt{\left(c_2\sqrt{J_2^{\varepsilon}} + c_3\varepsilon_1 + c_4 I_1^{\varepsilon}\right)^2 + 4c_1 J_2^{\varepsilon}} \right). \tag{3.39}$$

where ε_1 is the maximum principal total strain, c_1, c_2, c_3 and c_4 are the coefficients depending on $\alpha_1=f_t/f_c=k$, $\alpha_2=f_{bc}/f_c=r_{bc}^{\sigma}$ and α_3 and γ are the multipliers of the material strength in triaxial compression. The other definition of the equivalent strain measure $\tilde{\varepsilon}$ was used for concrete by Mazars and Pijaudier-Cabot (1989) using principal strains.

To describe the evolution of the damage parameter D in the tensile regime, the exponential softening law by (Peerlings et al. 1998) was mainly used (Fig. 3.4)

$$D = 1 - \frac{\kappa_0}{\kappa}\left(1 - \alpha + \alpha e^{-\beta(\kappa-\kappa_0)}\right), \tag{3.40}$$

where α and β are the material constants. The alternative forms of the damage evolution law were proposed by Geers et al. (1998), Zhou et al. (2002), Huerta et al. (2003) and Jirásek (2004a).

The damage evolution law determines the shape of the softening curve, i.e. material brittleness. The material softening starts when the when the equivalent strain measure reaches the initial threshold κ_0 (material hardening is neglected). The parameter β determines the rate of the damage growth (larger value of β causes a faster damage growth). In one dimensional problems, for $\varepsilon \to \infty$ (uniaxial tension), the stress approaches the value of $(1-\alpha)E\kappa_0$ (Fig. 3.4b).

The constitutive isotropic damage model for concrete requires the following 5 material constants: E, υ, κ_0, α and β (Eq. 3.35 and 3.40), 6 material constants: E, υ, κ_0, α, β and c (Eq. 3.37 and 3.40), 6 material constants: E, υ, k, κ_0, α and β (Eq. 3.38 and 3.40) or 9 material constants E, υ, κ_0, α, β, α_1, α_2, α_3 and γ (Eq. 3.39 and 3.40). The model is suitable for tensile failure (Marzec et al. 2007, Skarżyński et al. 2011) and mixed tensile-shear failure (Bobiński and Tejchman 2010). However, it cannot realistically describe irreversible deformations, volume changes and shear failure (Simone and Sluys 2004).

3.1.3 Anisotropic Smeared Crack Model

In a smeared crack approach, a discrete crack is represented by cracking strain distributed over a finite volume (Rashid 1968, Cope et. al. 1980, Willam et al. 1986, de Borst and Nauta 1985, de Borst 1986, Rots 1988, Rots and Blaauwendraad 1989). The model is capable of properly combining crack formation and a non-linear behaviour of concrete between cracks and of handling secondary cracking owing to rotation of the principal stress axes after primary crack formation. A secondary crack is allowed if the major principal stress exceeds tensile strength and/or if the angle between the primary crack and secondary crack exceeds a threshold angle. Since the model takes into account the crack orientation, it reflects the crack-induced anisotropy.

The total strain rate $\dot{\varepsilon}_{ij}$ is composed of a concrete strain rate $\dot{\varepsilon}_{ij}^{con}$ and several cracks strain rates $\dot{\varepsilon}_{ij}^{I}$, $\dot{\varepsilon}_{ij}^{II}$ etc. (de Borst 1986)

$$\dot{\varepsilon}_{ij} = \dot{\varepsilon}_{ij}^{con} + \dot{\varepsilon}_{ij}^{I} + \dot{\varepsilon}_{ij}^{II} . \tag{3.41}$$

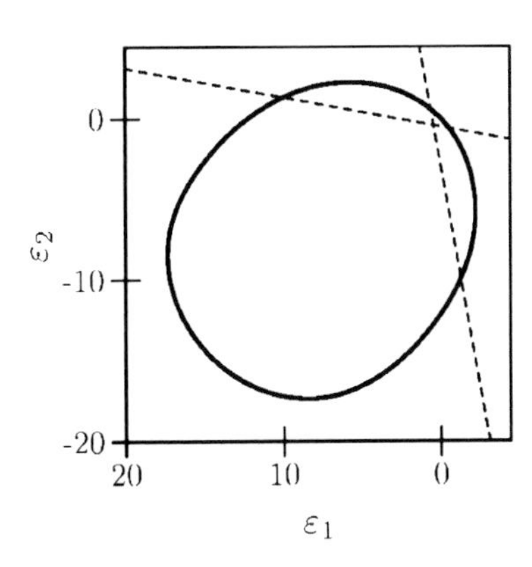

Fig. 3.3 Equivalent strain definition in principal strain space (dashed lines represent uniaxial stress paths) (Peerlings et al. 1998)

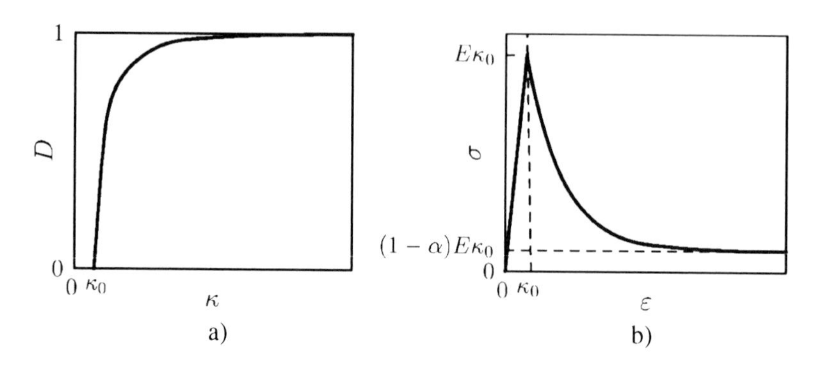

Fig. 3.4 Damage model: a) damage variable as a function of κ, b) homogeneous stress-strain behaviour during uniaxial tension (E –modulus of elasticity) (Peerlings et al. 1998)

The concrete strain rate is assumed to be related to some stress rate

$$\dot{\sigma}_{ij} = C^{con}_{ijkl}\, \dot{\varepsilon}^{con}_{kl} \ . \tag{3.42}$$

It may take into account elastic and plastic stress rates. The relation between the stress rate in the crack $\dot{\sigma}'_{ij}$ and the crack strain rate $\dot{\varepsilon}'_{kl}$ in the primary crack is given by

$$\dot{\sigma}'_{ij} = C'_{ijkl}\, \dot{\varepsilon}'_{kl} \ . \tag{3.43}$$

where the primes signify that the stress rate and the crack strain components of the primary crack are taken with respect to the coordinate system of the crack. The tensor C'_{ijkl} represents the stress-strain relation within the primary crack. Analogously, we have for a secondary crack

$$\dot{\sigma}''_{ij} = C''_{ijkl}\, \dot{\varepsilon}''_{kl} \ . \tag{3.44}$$

The double primes signify mean that the stress rate and the crack strain components of the secondary crack are taken with respect to the coordinate system of the crack. If α_{ik} are the direction cosines of the global coordination system with respect to the coordinate system of the primate crack, and if β_{ik} are the direction cosines of the global coordination system with respect to the coordinate system of the secondary crack

$$\dot{\varepsilon}_{ij}^{I} = \alpha_{ik}\alpha_{jl}\dot{\varepsilon}_{kl}^{'}, \tag{3.45}$$

$$\dot{\varepsilon}_{ij}^{II} = \beta_{ik}\beta_{jl}\dot{\varepsilon}_{kl}^{''}, \tag{3.46}$$

$$\dot{\sigma}_{ij} = \alpha_{ki}\alpha_{lj}\dot{\sigma}_{kl}, \tag{3.47}$$

$$\dot{\sigma}_{ij}^{''} = \beta_{ki}\beta_{lj}\dot{\sigma}_{kl}. \tag{3.48}$$

After a transformation of crack strain rates in global coordinates, one obtains the following relationship

$$\dot{\sigma}_{kl} = D_{klmn}^{con}(\dot{\varepsilon}_{mn} - \alpha_{mo}\alpha_{np}\dot{\varepsilon}_{op}^{'} - \beta_{mo}\beta_{np}\dot{\varepsilon}_{op}^{''}). \tag{3.49}$$

Next, after some arrangements, a relationship between $\dot{\sigma}_{ij}$ and $\dot{\varepsilon}_{kl}$ is derived. A crack is initiated if the major principal stress exceeds the tensile strength. The crack direction is usually assumed to be orthogonal to the principal tensile major stress. Between stresses and strains in the crack plane z'-y' (Fig. 3.5), we have the following relationship during loading

$$\begin{bmatrix} \dot{\sigma}_{xx}^{'} \\ \dot{\sigma}_{xy}^{'} \\ \dot{\sigma}_{xz}^{'} \end{bmatrix} = \begin{bmatrix} C & 0 & 0 \\ 0 & \beta G & 0 \\ 0 & 0 & \beta G \end{bmatrix} \begin{bmatrix} \varepsilon_{xx}^{'} \\ \varepsilon_{xy}^{'} \\ \varepsilon_{xz}^{'} \end{bmatrix} \tag{3.50}$$

and during unloading

$$\begin{bmatrix} \dot{\sigma}_{xx}^{'} \\ \dot{\sigma}_{xy}^{'} \\ \dot{\sigma}_{xz}^{'} \end{bmatrix} = \begin{bmatrix} S & 0 & 0 \\ 0 & \beta G & 0 \\ 0 & 0 & \beta G \end{bmatrix} \begin{bmatrix} \varepsilon_{xx}^{'} \\ \varepsilon_{xy}^{'} \\ \varepsilon_{xz}^{'} \end{bmatrix}, \tag{3.51}$$

where the tangent modulus C represents the relation between the normal crack strain increment and normal stress increment during loading, S is the secant modulus of the unloading branch (Fig. 3.6), G is the elastic shear modulus and β is the shear stiffness reduction factor (the term βG account for effects like aggregate interlock). In addition, a threshold angle is introduced which allows new cracks to form only when the angle between the current direction of the major principal stress and the normal to the existing cracks is exceeded. When a crack fully closes, the stiffness of the uncracked concrete is again inserted.

The model has two variants: 1) the so-called fixed crack model, in which the crack orientation is fixed when the maximum principal stress attains the tensile strength (de Borst 1986), and 2) the so-called rotating crack model, in which the crack orientation is rotated so as to always remain perpendicular to the maximum principal strain direction (Rots and Blaauwendraad 1989).

In our calculations, we assumed a simplified smeared crack approach. The total strains ε_{ij} were decomposed into the elastic ε_{ij}^{e} and inelastic crack strains ε_{ij}^{cr}

$$\varepsilon_{ij} = \varepsilon_{ij}^{e} + \varepsilon_{ij}^{cr} . \tag{3.52}$$

The concrete stresses were related to the elastic strains via

$$\sigma_{ij} = C_{ijkl}^{e} \varepsilon_{kl}^{e} , \tag{3.53}$$

Between the concrete stresses and cracked strains, the following relationship was valid (in a local coordinate system)

$$\sigma_{ij} = C_{ijkl}^{cr} \varepsilon_{kl}^{cr} \tag{3.54}$$

with the secant cracked stiffness matrix C_{ijkl}^{cr} (defined only for open cracks).

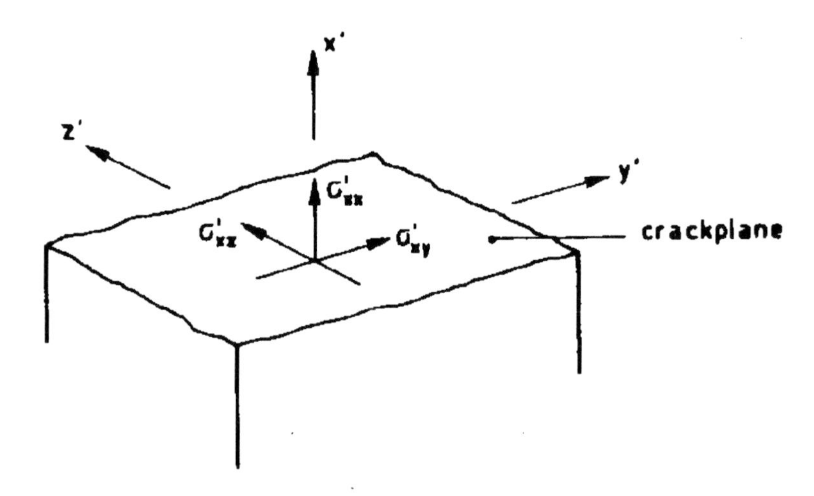

Fig. 3.5 Local coordinate system of a crack (de Borst 1986)

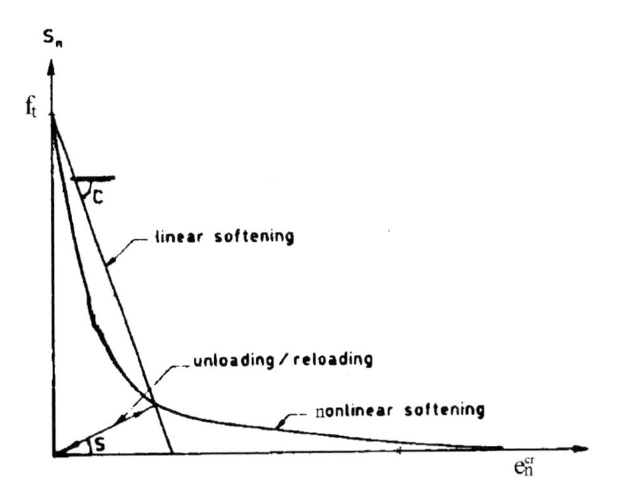

Fig. 3.6 Relationships between normal crack stress versus normal crack strain in softening range during loading, unloading and reloading (de Borst 1986)

The matrix C_{ijkl}^{cr} was assumed to be diagonal. A crack was created when the maximum tensile stress exceeded the tensile strength f_t. To define softening in a normal direction under tension, a curve by Hordijk (1991) was adopted

$$\sigma_t(\kappa) = f_t\left(\left(1 + A_1 \kappa \varepsilon_i^{cr^3}\right)\exp\left(-A_2 \varepsilon_i^{cr}\right) - \left(A_3\right)\varepsilon_i^{cr}\right) \tag{3.55}$$

with

$$A_1 = \frac{b_1}{\varepsilon_{nu}}, \qquad A_2 = \frac{b_2}{\varepsilon_{nu}}, \qquad A_3 = \frac{1}{\varepsilon_{nu}}(1 + b_1^3)\exp(-b_2), \tag{3.56}$$

where ε_i^{cr} is the normal cracked strain in a local i-direction, ε_{nu} denotes the ultimate cracked strain in tension and the material constants are b_1=3.0 and b_2=6.93, respectively.

The shear modulus G was reduced by the shear reduction factor β according to Rots and Blaauwendraad (1989)

$$\beta = \left(1 - \frac{\varepsilon_i^{cr}}{\varepsilon_{su}}\right)^p, \tag{3.57}$$

where ε_{su} is the ultimate cracked strain in shear and p is the material parameter. Combining Eqs. 3.51-3.54, the following relationship between stresses and total strains (in a local coordinate system) was derived

$$\sigma_{ij} = C^{s}_{ijkl}\varepsilon_{kl} \tag{3.58}$$

with the secant stiffness matrix C^{s}_{ijkl} as

$$C^{s}_{ijkl} = C^{e}_{ijkl} - C^{e}_{ijrs}(C^{e}_{rstu} + C^{cr}_{rstu})^{-1}C^{e}_{tukl}\,. \tag{3.59}$$

After cracking, the isotropic elastic stiffness matrix was replaced by the orthotropic one (in a local coordinate system). Two different formulations were investigated: a rotating crack model and a multi-fixed orthogonal crack model. In the first approach (rotating crack), only one crack was created which could rotate during deformation. To keep the principal axis of total strains and stresses aligned, the secant stiffness coefficient was calculated according to

$$C^{s}_{ijij} = \frac{\sigma_{ii} - \sigma_{jj}}{2\left(\varepsilon_{ii} - \varepsilon_{jj}\right)}\,. \tag{3.60}$$

The second formulation (fixed crack model) allowed one a creation of three mutually orthogonal cracks in 3D-problems (and two orthogonal cracks in 2D simulations, respectively). The orientation of the crack was described by its primary inclination at the onset, i.e. the crack did not rotate during loading.

The constitutive smeared crack model for concrete requires the following 8 material parameters: E, υ, p, c_1, c_2, f_t, ε_{su} and ε_{nu}.

3.2 Local Coupled Models for Cyclic Loading

An analysis of concrete elements under quasi-static cyclic loading under compression, tension and bending is complex mainly due to a stiffness degradation caused by fracture (Karsan and Jirsa 1969, Reinhardt et al. 1986, Hordijk 1991, Perdikaris and Romeo 1995). To take into account a reduction of both strength and stiffness, irreversible (plastic) strains and degradation of stiffness, a combination of plasticity and damage theories is in particular physically appealing since plasticity considers the first three properties and damage considers a loss of material strength and deterioration of stiffness. Within continuum mechanics, plasticity and damage couplings were analyzed by many researchers using different ideas (e.g. Lemaitre 1985, Mazars 1986, Simo and Ju 1987, Klisinski and Mróz 1988, Lubliner et al. 1989, Hansen and Schreyer 1994, Meschke et al. 1998, Pamin and de Borst 1999, Carol et al. 2001, Hansen and Willam 2001, Gatuingt and Pijaudier-Cabot 2002, Ibrahimbegovic et al. 2003, Salari et al. 2004, Bobiński and Tejchman 2006, Grassl and Jirásek 2006, Voyiadjis et al. 2009). An alternative to the cyclic concrete behaviour by elasto-plastic-damage models, is the application of an endochronic theory which deals with the plastic response of materials by means of memory integrals, expressed in terms of memory kernels (Bažant 1978, Khoei er al. 2003).

Below, the capability of 4 different coupled elasto-plastic damage continuum models to describe strain localization and stiffness degradation in a concrete beams subjected to quasi-static cyclic loading under tensile failure was investigated. The coupled elasto-plastic damage models proposed by Pamin and de Borst (1999), by Carol et al. (2001) and by Hansen and Willam (2001), by Meschke et al. (1998) and Marzec and Tejchman (2009, 2011) were taken into account.

The first model (Pamin and de Borst 1999) combines non-local damage with hardening plasticity based on effective stresses and a strain equivalence concept (Katchanov 1986, Simo and Ju 1987). The total strains are namely equal to strains in an undamaged skeleton between micro-cracks. Plastic flow can occur only in an undamaged specimen, therefore an elasto-plastic model is defined in terms of effective stresses. In the second model (Carol et al. 2001 and Hansen and Willam 2001) plasticity and damage are connected by two loading functions describing the behaviour of concrete in compression and tension. The onset and progression of material degradation is based upon the strain energy associated with the effective stress and strain. A damage approach (based on second-order tensors) simulates the behaviour of concrete under tension while plasticity describes the concrete behaviour under compression. A failure envelope is created by combining a linear Drucker-Prager formulation in compression with a damage formulation based on a conjugate force tensor and a pseudo-log damage rate in tension. In turn, in the third formulation (Meschke et al. 1998), an elasto-plastic criterion is enriched by new components including stiffness degradation. Degradation is written in the form of a Rankine'a criterion with hyperbolic softening. Following the partitioning concept of strain rates, an additional scalar internal variable is introduced into a constitutive formulation. Thus, the splitting of irreversible strains into components associated with plasticity and damage is obtained. Finally, based on an analysis of three initially presented formulations, an improved coupled formulation connecting plasticity and damage is presented using a strain equivalence hypothesis (Pamin and de Borst 1999). The plasticity is described with both a Drucker-Prager and a Rankine criterion in compression and tension, respectively. To describe the evolution of damage, a different definition is assumed for tension and compression. Finally to take into account a stiffness recovery at a crack closure and inelastic strains due to damage, combined damage in tension and compression based on stress weight factors is introduced.

Constitutive coupled model by Pamin and de Borst (1999)

The first formulation (called model '1') according to Pamin and de Borst (1999) combines elasto-plasticity with scalar damage assuming that total strains ε_{ij} are equal to strains in an undamaged skeleton (called effective strains ε_{ij}^{eff}). Elasto-plastic deformation occurs only in an undamaged specimen and is defined

$$\sigma_{ij}^{eff} = C_{ijkl}^{e}\varepsilon_{kl} , \tag{3.61}$$

The following failure criterion to describe a material response in an elasto-plastic regime is used

$$f_{ep} = F\left(\boldsymbol{\sigma}^{eff}\right) - \sigma_y\left(\kappa_p\right), \tag{3.62}$$

wherein σ_y - the yield stress and κ_p - the hardening parameter equal to plastic strain during uniaxial tension. As an elasto-plastic criterion in Eq. 3.62, the failure criterion by von Mises or Drucker-Prager may be used defined by effective stresses. Next, the material degradation is calculated with the aid of an isotropic damage model (3.38 and 3.40). The equivalent strain measure $\tilde{\varepsilon}$ can be defined in terms of total strains ε_{ij} or elastic strains ε_{ij}^{e}.

The local coupled elasto-plastic-damage model '1' requires the following 6 material constants to capture the cyclic tensile behaviour: E, ν, κ_0, α, β, k and one hardening yield stress function. In the case of linear hardening, 8 material constants are totally needed (in addition, the initial yield stress $\sigma_{yt}^{\,0}$ at κ_p=0 and hardening plastic modulus H_p).

Constitutive coupled model by Carol et al (2001) and Hansen and Willam (2001)
In the second model (called model '2'), a two-surface isotropic damage/plasticity model combining damage mechanics and plasticity in a single formulation is used (Carol et al. 2001 and Hansen and Willam 2001). A plastic region in compression is described with the aid of a linear Drucker-Prager criterion. The material experiences permanent deformation under sustained loading with no loss of the material stiffness. In turn in tension, damage is formulated in the spirit of plasticity by adopting the concept of a failure condition and a total strain rate decomposition into the elastic strain rate $\mathrm{d}\varepsilon_{ij}^{\,e}$ and degrading strain rate $\mathrm{d}\varepsilon_{ij}^{\,d}$ (as a result of the decreasing stiffness)

$$\mathrm{d}\varepsilon_{ij} = \mathrm{d}\varepsilon_{ij}^{e} + \mathrm{d}\varepsilon_{ij}^{d}. \tag{3.63}$$

The boundary between elastic and progressive damage is governed by a failure criterion

$$f_d = f\left(\sigma_{ij}, q_d\right), \tag{3.64}$$

where q_d is the damage history variable describing the evolution of the damage surface. The stress rate is equal to as

$$\mathrm{d}\sigma_{ij} = C_{ijkl}^{s}\left(\mathrm{d}\varepsilon_{kl} - \mathrm{d}\varepsilon_{kl}^{d}\right) \tag{3.65}$$

with C^s_{ijkl} as the secant stiffness matrix connected with a material damage parameter D via

$$C^s_{ijkl} = (1-D)C^e_{ijkl}. \tag{3.66}$$

The application of the secant stiffness is central to the idea that the degraded strains and stresses are reversible, since the material stiffness must degrade to make this idea possible (Carol et al. 2001, Hansen and Willam 2001). The degrading strain rate was defined as the excess strain rate beyond the value that corresponded to the stress increment according to the current secant stiffness.

The effective stress and effective strain are again experienced by the undamaged material between cracks. Assuming the energy equivalence, the mutual relationship between the nominal (observed externally) and effective stress and strain is taken as

$$\sigma_{ij} = \sqrt{1-D}\,\sigma^{eff}_{ij} \quad \text{and} \quad \varepsilon^{eff}_{ij} = \sqrt{1-D}\,\varepsilon_{ij} \tag{3.67}$$

and

$$\sigma^{eff}_{ij}\varepsilon^{eff}_{ij} = \sigma_{ij}\varepsilon_{ij}, \tag{3.68}$$

with

$$\sigma^{eff}_{ij} = C^e_{ijkl}\varepsilon^{eff}_{kl} \quad \text{and} \quad \sigma_{ij} = (1-D)C^e_{ijkl}\varepsilon_{kl}. \tag{3.69}$$

The loading function (Eq. 3.64) for the Rankine-type anistropic damage model is defined in terms of the modified principal tensile conjugate forces

$$f_d = \sum_i^3 f\left(-\hat{y}_{(i)}\right) - r(L), \tag{3.70}$$

where $-\hat{y}_{(i)}$ - the principal components of the tensile conjugate forces tensor and $r(L)$ - the resistance function as the complementary energy. The conjugate force $-\hat{y}_{(i)}$ is a second order energy tensor written with aid of the effective stresses and strains by assuming linear isotropic elasticity

$$-\hat{y}_{(i)} = \frac{1}{2}\left\langle \sigma^{eff}_i \right\rangle \left\langle \varepsilon^{eff}_i \right\rangle, \tag{3.71}$$

where $\langle \bullet \rangle$ is the Macauley bracket. Originally, Carol et al. (2001) and by Hansen and Willam (2001) proposed the following resistance function with two parameters G_f and r_o

$$r(L) = r_o e^{-\frac{r_o}{g_f}L}, \tag{3.72}$$

with g_f - the fracture energy and r_0 – the elastic strain energy at the peak of the uniaxial tension test (E - isotropic elastic modulus)

$$r_o = \frac{(f_t)^2}{2E}. \tag{3.73}$$

The parameter L in Eq. 3.72 denotes the pseudo-log damage variable and is calculated with the aid of Eqs. 3.71 and 3.72

$$L = \ln\frac{1}{1-D}, \qquad \left(D = 1 - e^{-L}\right). \tag{3.74}$$

The rate of L is

$$\dot{L} = \frac{\dot{D}}{1-D}. \tag{3.75}$$

However, Eq. 3.72 poorly influences the post-peak behaviour. Therefore, we proposed a new resistance function with also 2 parameters

$$r(L) = \frac{1}{2}E\kappa_0^2\exp\left(\frac{L(2-\beta)}{\beta}\right), \tag{3.76}$$

wherein κ_0 - the threshold strain value and β - the parameter describing softening. The resistance function adopted by Nguyen (2005) was used in numerical simulations as well

$$r(L) = \frac{1}{2}\frac{f_t^2}{E}\left(\frac{E + E_{pt}e^{-L\cdot n_t}}{Ee^{-L} + E_{pt}e^{-L\cdot n_t}}\right)^2, \tag{3.77}$$

with f_t - the tensile strength, E_{pt} - the damaged stiffness modulus and n_t - the rate of the stiffness modulus.

When simultaneously considering both damage and plasticity, the total strain rate becomes the sum of the elastic, damage and plastic rate

$$d\varepsilon_{ij} = d\varepsilon_{ij}^e + d\varepsilon_{ij}^d + d\varepsilon_{ij}^p. \tag{3.78}$$

The plastic strains are permanent while elastic and damage strains were reversible. Therefore, the elastic-damage strain $\mathrm{d}\varepsilon_{ij}^{ed}$ is introduced in the total value

$$\mathrm{d}\varepsilon_{ij} = \mathrm{d}\varepsilon_{ij}^{ed} + \mathrm{d}\varepsilon_{ij}^{p} . \tag{3.79}$$

The local coupled elasto-plastic-damage model requires the material constants: E, ν, ϕ, ψ, g_f and r_0 (Eq. 3.72), E, ν, ϕ, ψ, κ_0 and β (Eq. 3.76), E, ν, ϕ, ψ, f_t, E_{pt}, n_t (Eq. 3.77) and one compressive hardening/softening yield stress function.

Constitutive coupled model by Meschke et al. (1998)
In the third model (called model '3'), another concept of coupling was introduced. An elasto-plastic criterion is enhanced by a new component describing the stiffness degradation (Meschke et al. 1998). The permanent strain rate decomposition is assumed as

$$\mathrm{d}\varepsilon_{ij}^{pd} = \mathrm{d}\varepsilon_{ij}^{p} + \mathrm{d}\varepsilon_{ij}^{d} . \tag{3.80}$$

The plastic damage strain rate $\mathrm{d}\varepsilon_{ij}^{pd}$ is calculated as in classical plasticity. The component associated with degradation and plasticity is obtained by introducing a scalar constant γ between zero and one ($0\leq\gamma\leq 1$)

$$\mathrm{d}\varepsilon_{ij}^{p} = (1-\gamma)\,\mathrm{d}\varepsilon_{ij}^{pd} \qquad \text{and} \qquad \mathrm{d}\varepsilon_{ij}^{d} = \gamma\,\mathrm{d}\varepsilon_{ij}^{pd} . \tag{3.81}$$

The parameter γ enables one a simple splitting of effects connected with an inelastic slip process (which caused an increase of plastic strain) and a deterioration of microstructure (which contributed to an increase of the compliance tensor). The evolution law for the compliance tensor is ($d\lambda$ - proportionality factor)

$$\dot{\boldsymbol{D}} = \gamma \times d\lambda \frac{\dfrac{\partial f}{\partial \boldsymbol{\sigma}} \left(\dfrac{\partial f}{\partial \boldsymbol{\sigma}} \right)^T}{\left(\dfrac{\partial f}{\partial \boldsymbol{\sigma}} \right)^T \boldsymbol{\sigma}} . \tag{3.82}$$

The stresses are updated analogously to the standard plasticity theory. To simulate concrete softening in tension, a hyperbolic softening law is chosen

$$\sigma_t(\kappa) = \frac{f_t}{\left(1 + \dfrac{\kappa}{\kappa_0} \right)^2} , \tag{3.83}$$

where κ_0 - the parameter adjusted to the fracture energy.

This coupled elasto-plastic-damage model requires in tension the following 5 parameters: E, υ, f_t, κ_0 and γ.

Improved coupled elasto-plastic-damage model (Marzec and Tejchman 2010)
In order to describe the cyclic concrete behaviour under both tension and compression, an improved coupled model (called model '4') was proposed based on the model '1' by Pamin and de Borst (1999) (which combines elasto-plasticity with a scalar damage assuming a strain equivalence hypothesis). The elasto-plastic deformation is defined in terms of effective stresses according to Eq. 3.61. Two criteria are used in an elasto-plastic regime (Marzec et al. 2007, Majewski et al. 2008): a linear Drucker-Prager criterion with a non-associated flow rule in compression and a Rankine criterion with an associated flow rule in tension defined by effective stresses (Chapter 3.1). Next, the material degradation is calculated within damage mechanics, independently in tension and compression using one equivalent strain measure $\tilde{\varepsilon}$ proposed by Mazars (1986) (ε_i - principal strains)

$$\tilde{\varepsilon} = \sqrt{\sum_i \langle \varepsilon_i \rangle^2} . \tag{3.84}$$

In tension, the same damage evolution function by Peerlings et al. (1998) as in the model '1' is chosen (Eq. 3.40). In turn, in compression, the definition by Geers (1997) is adopted

$$D_c = 1 - \left(1 - \frac{\kappa_0}{\kappa}\right)\left(0.01\frac{\kappa_0}{\kappa}\right)^{\eta_1} - \left(\frac{\kappa_0}{\kappa}\right)^{\eta_2} e^{-\delta(\kappa-\kappa_0)} , \tag{3.85}$$

where η_1, η_2 and δ are the material constants. Equation 3.85 allows for distinguishing different stiffness degradation under tension and under compression. Damage under compression starts to develop later than under tension that is consistent with experiments. The damage term '1-D' (Eq. 3.33) is defined as in ABAQUS (1998) following Lubliner et al. (1989) and Lee and Fenves (1998a)

$$(1-D) = (1 - s_c D_t)(1 - s_t D_c) , \tag{3.86}$$

with two splitting functions s_t and s_c controlling the magnitude of damage

$$s_t = 1 - a_t w(\boldsymbol{\sigma}^{eff}) \qquad \text{and} \qquad s_c = 1 - a_c\left(1 - w(\boldsymbol{\sigma}^{eff})\right), \tag{3.87}$$

where a_t and a_c are the scale factors and $w(\boldsymbol{\sigma}^{eff})$ denotes the stress weight function which may be determined with the aid of principal effective stresses according to Lee and Fenves (1998a)

$$w\left(\boldsymbol{\sigma}^{eff}\right)=\begin{cases}0 & \text{if } \boldsymbol{\sigma}=0\\ \dfrac{\sum\left\langle\sigma_i^{eff}\right\rangle}{\sum\left|\sigma_i^{eff}\right|} & \text{otherwise}\end{cases}. \tag{3.88}$$

For relatively simple cyclic tests (e.g. uniaxial tension, bending), the scale factors a_t and a_c can be a_t=0 and a_c=1, respectively. Thus, the splitting functions are: $s_t=1.0$ and $s_c=w\left(\boldsymbol{\sigma}^{eff}\right)$. For uniaxial loading cases, the stress weight function becomes

$$w\left(\sigma^{eff}\right)=\begin{cases}1 \text{ if } \sigma^{eff}>0\\ 0 \text{ if } \sigma^{eff}<0\end{cases}. \tag{3.89}$$

Thus, under pure tension the stress weight function $w=1.0$ and under pure compression $w=0$.

Our constitutive model with a different stiffness in tension and compression and a positive-negative stress projection operator to simulate crack closing and crack re-opening is thermodynamically consistent. It shares main properties of the model by Lee and Fenves (1998a), which was proved to not violate thermodynamic principles (plasticity is defined in the effective stress space, isotropic damage is used and the stress weight function is similar). Moreover Carol and Willam (1996) showed that for damage models with crack-closing-re-opening effects included, only isotropic formulations did not suffer from spurious energy dissipation under non-proportional loading (in contrast to anisotropic ones).

Our local coupled elasto-plastic-damage model requires the following 10 material constants E, ν, κ_0, α, β, η_1, η_2, δ, a_t, a_c and 2 hardening yield stress functions (the function by Rankine in tension and by Drucker-Prager in compression). If the tensile failure prevails, one yield stress function by Rankine can be used only.

The quantities σ_y (in the hardening function) and κ_0 are responsible for the peak location on the stress-strain curve and a simultaneous activation of a plastic and damage criterion (usually the initial yield stress in the hardening function σ_y^0=3.5-6.0 MPa and κ_0=(8-15)×10^{-5} under tension). The shape of the stress-strain-curve in softening is influenced by the constant β in tension (usually β=50-800), and by the constants δ and η_2 in compression (usually δ=50-800 and η_2=0.1-0.8). The parameter η_2 influences also a hardening curve in compression. In turn, the stress-strain-curve at the residual state is affected by the constant α (usually α=0.70-0.95) in tension and by η_1 in compression (usually η_1=1.0-1.2). Since the parameters α and η_1 are solely influenced by high values of κ, they can arbitrarily be assumed for softening materials. Thus, the most crucial material constants are σ_y^0, κ_0, β, δ and η_2. In turn, the scale factors a_t and a_c influence the damage magnitude in tension and compression. In general, they vary between zero and one. There do not exist unfortunately the experimental data allowing for determining the magnitude of a_t and a_c. Since, the compressive stiffness is recovered upon the crack closure as the load changes from tension to compression and the tensile stiffness is not recovered due to compressive micro-cracks, the parameters a_c and a_t can be taken for the sake of simplicity as a_c=1.0 and a_t=0 for many different simple loading cases as e.g. uniaxial tension and bending. The equivalent strain measure $\tilde{\varepsilon}$ can be defined in terms of total strains or elastic strains. The drawback of our formulation is the necessity to tune up constants controlling plasticity and damage to activate an elasto-plastic criterion and a damage criterion at the same moment. As a consequence, the chosen yield stress σ_y may be higher than this obtained directly in laboratory simple monotonic experiments.

The material constants E, ν, κ_0, β, α, η_1, η_2, δ and two hardening yield stress functions can be determined for concrete with the aid of 2 independent simple monotonic tests: uniaxial compression test and uniaxial tension (or 3-point bending) test. However, the determination of the damage scale factors a_t and a_c requires one full cyclic compressive test and one full cyclic tensile (or 3-point bending) test.

Table 3.1 shows a short comparison between four coupled models. The major drawback of first 3 formulations is the lack of the damage differentiation in tension and compression, stiffness recovery associated with crack closing and relationship between the tensile and compression stiffness during a load direction change. To describe these phenomena, additional material constants have to be included.

The damage hardening/softening laws assumed in constitutive models have been fully based on experimental data from uniaxial compression and uniaxial tension tests which in turn strongly depend on the concrete nature, specimen size and boundary and loading conditions. It means that they are not physically based. This fact reveals the necessity to derive macroscopic laws in a softening regime from real micro-structure evolutions in materials during homogeneous tests using e.g. a discrete element model (Widulinski et al. 2011).

The coupled model '1' can be enriched by the crack-closure effect in a similar way as our model '4'. For the models '2' and '3' due to their different structure, the crack-closure effect can be incorporated by introducing a projection operator (model '2') or by modifying the evolution law for the compliance tensor (model '3').

Table 3.1 Comparison between four local coupled elasto-plastic damage formulations to describe concrete behaviour (Marzec and Tejchman 2009, 2011)

Nr.	Plastic strains in tension/compression	Stiffness degradation	Unique strain division	Stiffness recovery	Number of material parameters
Model '1'	Yes	Yes	No	No	Elastic: 2 Plastic: 1 (tens.) 3 (compr.) Damage: 4
Model '2'	Yes (only in compression)	Yes (only in tension)	Yes	No	Elastic: 2 Plastic: 3 Damage: 3-4
Model '3'	Yes	Yes	Yes	No	Elastic: 2 Plastic: 2 Damage: 1
Model '4'	Yes	Yes	No	Yes	Elastic: 2 Plastic: 1 (tension), 3 (compression) Damage:2 (tension), 3 (compression) Scale factors: 2

3.3 Regularization Techniques

Classical FE-simulations of the behaviour of materials with strain localization within continuum mechanics are not able to describe properly both the thickness of localization and distance between them. They suffer from mesh sensitivity (its size and alignment) and produce unreliable results. The strains concentrate in one element wide zones and the computed force-displacement curves are mesh-dependent (especially in a post-peak regime). The reason is that differential equations of motion change their type (from elliptic to hyperbolic in static problems) and the rate boundary value problem becomes ill-posed (de Borst et al. 1992). Thus, classical constitutive continuum models require an extension in the form of a characteristic length to properly model the thickness of localized zones. Such extension can be by done within different theories: a micro-polar (Mühlhaus

1986, Sluys 1992, Tejchman and Wu 1993, Tejchman et al. 1999), a strain gradient (Zbib and Aifantis 1989, Mühlhaus and Aifantis 1991, Pamin 1994, de Borst and Pamin 1996, Pamin 2004, Sluys and de Borst 1994, Peerlings et al. 1998, Meftah and Reynouard 1998, Pamin and de Borst 1998, Chen et al. 2001, Zhou et al. 2002, Askes and Sluys 2003), a viscous (Sluys 1992, Sluys and de Borst 1994, Neddleman 1988, Loret and Prevost 1990, Lodygowski and Perzyna 1997, Winnicki et al. 2001, Pedersen et al. 2008, Winnicki 2009) and a non-local (Pijaudier-Cabot and Bažant 1987, Bažant and Lin, 1988, Brinkgreve 1994, de Vree et al. 1995, Strömberg and Ristinmaa 1996, Marcher and Vermeer 2001, Maier 2002, 2003, di Prisco et al. 2002, Bažant and Jirásek 2002, Jirásek and Rolshoven 2003, Tejchman 2004).

Other numerical technique which also enables to remedy the drawbacks of a standard FE-method and to obtain mesh-independency during formation of cracks, are approaches with strong discontinuities which enrich continuous displacement modes of the standard finite elements with additional discontinuous displacements (Belytschko et al. 2001, 2009, Simone et al. 2002, Asferg et al. 2006, Oliver et al. 2006) or approaches with cohesive (interface) elements (Ortiz and Pandolfi 1999, Zhou and Molinari 2004) (Chapters 4.1 and 4.2). In the first approaches, discontinuity paths are placed inside the elements irrespective of the size and specific orientation. In the latter approaches, discontinuity paths are defined at the edges between standard finite elements. The most realistic approach to concrete (a continuous-discontinuous approach) was used by Moonen et al. (2008).

Below two different regularization methods (integral-type non-local and explicit second-gradient) are described in detail.

3.3.1 Integral-Type Non-local Approach

A non-local model of the integral type (so called "strongly non-local model") was used as a regularisation technique:

a) to properly describe strain localization (width and spacing),
b) to preserve the well-posedness of the boundary value problem,
c) to obtain mesh-independent results,
d) to take into account material heterogeneity and
e) to include a characteristic length of micro-structure for simulations of a deterministic size effect (Pijaudier-Cabot and Bažant 1987, Bažant and Jirásek 2002, Bobiński and Tejchman 2004).

It is based on a spatial averaging of tensor or scalar state variables in a certain neighbourhood of a given point (i.e. material response at a point depends both on the state of its neighbourhood and the state of the point itself). Thus, a characteristic length l_c can be incorporated and softening can spread over material points. It is in contrast to classical continuum mechanics, wherein the principle of local action holds (i.e. the dependent variables in each material point depend only upon the values of the independent variables at the same point), and softening at one material point does not affect directly the yield surfaces of other points. It has a physical motivation due to the fact the distribution of stresses in the interior of concrete is strongly non-uniform (Fig. 2.1b). Polizzotto et al. (1998) laid down a thermodynamic a consistent formulation of non-local plasticity. In turn, Borino et al. (2003) and Nguyen (2008) laid down a thermodynamic consistent formulation of non-local damage mechanics.

Usually it is sufficient to treat non-locally only one variable controlling material softening or degradation (Brinkgreve 1994, Bažant and Jirásek 2002, Huerta et al. 2003).

A full non-local model assumes a relationship between average stresses $\bar{\sigma}_{ij}$ and averaged strains $\bar{\varepsilon}_{ij}$ defined as

$$\bar{\sigma}_{ij}(x) = \frac{1}{\bar{V}} \iiint \omega \| x - \xi \| \sigma_{ij}(\xi) d\xi_1 d\xi_2 d\xi_3 \tag{3.90}$$

and

$$\bar{\varepsilon}_{ij}(x) = \frac{1}{\bar{V}} \iiint \omega \| x - \xi \| \varepsilon_{ij}(\xi) d\xi_1 d\xi_2 d\xi_3 \tag{3.91}$$

where $\bar{\sigma}_{ij}(\boldsymbol{x})$ and $\bar{\varepsilon}_{ij}(\boldsymbol{x})$ are the non-local softening parameters, $\boldsymbol{x}$ are the coordinates of the considered point, $\boldsymbol{\xi}$ are the coordinates of the surrounding points, ω denotes the weighting function and $\bar{V}$ denotes the weighed body volume

$$\bar{V} = \iiint \omega(\| x - \xi \|) d\xi_1 \, d\xi_2 d\xi_3 \, . \tag{3.92}$$

In general, it is required that the weighting function ω should not alter a uniform field which means that it must satisfy the normalizing condition (Bažant and Jirásek 2002).

As a weighting function ω (called also an attenuation function or a non-local averaging function), a Gauss distribution function was used which is in 2D calculations

$$\omega(r) = \frac{1}{l_c \sqrt{\pi}} e^{-\left(\frac{r}{l_c}\right)^2} \tag{3.93}$$

with

$$\int_{-\infty}^{\infty} \omega(r) \mathrm{d}r = 1 \, . \tag{3.94}$$

where the parameter l_c is a characteristic length of micro-structure and r is a distance between two material points. The averaging in Eq. 3.93 is restricted to a small representative area around each material point (the influence of points at the distance of $r=3l_c$ is only of 0.01%) (Fig. 3.7). A characteristic length is usually related to the micro-structure of the material (e.g. maximum aggregate size and crack spacing in concrete, pore and grain size in granulates, crystal size in metals). It is determined with an inverse identification process of experimental data (Geers et al. 1996, Mahnken and Kuhl 1999, Le Bellego et al. 2003).

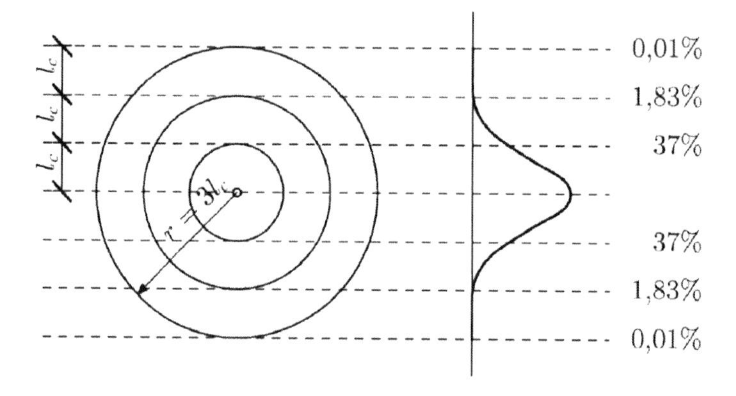

Fig. 3.7 Region of the influence of characteristic length l_c and weighting function ω (Bobiński and Tejchman 2004)

However, the determination of a representative characteristic length of micro-structure l_c is very complex in concrete since strain localization can include a mixed failure mode (cracks and shear zones) and a characteristic length (which is a scalar value) is related to the fracture process zone with a certain volume which changes during a deformation (the width of the fracture process zone increases according to e.g. Pijaudier-Cabot et al. 2004, but decreases after e.g. Simone and Sluys 2004). In turn, other researchers conclude that the characteristic length is not a constant, and it depends on the type of the boundary value problem and the current level of damage (Ferrara and di Prisco 2001). Thus, a determination of l_c requires further numerical analyses and measurements, e.g. using a Digital Image Correlation (DIC) technique (Bhandari and Inoue 2005). FE simulations of tests with measured load-displacement curves and widths of fracture process zones for different boundary value problems and specimen sizes are of importance. According to Pijaudier-Cabot and Bažant (1987), Bažant and Oh (1983), it is in concrete approximately $3\times d_a^{max}$, where d_a^{max} is the maximum aggregate size.

Other representations can be also used for the function ω (Ožbolt 1995, Akkermann 2000, Jirásek 2004a, di Prisco et al. 2002, Bažant and Jirásek 2002); e.g. the polynomial bell-shaped function reads

$$\omega = (1 - \frac{r^2}{R^2})^2 , \quad (3.95)$$

where R (interaction radius) is a parameter related to a characteristic length. To improve the behaviour of a non-local averaging in the vicinity of the boundary of a finite body, Polizzotto (2002) proposed the weight distribution preserving a uniform field and symmetry:

$$\omega = \omega(r) + [1 - \int_V \omega(r) dr] \delta(r) , \quad (3.96)$$

where δ denotes the Dirac distribution. This function is corrected by a suitable multiple of the local value to compensate for boundary effects. The FE-results by Jirásek et al. (2004b) show that the type of a non-local averaging near boundaries influences the peak of the load-displacement curve; the averaging with a symmetric local correction by Eq. 3.96 results in a lower resistance.

Our FE calculations were carried out mainly with the characteristic length l_c=1.5 mm (for fine-grained concrete) and l_c=5 mm (usual concrete) based on DIC tests (Skarżynski et al. 2011, Syroka 2011).

Monotonic loading

In the calculations within elasto-plasticity (Eqs. 3.27-3.32), the softening parameters κ_i (i=1, 2) were assumed to be a linear combination of the local and non-local values (independently for both yield surfaces f_i) (so called 'over-nonlocal' formulation, Brinkgreve 1994, Bobiński and Tejchman 2004)

$$\bar{\kappa}_i(\boldsymbol{x}) = (1-m)\kappa_i(\boldsymbol{x}) + m \frac{\int_V \omega(\|\boldsymbol{x}-\boldsymbol{\xi}\|)\kappa_i(\boldsymbol{\xi})d\boldsymbol{\xi}}{\int_V \omega(\|\boldsymbol{x}-\boldsymbol{\xi}\|)d\boldsymbol{\xi}} \qquad i=1, 2, \quad (3.97)$$

where $\bar{\kappa}_i(\boldsymbol{x})$ are the non-local softening parameters and m is the additional non-local parameter controlling the size of the localized plastic zone. For m=0, a local approach is obtained and for m=1, a classical non-local model is recovered (Pijaudier-Cabot and Bažant 1987, Bažant and Lin 1988). If the parameter m>1, the influence of the non-locality increases and the localized plastic region reaches a finite mesh-independent size (Brinkgreve 1994, Bažant and Jirásek 2002, Bobiński and Tejchman 2004). Brinkgreve (1994) derived an analytical formula for the thickness of a localized zone in an one-dimensional bar during tension with necking using a modified non-local approach by Eq. 3.97. According to this formula, if the non-local parameter was m=1, the thickness of the localized zone was equal to zero (similarly as in an usual local approach). The enhanced non-local elasto-plastic model has in addition two material parameters m and l_c.

The softening non-local parameters near boundaries were calculated also on the basis of Eqs. 3.93-3.95 (which satisfy the normalizing condition). During a FE-analysis, the integral in Eq. 3.96 was replaced by a summation operator

$$\bar{\kappa}_i(x_i) = (1-m)\kappa_i(x_i) + m\frac{\sum_j^{np}\omega(\|x_i-\xi_j\|)\kappa_i(\xi_j)V_j}{\sum_j^{np}\omega(\|x_i-\xi_j\|)V_j}, \tag{3.98}$$

where np is the number of all integration points in the whole body, x_j stand for co-ordinates of the integration point in each element and V_j is the actual element volume.

In the calculations within isotropic damage mechanics (Chapters 5 and 7), the equivalent strain measure $\tilde{\varepsilon}$ (Eqs. 3.35, 3.37, 3.38 and 3.39) was replaced by its non-local definition (Marzec et al. 2007)

$$\bar{\varepsilon} = \frac{\int_V \omega(\|x-\xi\|)\tilde{\varepsilon}(\xi)d\xi}{\int_V \omega(\|x-\xi\|)d\xi}. \tag{3.99}$$

It is to note, that in some other damage formulations, the use of a this non-local variable causes problems with energy dissipation can lead to an improper solution (Jirásek 1998, Jirásek and Rolshoven 2003, Borino et al. 2003). This case occurs in the coupled elasto-plastic-damage model '2'.

In the smeared crack approach, the secant matrix C^s_{ijkl} (Eqs. 3.52-3.60) was calculated with the non-local strain tensor $\bar{\varepsilon}_{kl}$ defined (independently for all tensor components) as (Jirásek and Zimmermann 1998)

$$\bar{\varepsilon}_{kl}(x) = \frac{\int \omega(\|x-\xi\|)\varepsilon_{kl}(\xi)d\xi}{\int \omega(\|x-\xi\|)d\xi}. \tag{3.100}$$

Thus, the resulting stresses were calculated from the relationship

$$\sigma_{ij} = C^s_{ijkl}(\bar{\varepsilon}_{kl})\varepsilon_{kl}. \tag{3.101}$$

Cyclic loading

In the first coupled elasto-plastic damage model (model '1'), non-locality was applied in damage (softening was not allowed in elasto-plasticity). The equivalent strain measure was replaced by its non-local counterpart (Eq. 3.94). In the second coupled model, non-locality was prescribed in tension to the energy release Y (Marzec and Tejchman 2009)

$$Y = \frac{1}{2}\varepsilon_{ij} C^{e}_{ijkl} \varepsilon_{kl}\,, \tag{3.102}$$

which is a component of the loading function in Eq. 3.70. The non-local damage energy was composed of a local and non-local term calculated in the current (i) and previous iteration (i-1) (Strömberg and Ristinmaa 1996, Rolshoven 2003)

$$\bar{Y}^{*}_{(i)} = \left(1 - m + mA_{kl}\right) Y_{(i)} + m\left(\bar{Y}_{(i-1)} - Y_{(i-1)} A_{kl}\right), \tag{3.103}$$

wherein m – the non-local parameter controlling the size of the localized plastic zone and distribution of the plastic strain and A_{kl} - the component of a non-local matrix

$$A_{kl} = \frac{\omega\left(\left\| x^{k} - x^{l} \right\|\right) V\left(x^{l}\right)}{\sum\limits_{j=1} \omega\left(\left\| x^{k} - x^{j} \right\|\right) V\left(x^{j}\right)}, \tag{3.104}$$

where $V(x^l)$ is the volume associated to the integration point l. In the third model, the rates of the softening parameter were averaged according to the Brinkreve's formula (Eq. 3.97) during both tension and compression. Finally, in the improved coupled formulation (model '4'), the non-locality was introduced similarly as in the model '1' i.e. local plasticity was combined with non-local damage. However another possibility non-local plasticity combined with local damage was also considered. In this case for both tension and compression, the non-locality was applied according to the Brinkreve's formula (Eq. 3.97).

A numerical problem in non-local elasto-plastic models is the way how to calculate non-local terms since the plastic rates are unknown in advance. The plastic strain rates can be approximated by the total strain rates $d\varepsilon$ (Brinkgreve 1994) or calculated iteratively in an exact way according to the algorithm given by Strömberg and Ristinmaa (1996), and Jirásek and Rolshoven (2003). To simplify the calculations, the non-local rates were replaced by their approximation $\Delta\kappa_i^{est}$ calculated on the basis of the known total strain increment values (Brinkgreve 1994):

$$\Delta\bar{\kappa}_i\left(\boldsymbol{x}\right) \approx \Delta\kappa_i(x) + m\left(\frac{\int_V \omega\left(\left\| x - \xi \right\|\right) \Delta\kappa_i^{est}(\xi)\, d\xi}{\int_V \omega\left(\left\| x - \xi \right\|\right) d\xi} - \Delta\kappa_i^{est}(x)\right), \quad i=1,2 \tag{3.105}$$

The plastic strain rates can be approximated by the total strain rates $d\varepsilon$. Eq. 3.105 enables one to 'freeze' the non-local influence of the neighbouring points and to determine the actual values of the softening parameters using the same procedures as in a local formulation. The strain rates can be calculated in all integration points of the specimen, in the integration points where only plastic strains occur or only in the integration points where both plastic strains and softening simultaneously occur. The FE-results show an insignificant influence of the calculation method of non-local plastic strain rates. An approximate method proposed by Brinkgreve (1994) is less time consuming (by ca.30%) (Bobiński and Tejchman 2004).

3.3.2 Second-Gradient Approach

Second gradient models have often been used for ductile materials (metals) (Fleck and Hutchinson 1993, Menzel and Steinmann 2000), quasi-brittle materials (rock, concrete) (Sluys 1992, and Pamin 1994) and granular materials (Vardoulakis and Aifantis 1991, Chambon et al. 2001, Maier 2002, Tejchman 2004). The gradient terms are thought to reflect the fact that below a certain size scale the interaction between the micro-structural carriers of deformation is non-local (Aifantis 2003). The constitutive models capture gradients in different ways. They usually involve the second gradient of a plastic strain measure (Laplacian) in the yield or potential function (plasticity) or in the damage function (damage mechanics). The plastic multiplier which is connected to the plastic strain measure is considered as a fundamental unknown and is computed at global level simultaneously with the displacement degrees of freedom (de Borst and Mühlhaus 1992) (in the classical theory of plasticity, the plastic multiplier is determined from a simple algebraic equation, Chapter 3). Such gradient model obviously requires a C^1-continuous interpolation of the plastic multiplier field. This requirement is fulfilled by e.g. element with the 8-nodal serendipity interpolation of displacements and 4-nodal Hermitian interpolation of plastic strain with 2×2 Gaussian integration (Pamin 1994). Alternatively, all strain gradients can be taken into account (Zervos et al 2001). The stress is conjugate to the strain rate, and the so-called double stress is conjugate to its gradient. To preserve that the derivatives are continuous across two-dimensional element boundaries, a triangular element of C^1 continuity with 36 degrees of freedom can be used (Maier 2002). The degrees of freedom at each node for each displacement are the displacement itself, its both first order derivatives and all three second order derivatives. The model requires a relationship between the double stress and strain gradient. The gradient terms can be evaluated not only by using additional complex shape functions but also by applying explicit method in the form of a standard central difference scheme.

Gradient–type regularization can be derived from non-local models. By expanding an arbitrary state variable $\kappa(x+r)$ into a Taylor series around the point $r=0$, choosing the error function ω as the weighting function and neglecting the terms higher than the second order, the following relationship is obtained for a non-local gradient of κ for one-dimensional problems (Pamin 1994):

$$\kappa^*(x)=\frac{1}{A}[\int_{-\infty}^{\infty}\frac{1}{l_c\sqrt{\pi}}e^{-(r/l)^2}\kappa(x)dr+\int_{-\infty}^{\infty}\frac{r}{l_c\sqrt{\pi}}e^{-(r/l_c)^2}\frac{\partial\kappa(x)}{\partial x}dr+$$
$$\int_{-\infty}^{\infty}\frac{r^2}{2l_c\sqrt{\pi}}e^{-(r/l_c)^2}\frac{\partial^2\kappa(x)}{\partial x^2}dr+\int_{-\infty}^{\infty}\frac{r^3}{6l_c\sqrt{\pi}}e^{-(r/l_c)^2}\frac{\partial^3\kappa(x)}{\partial x^3}dr+ \tag{3.106}$$
$$\int_{-\infty}^{\infty}\frac{r^4}{24l_c\sqrt{\pi}}e^{-(r/l_c)^2}\frac{\partial^4\kappa(x)}{\partial x^4}dr+....]=\kappa+l_c\frac{\partial\kappa}{\partial x}+\frac{l_c^2}{4}\frac{\partial^2\kappa}{\partial x^2}.$$

For two-dimensional problems, the enhanced variable κ^* is

$$\kappa^*(x,y)=\kappa+l_c(\frac{\partial\kappa}{\partial x}+\frac{\partial\kappa}{\partial y})+\frac{l_c^2}{4}(\frac{\partial^2\kappa}{\partial x^2}+\frac{\partial^2\kappa}{\partial y^2}+2\frac{\partial^2\kappa}{\partial x\partial y}). \tag{3.107}$$

The odd derivative can be canceled because of the implicit assumption of isotropy (de Borst et al. 1992). Thus, the enhanced variable κ^* is equal to

$$\kappa^*(x,y)=\kappa+\frac{l_c^2}{4}(\frac{\partial^2\kappa}{\partial x^2}+\frac{\partial^2\kappa}{\partial y^2}+2\frac{\partial^2\kappa}{\partial x\partial y}). \tag{3.108}$$

Instead of using complex shape functions to describe the evolution of the second gradient of κ, a central difference scheme was applied (di Prisco et al. 2002). The advantages of such approach are: simplicity of computation, little effort to modify each commercial FE-code and high computation efficiency. To take into account the effect of not only adjacent elements (as in the standard difference method), one assumed in the book a polynomial interpolation of the function κ of the fourth order in both directions:

$$\kappa(x)=Ax^4+Bx^3+Cx^2+Dx+E, \tag{3.109}$$

$$\kappa(y)=Ay^4+By^3+Cy^2+Dy+E, \tag{3.110}$$

where A, B, C, D and E are constants. From the theory of a finite difference method (when the difference steps dx and dy are infinitesimal), the second derivatives of the variable κ can be approximated in each triangular element of the quadrilateral composed of 4 triangles, e.g. in the triangle '13' of Fig. 3.8 (for a mesh regular in the vertical and horizontal direction) as

$$\frac{\partial^2 \kappa_{13}}{\partial x^2} = \frac{1}{dx^2}[-\frac{1}{12}\kappa_3 + \frac{16}{12}\kappa_8 - \frac{30}{12}\kappa_{13} + \frac{16}{12}\kappa_{18} - \frac{1}{12}\kappa_{23}], \quad (3.111)$$

$$\frac{\partial^2 \kappa_{13}}{\partial y^2} = \frac{1}{dy^2}[-\frac{1}{12}\kappa_{11} + \frac{16}{12}\kappa_{12} - \frac{30}{12}\kappa_{13} + \frac{16}{12}\kappa_{14} - \frac{1}{12}\kappa_{15}], \quad (3.112)$$

$$\begin{aligned}\frac{\partial^2 \kappa_{13}}{\partial x \partial y} &= \frac{1}{dxdy}[\frac{1}{12}(\frac{1}{12}\kappa_1 - \frac{8}{12}\kappa_2 + \frac{8}{12}\kappa_4 - \frac{1}{12}\kappa_5) - \\ &\frac{8}{12}(\frac{1}{12}\kappa_6 - \frac{8}{12}\kappa_7 + \frac{8}{12}\kappa_9 - \frac{1}{12}\kappa_{10}) + \\ &\frac{8}{12}(\frac{1}{12}\kappa_{16} - \frac{8}{12}\kappa_{17} + \frac{8}{12}\kappa_{19} - \frac{1}{12}\kappa_{20}) - \\ &\frac{1}{12}(\frac{1}{12}\kappa_{21} - \frac{8}{12}\kappa_{22} + \frac{8}{12}\kappa_{24} - \frac{1}{12}\kappa_{25})]\end{aligned}, \quad (3.113)$$

where the lower subscript at the variable κ denotes the number of the triangular element in the specified quadrilateral (Fig. 3.8), and dx and dy are the distances between the triangle centres in the neighbouring quadrilaterals in a horizontal and vertical direction, respectively. The calculations of the second derivatives of the variable κ in other triangles are similar. Thus, the effect of neighbouring elements near each element is taken into account. In FE calculations, the mixed derivative (Eq. 3.113) was neglected to reproduce the Laplacian of the variable κ only.

The advantage of a gradient approach is that it is suitable (as a non-local approach) for both shear and tension (decohesion) dominated applications. The explicit second-gradient strain isotropic damage approach (Eqs. 3.111 and 3.112) was used for reinforced concrete beams under monotonic loading.

The non-local and second-gradient model were implemented in the commercial finite element code ABAQUS (1998) for efficient computations. Such implementation can be performed with two methods. In the first one, two identical overlapping meshes are used. The first mesh allows to gather the information about coordinates of integration points in the entire specimen, area of all finite elements and total strain rates in each element. The elements in this mesh are defined by the user in the UEL procedure. They do not influence the results of stresses in the specimen body since they have no stiffness. The information stored is needed to calculate non-local variables with the aid of the second mesh which includes standard elements from the ABAQUS library (1998). The constitutive law is defined by the UMAT procedure. During odd iterations, the information is gathered in the elements of the first mesh. During even iterations, the stresses in the elements of the second mesh (including standard elements) are determined with taking into account non-local variables and a non-linear finite element equation is solved. Between odd and even iterations, the same element configuration is imposed. In the second method, only one mesh is used which contains user's elements (defined by the UEL procedure). During odd iterations, the information about the elements is stored, and during even iterations, the stresses within a non-local theory are determined. As compared to the first method, the second one consumes less time. However, it is less comfortable for the user due to the need of the definition of the stiffness matrix and out-of-balance load vector in finite elements.

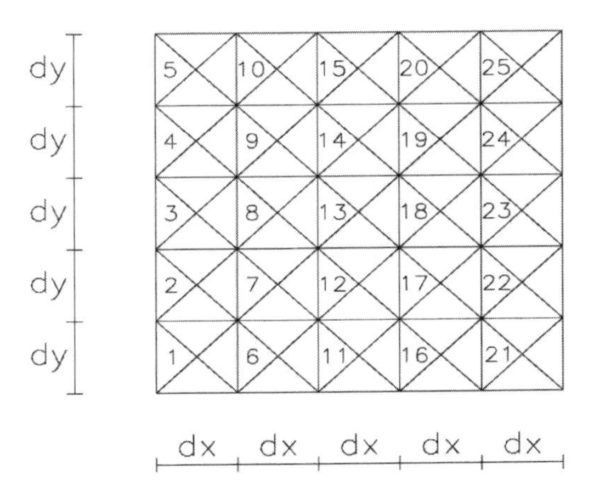

Fig. 3.8 Diagram for determination of the gradient of the constitutive variable κ in triangular finite elements using a central difference method (Tejchman 2004)

For the solution of the non-linear equation of motion governing the response of a system of finite elements, the initial stiffness method was used with a symmetric elastic global stiffness matrix instead of applying a tangent stiffness matrix (the choice was governed by access limitations to the commercial software ABAQUS (1998). To satisfy the consistency condition f=0 in elasto-plasticity, the trial stress method (linearized expansion of the yield condition about the trial stress point) using an elastic predictor and a plastic corrector with the return mapping algorithm (Ortiz and Simo 1986) was applied. The calculations were carried out using a large-displacement analysis (ABAQUS 1998). In this method, the current configuration of the body was taken into account. The Cauchy stress was taken as the stress measure. The conjugate strain rate was the rate of deformation. The rotations of the stress and strain tensor were calculated with the Hughes-Winget method (Hughes and Winget 1988). The non-local averaging was performed in the current configuration. This choice was governed again by the fact that element areas in this configuration were automatically calculated by ABAQUS (1998).

3.4 Bond-Slip Laws

Different bond-slip laws were assumed when modelling reinforced concrete elements. However, there does not exist an universal slip-bond law for reinforced concrete elements since it depends upon boundary conditions of the entire system (Chapter 2). To consider bond-slip, an interface with a zero thickness was assumed along a contact, where a relationship between the shear traction and slip was introduced. In the book, 4 different bond-slip laws were applied.

First, the simplest bond-slip proposed by Dörr (1980) without softening was used (Fig. 3.9)

$$\tau = f_t \left[0.5\left(\frac{u}{u_1}\right) - 4.5\left(\frac{u}{u_1}\right)^2 + 1.4\left(\frac{u}{u_1}\right)^3 \right] \quad \text{if} \quad 0<u\leq u_0, \tag{3.114}$$

$$\tau = \tau_{\max} = 1.9 f_t \quad \text{if} \quad u>u_1, \tag{3.115}$$

wherein τ denotes the bond stress, τ_{max} is the bond resistance, f_t is the tensile strength of concrete and u_1 is the displacement at which perfect slip occurs (u_1=0.06 mm).

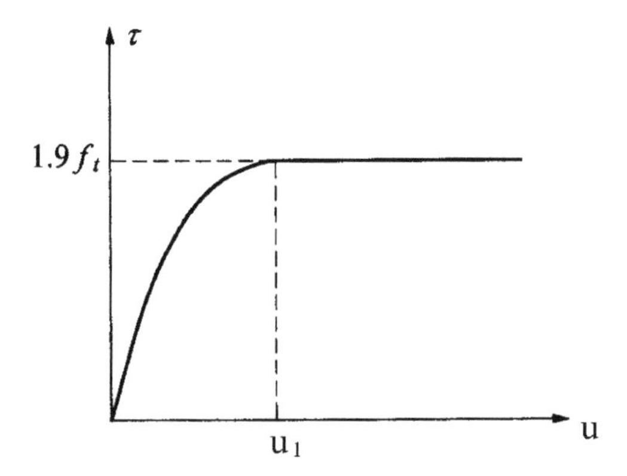

Fig. 3.9 Bond-slip law between concrete and reinforcement by Dörr (1980)

Second, the bond-slip law suggested by CEB-FIB (1992) was applied (Fig. 3.10)

$$\tau = \tau_{\max} \left(\frac{u}{u_1}\right)^{0.4}, \quad u < u_1, \tag{3.116}$$

$$\tau = \tau_{\max}, \quad u_1 \leq u \leq u_2, \tag{3.117}$$

$$\tau = \tau_{\max} - (\tau_{\max} - \tau_{res})\frac{u - u_2}{u_3 - u_2}, \qquad u_2 \le u \le u_3, \quad (3.118)$$

$$\tau = \tau_{res}, \qquad u_3 \le u, \quad (3.119)$$

where u_l=0.06 mm.

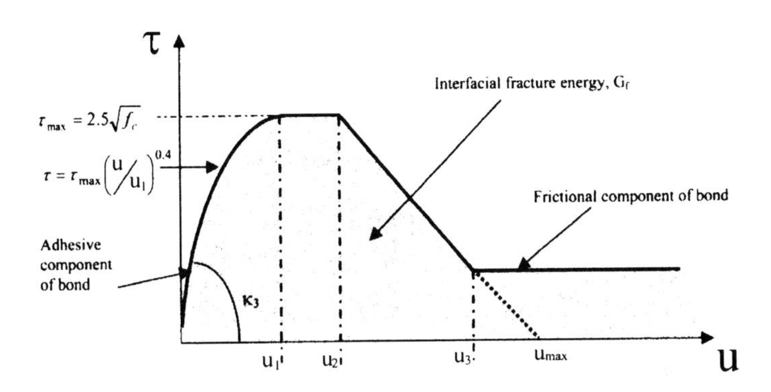

Fig. 3.10 Bond stress-slip relationship between concrete and reinforcement by CEB-FIP (1992)

Third, the bond-slip law by Haskett et al. (2008) on the basis of Eqs. 3.116-3.119 was used (Fig. 3.11)

$$\tau = \tau_{\max}\left(\frac{u}{u_1}\right)^{0.4}, \qquad 0 \le u \le u_1, \quad (3.120)$$

where u_l=1.5 mm is slip corresponding to the peak.

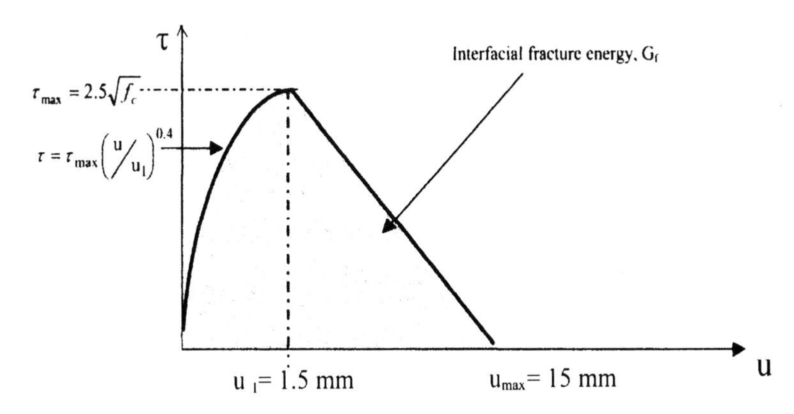

Fig. 3.11 Bond-slip relationship between concrete and reinforcement proposed by Haskett et al. (2008)

Finally, the bond-slip law was used which distinguishes pull-out failure where the bond strength is in the contact zone is exceeded, and a splitting failure which is caused by an insufficient concrete cover throughout (due to occurrence of radial cracks). Den Uijl and Bigaj (1996) and Akkermann (2000) proposed a bond model for ribbed bars based on concrete confinement. The bond model is formulated in the terms of a radial stress-radial strain relation (Fig. 3.12). The radial stresses are equal to the bond stresses. For the splitting failure, the radial strains are linear dependent on the slip, and for the pull-out failure, they are nonlinear dependent. If the radial stresses σ_r are smaller than the maximum slip stresses $\tau_{max}=5f_t$, a splitting failure takes places ($\tau_{max}/\sigma_r>1$), otherwise a pull-out failure takes place ($\tau_{max}/\sigma_r\leq 1$). The maximum radial stress and strain are at failure, respectively

$$\sigma_{r,\max} = 2f_t(\frac{c}{d_r})^{0.88}, \qquad \varepsilon_{r,\max} = 4.22\frac{f_t}{E_o}(\frac{c}{d_r})^{1.08}, \tag{3.121}$$

where E_o is the modulus of elasticity and c denotes the concrete cover. The bond stress is coupled with the radial stress by the friction angle.

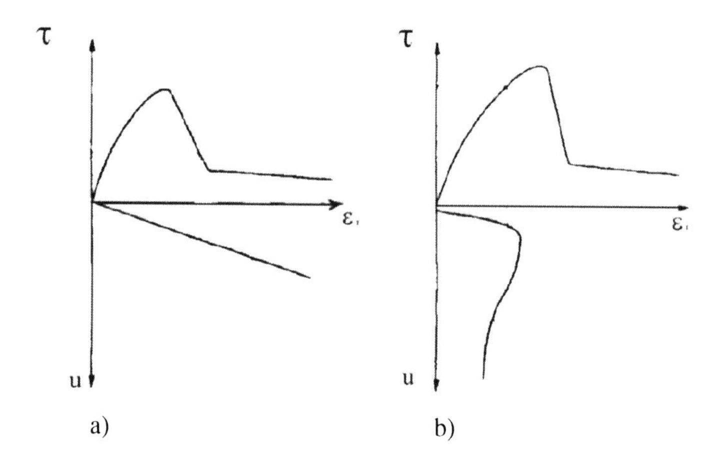

Fig. 3.12 Selected bond-slip laws between concrete and reinforcement: a) splitting failure, b) pull-out failure (τ – bond stress, u – slip, ε_r – radial strain) (den Uijl and Bigaj 1996)

References

ABAQUS. Theory Manual, Version 5.8, Hibbit, Karlsson & Sorensen Inc. (1998)

Aifantis, E.C.: Update on class of gradient theories. Mechanics of Materials 35(3), 259–280 (2003)

Akkermann, J.: Rotationsverhalten von Stahlbeton-Rahmenecken. PhD Thesis, Universität Fridericiana zu Karlsruhe, Karlsruhe (2000)

Asferg, P.N., Poulsen, P.N., Nielsen, L.O.: Modelling of crack propagation in concrete applying the XFEM. In: Meschke, G., de Borst, R., Mang, H., Bicanic, N. (eds.) Computational Modelling of Concrete Structures, EURO-C, pp. 33–42. Taylor and Francis (2006)

Askes, H., Sluys, L.J.: A classification of higher-order strain gradient models in damage mechanics. Archive of Applied Mechanics 53(5-6), 448–465 (2003)

Bažant, Z.P., Bhat, P.D.: Endochronic theory of inelasticity and failure of concrete. Journal of Engineering Mechanics Division ASCE 102(4), 701–722 (1976)

Bažant, Z.: Endochronic inelasticity and incremental plasticity. International Journal of Solids and Structures 14(9), 691–714 (1978)

Bažant, Z.P., Shieh, C.L.: Endochronic model for nonlinear triaxial behaviour of concrete. Nuclear Engineering and Design 47(2), 305–315 (1978)

Bažant, Z.P., Cedolin, L.: Blunt crack band propagation in finite element analysis. Journal of Engineering Mechanics Division ASCE 105(2), 297–315 (1979)

Bažant, Z.P., Oh, B.H.: Crack band theory for fracture of concrete. Materials and Structures RILEM 16(93), 155–177 (1983)

Bažant, Z.P., Lin, F.: Non-local yield limit degradation. International Journal for Numerical Methods in Engineering 26(8), 1805–1823 (1988)

Bažant, Z., Ožbolt, J.: Non-local microplane model for fracture, damage and size effect in structures. Journal of Engineering Mechanics ASCE 116(11), 2485–2505 (1990)

Bažant, Z., Jirásek, M.: Nonlocal integral formulations of plasticity and damage: survey of progress. Journal of Engineering Mechanics 128(11), 1119–1149 (2002)

Belytschko, T., Moes, N., Usui, S., Parimi, C.: Arbitrary discontinuities in finite elements. International Journal for Numerical Methods in Engineering 50(4), 993–1013 (2001)

Belytschko, T., Gracie, R., Ventura, G.: A review of extended/generalized finite element methods for material modeling. Modelling and Simulation in Material Science and Engineering 17(4), 1–24 (2009)

Bhandari, A.R., Inoue, J.: Experimental study of strain rates effects on strain localization characteristics of soft rocks. Soils and Foundations 45(1), 125–140 (2005)

Bobiński, J., Tejchman, J.: Numerical simulations of localization of deformation in quasi-brittle materials within non-local softening plasticity. Computers and Concrete 1(4), 433–455 (2004)

Bobiński, J., Tejchman, J.: Modelling of strain localization in quasi-brittle materials with a coupled elasto-plastic-damage model. International Journal for Theoretical and Applied Mechanics 44(4), 767–782 (2006)

Bobiński, J.: Implementacja i przykłady zastosowań nieliniowych modeli betonu z nielokalnym osłabieniem. PhD Thesis, Gdańsk University of Technology (2006)

Bobiński, J., Tejchman, J.: Continuous and discontinuous modeling of cracks in concrete elements. In: Bicanic, N., de Borst, R., Mang, H., Meschke, G. (eds.) Modelling of Concrete Structures, pp. 263–270. Taylor and Francis Group, London (2010)

Bolander, J.E., Sukumar, N.: Irregular lattice model for quasi-static crack propagation. Physical Review B 71(9), 094106(2005)

Borino, G., Failla, B., Parrinello, F.: A symmetric nonlocal damage theory. International Journal of Solids and Structures 40(13-14), 3621–3645 (2003)

Brinkgreve, R.B.J.: Geomaterial models and numerical analysis of softening. Phd Thesis, Delft University of Technology (1994)

Caballero, A., Carol, I., Lopez, C.M.: New results in 3d meso-mechanical analysis of concrete specimens using interface elements. In: Meschke, G., de Borst, R., Mang, H., Bicanic, N. (eds.) Computational Modelling of Concrete Structures, EURO-C, pp. 43–52. Taylor and Francis (2006)

Caballero, A., Carol, I., Lopez, C.M.: 3D meso-mechanical analysis of concrete specimens under biaxial loading. Fatigue and Fracture of Engineering Materials and Structures 30(9), 877–886 (2007)

Carol, I., López, C.M., Roa, O.: Micromechanical analysis of quasi-brittle materials using fracture-based interface elements. International Journal for Numerical Methods in Engineering 52(1-2), 193–215 (2001)

Carol, I., Willam, K.: Spurious energy dissipation/generation in stiffness recovery models for elastic degradation and damage. International Journal of Solids and Structures 33(20-22), 2939–2957 (1996)

Carol, I., Rizzi, E., Willam, K.: On the formulation of anisotropic elastic degradation. International Journal of Solids and Structures 38(4), 491–518 (2001)

CEB-FIP Model Code 90, London (1992)

Chambon, R., Caillerie, D., Matsuchima, T.: Plastic continuum with microstructure, local second gradient theories for geomaterials: localization studies. International Journal of Solids and Structures 38(46-47), 8503–8527 (2001)

Chen, E.: Non-local effects on dynamic damage accumulation in brittle solids. International Journal for Numerical and Analytical Methods in Geomechanics 23(1), 1–21 (1999)

Chen, J., Yuan, H., Kalkhof, D.: A non-local damage model for elastoplastic materials based on gradient plasticity theory. Report Nr.01-13, Paul Scherrer Institut, 1–130 (2001)

Cope, R.J., Rao, P.V., Clark, L.A., Norris, P.: Modeling of concrete reinforced behaviour for finite element analysis of bridge slabs. Numerical Methods for Non-linear Problems 1, 457–470 (1980)

Cusatis, G., Bažant, Z.P., Cedolin, L.: Confinement-shaer lattice model for concrete damage in tension and compression: I. Theory. Journal for Engineering Mechanics ASCE 129(12), 1439–1448 (2003)

de Borst, R., Nauta, P.: Non-orthogonal cracks in a smeared finite element model. Engineering Computations 2(1), 35–46 (1985)

de Borst, R.: Non-linear analysis of frictional materials. Phd Thesis, University of Delft (1986)

de Borst, R., Mühlhaus, H.-B., Pamin, J., Sluys, L.B.: Computational modelling of localization of deformation. In: Owen, D.R.J., Onate, H., Hinton, E. (eds.) Proc. of the 3rd Int. Conf. Comp. Plasticity, Swansea, pp. 483–508. Pineridge Press (1992)

de Borst, R., Mühlhaus, H.-B.: Gradient dependent plasticity: formulation and algorithmic aspects. International Journal for Numerical Methods in Engineering 35(3), 521–539 (1992)

de Borst, R., Pamin, J.: Some novel developments in finite element procedures for gradient-dependent plasticity. International Journal for Numerical Methods in Engineering 39(14), 2477–2505 (1996)

de Vree, J.H.P., Brekelmans, W.A.M., van Gils, M.A.J.: Comparison of nonlocal approaches in continuum damage mechanics. Computers and Structures 55(4), 581–588 (1995)

den Uijl, J.A., Bigaj, A.: A bond model for ribbed bars based on concrete confinement. Heron 41(3), 201–226 (1996) di Prisco, C., Imposimato, S., Aifantis, E.C.: A visco-plastic constitutive model for granular soils modified according to non-local and gradient approaches. International Journal for Numerical and Analytical Methods in Geomechanic 26(2), 121–138 (2002)

D'Addetta, G.A., Kun, F., Ramm, E.: On the application of a discrete model to the fracture process of cohesive granular materials. Granular Matter 4(2), 77–90 (2002)

Donze, F.V., Magnier, S.A., Daudeville, L., Mariotti, C.: Numerical study of compressive behaviour of concrete at high strain rates. Journal of Engineering Mechanics ASCE 125(10), 1154–1163 (1999)

Dörr, K.: Ein Beitag zur Berechnung von Stahlbetonscheiben unter Berücksichtigung des Verbundverhaltens. Phd Thesis, Darmstadt University (1980)

Dragon, A., Mróz, Z.: A continuum model for plastic-brittle behaviour of rock and concrete. International Journal of Engineering Science 17(2) (1979)

Etse, G., Willam, K.: Fracture energy formulation for inelastic behaviour of plain concrete. Journal of Engineering Mechanics ASCE 120(9), 1983–2011 (1994)

Ferrara, I., di Prisco, M.: Mode I fracture behaviour in concrete: nonlocal damage modeling. ASCE Journal of Engineering Mechanics 127(7), 678–692 (2001)

Fleck, N.A., Hutchinson, J.W.: A phenomenological theory for strain gradient effects in plasticity. Journal of Mechanics and Physics of Solids 41(12), 1825–1857 (1993)

Gatuingt, G., Pijaudier-Cabot, H.: Coupled damage and plasticity modelling in transient dynamic analysis of concrete. International Journal for Numerical and Analytical Methods in Geomechanics 26(1), 1–24 (2002)

Geers, M., Peijs, T., Brekelmans, W., de Borst, R.: Experimental monitoring of strain localization and failure behaviour of composite materials. Composites Science and Technology 56(11), 1283–1290 (1996)

Geers, M.G.D.: Experimental analysis and computational modeling of damage and fracture. PhD Thesis, Eindhoven University of Technology, Eindhoven (1997)

Geers, M., de Borst, R., Brekelmans, W., Peerlings, R.: Strain-based transient-gradient damage model for failure analyses. Computer Methods in Applied Mechanics and Engineering 160(1-2), 133–154 (1998)

Grassl, P., Jirásek, M.: Damage-plastic model for concrete failure. International Journal of Solids and Structures 43(22-23), 7166–7196 (2006)

Hansen, N.R., Schreyer, H.L.: A thermodynamically consistent framework for theories of elastoplasticity coupled with damage. International Journal of Solids and Structures 31(3), 359–389 (1994)

Hansen, E., Willam, K.: A two-surface anisotropic damage-plasticity model for plane concrete. In: de Borst, R. (ed.) Procceedings Int. Conf. Fracture Mechanics of Concrete Materials, Paris, Balkema, pp. 549–556 (2001)

Haskett, M., Oehlers, D.J., Mohamed Ali, M.S.: Local and global bond characteristics of steel reinforcing bars. Engineering Structures 30(2), 376–383 (2008)

Häu*β*ler-Combe, U., Prochtel, P.: Ein dreiaxiale Stoffgesetz fur Betone mit normalen und hoher Festigkeit. Beton- und Stahlbetonbau 100(1), 56–62 (2005)

Hentz, S., Donze, F.V., Daudeville, L.: Discrete element modelling of concrete submitted to dynamic loading at high strain rates. Computers and Structures 82(29-30), 2509–2524 (2004)

Herrmann, H.J., Hansen, A., Roux, S.: Fracture of disordered elastic lattices in two dimensions. Physical Review B 39(1), 637–647 (1989)

Hillerborg, A., Modeer, M., Peterson, P.E.: Analysis of crack propagation and crack growth in concrete by means of fracture mechanics and finite elements. Cem. Concr. Res. 6, 773–782 (1976)

Hillerborg, A.: The theoretical basis of a method to determine the fracture energy of concrete. Materials and Structures 18(1), 291–296 (1985)
Hordijk, D.A.: Local approach to fatigue of concrete, PhD dissertation, Delft University of Technology (1991)
Hsieh, S.S., Ting, E.C., Chen, W.F.: Plasticity-fracture model for concrete. International Journal of Solids and Structures 18(3), 181–187 (1982)
Huerta, A., Rodriguez-Ferran, A., Morata, I.: Efficient and reliable non-local damage models. In: Kolymbas, D. (ed.). Lecture Notes in Applied and Computational Mechanics, vol. 13, pp. 239–268. Springer, Heidelberg (2003)
Hughes, T.J.R., Winget, J.: Finite Rotation Effects in Numerical Integration of Rate Constitutive Equations Arising in Large Deformation Analysis. International Journal for Numerical Methods in Engineering 15(12), 1862–1867 (1980)
Ibrahimbegovic, A., Markovic, D., Gatuing, F.: Constitutive model of coupled damage-plasticity and its finite element implementatation. Revue Européenne des Eléments Finis 12(4), 381–405 (2003)
Jirásek, M., Zimmermann, T.: Analysis of rotating crack model. ASCE Journal of Engineering Mechanics 124(8), 842–851 (1998)
Jirásek, M.: Nonlocal models for damage and fracture: comparison of approaches. International Journal of Solids and Structures 35(31-32), 4133–4145 (1998)
Jirásek, M.: Comments on microplane theory. In: Pijaudier-Cabot, G., Bittnar, Z., Gerard, B. (eds.) Mechanics of Quasi-Brittle Materials and Structures, pp. 55–77. Hermes Science Publications (1999)
Jirásek, M., Rolshoven, S.: Comparison of integral-type nonlocal plasticity models for strain-softening materials. International Journal of Engineering Science 41(13-14), 1553–1602 (2003)
Jirásek, M.: Non-local damage mechanics with application to concrete. In: Vardoulakis, I., Mira, P. (eds.) Failure, Degradation and Instabilities in Geomaterials, Lavoisier, pp. 683–709 (2004a)
Jirásek, M., Rolshoven, S., Grassl, P.: Size effect on fracture energy induced by non-locality. International Journal for Numerical and Analytical Methods in Geomechanics 28(7-8), 653–670 (2004b)
Jirásek, M., Marfia, S.: Non-local damage model based on displacement averaging. International Journal for Numerical Methods in Engineering 63(1), 77–102 (2005)
Karsan, D., Jirsa, J.O.: Behaviour of concrete under compressive loadings. Journal of the Structural Division (ASCE) 95(12), 2543–2563 (1969)
Katchanov, L.M.: Introduction to continuum damage mechanics. Martimus Nijhoff, Dordrecht (1986)
Khoei, A.R., Bakhshiani, A., Mofid, M.: An implicit algorithm for hypoelasto-plastic and hypoelasto-viscoplastic endochronic theory in finite strain isotropic–kinematic-hardening model. International Journal of Solids and Structures 40(13-14), 3393–3423 (2003)
Klisiński, M., Mróz, Z.: Description of non-elastic deformation and damage of concrete. Script, Technical University of Poznan, Poznan (1988) (in polish)
Kozicki, J., Tejchman, J.: Effect of aggregate structure on fracture process in concrete using 2D lattice model. Archives of Mechanics 59(4-5), 1–20 (2007)
Krajcinovic, D., Fonseka, G.U.: The continuous damage theory of brittle materials. Journal of Applied Mechanics ASME 48(4), 809–824 (1981)
Kuhl, E., Ramm, E.: Simulation of strain localization with gradient enhanced damage models. Computational Material Sciences 16(1-4), 176–185 (2000)

Lade, P.V., Jakobsen, K.P.: Incrementalization of a single hardening constitutive model for frictional materials. International Journal for Numerical and Analytical Methods in Geomechanics 26(7), 647–659 (2002)

Le Bellego, C., Dube, J.F., Pijaudier-Cabot, G., Gerard, B.: Calibration of nonlocal damage model from size effect tests. European Journal of Mechanics A/Solids 22(1), 33–46 (2003)

Lee, J., Fenves, G.L.: Plastic-damage model for cyclic loading of concrete structures. Journal of Engineering Mechanics 124(8), 892–900 (1998a)

Lee, J., Fenves, G.L.: A plastic-damage concrete model for earthquake analysis of dams. Earthquake Engng. and Struct. Dyn. 27, 937–956 (1998b)

Lemaitre, J.: Coupled elasto-plasticity and damage constitutive equations. Computer Methods in Applied Mechanics and Engineering 51(1-3), 31–49 (1985)

Lemaitre, J., Chaboche, J.L.: Mechanics of solid materials. Cambridge University Press, Cambridge (1990)

Lodygowski, T., Perzyna, P.: Numerical modelling of localized fracture of inelastic solids in dynamic loading process. International Journal for Numerical Methods in Engineering 40(22), 4137–4158 (1997)

Loret, B., Prevost, J.H.: Dynamic strain localisation in elasto-visco-plastic solids, Part 1. General formulation and one-dimensional examples. Computer Methods in Applied Mechanics and Engineering 83(3), 247–273 (1990)

Lubliner, J., Oliver, J., Oller, S., Onate, E.: A plastic-damage model for concrete. International Journal of Solids and Structures 25(3), 229–326 (1989)

Majewski, T., Bobiński, J., Tejchman, J.: FE-analysis of failure behaviour of reinforced concrete columns under eccentric compression. Engineering Structures 30(2), 300–317 (2008)

Mahnken, T., Kuhl, E.: Parameter identification of gradient enhanced damage models. European Journal of Mechanics - A/Solids 18(5), 819–835 (1999)

Maier, T.: Numerische Modellierung der Entfestigung im Rahmen der Hypoplastizität. PhD Thesis, University of Dortmund (2002)

Maier, T.: Nonlocal modelling of softening in hypoplasticity. Computers and Geotechnics 30(7), 599–610 (2003)

Marcher, T., Vermeer, P.A.: Macro-modelling of softening in non-cohesive soils. In: Vermeer, P.A., et al. (eds.) Continuous and Discontinuous Modelling of Cohesive-Frictional Materials, pp. 89–110. Springer, Heidelberg (2001)

Marzec, I., Bobiński, J., Tejchman, J.: Simulations of crack spacing in reinforced concrete beams using elastic-plasticity and damage with non-local softening. Computers and Concrete 4(5), 377–403 (2007)

Marzec, I.: Application of coupled elasto-plastic-damage models with non-local softening to concrete cyclic behaviour. PhD Thesis, Gdańsk University of Technology (2009)

Marzec, I., Tejchman, J.: Modeling of concrete behaviour under cyclic loading using different coupled elasto-plastic-damage models with non-local softening. In: Oñate, E., Owen, D.R.J. (eds.) X International Conference on Computational Plasticity COMPLAS X, pp. 1–4. CIMNE, Barcelona (2009)

Marzec, I., Tejchman, J.: Enhanced coupled elasto-plastic-damage models to describe concrete behaviour in cyclic laboratory tests. Archives of Mechanics (2011) (in print)

Mazars, J.: A description of micro- and macro-scale damage of concrete structures. Engineering Fracture Mechanics 25(5-6), 729–737 (1986)

Mazars, J., Pijaudier-Cabot: Continuum damage theory – applications to concrete. Journal of Engineering Mechanics ASCE 115(2), 345–365 (1989)

Meftah, F., Reynouard, J.M.: A multilayered mixed beam element on gradient plasticity for the analysis of localized failure modes. Mechanics of Cohesive-Frictional Materials 3(4), 305–322 (1998)

Menétrey, P., Willam, K.J.: Triaxial failure criterion for concrete and its generalization. ACI Structural Journal 92(3), 311–318 (1995)

Menzel, A., Steinmann, P.: On the continuum formulation of higher gradient plasticity for single and polycrystals. Journal of the Mechanics and Physics of Solids 48(8), 1777–1796 (2000)

Meschke, G., Lackner, R., Mang, H.A.: An anisotropic elastoplastic-damage model for plain concrete. International Journal for Numerical Methods in Engineering 42(4), 702–727 (1998)

Moonen, P., Carmeliet, J., Sluys, L.J.: A continuous-discontinuous approach to simulate fracture processes. Philosophical Magazine 88(28-29), 3281–3298 (2008)

Mühlhaus, H.-B.: Scherfugenanalyse bei Granularen Material im Rahmen der Cosserat-Theorie. Ingen. Archiv. 56, 389–399 (1986)

Mühlhaus, H.B., Aifantis, E.C.: A variational principle for gradient plasticity. International Journal of Solids and Structures 28(7), 845–858 (1991)

Needleman, A.: A continuum model for void nucleation by inclusion debonding. Journal of Applied Mechanics 54(3), 525–531 (1987)

Needleman, A.: Material rate dependence and mesh sensitivity in localization problems. Computer Methods in Applied Mechanics and Engineering 67(1), 69–85 (1988)

Nguyen, G.D.: A thermodynamic approach to constitutive modelling of concrete using damage mechanics and plasticity theory, PhD Thesis, Trinity College, University of Oxford (2005)

Nguyen, G.D.: A thermodynamic approach to non-local damage modelling of concrete. International Journal of Solids and Structures 45(7-8), 1918–1932 (2008)

Oliver, J., Huespe, A.E., Sanchez, P.J.: A comparative study on finite elements for capturing strong discontinuities: E-FEM vs X-FEM. Computer Methods in Applied Mechanics and Engineering 195(37-40), 4732–4752 (2006)

Ortiz, M., Simo, I.C.: An analysis of a new class of integration algorithms for elastoplastic constitutive relation. International Journal for Numerical Methods in Engineering 23(3), 353–366 (1986)

Ortiz, M., Pandolfi, A.: Finite-deformation irreversible cohesive elements for three-dimensional crack-propagation analysis. International Journal for Numerical Methods in Engineering 44(9), 1267–1282 (1999)

Ottosen, N.S.: A failure criterion for concrete. Journal of Engineering Mechanics Division ASCE 103(4), 527–535 (1977)

Ožbolt, J.: Maßstabseffekt und Duktulität von Beton- und Stahlbetonkonstruktionen. Habilitation, Universität Stuttgart (1995)

Palaniswamy, R., Shah, S.P.: Fracture and stress-strain relationship of concrete under triaxial compression. Journal of the Structural Division (ASCE) 100(ST5), 901–916 (1974)

Pamin, J.: Gradient-dependent plasticity in numerical simulation of localization phenomena. PhD Thesis, University of Delft (1994)

Pamin, J., de Borst, R.: Simulation of crack spacing using a reinforced concrete model with an internal length parameter. Arch. App. Mech. 68(9), 613–625 (1998)

Pamin, J., de Borst, R.: Stiffness degradation in gradient-dependent coupled damage-plasticity. Archives of Mechanics 51(3-4), 419–446 (1999)

Pamin, J.: Gradient-enchanced continuum models: formulation, discretization and applications. Habilitation, Cracow University of Technology, Cracow (2004)

Pedersen, R.R., Simone, A., Sluys, L.J.: An analysis of dynamic fracture in concrete with a continuum visco-elastic visco-plastic damage model. Engineering Fracture Mechanics 75(13), 3782–3805 (2008)

Peerlings, R.H.J., de Borst, R., Brekelmans, W.A.M., Geers, M.G.D.: Gradient enhanced damage modelling of concrete fracture. Mechanics of Cohesive-Frictional Materials 3(4), 323–342 (1998)

Perdikaris, P.C., Romeo, A.: Size effect on fracture energy of concrete and stability issues in three-point bending fracture toughness testing. ACI Materials Journal 92(5), 483–496 (1995)

Pietruszczak, D., Jiang, J., Mirza, F.A.: An elastoplastic constitutive model for concrete. International Journal of Solids and Structures 24(7), 705–722 (1988)

Pijaudier-Cabot, G., Bažant, Z.P.: Nonlocal damage theory. Journal of Engineering Mechanics ASCE 113(10), 1512–1533 (1987)

Pijaudier-Cabot, G.: Non-local damage. In: Mühlhaus, H.-B. (ed.) Continuum Models for Materials and Microstructure, pp. 107–143. John Wiley & Sons Ltd. (1995)

Pijaudier-Cabot, G., Haidar, K., Loukili, A., Omar, M.: Ageing and durability of concrete structures. In: Darve, F., Vardoulakis, I. (eds.) Degradation and Instabilities in Geomaterials, Springer, Heidelberg (2004)

Polizzotto, C., Borino, G., Fuschi, P.: A thermodynamically consistent formulation of nonlocal and gradient plasticity. Mechanics Research Communications 25(1), 75–82 (1998)

Polizzotto, C., Borino, G.: A thermodynamics-based formulation of gradient-dependent plasticity. Eur. J. Mech. A/Solids 17, 741–761 (1998)

Polizzotto, C.: Remarks on some aspects of non-local theories in solid mechanics. In: Proc. 6th National Congr. SIMAI, Chia Laguna, Italy, CD-ROM (2002)

Pramono, E.: Numerical simulation of distributed and localised failure in concrete. PhD Thesis, University of Colorado-Boulder (1988)

Pramono, E., Willam, K.: Fracture energy-based plasticity formulation of plain concrete. Journal of Engineering Mechanics ASCE 115(6), 1183–1204 (1989)

Ragueneau, F., Borderie, C., Mazars, J.: Damage model for concrete-like materials coupling cracking and friction. International Journal for Numerical and Analytical Methods in Geomechanics 5(8), 607–625 (2000)

Rashid, Y.R.: Analysis of prestressed concrete pressure vessels. Nuclear Engineering and Design 7(4), 334–344 (1968)

Reinhardt, H.W., Cornelissen, H.A.W., Hordijk, D.A.: Tensile tests and failure analysis of concrete. Journal of Structural Engineering ASCE 112(11), 2462–2477 (1986)

Rolshoven, S.: Nonlocal plasticity models for localized failure. PhD Thesis, École Polytechnique Fédérale de Lausanne (2003)

Rots, J.G.: Computational modeling of concrete fracture. PhD Thesis, Delft University (1988)

Rots, J.G., Blaauwendraad, J.: Crack models for concrete, discrete or smeared? Fixed, Multi-directional or rotating? Heron 34(1), 1–59 (1989)

Sakaguchi, H., Mühlhaus, H.B.: Mesh free modelling of failure and localisation in brittle materials. In: Asaoka, A., Adachi, T., Oka, F. (eds.) Deformation and Progressive Failure in Geomechanics, pp. 15–21 (1997)

Salari, M.R., Saeb, S., Willam, K., Patchet, S.J., Carrasco, R.C.: A coupled elastoplastic damage model for geomaterials. Computers Methods in Applied Mechanics and Engineering 193(27-29), 2625–2643 (2004)

Schlangen, E., Garboczi, E.J.: Fracture simulations of concrete using lattice models: computational aspects. Engineering Fracture Mechanics 57, 319–332 (1997)

Simo, J.C., Ju, J.W.: Strain- and stress-based continuum damage models. Parts I and II. International Journal of Solids and Structures 23(7), 821–869 (1987)

Simone, A., Sluys, L.J.: The use of displacement discontinuities in a rate-dependent medium. Computer Methods in Applied Mechanics and Engineering 193(27-29), 3015–3033 (2004)

Skarżyński, L., Tejchman, J.: Mesoscopic modelling of strain localization in concrete. Archives of Civil Engineering LV(4), 521–540 (2009)

Skarżynski, L., Syroka, E., Tejchman, J.: Measurements and calculations of the width of the fracture process zones on the surface of notched concrete beams. Strain 47(s1), 319–322 (2011)

Simone, A., Wells, G.N., Sluys, L.J.: Discontinuous Modelling of Crack Propagation in a Gradient Enhanced Continuum. In: Proc. of the Fifth World Congress on Computational Mechanics WCCM V, Vienna, CDROM (2002)

Sluys, L.J.: Wave propagation, localisation and dispersion in softening solids, PhD thesis, Delft University of Technology, Delft (1992)

Sluys, L.J., de Borst, R.: Dispersive properties of gradient and rate-dependent media. Mechanics of Materials 18(2), 131–149 (1994)

Strömberg, L., Ristinmaa, M.: FE-formulation of nonlocal plasticity theory. Computer Methods in Applied Mechanics and Engineering 136(1-2), 127–144 (1996)

Syroka, E.: Investigations of size effects in reinforced concrete elements. Phd Thesis (2011) (under preparation)

Tejchman, J., Wu, W.: Numerical study on patterning of shear bands in a Cosserat continuum. Acta Mechanica 99(1-4), 61–74 (1993)

Tejchman, J., Herle, I., Wehr, J.: FE-studies on the influence of initial void ratio, pressure level and mean grain diameter on shear localisation. International Journal for Numerical and Analitycal Methods in Geomechanics 23(15), 2045–2074 (1999)

Tejchman, J.: Influence of a characteristic length on shear zone formation in hypoplasticity with different enhancements. Computers and Geotechnics 31(8), 595–611 (2004)

Widulinski, L., Tejchman, J., Kozicki, J., Leśniewska, D.: Discrete simulations of shear zone patterning in sand in earth pressure problems of a retaining wall. International Journal of Solids and Structures 48(7-8), 1191–1209 (2011)

Willam, K.J., Warnke, E.P.: Constitutive model for the triaxial behaviour of concrete. In: IABSE Seminar on Concrete Structures Subjected to Triaxial Stress, Bergamo, Italy, pp. 1–31 (1975)

Willam, K., Pramono, E., Sture, S.: Fundamental issues of smeared crack models. In: Proceedings SEM/RILEM International Conference of Fracture of Concrete and Rock (1986)

Winnicki, A., Pearce, C.J., Bicanic, N.: Viscoplastic Hoffman consistency model for concrete. Computers and Structures 79(1), 7–19 (2001)

Winnicki, A.: Viscoplastic and internal discontinuity models in analysis of structural concrete. Habilitation, University of Cracow (2009)

Vardoulakis, I., Aifantis, E.C.: A gradient flow theory of plasticity for granular materials. Acta Mechanica 87(3-4), 197–217 (1991)

Vervuurt, A., van Mier, J.G.M., Schlangen, E.: Lattice model for analyzing steel-concrete interactions. In: Siriwardane, Zaman (eds.) Comp. Methods and Advances in Geomechanics, Balkema, Rotterdam, pp. 713–718 (1994)

Voyiadjis, G.Z., Taqieddin, Z.N., Kattan, P.: Theoretical Formulation of a Coupled Elastic-Plastic Anisotropic Damage Model for Concrete using the Strain Energy Equivalence Concept. International Journal of Damage Mechanics 18(7), 603–638 (2009)

Zbib, H.M., Aifantis, C.E.: A gradient dependent flow theory of plasticity: application to metal and soil instabilities. Applied Mechanics Reviews 42(11), 295–304 (1989)

Zervos, Z., Papanastasiou, P., Vardoulakis, I.: A finite element displacement formulation for gradient plasticity. International Journal for Numerical Methods in Engineering 50(6), 1369–1388 (2001)

Zhou, W., Zhao, J., Liu, Y., Yang, Q.: Simulation of localization failure with strain-gradient-enhanced damage mechanics. International Journal for Numerical and Analytical Methods in Geomechanics 26(8), 793–813 (2002)

Zhou, F., Molinari, J.F.: Dynamic crack propagation with cohesive elements: a methodology to address mesh dependency. International Journal for Numerical Methods in Engineering 59(1), 1–24 (2004)

Yang, Z., Xu, X.F.: A heterogeneous cohesive model for quasi-brittle materials considering spatially varying random fracture properties. Computer Methods in Applied Mechanics and Engineering 197, 4027–4039 (2008)

Chapter 4
Discontinuous Approach to Concrete

Abstract. The Chapter discusses discontinuous approaches to simulate cracks in concrete. Two approaches are described: a cohesive crack model using interface elements defined along finite element boundaries and eXtended Finite Element Method (XFEM) wherein cracks can occur arbitrarily in the interior of finite elements.

4.1 Cohesive Crack Model

A cohesive crack model for simulating macro-cracks as discontinuities was initiated from the Hilleborg's fictitious crack model (Hilleborg et al. 1976) based on the idea of Dugdale (1960) and was further applied in conjunction with interface elements (cohesive elements) by Camacho and Ortiz (1996). No information on an initial crack needs to be known and the onset of crack initiation can be predicted within a preset cohesive zone, which is considered to be a potential crack propagation path. The cohesive crack model describes highly localized inelastic processes by traction-separation laws that link the cohesive traction transmitted by a discontinuity or surface to the displacement jump characterized by the separation vector (Needleman 1987, Camacho and Ortiz 1996, Ortiz and Pandolfi 1999, Chandra et al. 2002, Gálvez et al. 2002, Scheider and Brocks 2003, Zhou and Molinari 2004, de Lorenzis and Zavarise 2009). Cohesive elements are defined at the edges (interface) between standard finite elements to nucleate cracks and propagate them following the deformation process. They govern the separation of crack flanks in accordance with an irreversible cohesive law. Branching, crack coalescence, kinking and tortuousness (any material separation) are naturally handled by this approach. If the crack path is not known a priori, cohesive surfaces are placed between all finite elements. Interfacial normal and tangential tractions are non-linearly connected to the normal (mode-I) and tangential (mode-II) relative displacements on the interface. As the cohesive interface gradually separates, the magnitude of interfacial stresses at first increases, reaches a maximum, and then decreases with increasing separation, finally approaching zero. Thus, depending on the level of the interfacial relative displacements, the cohesive interface represents

J. Tejchman, J. Bobiński: Continuous & Discontinuous Modelling of Fracture, SSGG, pp. 95–107.
springerlink.com

the entire spectrum of the behaviour ranging from perfect bonding to complete separation. A shape of post-peak traction-opening is linked with the development of the so-called fracture process zone where many complex phenomena occur such as micro-cracking, interlocking, bridging, friction between surfaces and aggregates, etc. The fundamental material parameters in cohesive models for concrete are the fracture energy and the shape of the traction versus crack opening. The cohesive fracture energy is the external energy required to create and fully break a unit surface area of a cohesive crack and coincides with the area under the softening function. The crack models can be based on formulations of the classical plasticity (Gálvez et al. 2002) or damage mechanics (Omiya and Kishimoto 2010). The model is attractive since it is straightforward in implementation.

The cohesive crack model provides an objective description of fully localized failure if the mesh is fine. The cohesive traction-separation law with softening does not need any adjustment for the element size because mesh refinement does not change the resolved crack pattern. However, the model possesses some restrictions. The crack paths are dominated by preferred mesh orientations (Zhou and Molinari 2004) and the mesh independence is questionable if the cracking pattern is diffuse (Bažant and Jirásek 2002). Moreover, stress multiaxiality in the fracture process zone is not captured (Bažant and Jirásek 2002). In models with interface elements inserted a priori, spurious elastic deformation occurs prior to cracking onset, so too high initial elastic normal stiffness can lead to spurious traction oscillations in the pre-cracking phase (de Borst and Remmers 2006). A dummy stiffness (theoretically infinite) is usually required to keep the inactive interface elements closed. The mesh dependency can be improved if the mesh is very fine, inertia forces and viscosity are included, a non-local formulation for the interface mode is used, a separation approximation in the process zone is enriched (de Borst and Remmers 2006, Samimi et al. 2009) or cohesive elements strength follows a stochastic distribution (Zhou and Molinari 2004, Yang and Xu 2008). Recently, Cazes et al. (2009) proposed a thermodynamic method for the construction of a cohesive law from a nonlocal damage model.

The cohesive zone model includes 3 main steps:

- the constitutive continuum modelling (usually by means of linear elasticity if tensile loading prevails),
- the introduction of an initiation criterion for crack opening/growth (loading function),
- the evolution equation for softening of normal/shear tractions.

To take into account mixed mode loading conditions Camacho and Ortiz (1996) defined the effective crack opening displacement as

$$\delta_{eff} = \sqrt{\delta_n^2 + \eta^2\,\delta_s^2}\ , \tag{4.1}$$

wherein δ_n and δ_s are the normal crack opening displacement and tangential relative displacement (sliding), respectively, while the coefficient η takes into account the coupling between the failure mode I and failure mode II. The loading function is defined as

$$f\left(\delta_{eff}, \kappa\right) = \delta_{eff} - \kappa \tag{4.2}$$

with the history parameter κ equal to the maximum value of the effective displacement δ_{eff} obtained during loading. The effective traction is

$$t_{eff} = \sqrt{t_n^2 + \eta^{-2} t_s^2}\ . \tag{4.3}$$

The effective traction t_{eff} is calculated with the aid of an exponential, bilinear or linear softening relationship. Finally, the normal and shear tractions are evaluated as

$$t_n = \frac{t_{eff}}{\delta_{eff}} \delta_n \quad \text{and} \quad t_s = \eta^2 \frac{t_{eff}}{\delta_{eff}} \delta_s\ . \tag{4.4}$$

Unloading takes place to the origin. In compression, the penalty stiffness is applied.

We used a simple version of a cohesive crack model with interface elements placed 'a priori' between all finite elements of the FE mesh (Fig. 4.1). The bulk finite elements were modeled as linear elastic. In turn, in the interface elements, a damage constitutive relationship between the traction vector $\boldsymbol{t}=[t_n, t_s]$ and relative displacement vector $\boldsymbol{\delta}=[\delta_n, \delta_s]$ was assumed

$$t = (1 - D) E_0 \boldsymbol{I} \boldsymbol{\delta} \tag{4.5}$$

with the penalty (dummy) stiffness E_0 and unit tensor $\boldsymbol{I}$. To take into account both the normal and shear terms in the separation vector, an effective opening displacement was used by Eq. 4.1. To describe softening after cracking, an exponential law was assumed following Camacho and Ortiz (1996)

$$t_{eff}(\kappa) = f_t \exp\left(-\beta\left(\kappa - \frac{f_t}{E_0}\right)\right), \tag{4.6}$$

where β is the model parameter. The crack was initiated if (Fig. 4.2)

$$\kappa = \delta_0 = \frac{f_t}{E_0}\ . \tag{4.7}$$

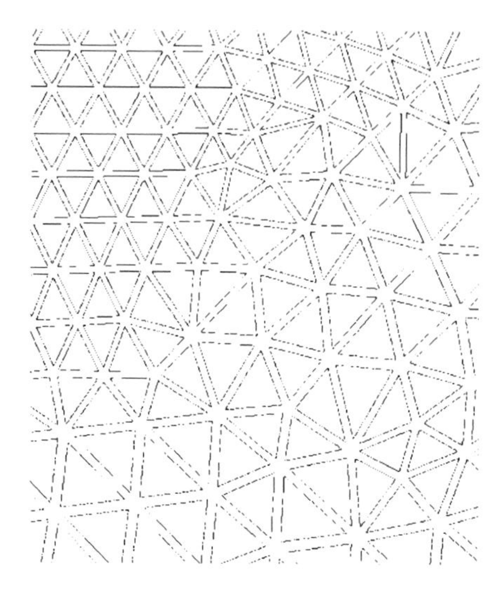

Fig. 4.1 Bulk and cohesive (interface) elements in FE mesh (Bobiński and Tejchman 2008)

The damage parameter was equal to

$$D = 1 - \frac{1}{E_0}\frac{t_{eff}}{\delta_{eff}}. \tag{4.8}$$

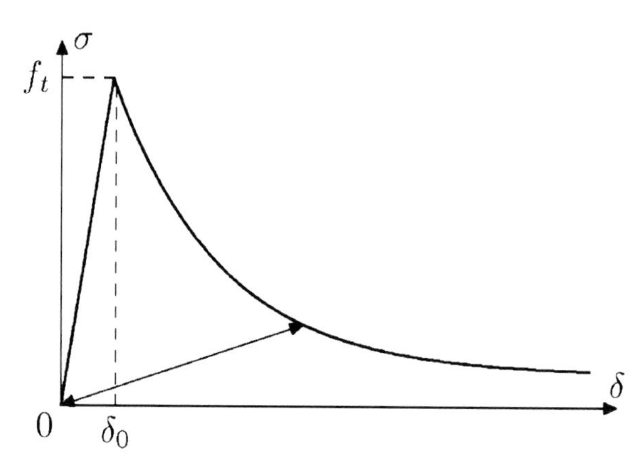

Fig. 4.2 Traction-separation cohesive law assumed for numerical calculations (Bobiński and Tejchman 2008)

4.2 Extended Finite Element Method

The Extended Finite Element Method (XFEM) is based on the Partition of Unity concept (Melenk and Babuska 1996) that allows for adding locally extra terms to the standard FE displacement field approximation in order to capture displacement discontinuities. These extra terms are defined based on a known analytical solution of the problem. The idea is to enrich only selected nodes with additional terms. There is no need to modify the original FE mesh (Belytschko and Black 1999) used to model cracks in elastic-brittle materials. To describe the stress field around the crack tip, Moes and Belytschko (1999) formulated a model with branch functions for elements with crack tip and Heaviside jump function for elements cut entirely by a crack. Later XFEM was extended to deal with branching and intersecting cracks (Daux et al. 2000) and to simulate three-dimensional problems (Sukumar et al. 2000). In turn, Samaniego and Belytschko (2005) simulated dynamic propagation of shear zones. The enrichment was defined only in tangential direction and no separation was allowed in normal direction. XFEM was also used to analyze problems with weak discontinuities (like material interfaces), and in fluid mechanics by modelling voids and holes, phase transformations, biofilms and dislocations. Wells and Sluys (2001) were the first to couple XFEM with cohesive cracks (only Heaviside jump function was defined to describe the displacement jump across the discontinuity). Moes and Belytschko (2002) used XFEM to simulate cohesive cracks. Zi and Belytschko (2003) formulated a new crack tip element using linear ramp functions for the description of the crack tip location. Mergheim et al. (2005) adopted the idea of Hansbo and Hansbo (2004) with no extra degrees of freedom in nodes. Any element with a crack was described by two overlapping standard finite elements with zero shape functions either on the left and on the right side of a discontinuity. Only displacement degrees of freedom were used, but extra phantom nodes had to be added in cracked elements to double standard nodes at the moment of cracking. This phantom node method turned out to be equivalent with the XFEM method. This approach later has been used by Song et al. (2006) to simulate cohesive shear zones. Rabczuk et al. (2008) extended the phantom node method for handling crack tips also inside of elements. To simulate shear zones in soils, a discrete Mohr-Coulomb law with softening was used by Bobiński and Brinkgreve (2010).

The formulation used here follows (with some minor modifications) the original model proposed by Wells and Sluys (2001). In a body Ω crossed by a discontinuity Γ_u (Fig. 4.3), a displacement field $\boldsymbol{u}$ can be decomposed into a continuous part $\boldsymbol{u}_{cont}$ and discontinuous part $\boldsymbol{u}_{disc}$. A displacement field can be defined as (Belytschko and Black 1999, Wells and Sluys 2001)

$$\boldsymbol{u}(\boldsymbol{x},t)=\hat{\boldsymbol{u}}(\boldsymbol{x},t)+\Psi(\boldsymbol{x})\tilde{\boldsymbol{u}}(\boldsymbol{x},t) \tag{4.9}$$

with the continuous functions $\hat{u}$ and $\tilde{u}$ and the generalized step function Ψ

$$\Psi(\boldsymbol{x})=\begin{cases}1 & \boldsymbol{x}\in\Omega^{+}\\ -1 & \boldsymbol{x}\in\Omega^{-}\end{cases}. \tag{4.10}$$

Other definitions can be also used here, e.g. the Heaviside step function. A collection of functions ϕ_i associated with set of discrete points i (i=1, 2, …, n) constitutes the partition of unity if

$$\sum_{i=1}^{n} \phi_i(\boldsymbol{x}) = 1, \qquad \forall \boldsymbol{x} \in \Omega. \tag{4.11}$$

A field $\boldsymbol{u}$ over body Ω can be interpolated as

$$\boldsymbol{u} = \sum_{i=1}^{n} \phi_i \left(a_i + \sum_{i=1}^{m} b_{ij} \gamma_j \right), \tag{4.12}$$

where a_i and b_{ij} are the discrete nodal values, γ_j – the enhanced basis and m – the number of enhanced terms for a particular node. The finite element shape functions N_i also define the partition of unity concept since

$$\sum_{i=1}^{n} N_i(\boldsymbol{x}) = 1, \qquad \forall \boldsymbol{x} \in \Omega. \tag{4.13}$$

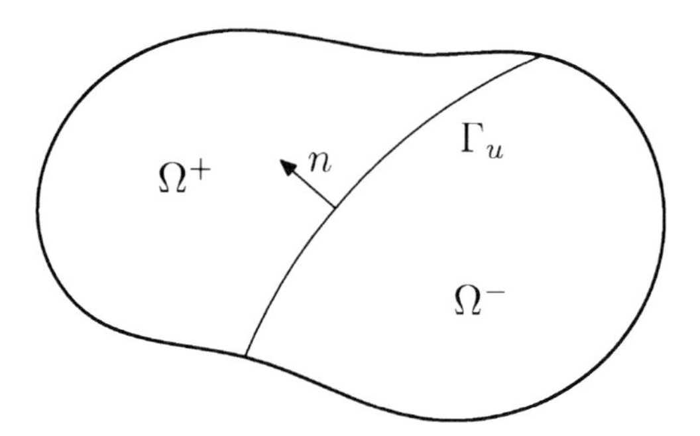

Fig. 4.3 Body crossed by a discontinuity

In a finite element format, Eq. 4.9 can be written as

$$\boldsymbol{u}(\boldsymbol{x}) = \boldsymbol{N}(\boldsymbol{x})\boldsymbol{a} + \boldsymbol{\Psi}(\boldsymbol{x})\boldsymbol{N}(\boldsymbol{x})\boldsymbol{b}, \tag{4.14}$$

where $\boldsymbol{N}$ contains shape functions, $\boldsymbol{a}$ – standard displacements at nodes and $\boldsymbol{b}$ – enriched displacements (jumps) at nodes. Only nodes belonging to 'cracked' elements are enriched. Here a formulation by Belytschko et al. (2001) called the shifted-basis enrichment is used that assumes the following definition of the displacement field

$$u(x) = N(x)a + (\Psi(x) - \Psi(x_I))N(x)b \tag{4.15}$$

with the diagonal matrices $\Psi(x)$ and $\Psi(x_I)$ containing $\Psi(x)$ and $\Psi(x_I)$, respectively (x_I is the position of the node I). The strain rate in the bulk continuum can be calculated as

$$\dot{\varepsilon} = B\dot{a} + (\Psi - \Psi_I)B\dot{b}\,, \tag{4.16}$$

whereas the rate of the displacement jump $[[u]]$ at the discontinuity is defined as

$$[[\dot{u}]] = 2N\dot{b}\,. \tag{4.17}$$

This formulation has two main advantages over the standard version (Eq. 4.9); the total displacements in nodes are equal to the standard displacements a and the implementation is simpler since two types of elements exist only.

The weak form of the equilibrium equation

$$\int_{\Omega} \nabla^s \eta : \sigma \mathrm{d}\Omega - \int_{\Gamma_u} \eta \cdot \bar{t} \mathrm{d}\Gamma = 0 \tag{4.18}$$

holds for all admissible displacement variations η (body forces are neglected and $\bar{t}$ stands for tractions applied on the boundary Γ_u). After several transformations (Wells 2001, Bobiński and Brinkgreve 2010), the following discrete weak equations are obtained

$$\begin{aligned} &\int_{\Omega} B^T \sigma \mathrm{d}\Omega = \int_{\Gamma_u} N^T \bar{t} \mathrm{d}\Gamma \\ &\int_{\Omega} (\Psi - \Psi_I) B^T \sigma \mathrm{d}\Omega + 2\int_{\Gamma_d} N^T t \mathrm{d}\Gamma = \int_{\Gamma_u} (\Psi - \Psi_I) N^T \bar{t} \mathrm{d}\Gamma \end{aligned} \tag{4.19}$$

with the strain-nodal displacement matrix B. The linearized equations of the total system are

$$\begin{bmatrix} K_{aa} & K_{ab} \\ K_{ba} & K_{bb} \end{bmatrix} \begin{bmatrix} \mathrm{d}a \\ \mathrm{d}b \end{bmatrix} = \begin{bmatrix} f_a^{ext} \\ f_b^{ext} \end{bmatrix} - \begin{bmatrix} f_a^{int} \\ f_b^{int} \end{bmatrix} \tag{4.20}$$

with the blocks of the global stiffness matrix $\boldsymbol{K}$ defined as

$$\begin{aligned} \boldsymbol{K}_{aa} &= \int_{\Omega} \boldsymbol{B}^T \boldsymbol{D} \boldsymbol{B} \mathrm{d}\Omega & \boldsymbol{K}_{ab} &= \int_{\Omega} \boldsymbol{B}^T \boldsymbol{D} \boldsymbol{B} (\boldsymbol{\Psi} - \boldsymbol{\Psi}_I) \mathrm{d}\Omega \\ \boldsymbol{K}_{ba} &= \int_{\Omega} (\boldsymbol{\Psi} - \boldsymbol{\Psi}_I) \boldsymbol{B}^T \boldsymbol{D} \boldsymbol{B} \mathrm{d}\Omega & \boldsymbol{K}_{bb} &= \int_{\Omega} (\boldsymbol{\Psi} - \boldsymbol{\Psi}_I) \boldsymbol{B}^T \boldsymbol{D} \boldsymbol{B} (\boldsymbol{\Psi} - \boldsymbol{\Psi}_I) \mathrm{d}\Omega + 4 \int_{\Gamma_d} \boldsymbol{N}^T \boldsymbol{T} \boldsymbol{N} \mathrm{d}\Gamma \end{aligned}, \tag{4.21}$$

where $\boldsymbol{T}$ is the stiffness matrix at the discontinuity. The force vectors are equal to

$$\begin{aligned} f_a^{ext} &= \int_{\Gamma_u} \boldsymbol{N}^T \bar{\boldsymbol{t}} \mathrm{d}\Gamma & f_a^{int} &= \int_{\Omega} \boldsymbol{B}^T \sigma \mathrm{d}\Omega \\ f_b^{ext} &= \int_{\Gamma_u} (\boldsymbol{\Psi} - \boldsymbol{\Psi}_I) \boldsymbol{N}^T \bar{\boldsymbol{t}} \mathrm{d}\Gamma & f_b^{int} &= \int_{\Omega} (\boldsymbol{\Psi} - \boldsymbol{\Psi}_I) \boldsymbol{B}^T \sigma \mathrm{d}\Omega + 2 \int_{\Gamma_d} \boldsymbol{N}^T t \mathrm{d}\Gamma \end{aligned}. \tag{4.22}$$

In un-cracked continuum, usually a linear elastic constitutive law between stresses and strains is assumed under tension. To activate a crack, the Rankine condition has to be fulfilled at least in one integration point in the element at the front of the crack tip

$$max\{\sigma_1, \sigma_2, \sigma_3\} > f_t, \tag{4.23}$$

where σ_i are the principal stresses and f_t is the tensile strength. This inequality can be also verified at the crack tip directly (Mariani and Perego 2003). A very important issue is a determination of the crack propagation direction. If this direction is known in advance, it can be assumed (fixed) directly. Otherwise a special criterion has to be used. The most popular criterion assumes that the direction of the crack extension is perpendicular to the direction of the maximum principal stress. To smooth the stress field around the crack tip, non-local stresses σ^* instead of local values can be taken to determine the crack direction (Wells and Sluys 2001)

$$\boldsymbol{\sigma}^* = \int_V \sigma\, w \mathrm{d}V, \tag{4.24}$$

where the domain V is a semicircle at the front of the crack tip and a weight function w is defined as

$$w = \frac{1}{(2\pi)^{3/2} l^3} \exp\left(-\frac{r^2}{2l^2}\right). \tag{4.25}$$

Here the length l is the averaging length (usually equal to 3 times the average element size) and r denotes the distance between the integration point and crack tip. This operation does not introduce non-locality connected to material microstructure into the model. Mariani and Perego (2003) used higher order polynomials for a better description of the stress state (and also the displacement state) around the crack tip. Stresses in the crack tip were determined using an interpolation of nodal values. Oliver et al. (2004) formulated a global tracking algorithm, where propagation directions of cracks were determined globally by solving a stationary anisotropic heat conduction type problem. Moes and Belytschko (2002) assumed that cohesive tractions had no influence on the crack propagation direction and used the maximum circumferential stress criterion from Linear Elastic Fracture Mechanics (LEFM). Another important item of the formulation is a discrete cohesive law which links tractions $\boldsymbol{t}$ with displacement jumps $[[u]]$ at a discontinuity. The simplest one assumes the following format of the loading function (Wells and Sluys 2001)

$$f\left([[u_n]],\kappa\right) = [[u_n]] - \kappa \tag{4.26}$$

with the history parameter κ equal to the maximum value of the displacement jump $[[u_n]]$ achieved during loading. Softening of the normal component of the traction vector can be described using an exponential

$$t_n = f_t \exp\left(-\frac{f_t \kappa}{G_f}\right) \tag{4.27}$$

or a linear relationship

$$t_n = f_t\left(1 - \frac{\kappa}{\kappa_u}\right), \qquad \kappa_u = \frac{2G_f}{f_t}, \tag{4.28}$$

where G_f denotes the fracture energy. During unloading, the secant stiffness is used with a return to the origin (damage format). In a compressive regime, a penalty elastic stiffness matrix is assumed. In a tangent direction, a linear relationship between a displacement jump and traction is defined with the stiffness T_s. Similar constitutive models were used by Remmers at al. (2003) and Mergheim et al. (2005). Alternatively, formulations based on effective displacements described in Chapter 4.1 may be used (Mariani and Perego 2003, Comi and Mariani 2007). Note that the cohesive crack formulation with the coefficient η=0 (Eq. 4.1) is equivalent with the discontinuity model described above, if the stiffness T_s=0. To overcome convergence problems in situations when $[[u_n]]$ changes its sign, Cox (2009) modified a linear softening curve in normal direction as

$$t_n = f_t\left(1 - \frac{\kappa}{\kappa_u}\right)\left(1 - \exp\left(-d_f \frac{\kappa}{\kappa_u}\right)\right), \tag{4.29}$$

where d_f is a drop factor. With increasing the value of d_f, the influence of the second term diminishes. The same modification was applied to an exponential softening curve.

The inclusion of enriched displacements $\boldsymbol{b}$ requires several modifications in the standard FE code. The final number of extra degrees of freedom $\boldsymbol{b}$ is unknown at the beginning and it may grow during calculations. Therefore special techniques are required to handle the extra data. If an essential boundary condition has been specified at a node with enriched degrees of freedom, the additional condition b=0 has to be added at this node. A new crack segment can be defined in the converged configuration only. After defining a new segment, a current increment has to be restarted. Moreover, nodes that share the edge with a crack tip may not be enriched. A definition of the crack segment geometry obeys the following rules:

- a new crack segment is defined from one element side to another one (a crack tip cannot be placed inside elements),
- segment end points cannot be placed at element vertices,
- a crack segment is straight inside one element,
- a crack is continuous across elements and adjacent segments share the same point.

To avoid placing cracks at element's vertices, three minimal distances are declared (Fig. 4.4):

- minimum distance between the vertex and crack segment $v_{\min}$,
- minimum distance between the vertex and crack segment end point along the side $l_{\min}$,
- minimum distance between the triangle side not touched by a discontinuity and the crack segment end point $s_{\min}$.

With the same values of $v_{\min}$ and $l_{\min}$, the first condition is stronger, since it takes into account also the distance between the vertex and segment. Finally, a new scheme for calculating strains, stresses, internal forces and stiffness in a cracked

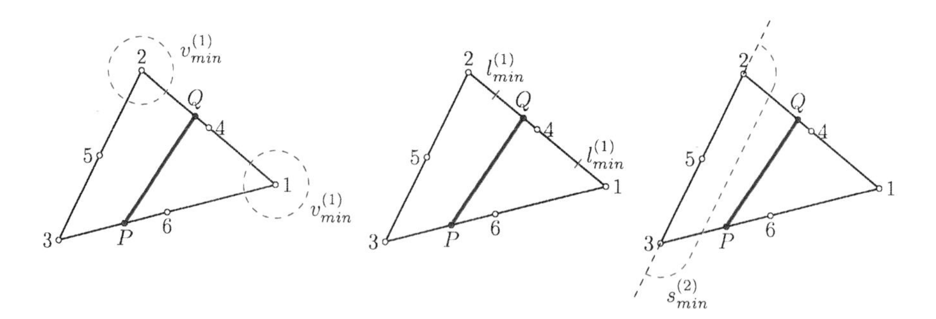

Fig. 4.4 Minimal distances between crack segment and triangle vertices/sides

element is required. Due to an arbitrary location of a discontinuity segment inside an element, new coordinates of integration points have to be defined. To determine these coordinates a sub-division algorithm is proposed. The triangle sub-region is divided into 3 triangles and quad sub-region into 4 triangles. In each triangle, 3 integration points are defined. In total, a numerical integration requires 21 integration points in the bulk and 2 points at the discontinuity (Wells and Sluys 2001, Bobiński and Brinkgreve 2010).

References

Bažant, Z., Jirásek, M.: Nonlocal integral formulations of plasticity and damage: survey of progress. Journal of Engineering Mechanics 128(11), 1119–1149 (2002)

Belytschko, T., Black, T.: Elastic crack growth in finite elements with minimal remeshing. International Journal for Numerical Methods in Engineering 45(5), 601–620 (1999)

Belytschko, T., Moes, N., Usui, S., Parimi, C.: Arbitrary discontinuities in finite elements. International Journal for Numerical Methods in Engineering 50(4), 993–1013 (2001)

Bobiński, J., Brinkgreve, R.: Objective determination of failure mechanisms in geomechanics, STW DCB.6368 Report, TU Delft (2010)

Camacho, G.T., Ortiz, M.: Computational modelling of impact damage in brittle materials. International Journal of Solids and Structures 33(20-22), 2899–2938 (1996)

Cazes, F., Coret, M., Combescure, A., Gravouil, A.: A thermodynamic method for the construction of a cohesive law from a nonlocal damage model. International Journal of Solids and Structures 46(6), 1476–1490 (2009)

Chandra, N., Li, H., Shet, C., Ghonem, H.: Some issues in the application of cohesive zone models for metal-ceramic interfaces. International Journal of Solids and Structures 39(10), 2827–2855 (2002)

Comi, C., Mariani, S.: Extended finite element simulation of quasi-brittle fracture in functionally graded materials. Computer Methods in Applied Mechanics and Engineering 196(41-44), 4013–4026 (2007)

Cox, J.V.: An extended finite element method with analytical enrichment for cohesive crack modelling. International Journal for Numerical Methods in Engineering 78(1), 48–83 (2009)

Daux, C., Moes, N., Dolbow, J., Sukumar, N., Belytschko, T.: Arbitrary branched and intersecting cracks with the extended finite element method. International Journal for Numerical Methods in Engineering 48(12), 1741–1760 (2000)

De Borst, T., Remmers, J.J.C.: Computational modelling of delamination. Composites Science and Technology 66(6), 713–722 (2006)

De Lorenzis, L., Zavarise, G.: Cohesive zone modeling of interfacial stresses in plated beams. International Journal of Solids and Structures 46(24), 4181–4191 (2009)

Dugdale, D.S.: Yielding of steel sheers containing slits. Journal of Mechanics and Physics of Solids 8(2), 100–108 (1960)

Gálvez, J.C., Červenka, J., Cendón, D.A., Saouma, V.: A discrete crack approach to normal/shear cracking of concrete. Cement and Concrete Research 32(10), 1567–1585 (2002)

Hansbo, A., Hansbo, P.: A finite element method for the simulation of strong and weak discontinuities in solid mechanics. Computer Methods in Applied Mechanics and Engineering 193(33-35), 3523–3540 (2004)

Hillerborg, A., Modeer, M., Peterson, P.E.: Analysis of crack propagation and crack growth in concrete by means of fracture mechanics and finite elements. Cement and Concrete Research 6(6), 773–782 (1976)

Mariani, S., Perego, U.: Extended finite element method for quasi-brittle fracture. International Journal for Numerical Methods in Engineering 58(1), 103–126 (2003)

Melenk, J.M., Babuška, I.: The partition of unity finite element method: basic theory and applications. Computer Methods in Applied Mechanics and Engineering 139(1-4), 289–314 (1996)

Mergheim, J., Kuhl, E., Steinman, P.: A finite element method for the computational modelling of cohesive cracks. International Journal for Numerical Methods in Engineering 63(2), 276–289 (2005)

Moës, N., Belytschko, T.: A finite element method for crack growth without remeshing. International Journal for Numerical Methods in Engineering 46(1), 131–150 (1999)

Moës, N., Belytschko, T.: Extended finite element method for cohesive crack growth. Engineering Fracture Mechanics 69(7), 813–833 (2002)

Needleman, A.: A continuum model for void nucleation by inclusion debonding. Journal of Applied Mechanics 54(3), 525–531 (1987)

Oliver, J., Huespe, A.E., Samaniego, E., Chaves, E.W.: Continuum approach to the numerical simulation of material failure in concrete. International Journal for Numerical and Analytical Methods in Geomechanics 28(7-8), 609–632 (2004)

Omiya, M., Kishimoto, K.: Damage-based cohesive zone model for rate-dependent interfacial fracture. International Journal of Damage Mechanics 19(4), 397–420 (2010)

Ortiz, M., Pandolfi, A.: Finite-deformation irreversible cohesive elements for three-dimensional crack-propagation analysis. International Journal for Numerical Methods in Engineering 44(9), 1267–1282 (1999)

Rabczuk, T., Zi, G., Gerstenberger, A., Wall, W.A.: A new crack tip element for the phantom-node method with arbitrary cohesive cracks. International Journal for Numerical Methods in Engineering 75(5), 577–599 (2008)

Remmers, J.J.C., de Borst, R., Needleman, A.: A cohesive segments method for the simulation of crack growth. Computational Mechanics 31(1-2), 69–77 (2003)

Samaniego, E., Belytschko, T.: Continuum-discontin0uum modelling of shear bands. International Journal for Numerical Methods in Engineering 63(13), 1857–1872 (2005)

Samimi, M., van Dommelen, J.A.W., Geers, M.G.D.: An enriched cohesive zone model for delamination in brittle interfaces. International Journal for Numerical Methods in Engineering 80(5), 609–630 (2009)

Scheider, I., Brocks, W.: Simulation of cup-cone fracture using the cohesive mode. Engineering Fracture Mechanics 70(14), 1943–1961 (2003)

Song, J.-H., Areias, P.M.A., Belytschko, T.: A method for dynamic crack and shear zone propagation with phantom nodes. International Journal for Numerical Methods in Engineering 67(6), 868–893 (2006)

Sukumar, N., Moës, N., Belytschko, T., Moran, B.: Extended finite element method for three-dimensional crack modelling. International Journal for Numerical Methods in Engineering 48(11), 1549–1570 (2000)

Wells, G.: Discontinuous modelling of strain localisation and failure. PhD Thesis, TU Delft (2001)

Wells, G.N., Sluys, L.J.: A new method for modelling cohesive cracks using finite elements. International Journal for Numerical Methods in Engineering 50(12), 2667–2682 (2001)

Yang, Z., Xu, X.F.: A heterogeneous cohesive model for quasi-brittle materials considering spatially varying random fracture properties. Computer Methods in Applied Mechanics and Engineering 197(45-48), 4027–4039 (2008)

Zhou, F., Molinari, J.F.: Dynamic crack propagation with cohesive elements: a methodology to address mesh dependency. International Journal for Numerical Methods in Engineering 59(1), 1–24 (2004)

Zi, G., Belytschko, T.: New crack-tip elements for XFEM and applications to cohesive cracks. International Journal for Numerical Methods in Engineering 57(15), 2221–2240 (2003)

Chapter 5
Continuous and Discontinuous Modelling of Fracture in Plain Concrete under Monotonic Loading

Abstract. This Chapter presents the FE results of continuous and discontinuous modelling of fracture in plain concrete under monotonic loading. Tests of uniaxial compression, uniaxial extension, bending and shear –extension were simulated.

5.1 Uniaxial Compression and Extension

FE results within elasto-plasticity and damage mechanics

The initial plane strain FE-calculations (Bobiński and Tejchman 2004) were performed with a specimen b=4 cm wide and h=14 cm high subjected to uniaxial compression and extension (Fig. 5.1) using the Drucker-Prager constitutive model with non-local softening (Eqs. 3.27-3.30, 3.93 and 3.97). The lower and upper edge were both smooth. All nodes along the lower edge of the specimen were fixed in a vertical direction. To preserve the stability of the specimen, the node in the middle of the lower edge was also kept fixed in a horizontal direction. The deformation was initiated through constant vertical displacement increments prescribed to nodes along the upper edge of the specimen.

Figure 5.2 shows a relationship between the yield stress σ_y and equivalent plastic strain ε_p in a compressive regime. In the simulations, localization was induced by one small material imperfection in the form of a weak element at mid-height of the specimen side (where the yield stress at peak σ_y^{max} of Fig. 5.2 was diminished by 2%). In addition, the calculations were carried out with three initial weak elements of a different size and spacing and with one initial strong element (where the cohesion yield stress at peak σ_y^{max} was increased by 2%).

To investigate the effect of the mesh size on the results, a various discretization was used: coarse (8×28), medium (16×56) and fine (24×84), where each quadrilateral was composed of four diagonally crossed triangular elements with linear shape functions (mesh inclination against the bottom was fixed and equal to

J. Tejchman, J. Bobiński: Continuous & Discontinuous Modelling of Fracture, SSGG, pp. 109–162.
springerlink.com

θ=45°). To examine the influence of the mesh alignment on the results, two other types of meshes were used: with a mesh inclination θ smaller than 45°: θ=26.6° (8×56), θ=36.9° (12×56), θ=33.7° (16×84) and θ=39.8° (20×84), and with a mesh inclination θ greater than 45° θ=56.3° (12×28), θ=63.5° (16×28), θ=51.3° (20×56) and θ=56.3° (24×56).

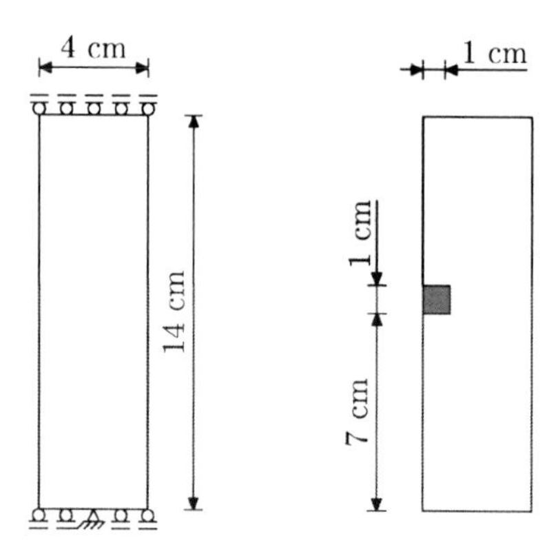

Fig. 5.1 Geometry of the specimen, boundary conditions and location of the imperfection for elasto-plastic calculations (Bobiński and Tejchman 2004)

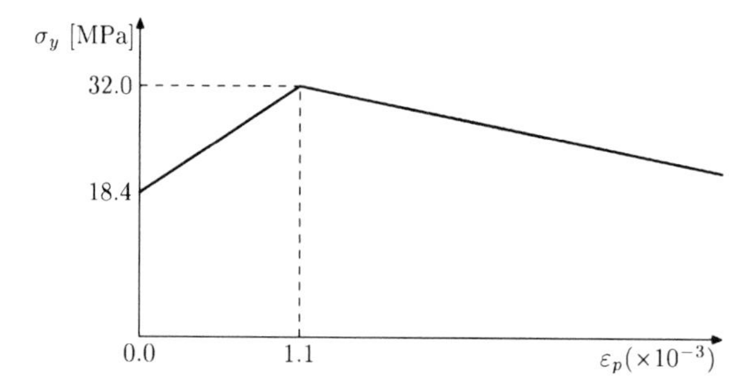

Fig. 5.2 Relationship between cohesion yield stress σ_y and equivalent plastic strain ε_p in a compressive regime for elasto-plastic calculations (Bobiński and Tejchman 2004)

First, an elasto-plastic analysis was carried out without non-local terms (local formulation) during uniaxial compression. Figure 5.3 shows deformed meshes for a various mesh refinement (θ=45°) with φ=0° and ψ=0° (Fig. 5.3a) and φ=25° and ψ=10° (Fig. 5.3b). The deformations localize always in one element wide shear zone with the inclination of 45° (equal to the mesh inclination). The load-displacement curves are strongly dependent upon the discretization size in the softening regime.

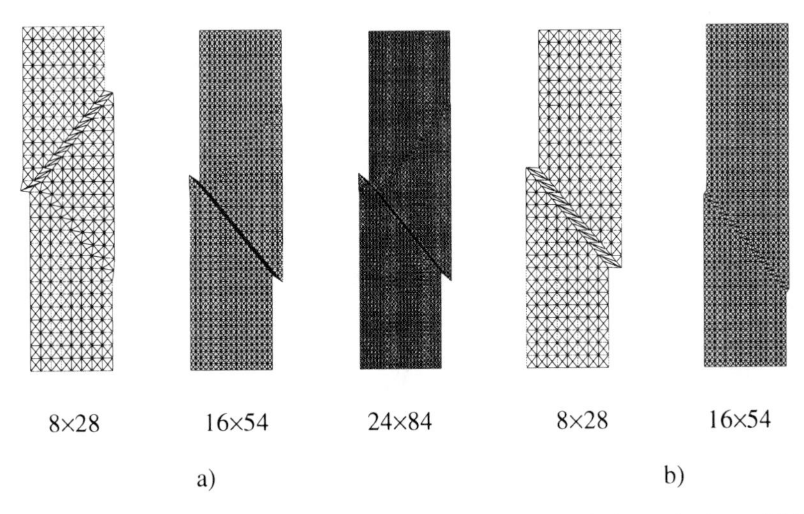

Fig. 5.3 Deformed meshes during uniaxial compression with local elasto-plastic model (non-locality parameter m=0): a) φ=0° and ψ=0°, b) φ=25° and ψ=10° (Bobiński and Tejchman 2004)

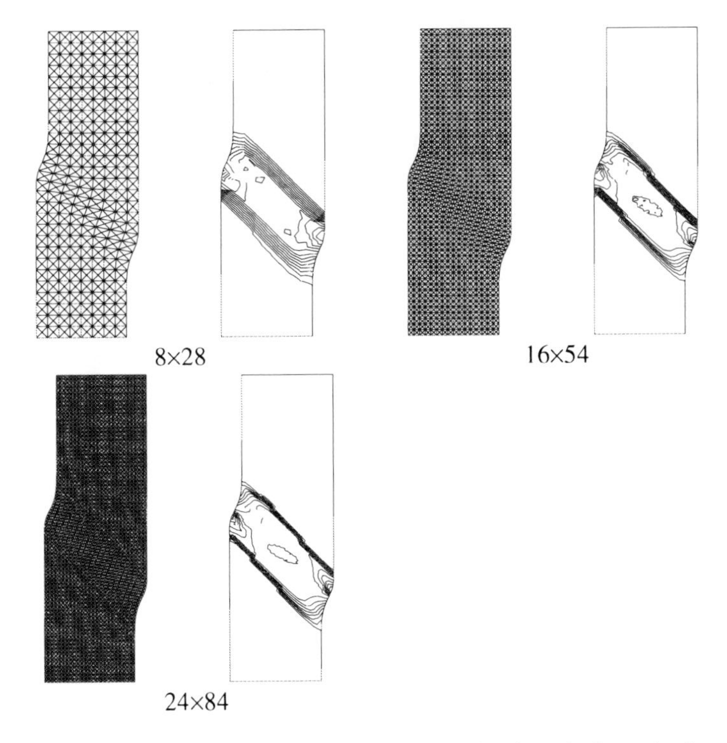

Fig. 5.4 Deformed meshes and contours of non-local equivalent plastic strains (elasto-plastic model with non-local softening, m=2, l_c=7.5 mm, φ=25° and ψ=10°) (Bobiński and Tejchman 2004)

Figures 5.4-5.7 depict the FE-results with an elasto-plastic model with non-local softening during uniaxial compression (with the mesh alignment of θ=45°). As compared to the conventional elasto-plastic analysis, two additional constants were taken into account: a non-locality parameter m=2 and a characteristic length l_c=7.5 mm. The calculated deformed meshes and non-local equivalent plastic strains are shown in Fig. 5.4. The load-displacement diagrams are presented in Fig. 5.5. In turn, Fig. 5.6 demonstrates the distribution of the equivalent plastic strains across the shear zone. The evolution of the equivalent non-local plastic strain is shown in Fig. 5.7.

During compression, two shear zones are simultaneously created expanding outward from the weak element on the left side (Fig. 5.7). They occur directly before the peak of the resultant vertical force on the top. After the peak, and up to the end, only one shear zone dominates. The complete shear zone is noticeable shortly after the peak. The shear zone is wider than one finite element (Fig. 5.4). The thickness of a shear zone is approximately t_s=2.4 cm (3.2×l) and does not depend upon the mesh size. The inclination of the shear zone against the bottom is equal to θ=44.8°, 44.8° and 46.8° with a coarse, medium and fine mesh, respectively. These values are in a good agreement with an inclination of a shear zone (θ=46°) obtained from an analytical formula based on a bifurcation theory (Sluys 1992)

$$\tan^2\Theta = -\frac{9(s_1 + \nu s_3) + (1+\nu)\sqrt{3J_2}\,(\varphi+\psi)}{9(s_2 + \nu s_3) + (1+\nu)\sqrt{3J_2}\,(\varphi+\psi)}, \tag{5.1}$$

where s_i are the components of the deviatoric principle stress tensor, J_2 denotes the second invariant of the deviatoric stress tensor and ν is the Poisson's ratio.

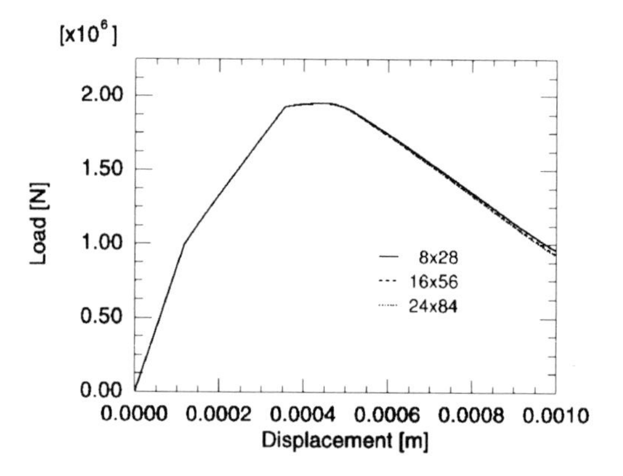

Fig. 5.5 Load–displacement curves for different mesh discretizations (elasto-plastic model with non-local softening, m=2, l_c=7.5 mm, φ=25° and ψ=10°) (Bobiński and Tejchman 2004)

The calculated inclinations of the shear zone against the bottom (using a medium mesh) with other values of φ and ψ: φ=40° and ψ=10°, and φ=40° and ψ=25° were equal to 48.6° and 52.0°, respectively. These values again agree well with Eq. 5.1 (48.1° and 50.2°, respectively). The growth of the internal friction angle and dilatancy angle obviously increases the inclination of the shear zone against the bottom.

The evolution of the vertical force along the top before and after the peak is the same for a various discretization (Fig. 5.5). The maximum vertical force, 1.95 MN (corresponding to the compressive strength 49 MPa), occurs at the vertical displacement of 0.4 mm. The distribution of non-local equivalent plastic strains in a section perpendicular to the shear zone is fully uniform (Fig. 5.6). Due to that a non-locality parameter is larger than 1, the difference between the non-local and local plastic strains results in a constant value of the plastic strain (Jirásek and Rolshoven 2003). Separately, the non-local and local plastic strains have an exponential shape.

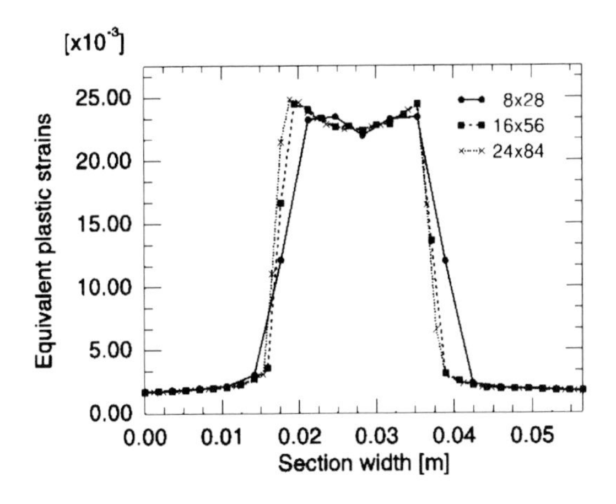

Fig. 5.6 Non-local equivalent plastic strains in a shear zone different mesh discretization (elasto-plastic model with non-local softening, m=2, l_c=7.5 mm, φ=25° and ψ=10°) (Bobiński and Tejchman 2004)

The FE-calculations were also performed with a mesh alignment lower than 45°. The width of the shear zone was equal to t_s=2.4 cm for all meshes except for the coarsest one (8×56) where the localization zone was equal to t_s=3.6 cm. The inclination of the shear zone was: 44.0° (8×56, θ=26.6°), 43.9° (12×56, θ=36.9°), 47.6° (16×84, θ=33.7°) and 45.2° (20×84, θ=39.8°). The load-displacement diagrams were similar using all mesh discretizations. In the FE-studies assuming a mesh inclination greater than 45°, the width of the shear zone was again 2.4 cm with the meshes 20×56 and 24×56, and 2.9 cm with the meshes 12×28 and 16×28.

Fig. 5.7 Evolution of non-local equivalent plastic strains in the specimen during uniaxial compression (elasto-plastic model with non-local softening, m=2, l_c=7.5 mm, φ=25° and ψ=10°) (Bobiński and Tejchman 2004)

The inclination of the shear zone was 50.3° (12×28, θ=56.3°), 47.2° (16×28, θ=47.2°), 50.1° (20×56, θ=50.1°) and 49.1° (24×56, θ=56.3°). The FE-results showed that the evolution of the vertical force along the top and thickness of the spontaneous shear zone did not depend upon mesh refinement within a modified non-local continuum. The effect of the mesh alignment on the inclination of the shear zone was negligible.

The FE-results with a classical non-local model, m=1, are demonstrated in Figs. 5.8-5.10 (using meshes with an inclination lower θ than 45°), and in Figs. 5.11-5.13 (using meshes with an inclination θ greater than 45°). The calculations show that the results are only partly mesh-independent. The evolution of the vertical force on the top edge was only slightly different in a softening regime (Figs. 5.9 and 5.12). However, the width of the localized zone was different: t_s=2.7 cm (12×56 and 16×84) and t_s=1.9 cm (20×84) at θ<45° (Fig. 5.10). The shear zone using a mesh 8×56 was very wide and had an opposite direction. The inclination of the shear zone was similar: 43.6° (12×56, θ=36.9°), 43.8° (16×84, θ=33.7°) and 45.5° (20×84, θ=39.8°). When the mesh alignment was larger than 45°, the width of the shear zone was again different: L=2.7 cm, 3.3 cm, 1.2 cm and 1.7 cm for meshes 12×28, 16×28, 20×56 and 24×56 mesh, respectively (Fig. 5.12). Its inclination was varied: 55.2° (12×28, θ=56.3°), 47.5° (16×28, θ=47.2°), 51.3° (20×56, θ=50.1°) and 50.6° (24×56, θ=56.3°). The results confirm that classical non-local model (m=1) is able to regularize force–displacement diagram only.

Effect of imperfections

To investigate the effect of imperfections on the material behaviour during uniaxial plane strain compression, in addition, the simulations were carried out with one initial strong element and few initial weak elements with a different size and spacing.

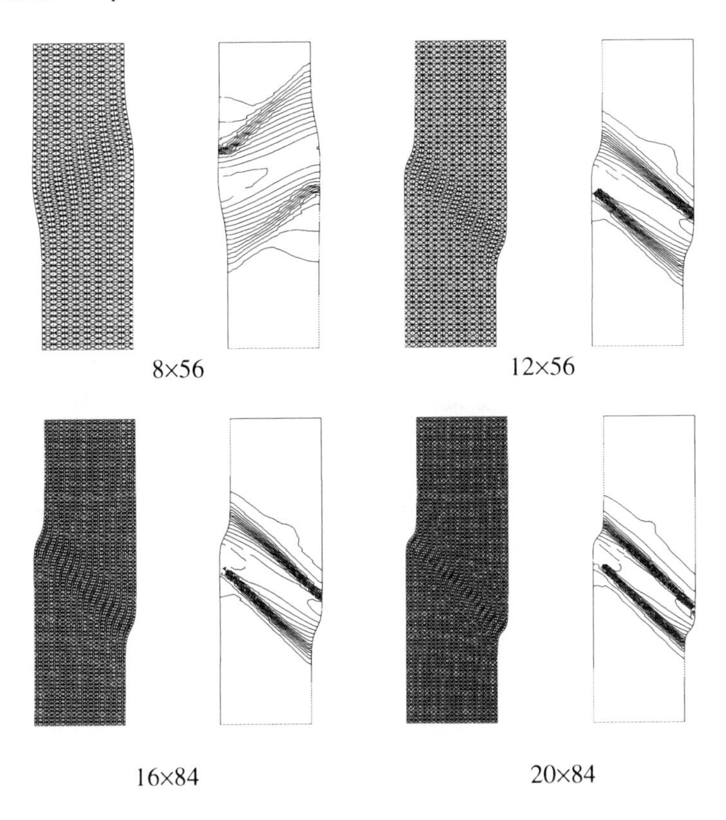

Fig. 5.8 Deformed meshes and contours of non-local equivalent plastic strains (elasto-plastic model with non-local softening, θ<45°, m=1, l_c=20 mm, φ=25° and ψ=10°) (Bobiński and Tejchman 2004)

Figure 5.14 presents the results with one strong element in the middle of the left side of the specimen. The location of the shear zone is not connected with the position of the imperfection. The shear zone (reflected from the rigid bottom) is created in the lower part of the specimen.

The evolution of non-local plastic strains in the specimen with three initial weak elements distributed at the same distance along the left side is demonstrated in Fig. 5.15. Before the peak, two shear zones are created at each week element. The shear zones propagate towards both the top and bottom of the specimen. After the peak, only one shear zone dominates in the whole specimen. The location of this shear zone is, however, different as compared to the results with one weak element at mid-height. The other calculations showed that the size and number of weak elements and distance between them did not influence the thickness and inclination of the shear zone, and the load-displacement curve. Only the location of the shear zone was affected. This result is in contrast to FE-calculations by Shi and Chang (2003) wherein the thickness of a shear zone was influenced by the imperfection spacing.

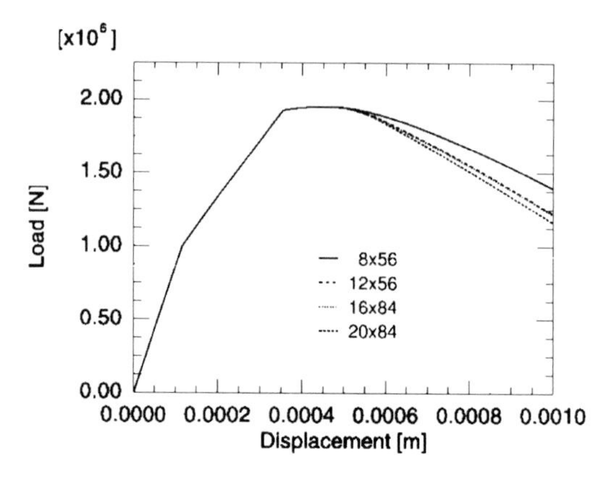

Fig. 5.9 Load–displacement curves (elasto-plastic model with non-local softening, θ<45°, m=1, l_c=20 mm, φ=25° and ψ=10°) (Bobiński and Tejchman 2004)

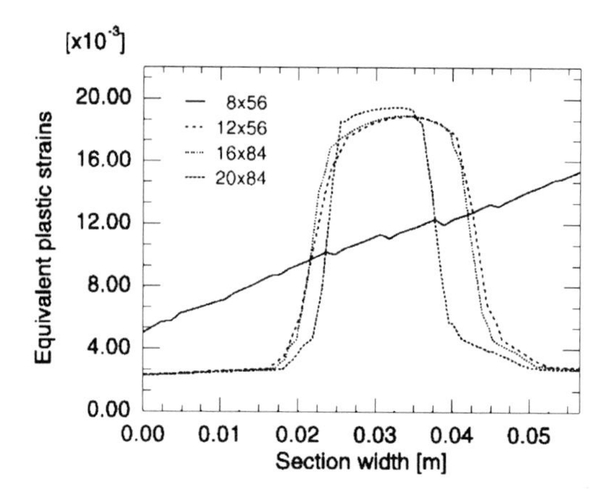

Fig. 5.10 Distribution of non-local equivalent plastic strains in a shear zone (elasto-plastic model with non-local softening, θ<45°, m=1, l_c=20 mm, φ=25° and ψ=10°) (Bobiński and Tejchman 2004)

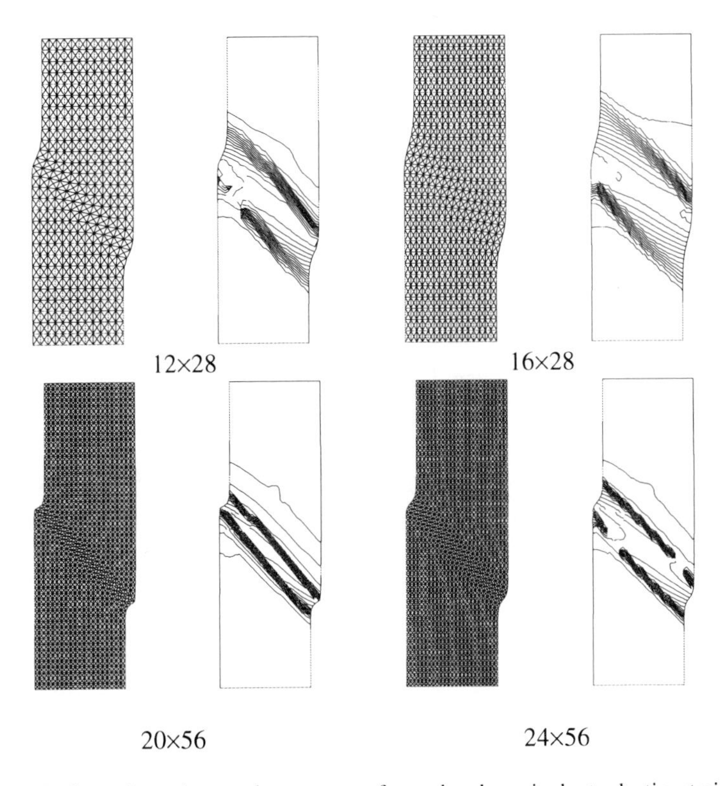

Fig. 5.11 Deformed meshes and contours of non-local equivalent plastic strains with various meshes (elasto-plastic model with non-local softening, θ>45°, m=1, l_c=20 mm, φ=25° and ψ=10°) (Bobiński and Tejchman 2004)

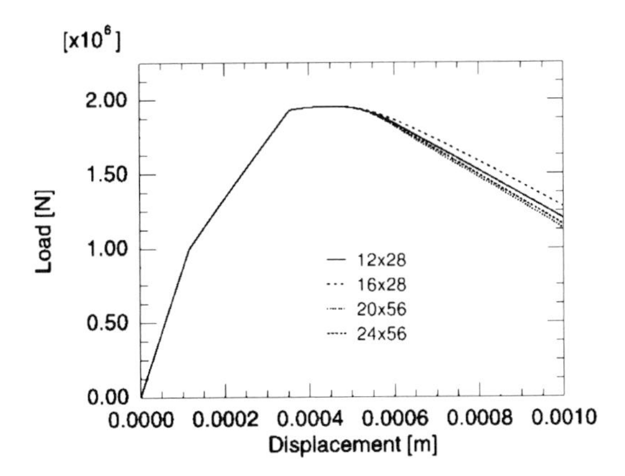

Fig. 5.12 Load–displacement curves (elasto-plastic model with non-local softening, θ>45°,m=1, l_c=20 mm, φ=25° and ψ=10°) (Bobiński and Tejchman 2004)

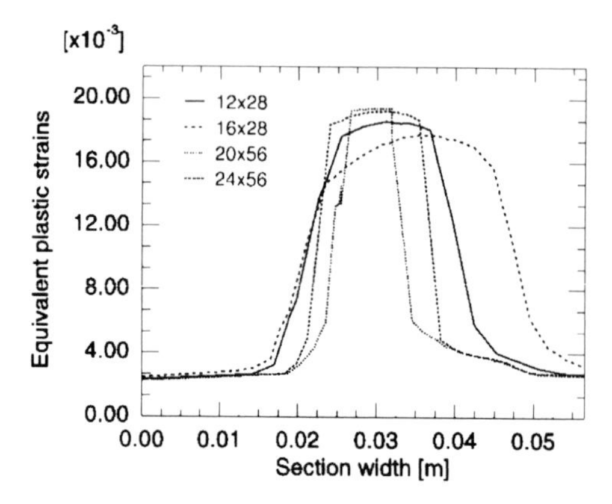

Fig. 5.13 Distribution of non-local equivalent plastic strains in a shear zone (non-local elasto-plastic model, θ>45°, m=1, l_c=20 mm, φ=25° and ψ=10°) (Bobiński and Tejchman 2004)

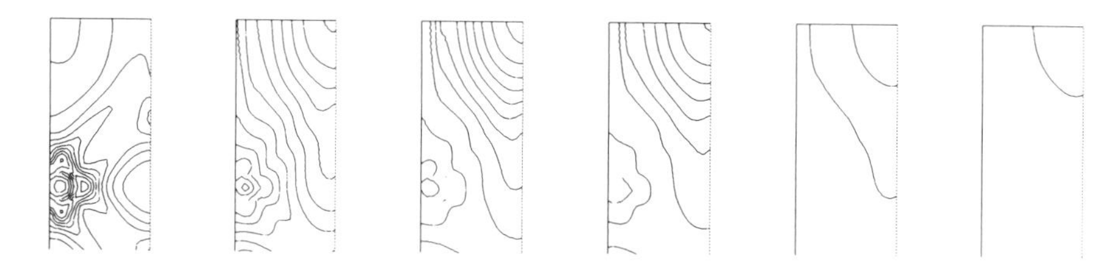

Fig. 5.14 Evolution of non-local equivalent plastic strains in the specimen during uniaxial compression (elasto-plastic model with non-local softening, one initial strong element, m=2, l_c=7.5 mm, φ=25° and ψ=10°)

Fig. 5.15 Evolution of non-local equivalent plastic strains in the specimen during uniaxial compression (elasto-plastic model with non-local softening, three initial weak elements, m=2, l_c=7.5 mm, φ=25° and ψ=10°) (Bobiński and Tejchman 2004)

Effect of direction of deformation

Figure 5.16 shows the numerical results during uniaxial plane strain extension of the specimen with one weak element at mid-height. Similarly, as during uniaxial compression, the results do not depend on the mesh discretization. The thickness of the shear zone is approximately equal to t_s=1.7 cm (2.2×l). Thus, the thickness of the shear zone during extension is smaller by 30% than during compression. The inclination of the shear zone against the bottom is 45.4° and 45.2° with a coarse and fine mesh, respectively.

Effect of characteristic length and non-locality parameter

The effect of the parameters m (m=2-5) and l_c (l_c=2-12 mm) on the width of shear localization was investigated using the medium mesh (16×56, θ=45°) during uniaxial compression and extension.

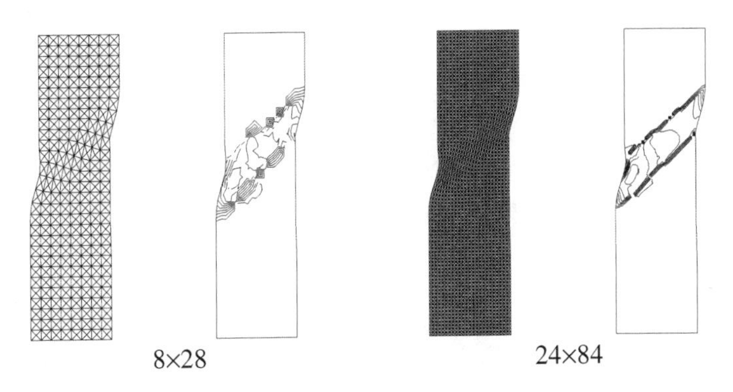

Fig. 5.16 Deformed meshes and contours of non-local equivalent plastic strains during uniaxial extension (elasto-plastic model with non-local softening, m=2, l_c=7.5 mm, φ=25°, ψ=10°) (Bobiński and Tejchman 2004)

The results show that the larger the parameters l_c and m, the wider the shear zone t_s (Tab. 5.1). The parameters m and l_c strongly influence the load-displacement curves in a post-peak regime (Fig. 5.17). The larger l, the smaller is the drop of the curves after the peak. However, they have no influence on the maximum vertical force on the top due to the presence of very smooth horizontal boundaries. The width of the shear zone t_s during uniaxial compression (Tab. 5.1) can be approximately co-related with the parameters m and l_c by a following relation

$$t_s \cong (1.2-1.6)m \times l_c \,. \tag{5.2}$$

However, during uniaxial extension, the width of the shear zone is described by a different relation:

$$t_s \cong (0.8-1.2)m \times l_c \,. \tag{5.3}$$

The thickness of the shear zone is obviously dependent upon the boundary conditions of the entire system expressed by e.g. the specimen geometry and loading conditions.

Table 5.1 The calculated width within enhanced elasto-plasticity of the shear zone t_s [cm] during uniaxial compression (Bobiński and Tejchman 2004)

l_c [mm]	m			
	2.0	3.0	4.0	5.0
2	1 element	1.2	1.2	1.5
4	1.4	2.0	2.3	2.5
6	2.1	2.7	3.2	3.6
8	2.7	3.5	4.0	4.7
10	3.2	4.6	diffuse	diffuse
12	3.8	diffuse	diffuse	diffuse

Finally, the effect of the boundary roughness was investigated in concrete elements (a weak element was inserted at the lower left corner) (Fig. 5.18). In this case, the horizontal displacements along horizontal boundaries were fixed. With smooth boundaries, the strength is always the same. When the boundaries are very rough, the strength increases with increasing characteristic length. The brittleness increases as usually with decreasing l_c independently of the boundary roughness.

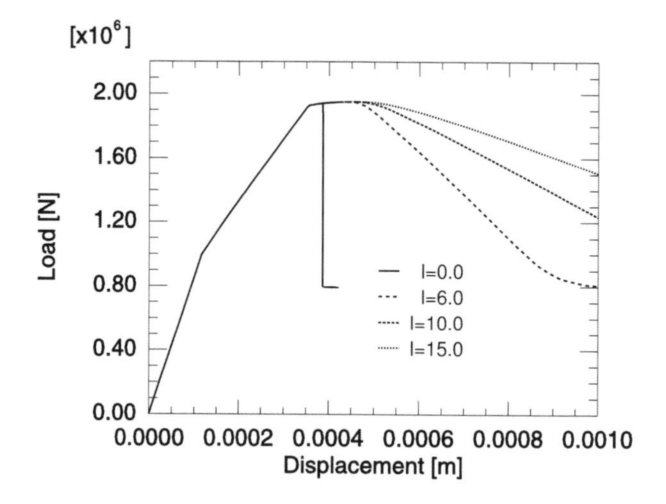

Fig. 5.17 Load–displacement curves with a different characteristic length (elasto-plastic model with non-local softening, m=2, φ=25° and ψ=10°) (Bobiński and Tejchman 2004)

Next, the problem of a symmetric double notched concrete specimen under uniaxial tension was numerically studied with a Rankine'a constitutive model (Eqs. 3.32, 3.93 and 3.97) (Bobiński and Tejchman 2006). Experimentally it was investigated by Hordijk (1991). The geometry of the concrete specimen (width b=60 mm, height h=125 mm, thickness in the out-of-plane direction t=50 mm) and boundary conditions are presented in Fig. 5.19. Two symmetric notches 5×5 mm^2 were located at the mid-point of both sides of the specimen. Three different FE-meshes were used: coarse (1192 triangular elements), medium (1912 triangular elements) and fine (4168 triangular elements), Fig. 5.20. When calculating non-local quantities close to the notch, the so-called "shading effect" was considered (i.e. the averaging procedure considers the notch as an internal barrier that is shading the non-local interaction, Jirásek and Rolshoven 2003).

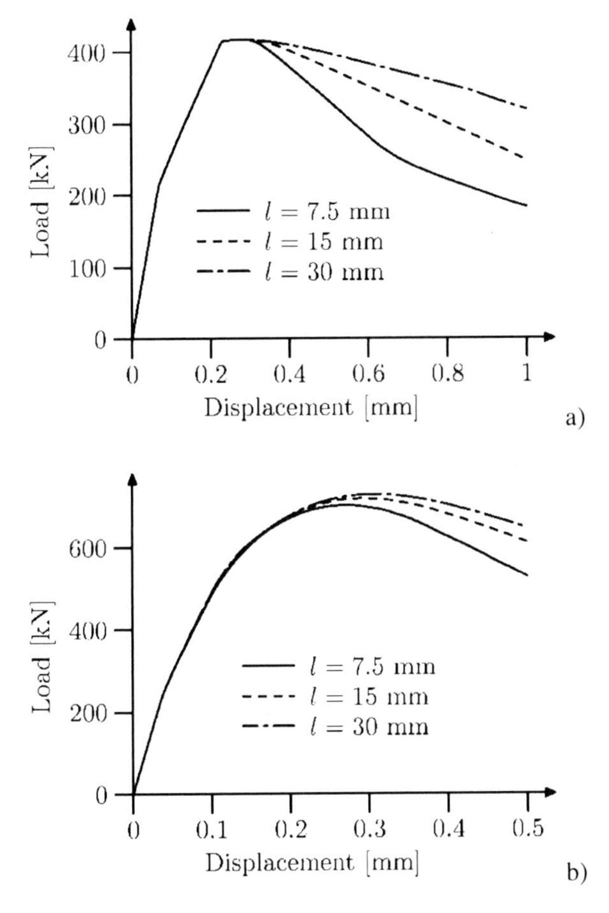

Fig. 5.18 Calculated load-displacement curves for a specimen under uniaxial compression for different characteristic length and roughness of horizontal boundaries (b=10 cm): (a) smooth boundaries with h=5 cm, b)-d) very rough boundaries with h=5 cm (b), h=10 cm (c) and h=20 cm (d) (elasto-plastic model with non-local softening) (Bobiński and Tejchman 2006)

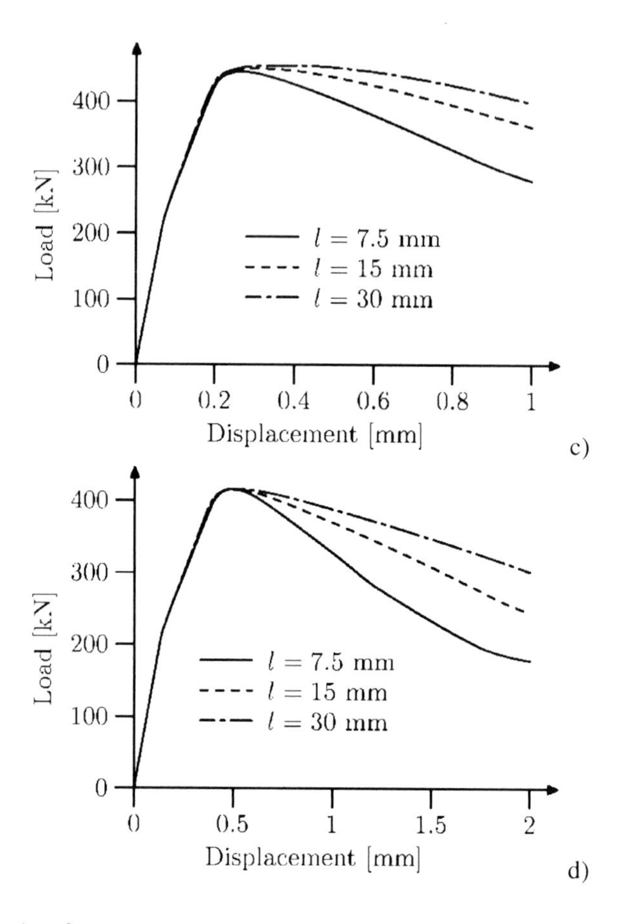

Fig. 5.18 (*continued*)

The modulus of elasticity was equal to E=18.0 GPa, the Poisson's ratio was υ=0.2 and the tensile strength was f_t=3.2 MPa. The elasto-plastic calculations were carried out with 3 different diagrams describing the tensile plastic stress σ_t versus the softening parameter κ_2 (Fig. 5.21). A linear and non-linear relationship σ_t=f(κ_2) was assumed in the softening tensile regime. In the case of linear softening, two different softening modules were used: H_t=3.2/(2.4×10^{-3})≈1300 MPa and H_t=3.2/(1.5×10^{-3})≈2100 MPa. In addition, a curvilinear exponential softening curve proposed by Hordijk (1991) was taken into account in the tensile regime (Eqs. 3.55 and 3.56) with κ_u=0.007 (κ_u – ultimate value of the softening parameter) and b_1=3.0 and b_2=6.93. The non-locality parameter was m=2 (Bobiński and Tejchman 2004). In turn, the following parameters were chosen for the damage model: κ_0=2.2×10^{-4}, α=0.96, β=600 and k=10 (von Mises failure criterion in terms of strains, Eqs. 3.38 and 3.40) and κ_0=1.7×10^{-4}, α=0.96 and β=900 (Rankine'a failure criterion, Eqs. 3.35 and 3.40). A characteristic length l_c was assumed to be 5 mm.

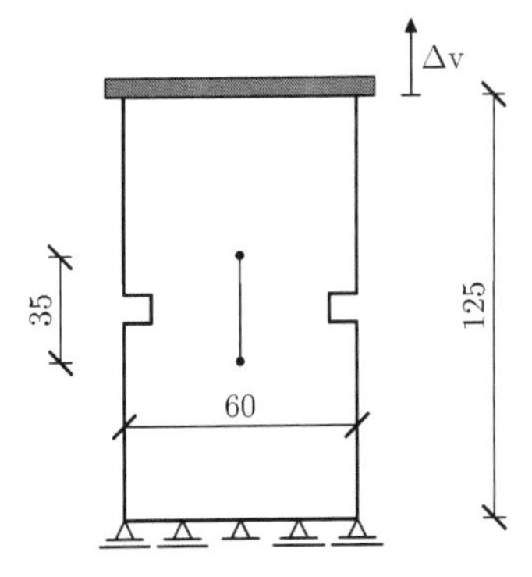

Fig. 5.19 Geometry and boundary conditions of a specimen with symmetric 2 notches under uniaxial tension (dimensions are given in mm) (Bobiński and Tejchman 2006)

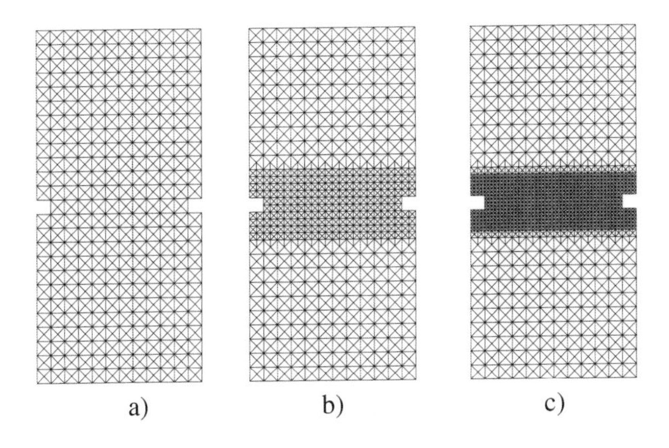

Fig. 5.20 FE-meshes used for elasto-plastic calculations of uniaxial tension: a) coarse, b) medium, c) fine (Bobiński and Tejchman 2006)

Figure 5.22 presents the nominal stress–elongation tensile curves for all meshes as compared to the experimental curve by Hordijk (1991). The elongation δ in Fig. 5.22 denotes the elongation of the specimen above and below both notches at the height of 35 mm, Fig. 5.19. It was measured experimentally by 4 pairs of extensometers with a gauge length of 35 mm. The vertical normal stress was calculated by dividing the calculated resultant vertical force by the specimen cross-section of 50×50 mm^2. The calculated load-displacement curves of Fig. 5.22 practically coincide for the different meshes and different models (in particular for damage models). They are also in a satisfactory agreement with the experimental curve (Hordijk 1991) (in particular, both damage curves and one elasto-plastic curve with linear softening using H_t=2100 MPa).

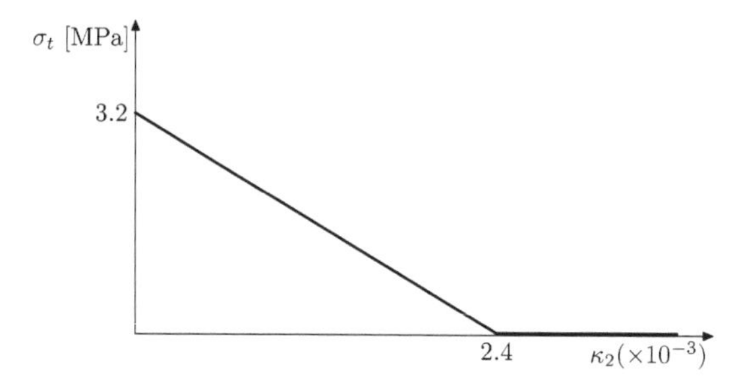

Fig. 5.21 Assumed curve $\sigma_t = f(\kappa_2)$ in tensile regime using the elasto-plastic model for uniaxial tension with linear softening modulus H_t=2300 MPa (σ_t – tensile plastic stress, κ_2 – softening parameter) (Bobiński and Tejchman 2006)

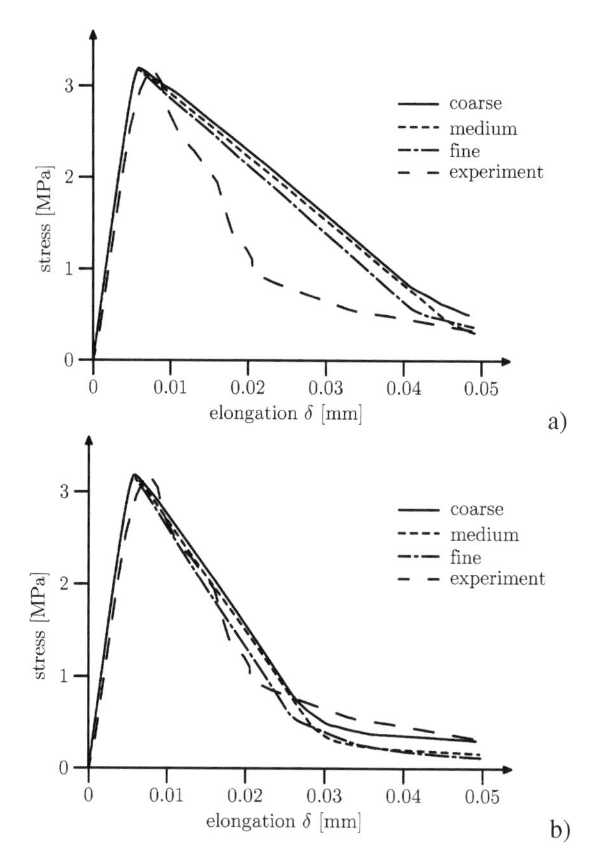

Fig. 5.22 Calculated stress–elongation diagrams for a specimen under uniaxial tension with different FE-meshes compared to the experimental diagram by Hordijk (1991): a) elasto-plastic model with non-local linear softening (softening modulus H_t=1300 MPa), b) elasto-plastic model with non-local linear softening (H_t=2300 MPa), c) elasto-plastic model with non-local exponential softening, d) damage model with non-local softening (Eq. 3.38), e) damage model with non-local softening (Eq. 3.35) (Bobiński and Tejchman 2006)

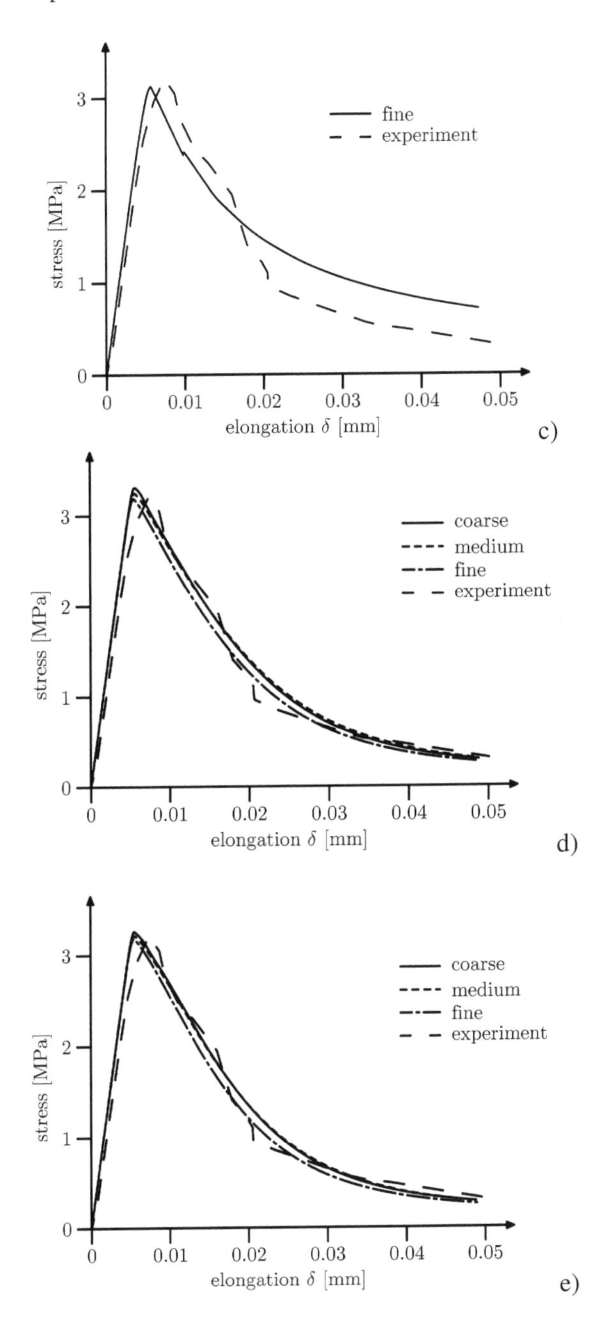

Fig. 5.22 *(continued)*

The calculated contours of the non-local parameter $\bar{\kappa}_2$ in the specimen are shown in Fig. 5.23 at residual state for δ=0.05 mm. The width of the localization zone between two notches w_{lz} (determined from the distribution of $\bar{\kappa}_2$) is similar for both damage models, w_{lz}=25 mm (5.0×l_c). In turn within elasto-plasticity, the width of the localization zone is approximately 25 mm for a coarse mesh, 20 mm for a medium mesh and 15 mm (3.0×l_c) for a fine mesh. The width of the localized zone is not influenced in elasto-plasticity by the rate of softening (Fig. 5.24).

In addition, the influence of the characteristic length l_c of micro-structure on the specimen behaviour was investigated. The FE-calculations were performed with l_c in the range from 2.5 mm up to 10.0 mm. The obtained load-displacement curves are presented in Fig. 5.25. The larger the characteristic length l_c (up to l_c=7.5 mm), the higher the maximum tensile stress (this effect is stronger within damage mechanics). The slope of all curves to the horizontal after the peak becomes smaller with increasing l_c (the material becomes more ductile with increasing l_c). The calculated contours of the non-local parameter $\bar{\kappa}$ in the specimen are shown in Fig. 5.26 at residual state for a fine mesh of Fig. 5.20c. In general, the width of the localized zone, w_{lz}, increased with increasing l_c; it was 10 mm (4.0×l_c for l_c=2.5 mm), and 15 mm (3.0×l_c for l_c=5 mm), 15 mm (2.0×l_c for l_c=7.5 mm) and 15 mm (1.5×l_c for l_c=10 mm) using the elasto-plastic model, and 15 mm (6.0×l_c for l_c=2.5 mm), 25 mm (5.0×l_c for l_c=5 mm), 35 mm (4.7×l_c for l_c=7.5 mm) and 35 mm (3.7×l_c for l_c=10 mm) using an isotropic damage model.

The results are qualitatively in a good accordance with the FE-results by Gutierrez and de Borst (2003) with the second-gradient elasto-plastic model and Peerlings et al. (1998) and Pamin (2004) using the second-gradient damage model in respect to the effect of a characteristic length on the width of strain localization and load-displacement curve.

The maximum normalized loads obtained from FE-simulations for notched concrete specimens within damage and elasto-plasticity during uniaxial tension were compared with a deterministic (energetic) size effect law given by Bažant for structures with pre-existing notches or large stress-free cracks growing in a stable manner prior to the maximum load (Bažant 2003) (Chapter 8)

$$\sigma_N = \frac{Bf_t}{\sqrt{1+D/D_0}}, \tag{5.5}$$

where σ_N – the nominal strength, B – the dimensionless geometry-dependent parameter (depending on the geometry of the structure and of the crack), D – the specimen size (equal to the specimen height h) and D_0 – the size-dependent

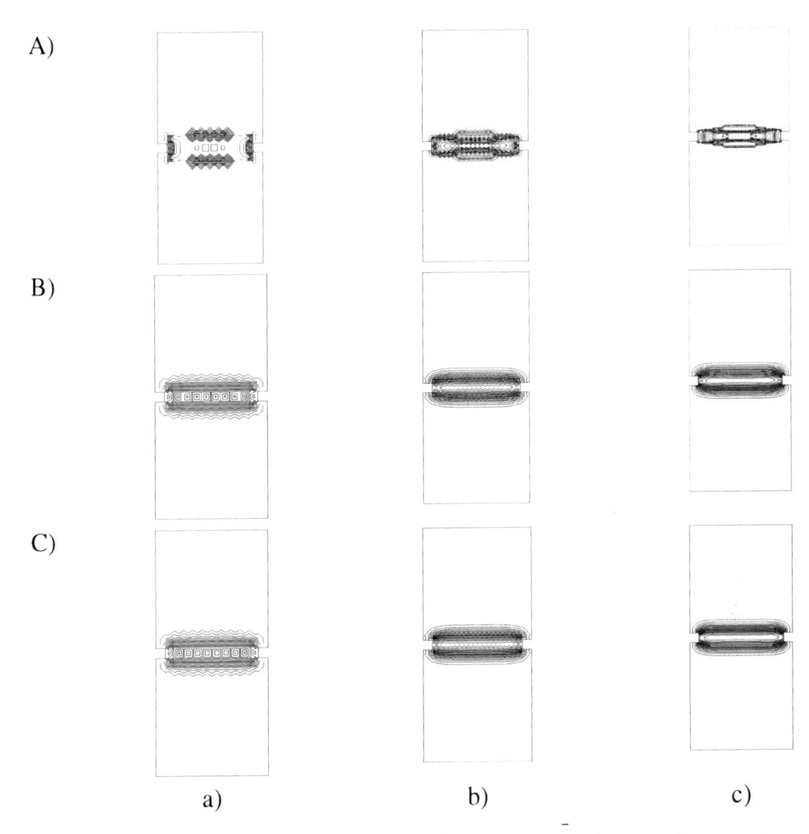

Fig. 5.23 Calculated contours of the non-local parameter $\bar{\kappa}_2$ in a specimen under uniaxial tension for a) coarse, b) medium and c) fine mesh: A) elasto-plastic model with non-local softening (softening modulus H_r=1300 MPa), B) damage model with non-local softening (Eq. 3.38), C) damage model with non-local softening (Eq. 3.35) (Bobiński and Tejchman 2006)

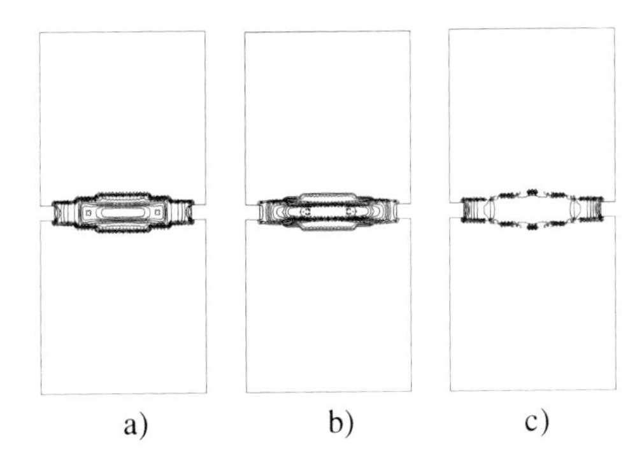

Fig. 5.24 Calculated contours of the non-local parameter $\bar{\kappa}_2$ in a specimen under uniaxial tension for fine mesh within elasto-plasticity with non-local softening: a) linear softening (H_r=1300 MPa), b) linear softening (H_r=2300 MPa), c) curvilinear softening by Eq. 3.55 (Bobiński and Tejchman 2006)

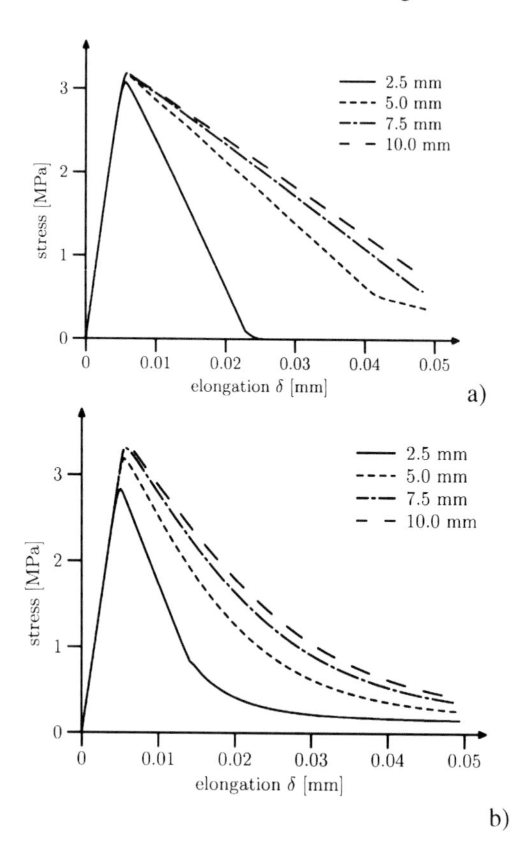

Fig. 5.25 Calculated vertical stress–elongation diagrams using different characteristic lengths l_c for a specimen under uniaxial tension using a fine mesh: a) elasto-plastic model with non-local linear softening (H_t=1300 MPa), b) damage model with non-local softening (Eq. 3.38) (Bobiński and Tejchman 2006)

parameter called transitional size. To find the parameters B and D_o from FE-analyses, a linear regression was used. Figure 5.27 presents a comparison between FE-result and the size effect law by Bažant (2003). A good agreement was obtained. The normalized strength decreases almost linearly with increasing size ratio h/l_c in the considered range.

In addition, similar elasto-plastic calculations (using the Rankine's failure criterion, Eq. 3.32) were carried out with a concrete specimen under uniaxial tension possessing one non-symmetric notch (Fig. 5.28). The tensile strength was equal to f_t=3 MPa, the softening modulus in tension H_t=1.0 GPa, the non-local parameter m=2 and the characteristic length l_c=15 mm. Figure 5.29 presents a normalized load-displacement diagrams for 3 different geometrically similar

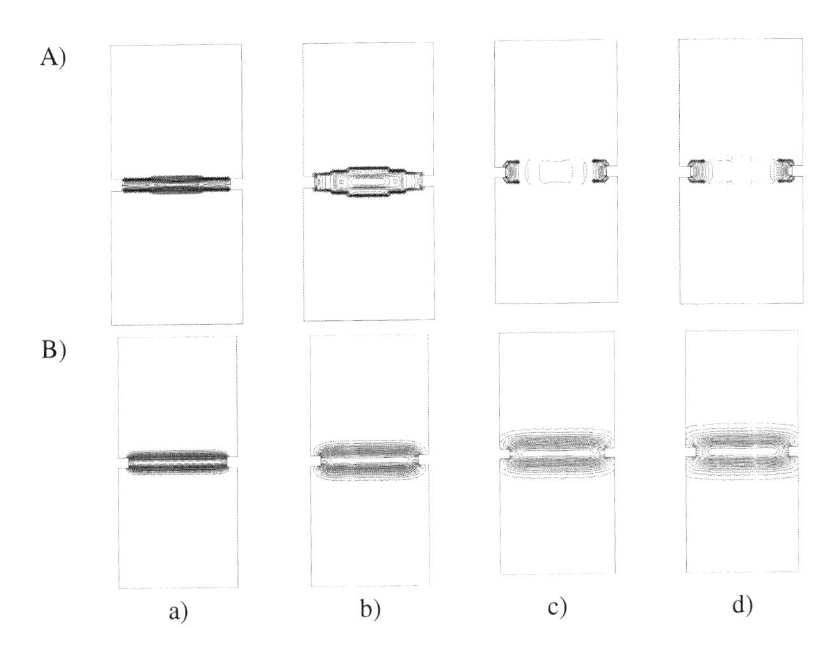

Fig. 5.26 Calculated contours of the parameter $\bar{\kappa}_2$ in a specimen under uniaxial tension for fine mesh: a) l_c=2.5 mm, b) l_c=5 mm, c) l_c=7.5 mm, d) l_c=10 mm: A) elasto-plastic model with non-local softening, B) damage model with non-local softening (Eq. 3.38) (Bobiński and Tejchman 2006)

concrete specimens. The strength and ductility increase with decreasing size. The contours of a nonlocal hardening tensile parameter near the notch are shown in Fig. 5.30. The calculated width of the localized zone increased with increasing specimen size and was equal to 5.0 cm ($3\times l$), 6.0 cm ($4\times l$) and 7.0 cm ($4.6\times l$) for a small, medium and large specimen, respectively.

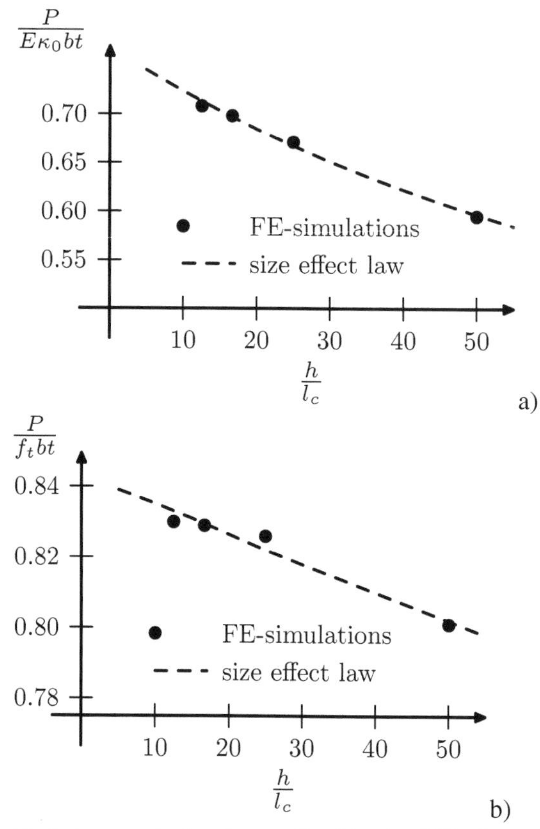

Fig. 5.27 Relationship between calculated normalized loads: $P/(E\kappa_o bt)$ and $P/(f_t bt)$ during uniaxial tension (with l_c=5 mm) and ratio h/l_c as compared to size effect law by Bažant (2003) within: a) damage mechanics, b) elasto-plasticity (Bobiński and Tejchman 2006)

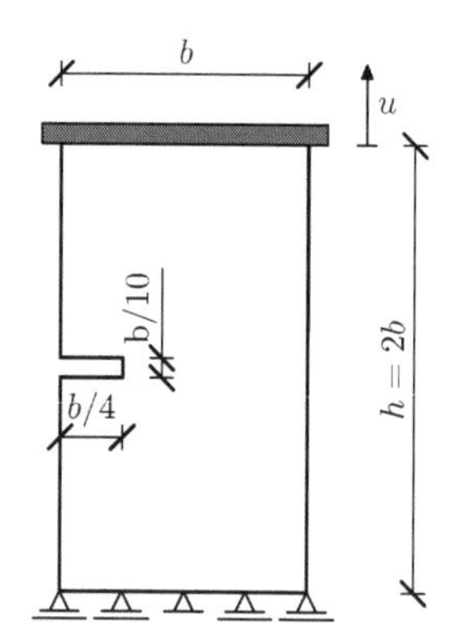

Fig. 5.28 Geometry and boundary conditions of specimen with one notch under uniaxial tension for elasto-plastic calcualtions (Bobiński and Tejchman 2006)

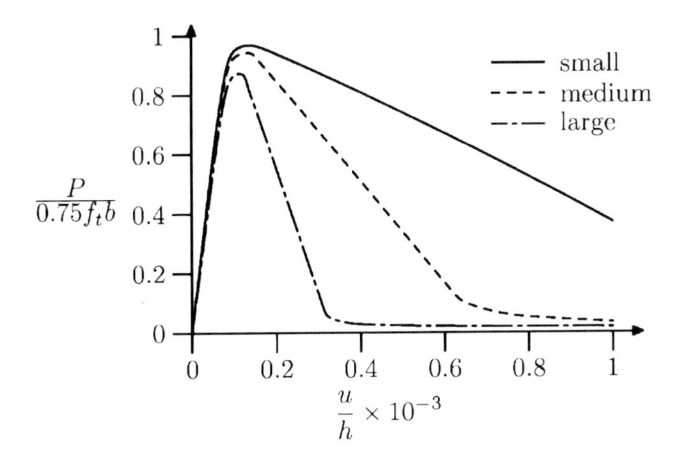

Fig. 5.29 Normalized load-displacement diagrams for a specimen with one notch under uniaxial tension (specimen $b\times h$: small: 5×10 cm^2, medium: 10×20 cm^2, large: 20×40 cm^2) (elasto-plastic model with non-local softening) (Bobiński and Tejchman 2006)

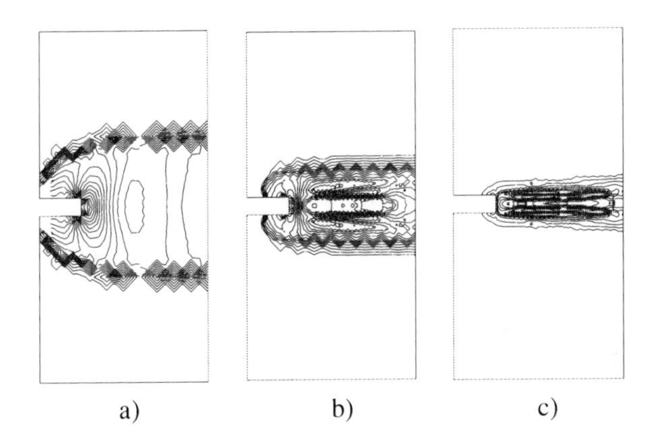

Fig. 5.30 Calculated contours of the nonlocal parameter $\bar{\kappa}_2$ near the notch during uniaxial tension for small (a), medium (b) and large specimen (c) (increased by a factor 4 (small), 2 (medium) and 1 (large)) (elasto-plastic model with non-local softening) (Bobiński and Tejchman 2005)

FE results with XFEM

In the calculations using XFEM (Chapter 4.2), the size of the specimen subjected to uniaxial tension was 100 mm (width) and 150 mm (height), Fig. 5.31. All nodes along the bottom were fixed in a vertical direction. The vertical tensile deformation was imposed by enforcing the vertical displacement of all nodes along the upper edge by the amount of u=0.1 mm. To preserve the stability of the specimen, the node in the middle of bottom was fixed in a horizontal direction. The starting point of the crack propagation was defined in the middle of the left edge. The modulus of elasticity was equal to E=30 GPa and the Poisson's ratio was ν=0.2. The tensile strength was taken as f_t=3 MPa and fracture energy as G_f=100 N/m. The exponential softening in a normal direction was assumed. A fixed horizontal crack propagation was assumed in advance. To investigate the mesh insensitivity, three different FE meshes were defined: coarse, medium and fine with 600, 2400 and 5400 3-node triangular elements, respectively. Plane stress state was assumed. The calculated force – displacement diagrams are shown in Fig. 5.32. The curves are almost identical, only a small discrepancy for a fine mesh can be seen. The deformed meshes with a discrete horizontal crack are depicted in Fig. 5.33.

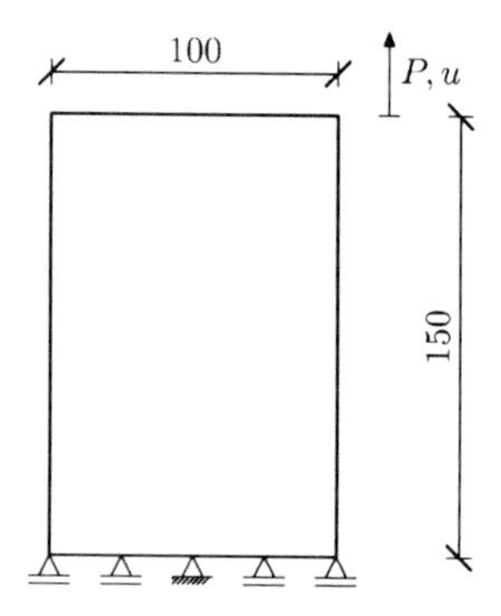

Fig. 5.31 Geometry and boundary conditions of concrete specimen during uniaxial tension test in calculations with XFEM of Chapter 4.2 (Bobiński and Tejchman 2011)

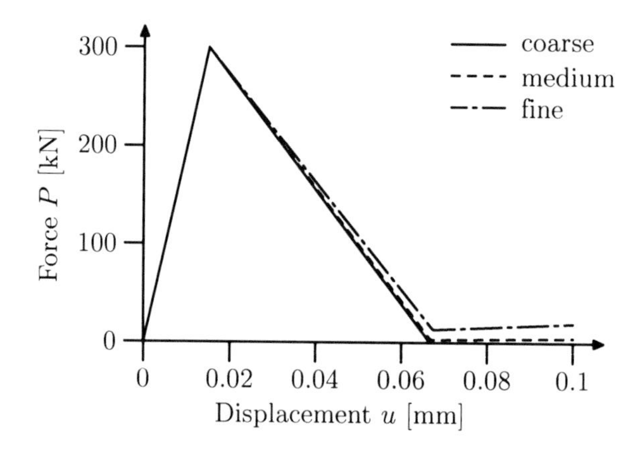

Fig. 5.32 Calculated force–displacement curves during uniaxial tension (using XFEM of Chapter 4.2) (Bobiński and Tejchman 2011)

Fig. 5.33 Calculated horizontal crack for different meshes during uniaxial tension using XFEM of Chapter 4.2 (Bobiński and Tejchman 2011)

The following conclusions can be derived:

- The FE-calculations on strain localization demonstrate that conventional elasto-plastic models suffer from a mesh-dependency when material softening is included. The thickness and inclination of localized zones inside a specimen, and load-displacement diagram in a post-peak regime depend strongly upon the mesh discretisation.
- An elasto-plastic model with modified non-local softening and a damage model with usual non-local softening cause a full regularisation of the boundary value problem during uniaxial compression and extension. A finite and the same size of the strain localization zone is obtained upon mesh refinement. The load-displacement curves are similar. The effect of the mesh alignment on the inclination of localized zone is negligible.
- The thickness of localized zones increases with increasing characteristic length and non-local parameter. The thickness of a shear zone during uniaxial compression is larger than during uniaxial extension.
- The size and number of imperfections, and the distance between them do not influence the thickness and inclination of the localized zone.
- The vertical force on the top during uniaxial compression and tension increases in the softening regime with increasing characteristic length and non-local parameter. The effect of a characteristic length on the maximum vertical force is noticeable only in the case of very rough boundaries (using a weak element). The influence of a characteristic length on the maximum vertical force is stronger with non-symmetric notches than with symmetric ones in the case of smooth boundaries.
- The larger the ratio between the characteristic length of micro-structure and the specimen size, the higher usually both the specimen strength and the ductility of the specimen during extension and bending.

• The width of the localized zone is larger in FE-analyses with a damage model than with an elasto-plastic model using a similar l_c. For uniaxial tension, the width of the localized strain zone is about (1.5-4.0)$\times l_c$ within elasto-plasticity, whereas it is (3.7-6.0)$\times l_c$ within damage mechanics. The width of the localized zone does not depend on the rate of softening in elasto-plasticity.
• The width of the localized zone grows during the entire deformation process within damage mechanics, whereas it is almost constant within elasto-plasticity.
• The size effect decreases almost linearly with decreasing ratio between the specimen size and characteristic length. The calculated deterministic size effect in concrete elements during tension is in agreement with the corresponding size effect law by Bažant (2003).
• The results with XFEM are mesh objective.

5.2 Bending

FE results within elasto-plasticity and damage mechanics

The behaviour of a symmetric concrete beam with a notch at the bottom at mid-span and free ends during three-point bending was simulated. This behaviour was experimentally investigated by Le Bellego at al. (2003), and later numerically simulated by Le Bellego et al. (2003) and Rodriguez-Ferran et al (2002) with a non-local damage approach. Three different beams were used in laboratory tests: small (h=8 cm), medium (h=16 cm) and large one (h=32 cm). The beam span was L=3h. The geometry and boundary conditions of the beam are presented in Fig. 5.34. The loading was prescribed at the top edge in the mid-span via the vertical displacement.

Three different FE-meshes were assumed: with 1534, 2478 and 4566 triangular elements for a small, medium and large specimen, respectively (Fig. 5.35). Due to the symmetry of the problem, only the left half of the beam was modelled.

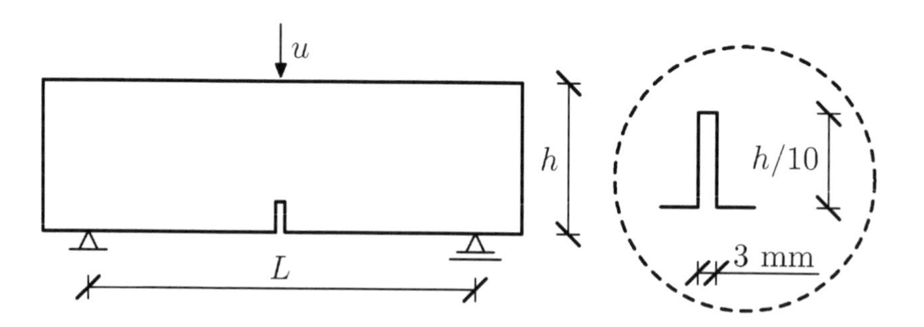

Fig. 5.34 Geometry of notched beam and boundary conditions in laboratory tests by Le Bellego et al. (2003)

In the elasto-plastic simulations, the modulus of elasticity was taken as E=38.5 GPa and the Poisson ratio as ν=0.2. In the tensile regime of the elasto-plastic model, the Rankine criterion with the exponential curve by Hordijk (1991) was used (Eqs. 3.32, 3.55, 3.93 and 3.97). Two different internal lengths were chosen in the FE-analysis: l_c=5 mm and l_c=10 mm. Due to two different l_c, two sets of the material parameters were also chosen in Eq. 3.55 (with m=2): f_t=3.6 MPa, κ_u=0.005 for l_c=5 mm, and f_t=3.3 MPa, κ_u=0.003 for l_c=10 mm to obtain the best agreement between the load-displacement diagrams from FE-analyses and laboratory tests (Le Bellego et al. 2003). The internal friction angle was equal to φ=10° (Eq. 3.27) and the dilatancy angle ψ=5° (Eq. 3.30). The compressive strength was equal to f_c=40 MPa. A linear softening modulus in compression was H_c=0.8 MPa. The effect of material parameters in compression was found to be insignificant on the FE-results. In the case of a damage model with the Rankine criterion (Eqs. 3.35, 3.40, 3.93 and 3.99), two sets of material parameters were again chosen for two different l_c: κ_0=7·10^{-5}, α=0.99, β=600 for l_c=5 mm, and κ_0=6.25·10^{-5}, α=0.99, β=1000 for l_c=10 mm.

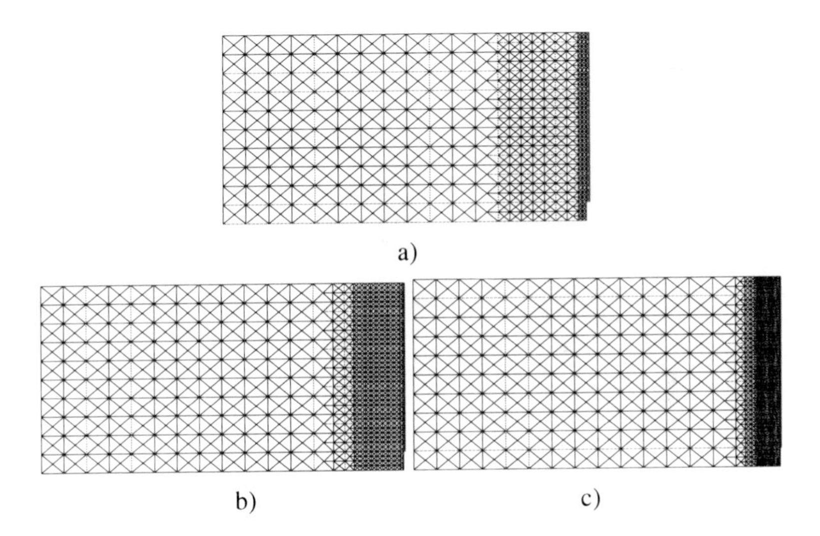

Fig. 5.35 FE meshes used for calculations of bending: a) coarse, b) medium, c) fine (elasto-plastic model with non-local softening) (Bobiński and Tejchman 2006)

Figure 5.36 presents load-displacement curves for all beams obtained from FE-calculations using the characteristic length l_c=5 mm and l_c=10 mm, respectively (compared to experiments). A satisfactory agreement with experiments was achieved. The FE-results overestimate slightly the load bearing capacity of the small and medium beam and underestimate the maximum load for the large beam. The ratio between the specimen strength of the large and medium beam and of the medium and small beam is similar. The same numerical results were obtained by

Le Bellego et al. (2003) and Rodriguez-Ferran et al. (2002), although they used different definitions of the equivalent strain $\tilde{\varepsilon}$ and evolution laws. Other calculations demonstrate that the larger the characteristic length, the higher the beam strength (in particular in damage mechanics) (Bobiński and Tejchman 2005).

Figures 5.37 and 5.38 show the distribution of a non-local parameter $\bar{\kappa}_2$ above the notch. The width of the localization zone ***w*** was at the residual state: within elasto-plasticity about 25 mm ($5\times l_c$) for l_c=5 mm and all beams, and 50 mm ($5\times l_c$) (small beam), 50 mm ($5\times l_c$) (medium beam) and 45 mm ($4.5\times l_c$) (large beam) for l_c=10 mm, respectively. In the case of the damage model, it was equal to 40 mm ($8\times l_c$) for l_c=5 mm, and 65 mm ($6.5\times l_c$) (small beam), 75 mm ($7.5\times l_c$) (medium beam) and 90 mm ($9\times l_c$) (large beam) for l_c=10 mm, respectively. It did not depend on the mesh size.

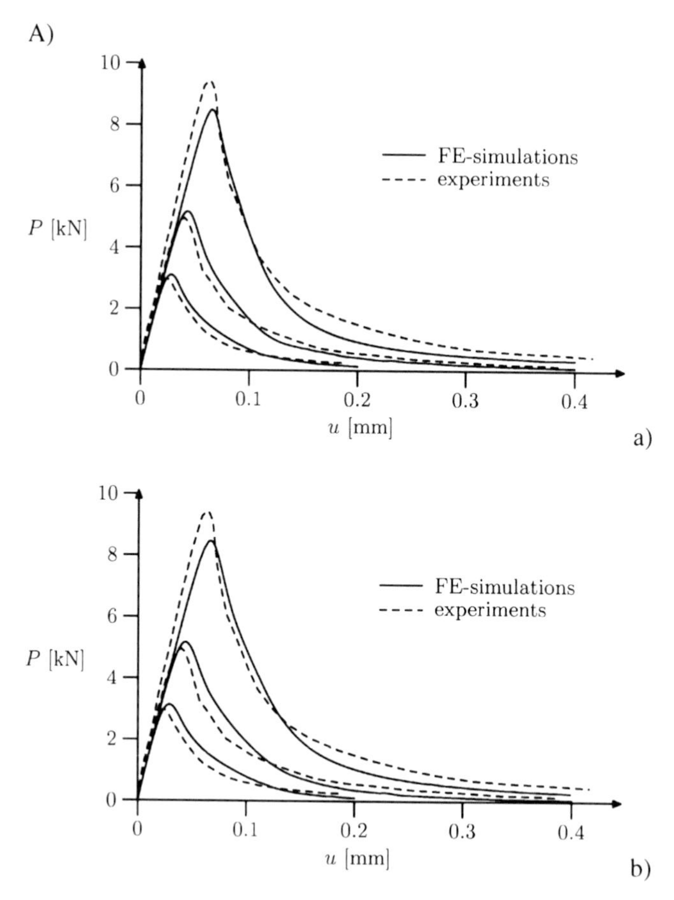

Fig. 5.36 Load-displacement curves from experiments (Le Bellego et al. 2003) and FE-simulations: a) l_c=5 mm, b) l_c=10 mm, A) elasto-plastic model with non-local softening, B) damage model with non-local softening (Bobiński and Tejchman 2006)

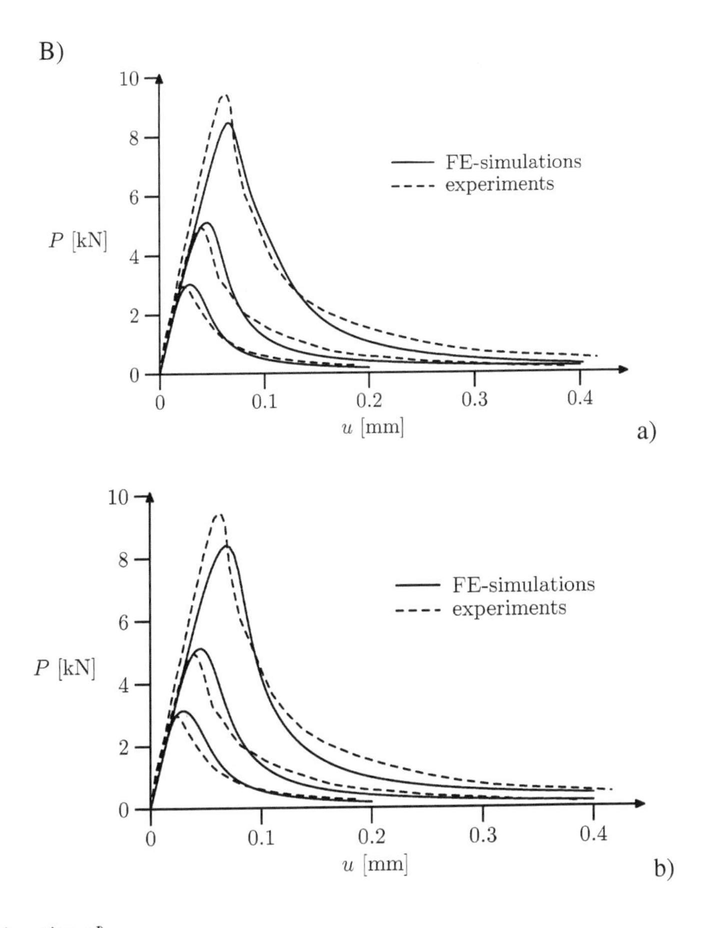

Fig. 5.36 *(continued)*

Figure 5.39 presents the evolution of the localized zone during deformation. The width of the localized zone increases at the beginning of deformation after the peak, and next it remains almost the same with advanced deformation within elasto-plasticity, however, its continuous increase takes place within damage mechanics. This outcome is in accordance with FE-calculations by Pamin (2004, 2005) using a second–gradient elasto-plastic and second-gradient damage model.

Finally, Fig. 5.40 shows the influence of the parameters α, β and κ_o on the force-displacement curve using an isotropic damage model (Eqs. 3.35, 3.40, 3.93 and 3.99). The parameter κ_o influences strongly the peak on the load-displacement. In turn, the parameter β affects the slope of the curve in the softening regime and the parameter α has a large effect on the residual value.

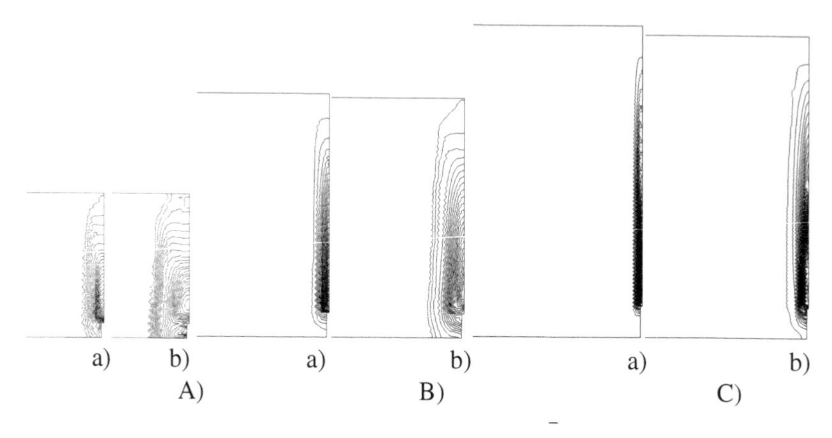

Fig. 5.37 Calculated contours of the non-local parameter $\bar{\kappa}_2$ along the beam height at the left side of notch: a) l_c=5 mm, b) l_c=10 mm, A) h=80 mm, B) h=160 mm, C) h=320 mm (elasto-plastic model with non-local softening) (Bobiński and Tejchman 2006)

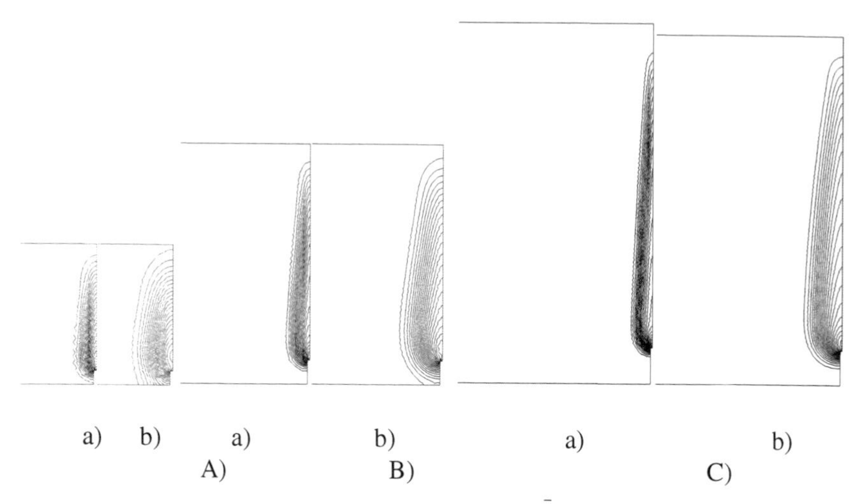

Fig. 5.38 Calculated contours of the non-local parameter $\bar{\kappa}_2$ along the beam height at the left side of notch: a) l_c=5 mm, b) l_c=10 mm, A) h=80 mm, B) h=160 mm, C) h=320 mm (damage model with non-local softening, Eqs. 3.35 and 3.40) (Bobiński and Tejchman 2006)

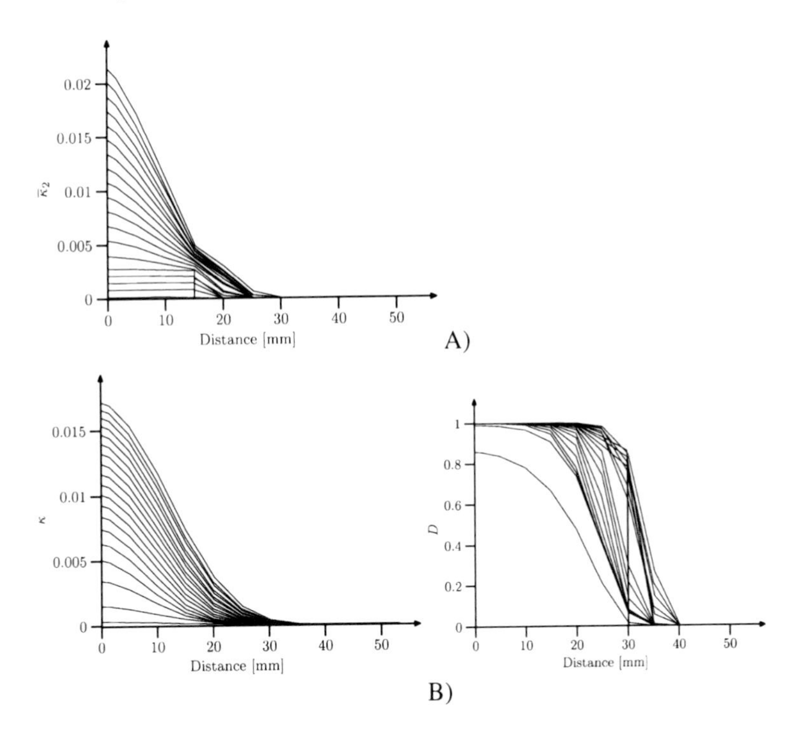

Fig. 5.39 Evolution of non-local parameter $\bar{\kappa}_2$ (a) and damage parameter D above the notch (right side of the beam) (b) (l_c=10 mm, h=160 mm): A) elasto-plastic model with non-local softening, B) damage model with non-local softening (Eqs. 3.35 and 3.40) (Bobiński and Tejchman 2006)

Figure 5.41 presents a comparison between FE-result and the size effect law by Bažant (2003) - a good agreement was obtained. The normalized strength decreases almost linearly with increasing size ratio h/l_c in the considered range.

The problem of a notched beam under three-point bending was experimentally studied by Kormeling and Reinhardt (1983) and numerically simulated by Jirásek (2004) using a non-local damage continuum model. The geometry of the beam with a depth of t=100 mm is shown in Fig. 5.42 (span length L=450 mm, height h=100 mm). The deformation was obtained by imposing the vertical displacement to the top at the mid-span of the beam. The elastic properties were: E=20 GPa and υ=0.2. The following damage parameters were chosen: $\kappa_0=1.2\times10^{-4}$, α=0.96, β=200, and k=10 (Eqs. 3.38, 3.40, 3.93 and 3.99). A characteristic length was l_c=5 mm. The calculations were carried out also with β=500. A coarse (2120 elements), medium (2300 elements) and fine (4380 elements) mesh was used.

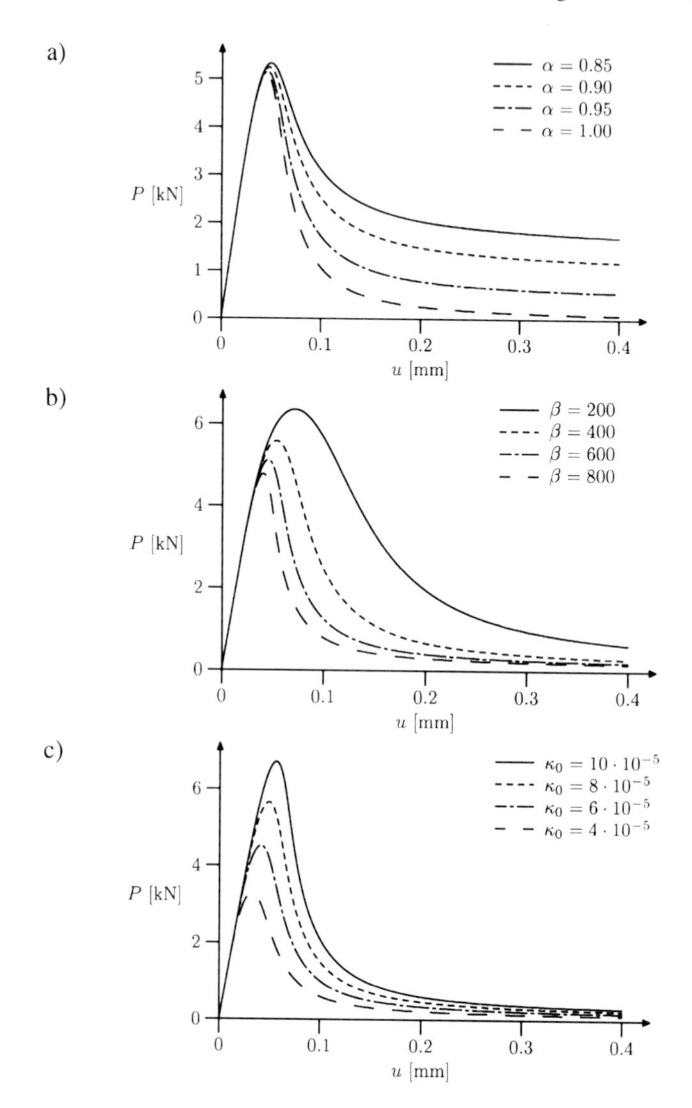

Fig. 5.40 Influence of the parameters α, β and κ_o on load-displacement curve using damage model (l_c=5 mm, h=160 mm) (Eqs. 3.35and 3.40) (Bobiński and Tejchman 2006)

The calculated contours of the damage parameter κ at residual state are shown in Fig. 5.43. The results are mesh-independent. The width of the localisation zone is approximately equal to 23 mm (4.5×l_c). In turn, Fig. 5.44 presents the calculated load–displacment curves for different meshes which are almost indentical and match well with the experimental curves (Kormeling and Reinhardt 1983) (they lie between them) for the parameter β=200. The effect of β on the load-displacement diagram is strong. The larger β, the smaller the beam strength. The results with different characteristic lengths l_c (l_c=2-10 mm), indicate that the beam strength increases with increasing l_c (Fig. 5.45). Thus, a pronounced size effect caused by the ratio l_c/L (L - specimen size) occurs. With increasing characteristic length, the behaviour of the material after the peak becomes more ductile. The results are in a good agreement with the FE-results by Jirásek (2004).

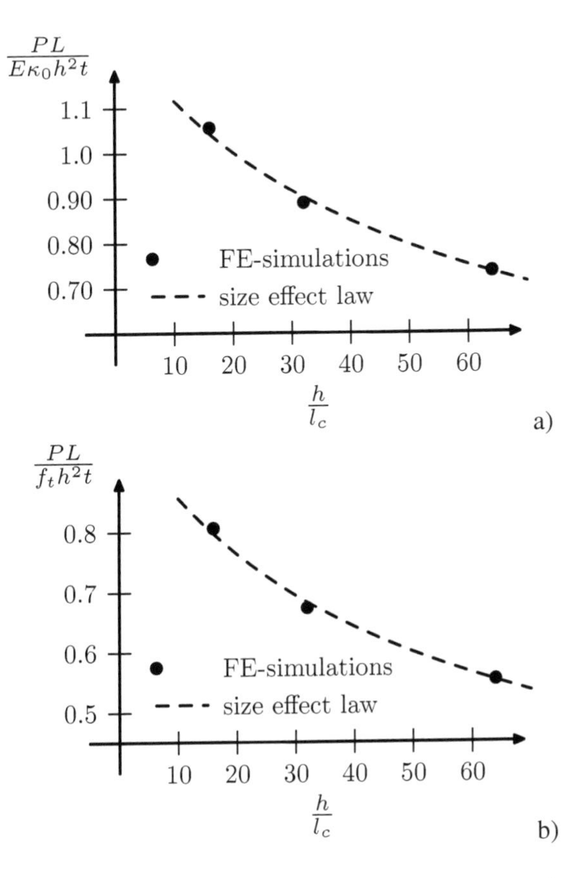

Fig. 5.41 Relationship between the calculated normalized loads: $(PL)/(E\kappa_o h^2 t)$ and $(PL)/(f_t h^2 t)$ during bending (with l_c=5mm) and the ratio h/l_c as compared to the size effect law by Bažant (2003) within: a) damage mechanics, b) elasto-plasticity (Bobiński and Tejchman 2006)

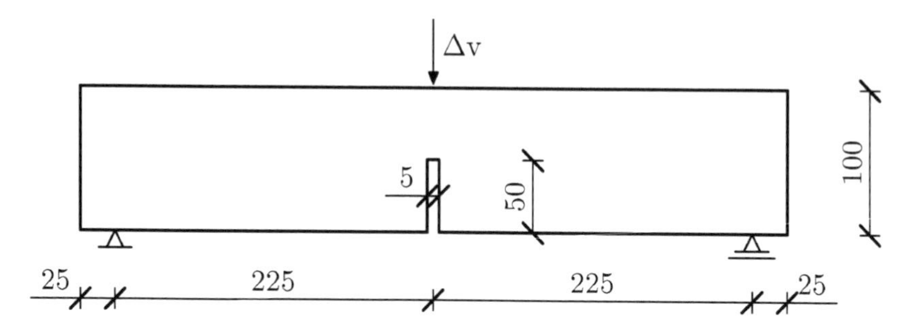

Fig. 5.42 Geometry and boundary conditions of the beam under three-point bending (dimensions are given in mm) (Bobiński and Tejchman 2005)

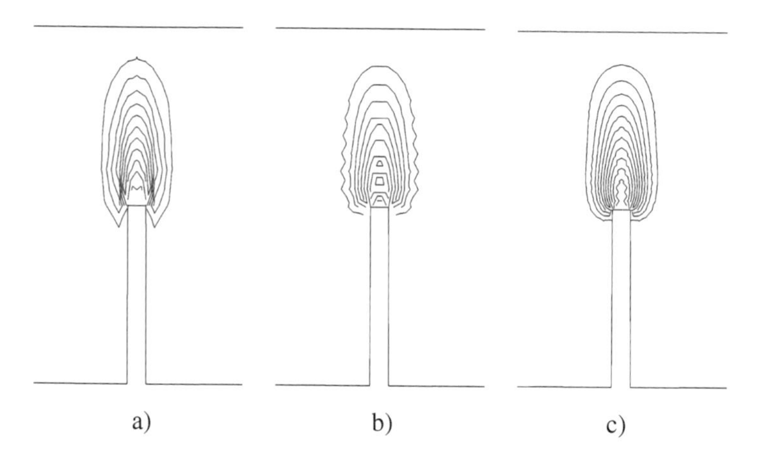

Fig. 5.43 Calculated contours of damage parameter near the notch of the beam under three-point bending for: a) coarse, b) medium and c) fine mesh (Bobiński and Tejchman 2005)

In turn, the problem of a notched beam under four-point bending was experimentally investigated by Hordijk (1991) and numerically simulated by both Pamin (2004) with second-gradient plasticity and Simone et al. (2002) with a second-gradient damage model. The geometry of the specimen is given in Fig. 5.46 (span length L=450 mm, height h=100 mm). The beam had the 5×10 mm^2 notch at the mid-span. The thickness of the beam in out-of-plane direction was t=50 mm. The deformation was induced by imposing a vertical displacement in two nodes at the top in the central part of the beam. The modulus of elasticity was E=40 GPa and the Poisson's ratio υ=0.2. The remaining material parameters were: κ_0=0.75×10^{-4}, α=0.92, β=200, k=10 and l_c=5 mm (Eqs. 3.32, 3.55, 3.93 and 3.99). The FE-analyses were carried out with a coarse (2152 elements), medium (2332 elements) and fine (4508 elements) mesh.

The calculated contours of the damage parameter are shown in Fig. 5.47. The obtained results do not depend on the mesh size. The width of the localisation zone is approximately equal to 26 mm (5.2×l_c). Figure 5.48 depicts the load–displacment curves for various FE-meshes which are almost indentical. They are also close to the experimental curve by Eq. 3.55 (Hordijk 1991). In addition, the influence of the characteristic length in the range of l_c=2.5-10.0 mm was analyzed (Fig. 5.49). The results of Fig. 5.49 show that the effect of the characteristic length on the load-displacement diagrams is pronounced. The results are close to those by Pamin (2004) and Simone et al. (2002).

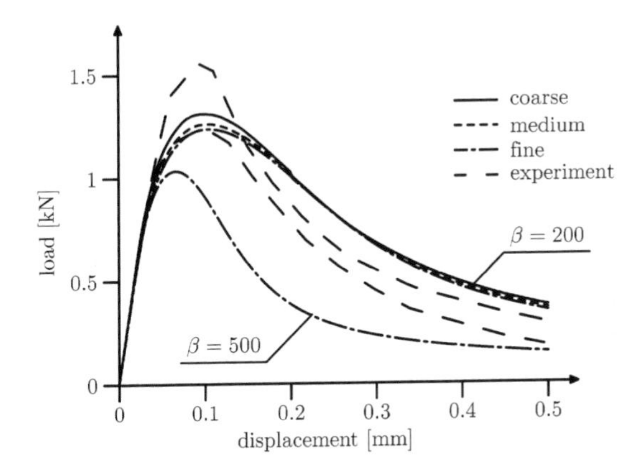

Fig. 5.44 Calculated load–displacement diagrams within damage mechanics for beam under three-point bending with different FE-meshes compared to the experimental curves by Kormeling and Reinhardt (1983) (Bobiński and Tejchman 2005)

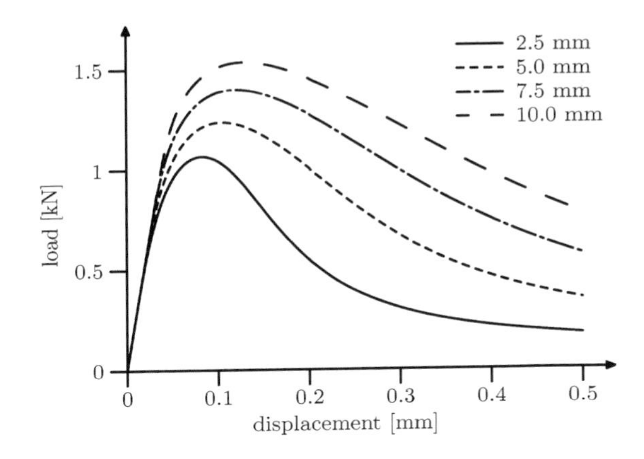

Fig. 5.45 Calculated load–displacement diagrams for different characteristic lengths l_c within damage mechanics (beam under three-point bending) (Bobiński and Tejchman 2005)

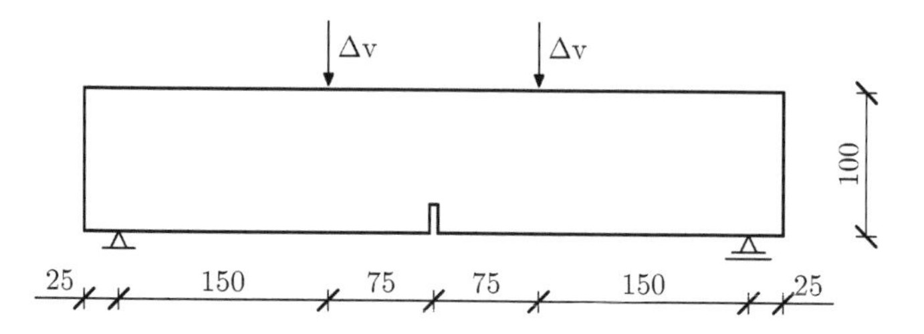

Fig. 5.46 Geometry and boundary conditions of beam under four-point bending (dimensions are given in mm) (Bobiński and Tejchman 2005)

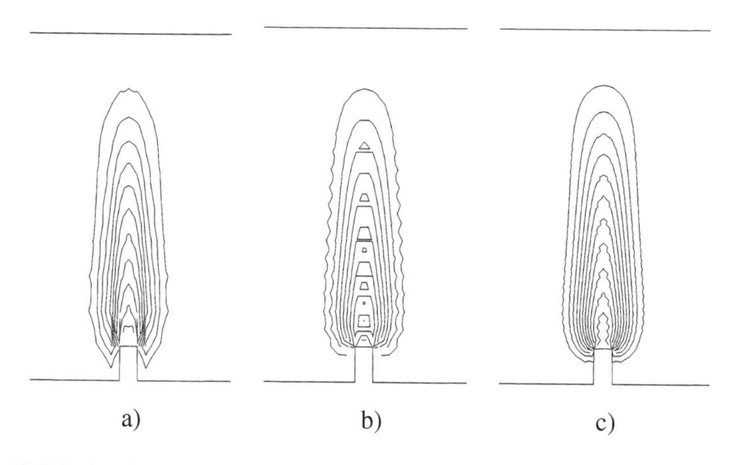

Fig. 5.47 Calculated contours of damage parameter near the notch for a beam under four-point bending for different FE-meshes: a) coarse, b) medium and c) fine mesh (Bobiński and Tejchman 2005)

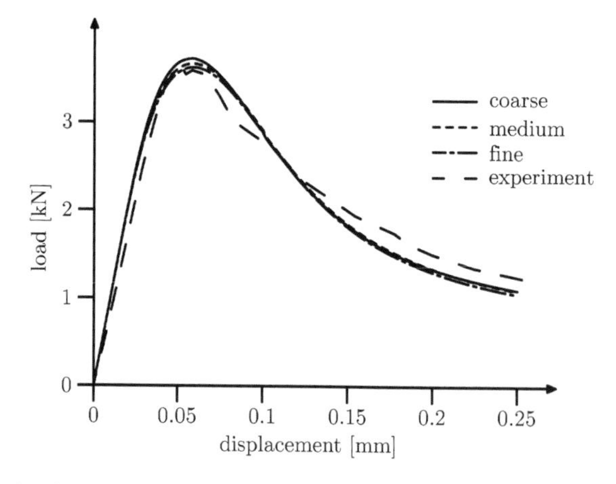

Fig. 5.48 Calculated load displacement diagrams for beam under four-point bending within damage mechanics for different meshes compared to experiment (Bobiński and Tejchman 2005)

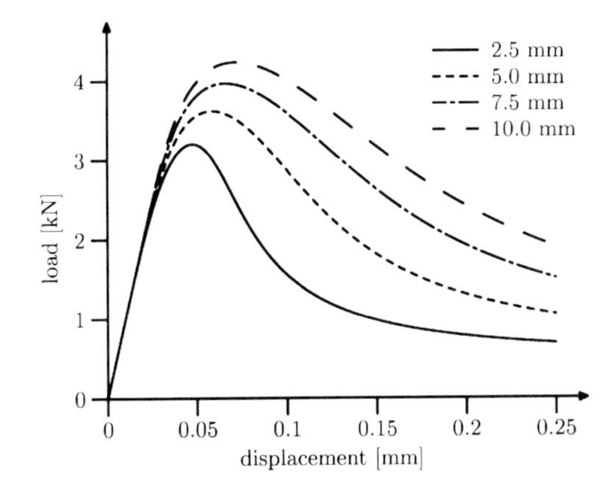

Fig. 5.49 Calculated load–displacement diagrams for different characteristic lengths l_c within damage mechanics (beam under four-point bending) (Bobiński and Tejchman 2005)

Figure 5.50 shows the normalized strength $Pl/E\kappa_0 h^2 t$ against the ratio h/l_c from FE-studies. The force decreases almost linear with h/l_c. The relationship can be approximated by Bažant's size effect formula (Eq. 5.5), wherein the parameters were calculated using a linear regression.

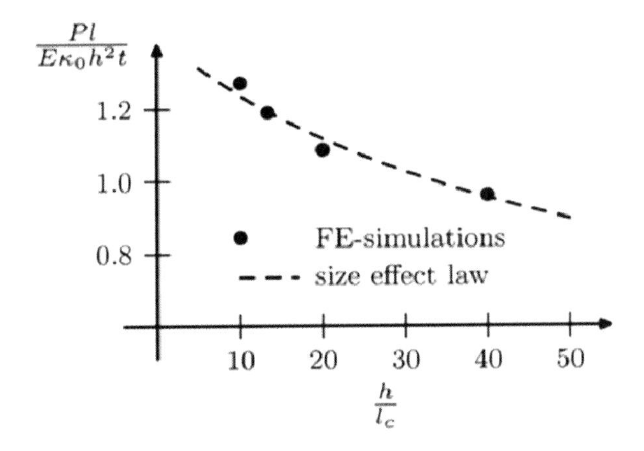

Fig. 5.50 Relationship between strength $Pl/E\kappa_0 h^2 t$ and ratio l_c/h during four-point bending within damage mechanics as compared to size effect law by Bažant (2003) (Bobiński and Tejchman 2005)

FE results with cohesive elements

The three-point bending tests with notched beams by Le Bellego et al. (2003) were investiagated with cohesive elements of Chapter 4.1 (Bobiński and Tejchman 2008). The elastic material patameters were: E=38.5 GPa and ν=0.24 and the cohesive data were: f_t=3.5 MPa, η=0.0 and β=45000. The FE meshes consisted of 3068, 4956 and 9132 3-node triangles. Interface elements were placed only along the expected crack trajectory, i.e. along the vertical symmetry line.

Figures 5.51 and 5.52 show the numerical results with cohesive elements for three different beams from laboratory experiments by Le Bellego et al. (2003). The outcomes indicate that the cohesive elements are able to realistically describe a deterministic size effect and a vertical crack above the beam notch.

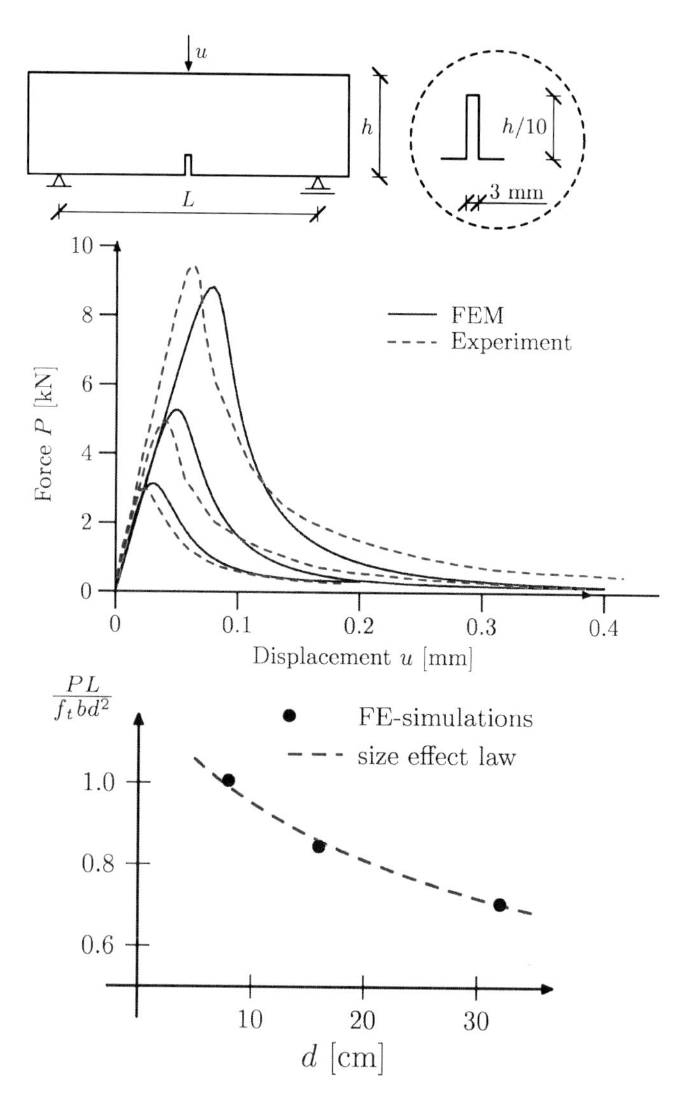

Fig. 5.51 Calculated load-displacement curves from experiments by Le Bellego et al. (2003) with 3 different notched beams and calculated size effect in FE-simulations with cohesive elements of Chapter 4.1 (Bobiński and Tejchman 2008)

Fig. 5.52 Defomed medium-size notched beam with vertical crack in FE-simulations with cohesive elements of Chapter 4.1 (Bobiński and Tejchman 2008)

FE results with XFEM

The geometry of the specimen during three-point bending was taken from Häussler-Combe (2002). The beam was deformed by prescribing a vertical displacement at the upper edge of the beam at its mid-span up to the final value u=0.4 mm (Fig. 5.53). The starting point of the crack was defined in the middle of the bottom edge. The Young modulus was equal to E=30 GPa and the Poisson's ratio was ν=0.2. The tensile strength was taken as f_t=3 MPa. The fracture energy G_f=120 N/m with exponential softening was defined. The fixed vertical crack was assumed (Chapter 4.2). The simulations were performed with three FE meshes with 410, 1620 and 3630 3-node triangles (Bobiński and Tejchman 2011). A plane stress state was assumed. The odd number of finite elements between supports was defined to locate the crack starting point in the middle of the elements' edge. Figure 5.54 shows the calculated force – displacement curves which are similar. For a coarse mesh, several drops in the force evolution were obtained which are caused by an edge-like crack propagation (crack tip can be placed at finite element edge only). With decreasing element size this phenomenon disappears. The obtained deformed coarse mesh is shown in Fig. 5.55 (displacements were scaled 20 times). It should be noted here that the use of the crack direction propagation criterion based on the maximum principal stress resulted in a sudden unrealistic change of the crack direction at the certain stage of deformation. The reason was an isotropic stress stage (biaxial tension) attained at the front of the crack tip. Next, the vertical tensile stresses become larger than horizontal ones and the crack turned by 90 degrees in the left (or right) direction.

A beam with a notch under three-point bending was also analyzed according to experiments by Le Bellego et al. (2003). The Young modulus and the Poisson's ratio were taken as E=38.5 GPa and ν=0.24, respectively. The tensile strength was equal to f_t=3.2 MPa. The exponential softening with fracture energy G_f=80 N/m was defined. A fixed vertical crack was assumed again in advance. Three different meshes with 3068, 4956 and 9132 triangles were defined for a small, medium and large beam, respectively. The crack starting points were located at the left side near the node at the symmetry of the each beam.

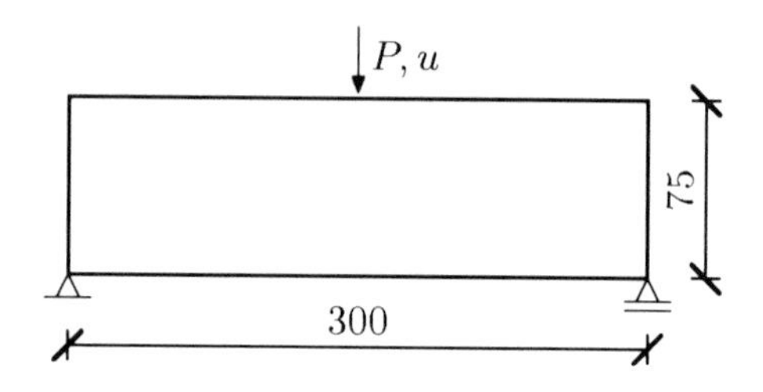

Fig. 5.53 Geometry and boundary conditions of concrete beam during bending test using XFEM of Chapter 4.2 (Bobiński and Tejchman 2011)

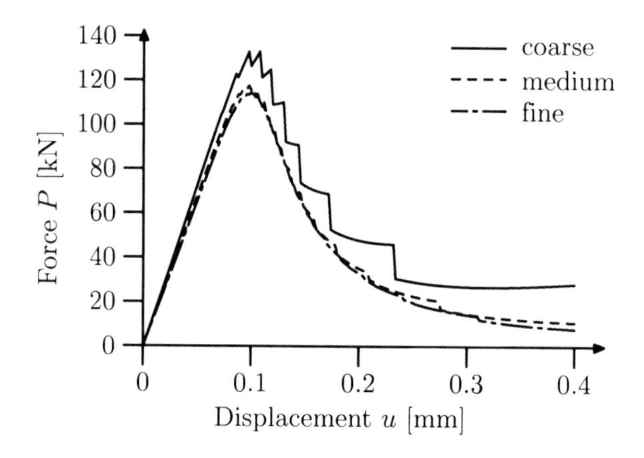

Fig. 5.54 Calculated force–displacement curves during bending test using XFEM of Chapter 4.2 (Bobiński and Tejchman 2011)

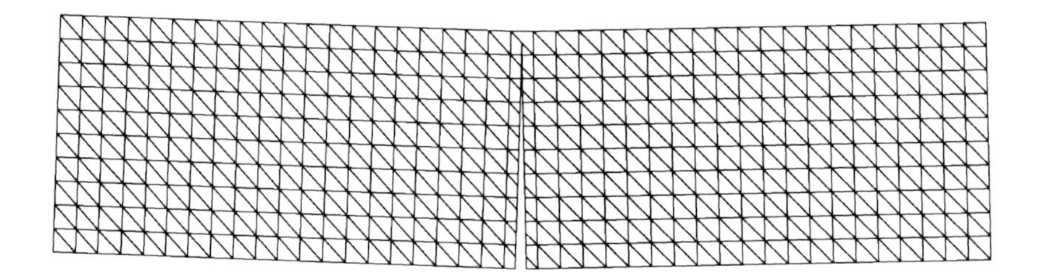

Fig. 5.55 Calculated crack for coarse mesh during bending test using XFEM of Chapter 4.2 (Bobiński and Tejchman 2011)

Figure 5.56 shows the calculated force-displacement diagrams as compared with experimental curves. A satisfactory agreement was achieved and the experimental size effect was well reproduced. The obtained mesh for a small beam with a propagating vertical crack is presented in Fig. 5.57.

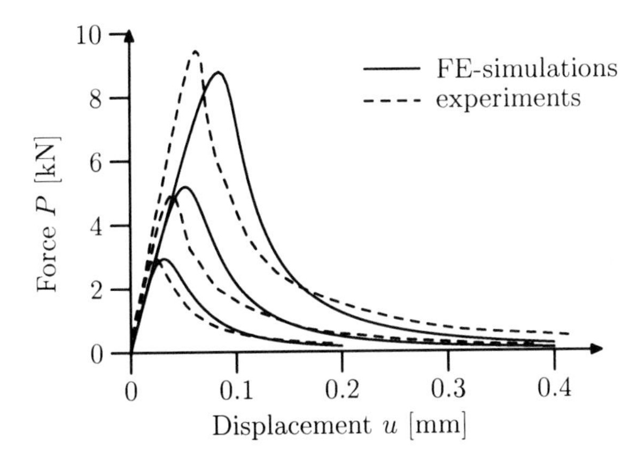

Fig. 5.56 Calculated force–displacement curves during bending test using XFEM of Chapter 4.2 compared to experiments by Le Bellego et al. (2003) (Bobiński and Tejchman 2011)

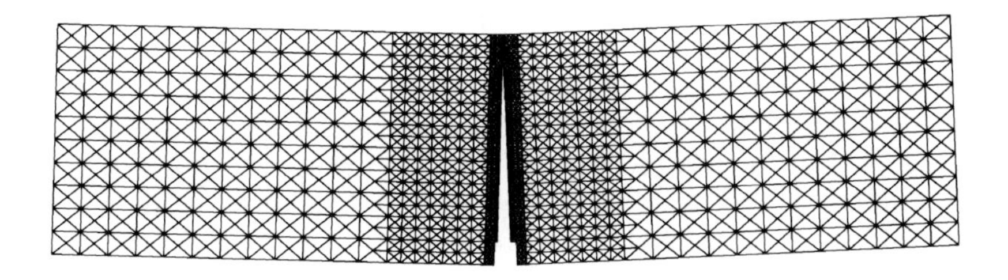

Fig. 5.57 Calculated vertical crack for small beam during bending test using XFEM of Chapter 4.2 (Bobiński and Tejchman 2011)

The following conclusions can be derived from FE simulations of beam bending:

- The larger the ratio between the characteristic length of micro-structure and the specimen size, the higher both the specimen strength and the ductility of the specimen during bending.
- The width of the localized strain zone is about (4.5-5.0)×l_c within elasto-plasticity and (6.5-9.0)×l_c within damage mechanics.
- The width of the localized zone grows during the entire deformation process within damage mechanics, whereas it is almost constant within elasto-plasticity.
- The size effect decreases almost in a hyperbolic way with decreasing ratio between the specimen size and characteristic length. The calculated deterministic size effect in notched concrete elements during bending is in agreement with the corresponding size effect law by Bažant (2003).
- The size effect is more pronounced during beam bending than during uniaxial tension.

- The larger value of the damage parameter β, the faster the damage growth. The parameter β is not uniated versal for all boundary value problems.
- The cohesive crack model and XFEM are able to describe a deterministic size effect and a propagating crack. The FE results do not depend on the mesh size.

5.3 Shear-Extension

In this chapter, two different boundary value problems were numerically studied under combined shear and tension: a double-edge notched concrete specimen and a single-edge notched concrete beam.

Double-edge notched

A double-edge notched (DEN) specimen under various different loading paths of combined shear and tension was experimentally investigated by Nooru-Mohamed (1992). The dimensions of the largest specimen and boundary conditions are presented in Fig. 5.58. The length and height of the element was 200 mm. The thickness was 50 mm. Two notches with dimensions of 25×5 mm^2 were placed in the middle of the vertical edges. The loading was prescribed by rigid steel frames glued to concrete. During one of the loading paths (called '4'), a shear force P_s was applied until it reached a specified value, while the horizontal edges were free. At the second stage, the shear force remained constant and the vertical tensile displacement was prescribed. In the experiment, two curved cracks with an inclination depending on the shear force (for a small value of P_s – almost horizontal, for a large value of P_s – highly curved, Fig. 5.58) were obtained.

The following elastic material parameters were chosen in the FE-analyses: E=32.8 GPa and ν=0.2. A FE-mesh was composed of 12600 3-node triangular finite elements.

FE-results within elasto-plasticity

The tensile strength was assumed as f_t=2.6 MPa and the parameter κ_u=0.033 when using linear softening (Eqs. 3.32, 3.93 and 3.97). A characteristic length was equal to l_c=1 mm and the non-local parameter was m=2. Figure 5.59 presents the obtained FE results with the shear force P_s=5 kN (path '4a'). A very good agreement was achieved with respect to both the force-displacement curve and geometry of localized zones, although the calculated maximum force P was too large. The FE-results for the path '4b' (P_s=10 kN) are shown in Fig. 5.60. The force-displacement curve is satisfactorily reproduced. Two curved localized zones were numerically obtained again, but they were too flat as compared to the experiment (wherein they were more curved and the distance between them was larger).

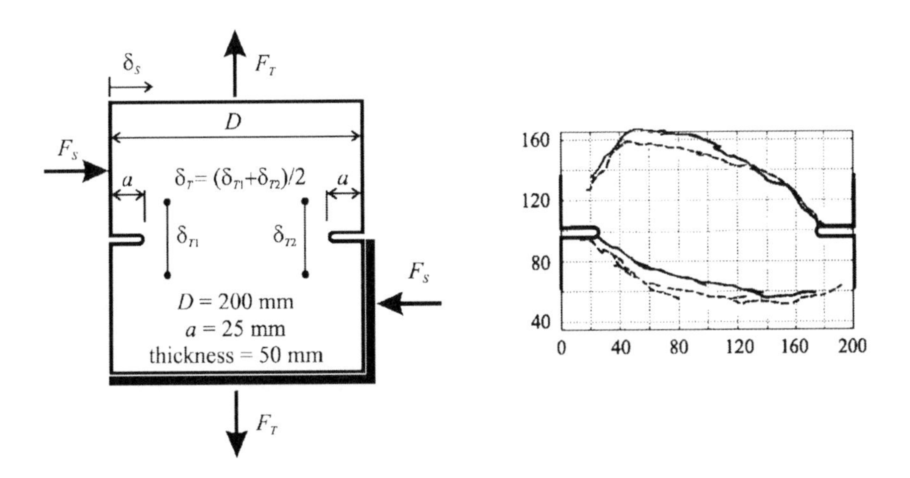

Fig. 5.58 Geometry of DEN specimen (dimensions in mm) and experimental curved cracks during loading path '4c' (Nooru–Mohamed 1992)

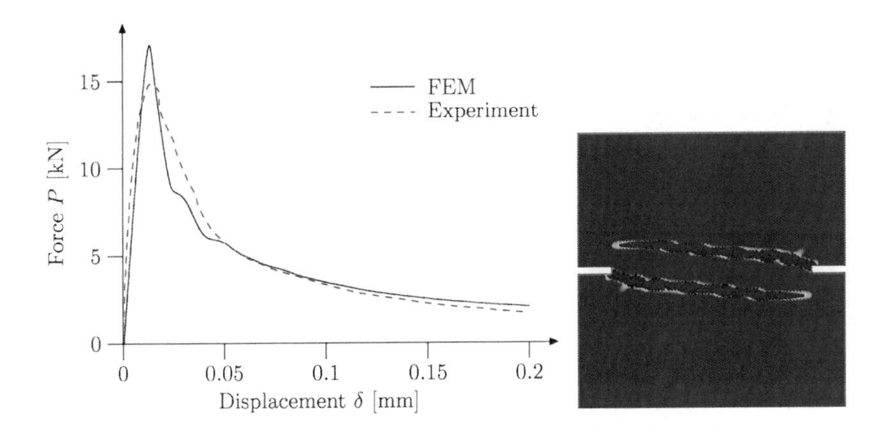

Fig. 5.59 The force-displacement curves and the contour map of non-local parameter $\bar{\kappa}$ for shear force P_s=5 kN within elasto-plasticity (l_c=1 mm) (Bobiński and Tejchman 2010)

FE-results within damage mechanics

First, the Rankine definition of the equivalent strain measure $\tilde{\varepsilon}$ was used (Eqs. 3.35, 3.40, 3.93 and 3.99). The following material parameters were assumed: $\kappa_0=7\times10^{-5}$, α=0.92, β=100 and l_c=0.5 mm. Figure 5.61 presents results at the shear force P_s=10 kN. The force-displacement curve overestimated the maximum value, and its slope after the peak was too high. Two localized zones were numerically obtained, however their curvature was not smooth.

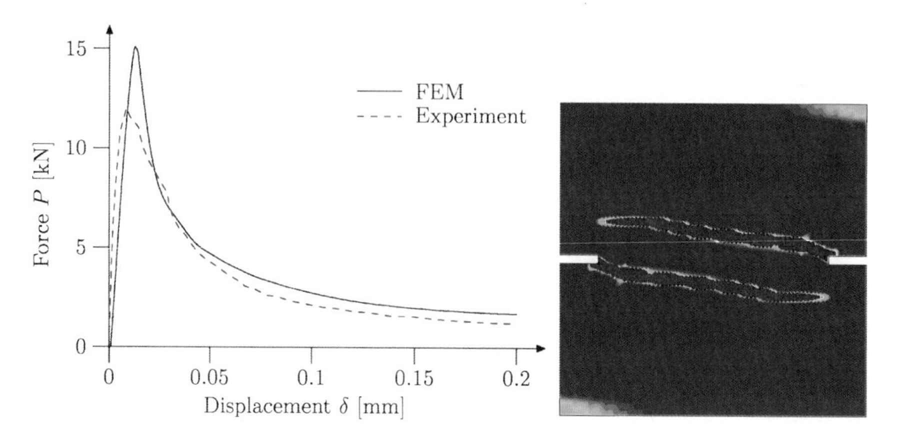

Fig. 5.60 Calculated force-displacement curves and contour map of non-local parameter $\bar{\kappa}$ for shear force P_s=10 kN within elasto-plasticity (l_c=1 mm) (Bobiński and Tejchman 2010)

To improve the behaviour of the damage model, the modified Rankine definition (Eq. 3.37) was also used. The coefficient c was taken as 0.15. The pattern of localized zones and force-displacement diagram (Fig. 5.62) reflect the experimental results much better than those with a standard Rankine definition by Eq. 3.35 (Fig. 5.61).

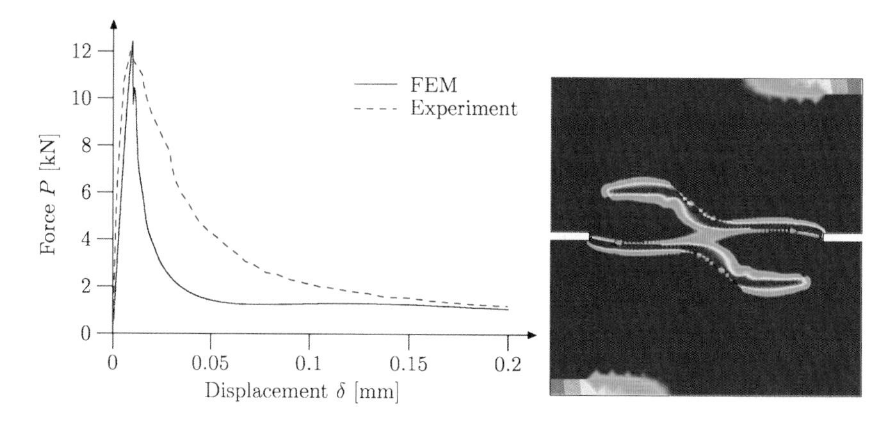

Fig. 5.61 Calculated force-displacement curves and contour map of damage parameter D for shear force P_s=10 kN within damage mechanics (Eq. 3.35) (Bobiński and Tejchman 2010)

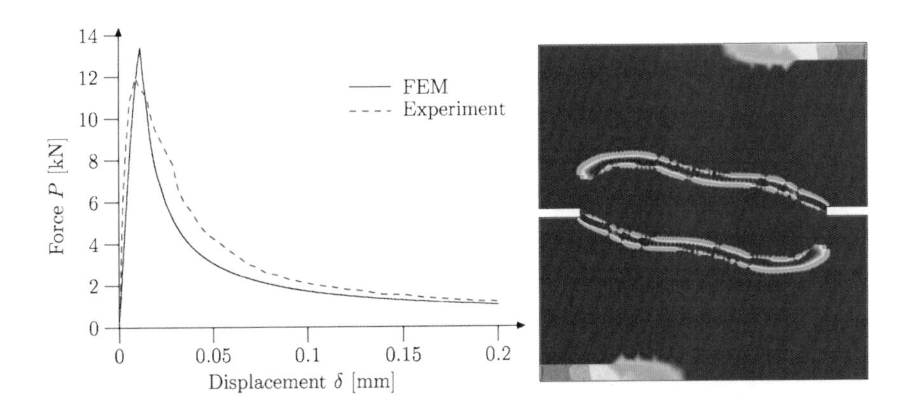

Fig. 5.62 Calculated force-displacement curves and contour map of damage parameter D for shear force P_s=10 kN within damage mechanics (modified Rankine definition by Eq. 3.37) (Bobiński and Tejchman 2010)

Afterwards, the modified von Mises definition was used (Eqs. 3.38, 3.40, 3.93 and 3.99) with the same set of parameters as for the Rankine definition (k=10). Both the force-displacement curve and pattern of localized zones are in good agreement with the experimental results (Fig. 5.63).

FE results with smeared crack model

A characteristic length was equal to l_c=2 mm. The tensile strength was taken as f_t=2.6 MPa and the ultimate normal crack strain ε_{nu}=0.02. For a fixed crack model, the shear retention parameters were assumed as: ε_{su}=0.02 and p=8 (Eqs. 3.52-3.60, Chapter 3.1.3). Figure 5.64 shows results obtained with a multi-orthogonal fixed crack model. The force-displacement curve is reproduced quite well. One obtains two straight localized zones similarly as in the experiment. The calculations with a rotating crack model were also carried out, but they were not successful. Serious numerical problems with convergence took place shortly after the peak. The obtained localization pattern was similar to that obtained with a damage model using a Rankine'a definition of the equivalent strain measure (one horizontal localized zone or two almost horizontal localized zones were created).

FE-results with cohesive elements

The cohesive cracks were described by the tensile strength f_t=2.2 MPa and parameters η=0.0 and β=30000 (Chapter 4.1). The FE mesh with 10184 3-node triangles was defined. Cohesive elements were placed between elements in the central horizontal region of the specimen.

Figures 5.65 and 5.65 show the results obtained for the shear force P_s of 5 kN and 10 kN, respectively. In both cases, a very good agreement was obtained between experimental and numerical crack patterns and force-displacement curves (it should be noted that the tensile strength was slightly decreased as compared to the experimental value).

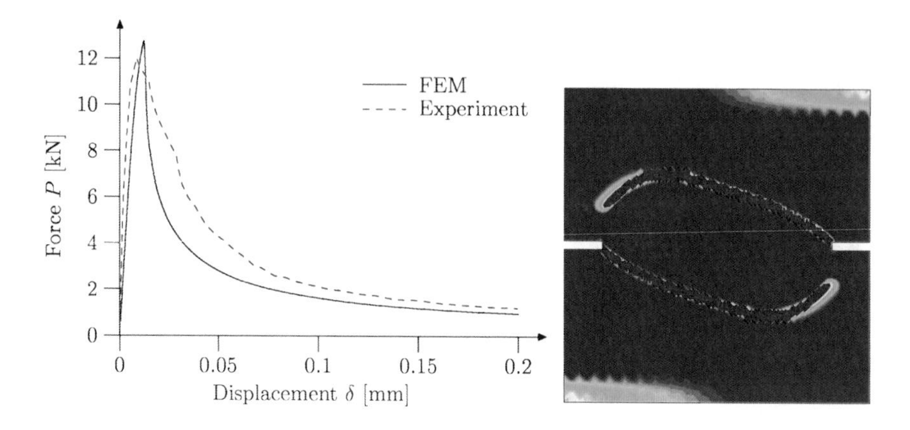

Fig. 5.63 Calculated force-displacement curves and contour map of parameter D for shear force P_s=10 kN within damage mechanics (Eq. 3.38) (Bobiński and Tejchman 2010)

The influence of cohesive parameters η and β is indicated in Figs. 5.67 and 5.68 (at P_s=5 kN). The larger the value of η and β, the larger softening of the force – displacement curve was obtained (Fig. 5.67). The effect of these parameters on the maximum force was not significant. With increasing value of the parameter η, calculated cracks became more horizontal (Fig. 5.68).

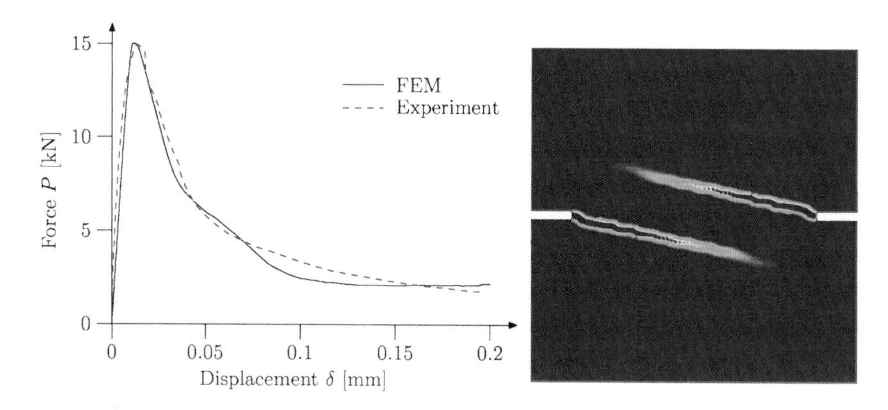

Fig. 5.64 Calculated force-displacement curves and the contour map of strain for shear force P_s=5 kN with smeared crack model of Chapter 3.1.3 (Bobiński and Tejchman 2010)

FE-results with XFEM

The following constants were assumed in elasticity: Young modulus E=38.2 GPa and Poisson's ratio ν=0.2. The tensile strength was equal to f_t=3 MPa. The exponential softening and the fracture energy G_f=100 N/m were assumed for cohesive cracks. Simulations were performed in the plane stress conditions. Two crack starting points were defined near the notch corners. To calculate the crack propagation direction, the criterion based on a direction of the maximum principal stress was used. The stress averaging length was taken as l=1 cm (Chapter 4.2). The FE mesh included 3840 3-node triangles.

Figure 5.69 shows the results obtained for the loading path '4a' (shear force P_s=5 kN). Two inclined cracks were obtained (too strongly curved as compared to the experiment). The calculated force – displacement curve indicated unphysical rehardening caused by a self-locking of both cracks due to their sudden direction change. Crack tips were located at the edges of the earlier cracked elements and the crack evolution was stopped.

The results for the path '4b' (shear force P_s=10 kN) are presented in Fig. 5.70. Too slightly curved cracks were again obtained. Moreover, a strange jump in the crack trajectory was observed in each crack. Despite this fact a good agreement was observed with respect to the force – displacement diagram.

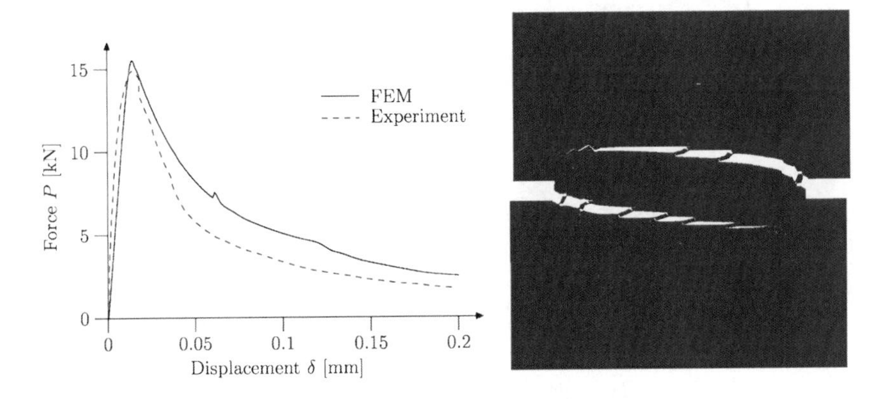

Fig. 5.65 The force-displacement curves and deformed specimen for shear force P_s=5 kN using cohesive elements of Chapter 4.1 (Bobiński and Tejchman 2010)

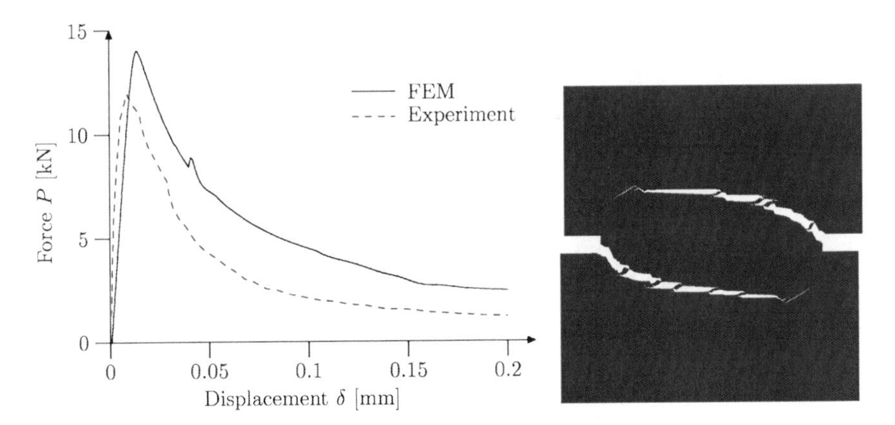

Fig. 5.66 Calculated force-displacement curves and cracks for shear force P_s=10 kN using cohesive elements of Chapter 4.1 (Bobiński and Tejchman 2010)

Single-edge notched (SEN) concrete beam

A single-edge notched (SEN) concrete beam under four-point shear loading (anti-symmetric loading) was analyzed (Schlangen 1993) (Fig. 5.71). The length and height of the beam were equal to 440 mm and 100 mm, respectively. The depth of the notch was equal to 20 mm and its thickness was 5 mm. In the experiments, a curved crack starting from the lower-right part of the notch towards a point to the right of the lower right support was obtained.

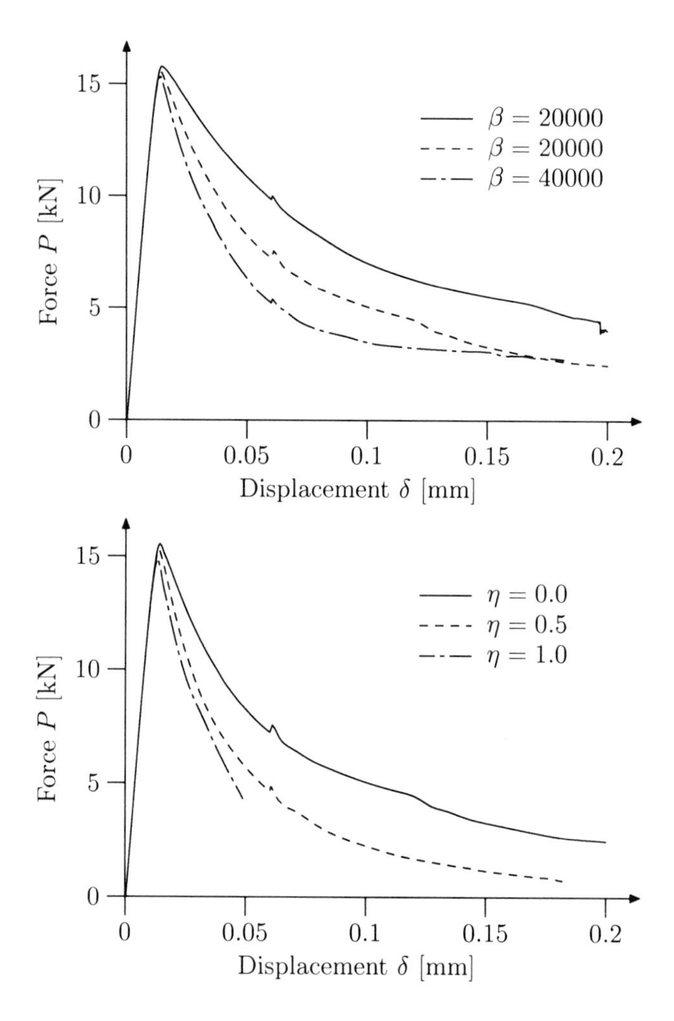

Fig. 5.67 Calculated force – displacement curves for different values of η and β using cohesive elements of Chapter 4.1 at shear force P_s=5 kN

FE results within elasto-plasticity

Figure 5.72 demonstrates the results obtained with the elasto-plastic model (f_t=3 MPa, κ_u=0.040 and l_c=1 mm), Eqs. 3.32, 3.40, 3.93 and 3.97. The localized zone was curved and its shape matched well the experiment. A satisfactory agreement between a numerical and experimental force-displacement diagram was also achieved.

A FE-mesh consisted of 6556 3-node triangle finite elements. The modulus of elasticity was taken as E=35 GPa and the Poisson ratio as ν=0.2. The deformation was induced by linearly increasing the distance δ_2 (due to the snap-back behaviour of vertical displacements at the points where the forces were applied) (Fig. 5.72) (Chapter 8).

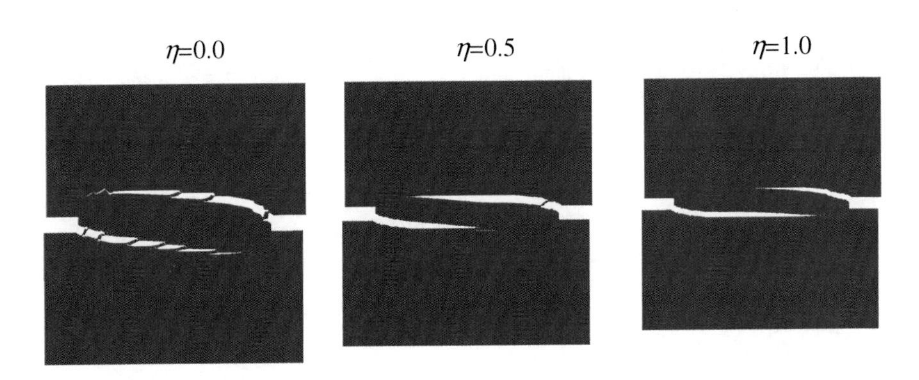

Fig. 5.68 Crack patterns for different values of η using cohesive elements of Chapter 4.1 at shear force P_s=5 kN

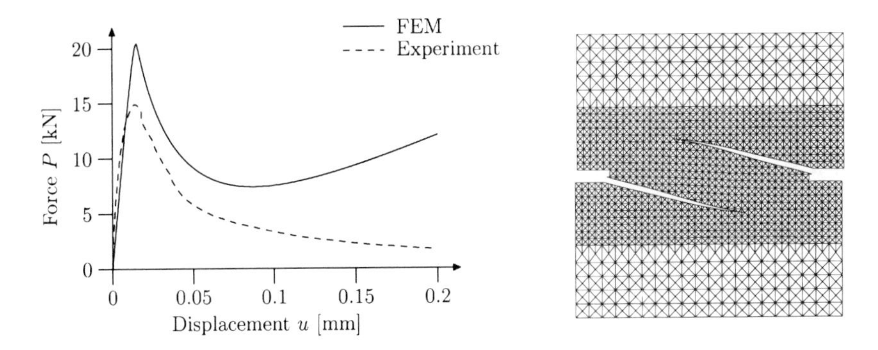

Fig. 5.69 Calculated force – displacement curves and deformed mesh for shear force P_s=5 kN (displacements were scaled 20 times) using XFEM of Chapter 4.2 (Bobiński and Tejchman 2011)

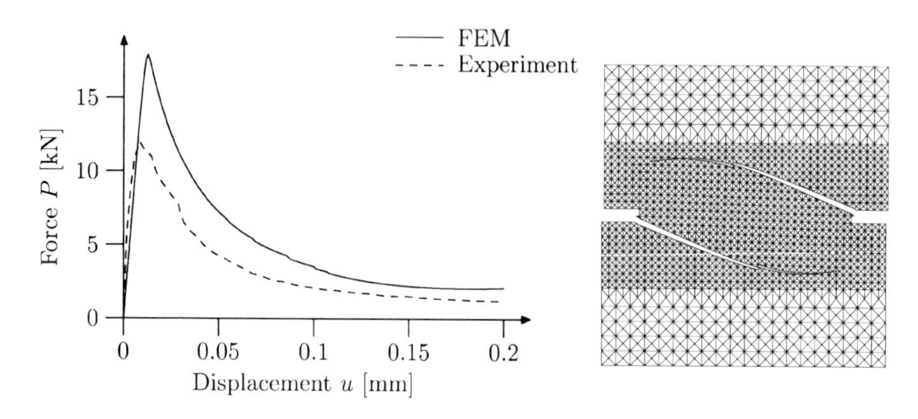

Fig. 5.70 Calculated force – displacement curves and deformed mesh for shear force P_s=10 kN (displacements were scaled 20 times) using XFEM of Chapter 4.2 (Bobiński and Tejchman 2011)

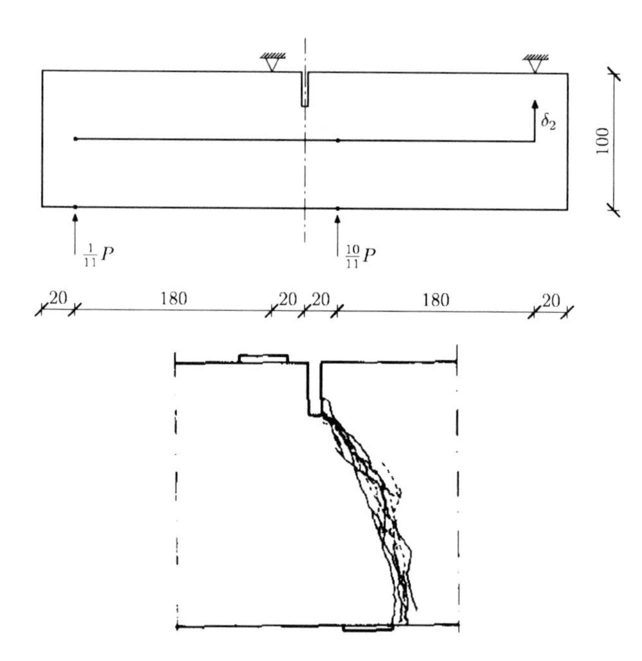

Fig. 5.71 Geometry of SEN specimen (dimensions in mm) and experimental curved crack (Schlangen 1993)

FE results within damage mechanics

The numerical calculations were performed with the Rankine definition of the equivalent strain measure (Eqs. 3.35, 3.40, 3.93 and 3.99). The following material parameters were assumed: l_c=1 mm, κ_0=8.5×10^{-5}, α=0.92 and β=150. Both the force displacement-diagram and strain localization differ significantly from the experimental outcome (Fig. 5.73). To obtain a better agreement with the experiment, the modified von Mises definition (Eqs. 3.38, 3.40, 3.93 and 3.99) of the equivalent strain measure was used with the following parameters: l_c=1 mm, κ_0=8.0×10^{-5}, α=0.92, β=150 and k=10. Although, the slope of the calculated force-displacement curve was too sharp, the shape of the localized zone was properly reproduced (Fig. 5.74).

The results of our FE simulations within continuum mechanics of the concrete behaviour under mixed mode conditions have shown that a proper choice of a constitutive law is a very important issue. The models show a different capability to capture the localized zone phenomenon. In general, an elasto-plastic model with the Rankine failure criterion is the most effective among continuous models. The usefulness of an isotropic damage model depends on the definition of the equivalent strain measure. The influence of the material description in the tensile-compression regime had to be taken into account (e.g. by a non-realistic increase

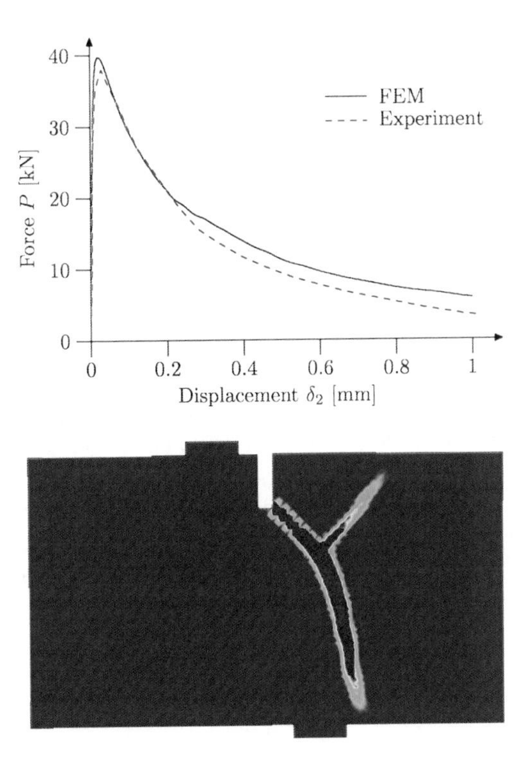

Fig. 5.72 The force-displacement curves and contour map of non-local parameter $\overline{\kappa}$ in central part of beam within elasto-plasticity (l_c=1 mm) (Bobiński and Tejchman 2010)

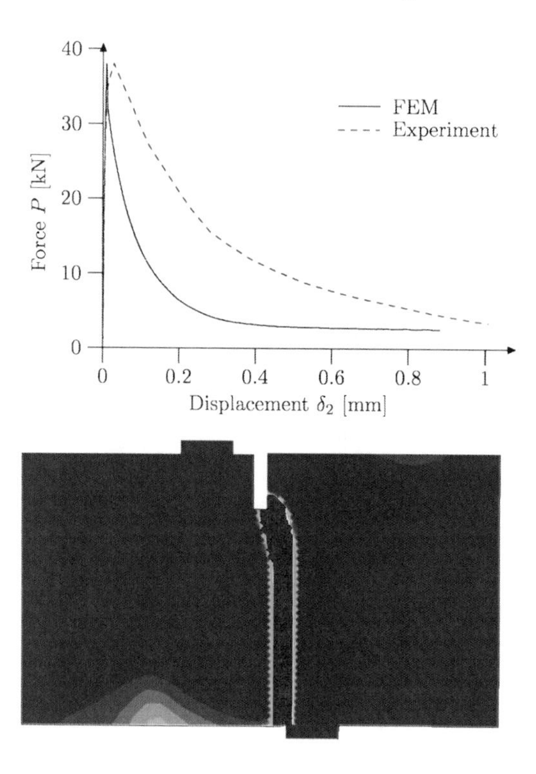

Fig. 5.73 The force-displacement curves and contour map of damage parameter D in central part of beam within damage mechanics (with Rankine equivalent strain definition by Eqs. 3.35) (Bobiński and Tejchman 2010)

of strength). A fixed smeared crack model was not able to reproduce curved localized zones (the worst results were obtained with a rotating smeared crack model). In general, the approach with cohesive elements provided the best approximation of experiments. The calculations using XFEM indicated the importance of the crack propagation criterion defined to realistically model a crack trajectory (the crack evolution was sometimes blocked using an conventional criterion based on a direction of the maximum principal stress).

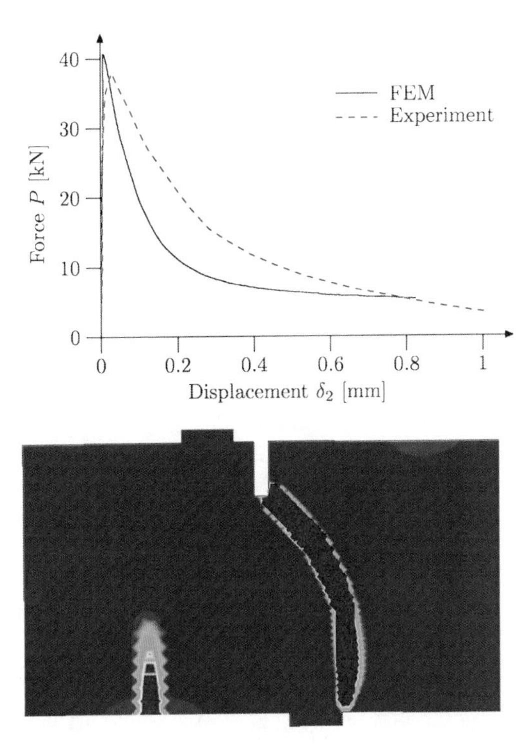

Fig. 5.74 The force-displacement curves and contour map of damage parameter D in central part of beam within damage mechanics (with modified von Mises equivalent strain definition by Eqs. 3.38) (Bobiński and Tejchman 2010)

References

Bažant, Z.P.: Scaling of Structural Strength. Hermes-Penton, London (2003)

Bobiński, J., Tejchman, J.: Numerical simulations of localization of deformation in quasi brittle materials within non-local softening plasticity. Computers and Concrete 1(4), 433–455 (2004)

Bobiński, J., Tejchman, J.: Modelling of concrete behaviour with a non-local continuum damage approach. Archives of Hydro-Engineering and Environmental Mechanics 52(3), 243–263 (2005)

Bobiński, J., Tejchman, J.: Modelling of size effects in concrete using elasto-plasticity with non-local softening. Archives of Civil Engineering. LII(1), 7–35 (2006)

Bobiński, J., Tejchman, J.: Quasi-static crack propagation in concrete with cohesive elements under mixed-mode conditions. In: Schrefler, B.A., Perego, U. (eds.) Proc. 8th Worls Congress on Computational Mechanics, WCCM 2008, Venice, June 29-July 1 (2008)

Bobiński, J., Tejchman, J.: Continuous and discontinuous modeling of cracks in concrete elements. In: Bicanic, N., de Borst, R., Mang, H., Meschke, G. (eds.) Modelling of Concrete Structures, pp. 263–270. Taylor and Francis Group, London (2010)

Bobiński, J., Tejchman, J.: Modeling cracks in concrete elements with XFEM. In: Proc. 19th Int. Conf. on Computer Methods in Mechanics, Warsaw, May 9-12, pp. 139–140 (2011)

Gutierrez, M.A., de Borst, R.: Simulation of size-effect behaviour through sensitivity analysis. Engineering Fracture Mechanic 70(16), 2269–2279 (2003)

Häussler-Combe, U.: Size effect for plain and reinforced concrete structures. In: Proceedings WCCM V, Vienna, Austria (2002)

Hordijk, D.A.: Local approach to fatigue of concrete. PhD Thesis, Delft University of Technology (1991)

Jirásek, M.: Non-local damage mechanics with application to concrete. In: Vardoulakis, I., Mira, P. (eds.) Failure, Degradation and Instabilities in Geomaterials, Lavoisier, pp. 683–709 (2004)

Jirásek, M., Rolshoven, S.: Comparison of integral-type nonlocal plasticity models for strain-softening materials. Int. J. Engineering Science 41(13-14), 1553–1602 (2003)

Kormeling, H.A., Reinhardt, H.W.: Determination of the fracture energy of normal concrete and epoxy modified concrete. Report 5-83-18, Delft University of Technology (1983)

Le Bellego, C., Dube, J.F., Pijaudier-Cabot, G., Gerard, B.: Calibration of nonlocal damage model from size effect tests. European Journal of Mechanics A/Solids 22(1), 33–46 (2003)

Nooru-Mohamed, M.B.: Mixed mode fracture of concrete: an experimental approach. PhD Thesis, Delft University of Technology (1992)

Pamin, J.: Gradient-enchanced continuum models: formulation, discretization and applications. In: Habilitation Monography, Cracow University of Technology, Cracow (2004)

Pamin, J.: Gradient plasticity and damage models: a short comparison. Computational Material Science 32(3-4), 472–479 (2005)

Peerlings, R.H.J., de Borst, R., Brekelmans, W.A.M., Geers, N.G.D.: Gradient enhanced damage modelling of concrete fracture. Mech. Cohesion.-Friction. Materials 3(4), 323–342 (1998)

Rodriguez-Ferran, A., Morata, I., Huerta, A.: Numerical modelling of notched specimens. In: Proc. WCCM V, Vienna, Austria, CDROM (2002)

Schlangen, H.E.J.: Experimental and numerical analysis of fracture processes in concrete. PhD Thesis, Delft University of Technology (1993)

Shi, Q., Chang, C.S.: Numerical analysis for the effect of heterogeneity on shear band formation. In: Proc. 16th ASCE Engineering Mechanics Conference, University of Washington, Seattle, pp. 1–11 (2003)

Simone, A., Wells, G.N., Sluys, L.J.: Discontinuous Modelling of Crack Propagation in a Gradient Enhanced Continuum. In: Proc. of the Fifth World Congress on Computational Mechanics, WCCM V, Vienna, Austria, CDROM (2002)

Sluys, L.J.: Wave propagation, localisation and dispersion in softening solids. PhD Thesis, Delft University of Technology, Delft (1992)

Chapter 6
Continuous Modelling of Fracture in Plain Concrete under Cyclic Loading

Abstract. The enhanced coupled elasto-plastic damage models with non-local softening proposed by Pamin and de Borst (1999) (called model '1'), by Carol et al. (2001) and by Hansen and Willam (2001) (called model '2'), by Meschke et al. (1998) (called model '3') and Marzec and Tejchman (2009, 2010, 2011) (called model '4') described in detail in Chapter 3.2 were used in FE calculations. Quasi-static FE results were compared with corresponding laboratory tests on concrete specimens: dog-bone shaped specimen under monotonic uniaxial tension (van Vliet and van Mier 2000) and notched beams under cyclic loading (Hordijk 1991, Perdikaris and Romeo 1995).

Initial Results for Monotonic Uniaxial Tension

In the first step, the numerical calculations were carried out for concrete specimens under monotonic uniaxial tension. The main purpose was to check the effectiveness of a different non-local techniques used for each model. The experimental data presented by van Vliet and van Mier (2000) served as the reference data. In the experiments, a size effect in concrete with two-dimensional dog-bone shaped concrete specimens under quasi-static uniaxial tension (Fig. 6.1) was investigated. The five different specimen types (from 'A' to 'E') were used. Their height varied from 75 mm up to 2400 mm. In the numerical calculations, three different specimen sizes were considered only, namely: 'A', 'B' and 'C' (Tab. 6.1) with the height varying between 75 mm and 300 mm. The deformation was induced by imposing a vertical displacement at the node at the top part of the specimen. The number of triangular finite elements (with linear shape functions) was equal to 246, 1018 and 4102 for the specimen 'A', 'B' and 'C', respectively (with characteristic length l_c=5 mm). The modulus of elasticity was E=49.0 GPa and the Poisson's ratio was ν=0.2.

In the coupled model '1' by Pamin and de Borst (1999) (Chapter 3.2), the von Mises yield criterion with linear hardening was assumed in a plastic region (with the yield stress σ_{yt}^{0}=2.6 MPa and linear hardening modulus H_p=E/2). In a damage regime, the following material parameters were assumed: κ_0=7.9×10^{-5}, α=0.91, β=350 and k=10. The damage formulation was based on the total strain $\tilde{\varepsilon}\left(\varepsilon_{ij}\right)$.

J. Tejchman, J. Bobiński: Continuous & Discontinuous Modelling of Fracture, SSGG, pp. 163–182.
springerlink.com

In the coupled model '2' by Carol et al. (2001) and by Hansen and Willam (2001) (Chapter 3.2), the following parameters: E_{pr}=45 GPa, n_r=0.13 and f_t=2.54 MPa were assumed to describe the resistance function by Eq. 3.77. In turn the following parameters were chosen in the coupled model '3' by Meschke et al. (1998) (Chapter 3.2): f_t=2.85 MPa, $\kappa_0=3.0\times10^{-3}$ and β=0.15. For the coupled model '4' by Marzec and Tejchman (2010) (Chapter 3.2), the Rankine yield criterion with the yield stress σ^0_y=2.5 MPa was assumed in plasticity. A linear hardening parameter ($H_p=E/2$) was chosen. In a damage regime, the following material parameters were taken: $\kappa_0=9\times10^{-5}$, α=0.95 and β=230. The damage formulation was based on the total strains according to Eq. 3.84. The stiffness reduction factors were a_t=1 and a_c=1.

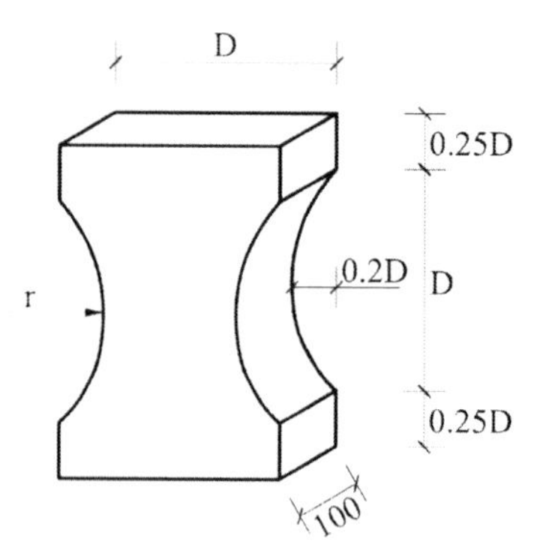

Fig. 6.1 Geometry of dog-bone shaped specimen (Vliet and van Mier 2000)

Table 6.1 Dimensions of dog-bone shaped specimens under uniaxial tension of Fig. 6.1 (van Vliet and van Mier 2000)

Specimen type (Fig. 6.1)	'A'	'B'	'C'
D [mm]	50	100	200
r [mm]	36.25	72.50	145

Figure 6.2 shows the calculated localized zone with four coupled models in the specimen A, B and C of Tab. 6.1. In turn, the calculated load-displacement diagrams for the concrete specimen 'B' of Tab. 6.1 with 4 coupled models compared to the experiment are demonstrated in Fig. 6.3. A satisfactory agreement with the experiment was obtained.

The calculated width of a localized zone is 2.2÷2.3 cm (4-5)×l_c (model '1' and '4'), 2.7 cm (5-6)×l_c (model '2') and 1.7 cm (3-4)×l_c (model '3'). The calculated evolution of the vertical force at the top is almost the same for each formulation and close to the experimental data.

Figure 6.4 presents a comparison of calculated and experimental values of the nominal strength σ_N versus the specimen size D (σ_N was calculated by dividing the ultimate vertical force at the top by the smallest specimen cross-section equal to 0.6 bD, b – specimen thickness, D – specimen width of Fig. 6.1. The numerical results were compared with the corresponding experimental mean values (and standard deviations).

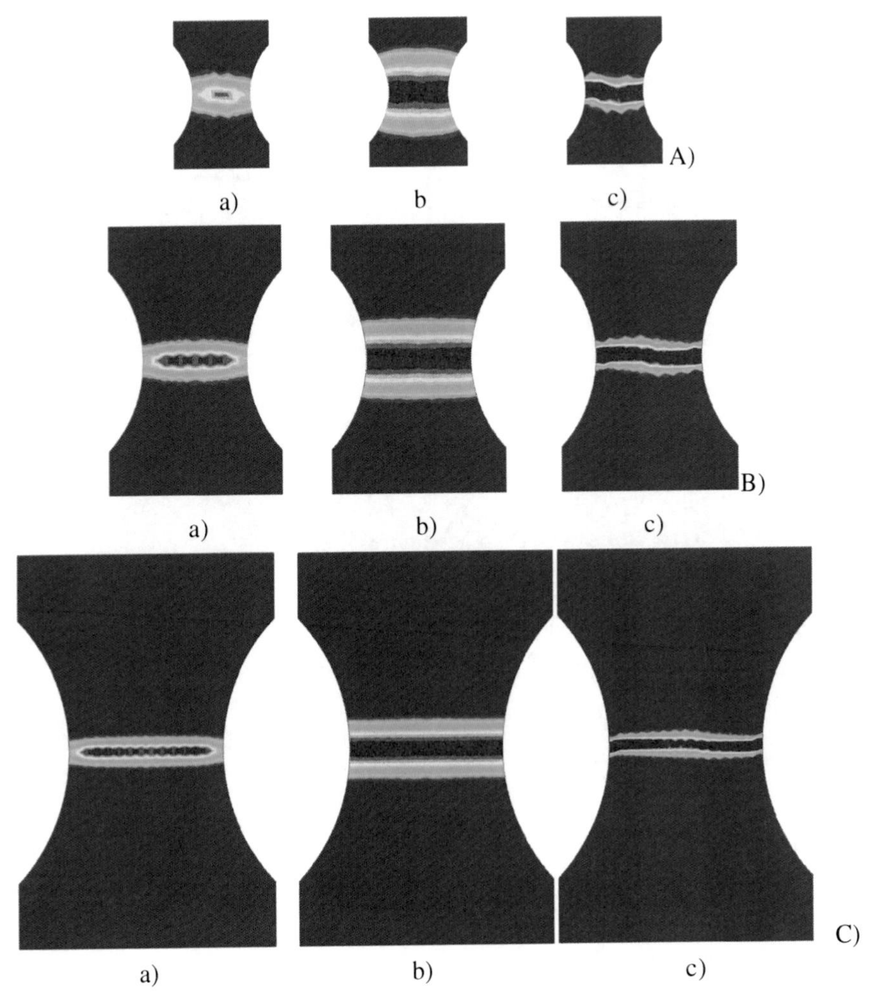

Fig. 6.2 Calculated contours of localized zones (for u=100 μm) with different enhanced coupled models: a) model '1' and '4', b) model '2' and c) model '3' for specimen sizes of Tab. 6.2: A) type A, B) type B and C) type C (Vliet and van Mier 2000) (specimens are not scaled)

Figure 6.5 shows the effect of a different resistance function (Eqs. 3.72, 3.76 and 3.77) in the coupled model '2' on results in a post-peak regime. The function of Eq. 3.72 gives limited possibilities to control the material behaviour in a softening regime (Fig. 6.5a), since a significant change of g_f/r_o slightly influences the load-displacement curve. The proposition in Eq. 3.36 describes a wider range of the post-peak behaviour with the same amount of parameters as the previous function (Fig. 6.5b). The third resistance function of Eq. 3.77 can the best control the rate of softening and the shape of the function in the post-peak regime with the help of one additional parameter (Fig. 6.5c).

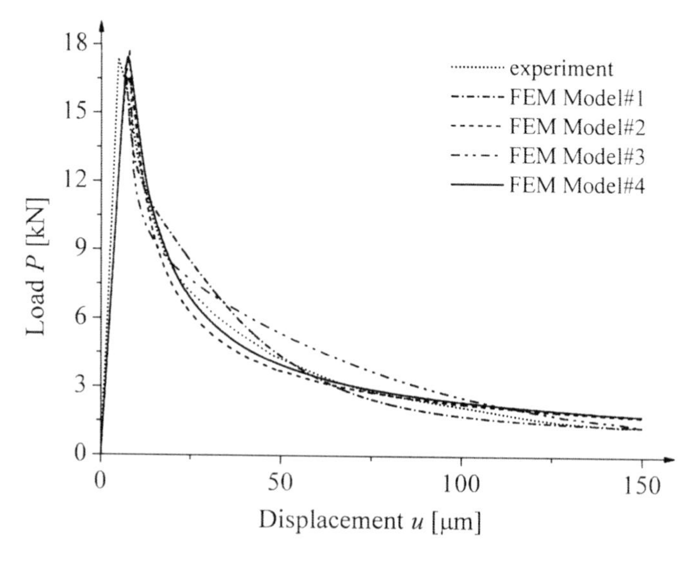

Fig. 6.3 Calculated load-displacement diagrams for four coupled elasto-plastic-damage models as compared with experimental data for specimen 'B' (Vliet and van Mier 2000)

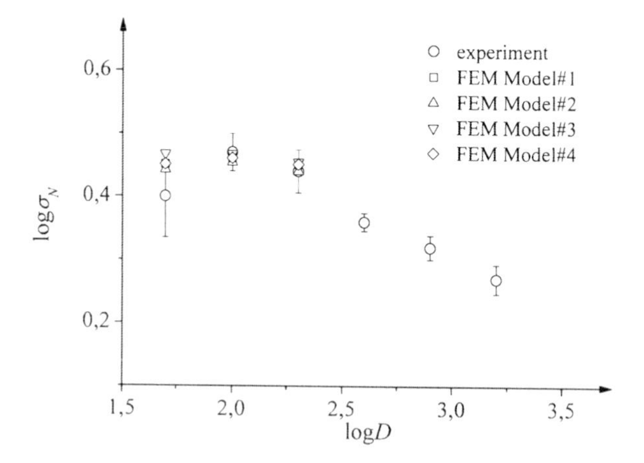

Fig. 6.4 Comparison between calculated and experimental values of nominal strength σ_N versus specimen size D of Fig. 6.1 for dog-bone shaped specimens (Vliet and van Mier 2000)

The experimental size effect on strength was well reflected in numerical calculations except the smallest specimen 'A' wherein a strong boundary effect took place in the experiment (the coarseness of the applied concrete mixture was simple too large in relation to the specimen dimensions, Vorechovsky 2007).

FE Results for Four-point Cyclic Bending of Notched Concrete Beams

The comparative numerical plane strain simulations were performed with a concrete notched beam under four-point cycling bending subjected to tensile failure (Hordijk 1991) (Fig. 6.6). The length of the beam was 0.5 m and the height 0.1 m. The deformation was induced by imposing a vertical displacement at two nodes at the top of the beam. In the calculations, the modulus of elasticity was E=40 GPa, Poisson ratio ν=0.2 and characteristic length l_c=5 mm. The tensile strength from experiments was varied between f_t=2.49 MPa and f_t=4.49 MPa. The calculations were performed with 7634 triangular finite elements. The size of elements was not greater than (2-3)$\times l_c$ to obtain objective FE results (Bobiński and Tejchman 2004, Marzec et al. 2007). The force-displacement diagrams P=f(u) are shown in Fig. 6.7. In turn, Fig. 6.8 presents the calculated contours of a localized zone above the notch. The evolution of non-local parameters: equivalent strain measure (model '1' and '4'), pseudo-log damage variable (model '2') and softening parameter (model '3') is demonstrated in Fig. 6.9.

For the first enhanced coupled model (model '1') with one surface in hardening plasticity, the von Mises criterion with the yield stress σ_{yt}^{0}=6.5 MPa (total strains) and σ_{yt}^{0}=5.9 MPa (elastic strains) was assumed with a linear hardening parameter (H_p=E/2). Since, an elasto-plastic model is not directly responsible for the evolution of the failure mechanism, the von Mises criterion was chosen for concrete in elasto-plasticity for the sake of simplicity (the application of the criterion by Drucker-Prager does not affect FE results). The following material constants were used: κ_0=9.5×10^{-5}, α=0.92 and β=140 with the total strains $\tilde{\varepsilon}\left(\varepsilon_{ij}\right)$, and κ_0=8.6×10^{-5}, α=0.92 and β=170 with the elastic strains $\tilde{\varepsilon}\left(\varepsilon_{ij}^{e}\right)$. The parameter set is different in both cases due to a varying coupling between plasticity and damage (via elastic or total strains).

Figure 6.7a shows the calculated load-displacement curves with a coupled elasto-plastic damage model using total strains. The load reversals exhibit a gradual decrease of the elastic stiffness, however calculated stiffness degradation is overestimate, especially for high values of κ. The calculated vertical force is close to experiment. The slope of the load-displacement curve is realistically reflected. The width of a localized zone above the notch in the beam is about 2.4 cm (4.8×l_c) (Fig. 6.9a).

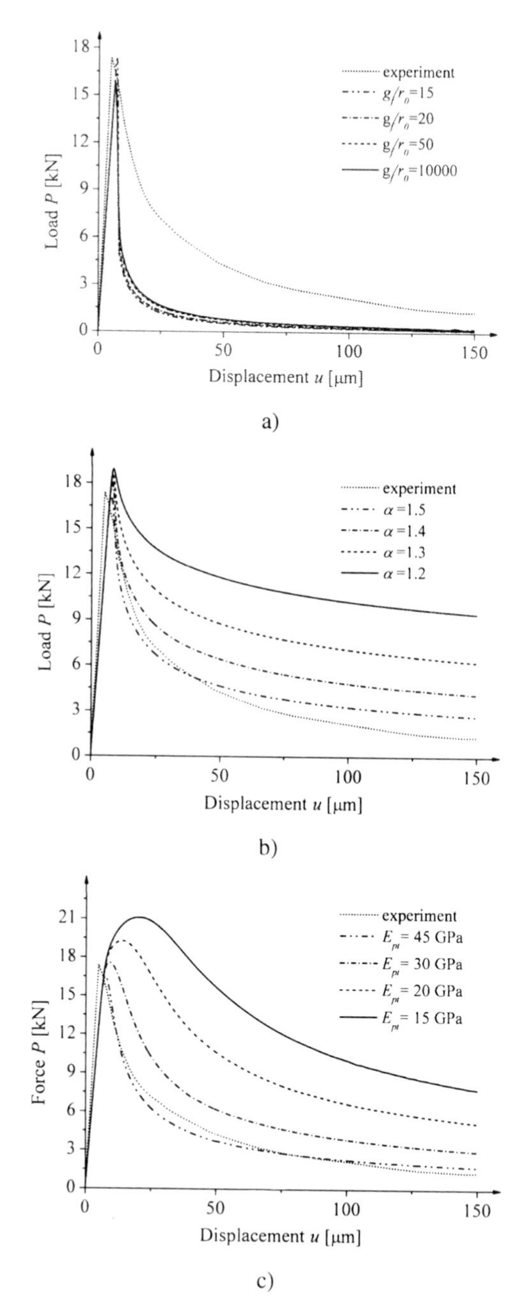

Fig. 6.5 Calculated load-displacement curves with resistance functions: a) of Eq. 3.72 for different ratios g_f/r_0, b) of Eq. 3.76 for different parameter α and c) of Eq. 3.77 for different parameters E_{pt} and n_t compared with experimental data for specimen 'B' (Vliet and van Mier 2000)

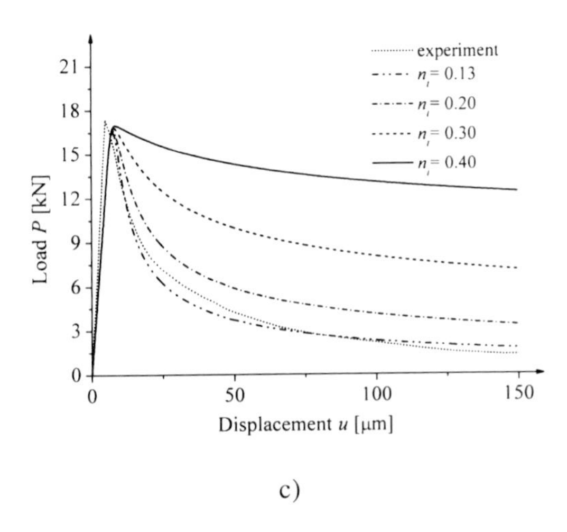

Fig. 6.5 (*continued*)

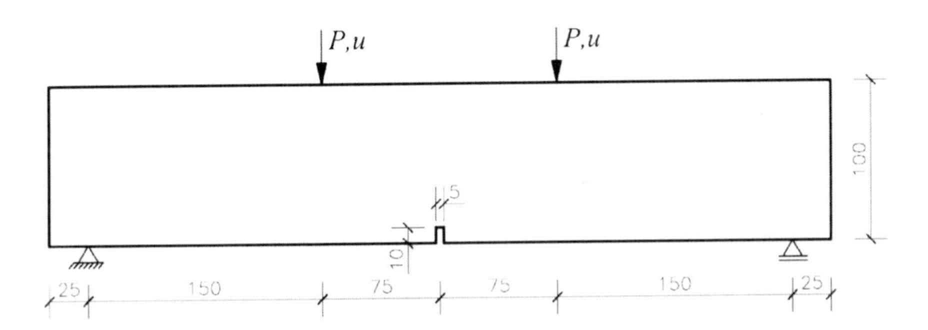

Fig. 6.6 Geometry and boundary conditions of a notched beam under four-point bending (Hordijk 1991)

Using the second enhanced coupled model, the resistance function by Nguyen (2005) was assumed with E_{pr}=37 GPa, n_r=0.175, f_t=2.85 MPa and m=1.2. The numerical results agree well with the experimental data only in the case of the ultimate vertical force and softening slope in the post-peak regime (Fig. 6.7b). The calculated stiffness degradation is significantly too high than in the experiment. As a consequence, the width of a localized zone increases up to 3.2 cm ($6.4\times l_c$) (Fig. 6.9b). The similar results are obtained with the resistance function by Eq. 3.76 (Marzec 2009).

In the third enhanced coupled model, the calculated ultimate vertical force (with the parameters: f_t=2.85 MPa, $\kappa_0=1.85\times10^{-3}$, γ=0.2 and m=2) again very similar as compared with the experimental value (Fig. 6.7c). Also the softening behaviour is realistically reflected. The slope of the experimental and numerical curve is almost the same. The calculated stiffness degradation exhibits a proper gradual decrease and it is close to experiment. The width of the localized zone above the notch is 1.4 cm ($2.8\times l_c$) (Fig. 6.9c).

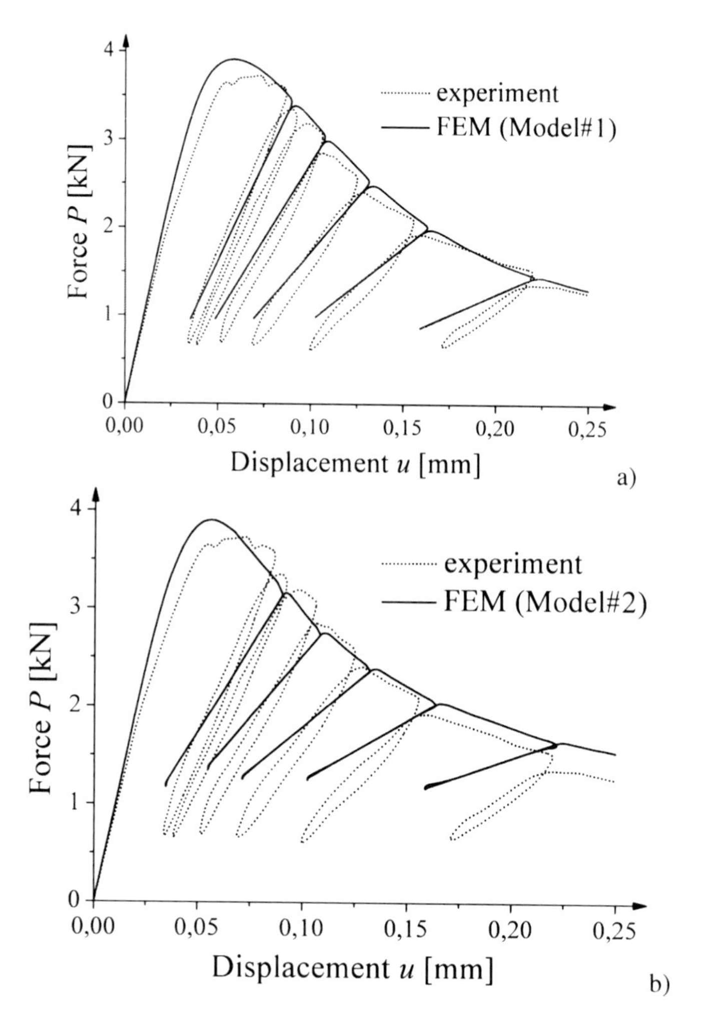

Fig. 6.7 Experimental and calculated force-displacement curves using 4 different coupled elasto-plastic-damage models with non-local softening during quasi-static four-point cyclic bending under tensile failure (Hordijk 1991): a) model '1' (damage based on total strains), b) model '2', c) model '3' and d) model '4' (damage based on elastic strains) (Marzec and Tejchman 2009, 2010)

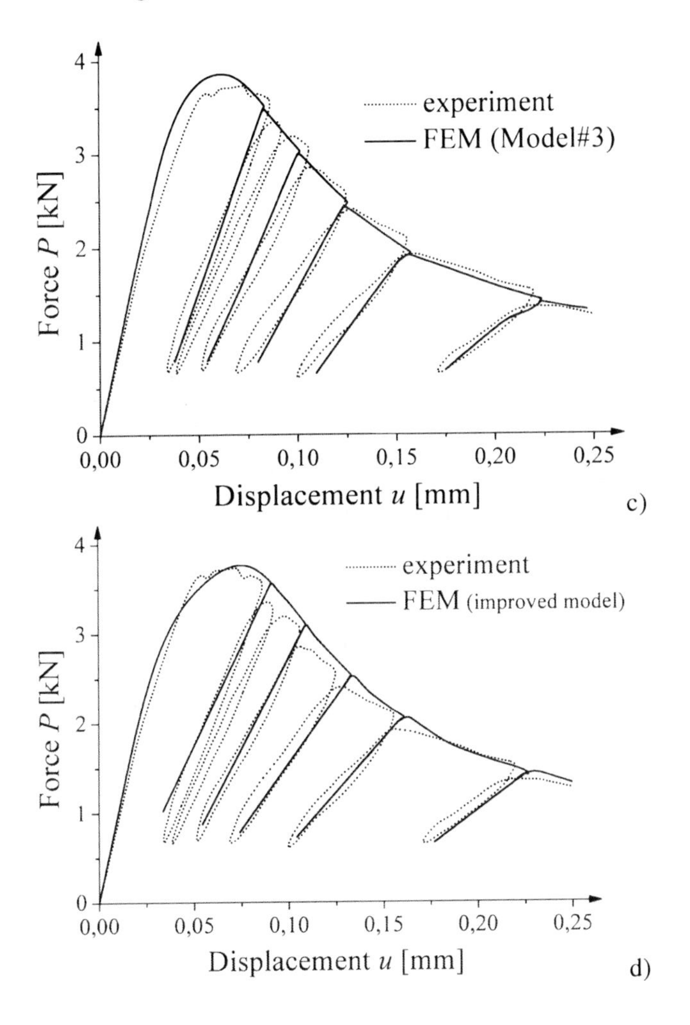

Fig. 6.7 (*continued*)

In the fourth enhanced model, the constants σ_{yt}^{0}=6.5 MPa, H_p=E/2, κ_0=4.3×10^{-5}, β=650, α=0.90, η_1=1.2, η_2=0.15, δ=450, a_t=0 and a_c=1 were used (damage was based on elastic strains). The calculated force-displacement curve exhibits good agreement with experimental outcomes (Fig. 6.7d). The bearing capacity of the beam is very well captured. The post-peak behaviour is close to experiment, however the softening slope is slightly worst reflected as in the model '3'. In turn a calculated stiffness decrease is almost the same as in the experiment. Thus, an evident improvement as compared to the model '1' with respect to the magnitude of the stiffness reduction was achieved. The calculated contours of a non-local variable describing the shape of a localized zone are similar as in the model '1' (Fig. 6.8d). The results of Figs. 6.8 and 6.9 demonstrate that the shape of a localized zone above the notch is different due to the material stiffness in a softening regime induced by the material formulation. The shape of a localized zone in the models '1' and '4' is the same due to a similar model formulation, and is typical for other solutions within damage mechanics (e.g. Peerlings 1999, Pamin and de Borst 1999).

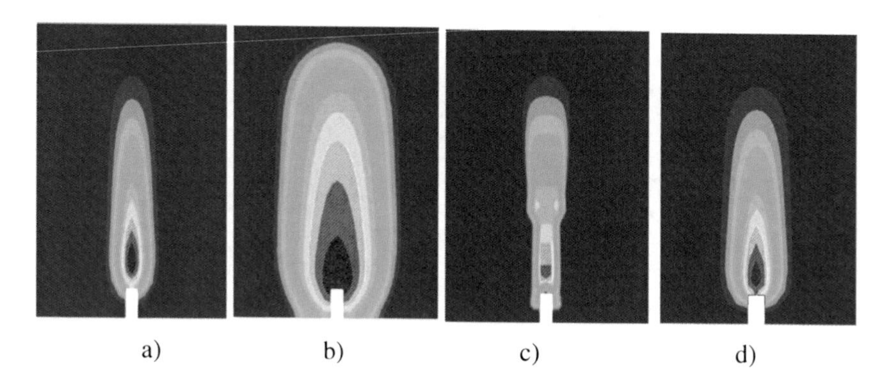

Fig. 6.8 Calculated contours of localized zone near notch in a beam under four-point bending with 4 different coupled elasto-plastic-damage models with non-local softening during four-point bending: a) model '1', b) model '2', c) model '3' and d) model '4' (at deflection u=0.15 mm) (Marzec and Tejchman 2010)

Summarized, the coupled models '1', '3' and '4' are capable to satisfactorily capture the cyclic concrete behaviour under tensile failure.

FE results for three-point cyclic bending of notched concrete beams

In order to check the capability of the improved coupled model '4' to simulate a deterministic size effect observed experimentally in brittle materials (van Vliet and van Mier, 2000), the FE-calculations were performed in addition with concrete notched beams under three-point cycling loading (Fig. 6.10 Tab. 6.2) (Perdikaris and Romeo 1995). The number of triangular finite elements was equal to 2292, 5213 and 9211 for a small-, medium- and large-size beam, respectively. The size of elements was again not greater than $3\times l_c$. The deformation was induced by imposing a vertical displacement at the mid-node at the beam top. The modulus of elasticity was E=45.6 GPa and the Poisson ratio was ν=0.2 and. To match the numerical results with the experimental ones, the same material constants for all three beams were chosen: σ^0_y=6.5 MPa, H_p=E/2, κ_0=9.0×10^{-5}, β=1550, α=0.99, η_1=1.2, η_2=0.15, δ=950, a_t=0 and a_c=1 and l_c=5 mm (equivalent strain measure based on elastic strains). As compared to FE calculations on four-point cyclic bending, the same constants σ_y, H_p, η_1, η_2, a_t and a_c were assumed.

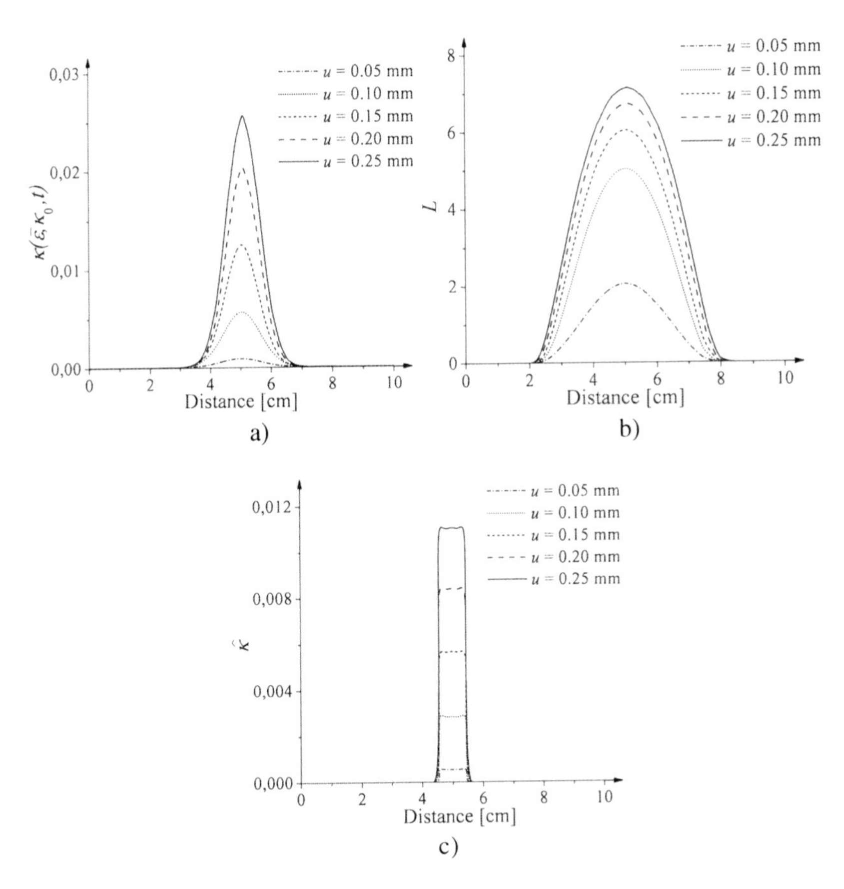

Fig. 6.9 Evolution of non-local parameter above notch in beam under four-point bending with 4 different coupled elasto-plastic-damage models with non-local softening: a) model '1', b) model '2' and c) model '3' (at deflection u=0.15 mm) (Marzec and Tejchman 2009)

Figures 6.11a and 6.11b demonstrate the calculated force-displacement diagrams for a small- and large-size beam compared with the experimental data. The stiffness degradation is again realistically captured by the model. The calculated ultimate force as compared to experiments is higher by 10-15%. To obtain a better agreement between ultimate forces and calculated stiffness, the material constant should be better calibrated (in particular κ_0 and parameters controlling the damage evolution β, δ and η_2).

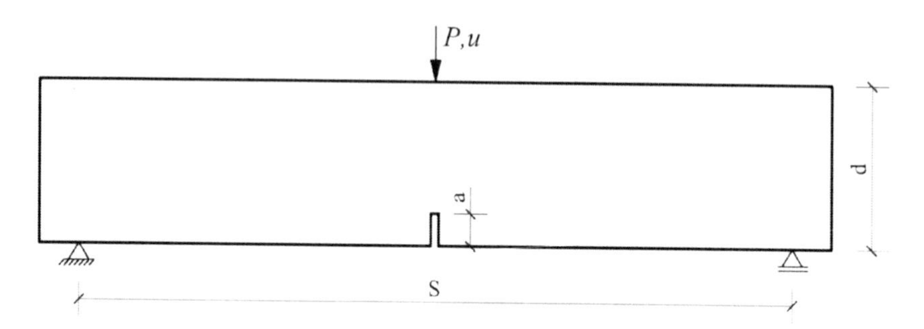

Fig. 6.10 Geometry and boundary conditions of a notched beam under three-point cyclic bending (Perdikaris and Romeo 1995)

Table 6.2 Beam dimensions in cyclic tests by Perdikaris and Romeo (1995)

Beam Size	Depth d [mm]	Width b [mm]	Span S [mm]	Notch a [mm]
Small	64	127	254	20
Medium	128	127	508	39
Large	254	127	1016	78

Next, the calculated results of a deterministic size effect with respect to the ultimate vertical force were confronted with the size effect law by Bažant (Eq. 5.5) for notched beams (Bažant and Planas 1998, Bažant 2003). The FE results show good agreement with the experimental data (Fig. 6.12).

First, simple cyclic uniaxial element tests were numerically performed to show the behaviour of the model '4' (with 4-node quadrilateral elements). Figure 6.13 shows the load-displacement diagrams under cyclic uniaxial tension and cyclic uniaxial compression for different influential material constants β, δ, η_2 and κ_0 (which were independently changed). The effect of the constant α (α=0.7-0.99) and η_1 (η_1=1.0-1.2) was negligible. The modulus of elasticity was E=40 GPa and the Poisson ratio was ν=0.18. In tension, the constants σ^0_y=4.0 MPa and $H_p=E/2$ (Rankine criterion), and in compression σ^0_y=40 MPa, $H_p=E/2$, ϕ=20° and ψ=10° (Drucker-Prager criterion) were chosen. The equivalent strain measure was based on total strains. The material constants varied in the following ranges: β=200-1100, δ=200-900, η_2=0.15-0.45 and κ_0=(15-25)×10^{-5} (with α=0.95, η_1=1.2, a_t=0.0 and a_c=1.0). The force-displacement results indicate that the effect of κ_0 is significant in tension and the effect of δ, η_2 and κ_0 in compression. The parameter κ_0 is responsible for a peak location and a simultaneous activation of a plastic and damage criterion. The parameters β, δ and η_2 affect a model response in softening

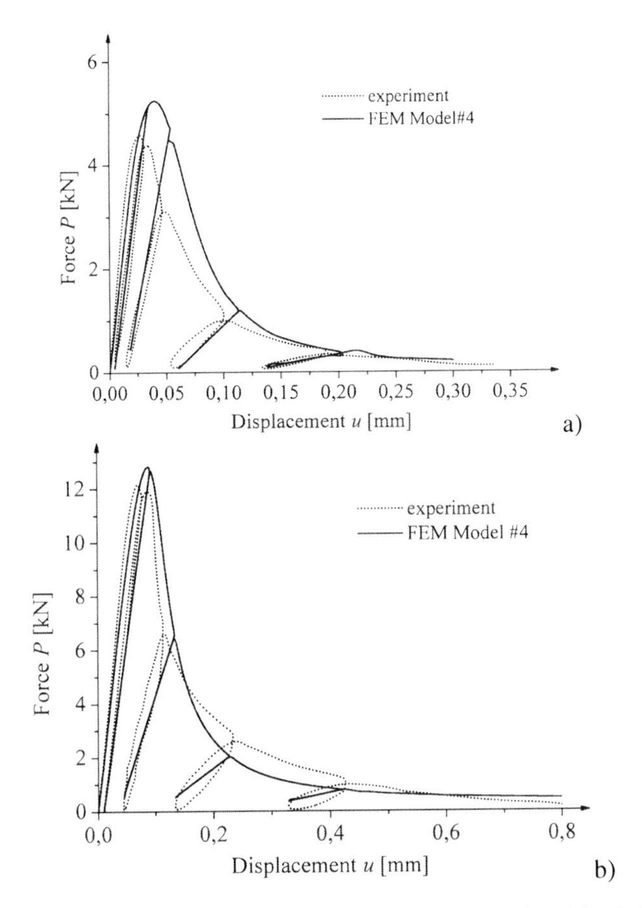

Fig. 6.11 Experimental (Perdikaris and Romeo 1995) and calculated load-displacement curves with enhanced coupled model '4' with damage based on elastic strains (quasi-static cyclic three-point bending): a) small-size beam, b) large-size beam (a_t=0.0 and a_c=1.0) (Marzec and Tejchman 2010)

during tension and compression, and the parameter η_2 influences a hardening curve in compression. The effect of two other parameters (α and η_1) describing the stress-strain curve at the residual state is negligible.

Cyclic behaviour of concrete under compression and tension

Next, a simple cyclic tension-compression-tension element test was calculated (Fig. 6.14) (σ_{yt}^0=4 MPa, σ_{yc}^0=40 MPa, H_p=E/2, ϕ=20°, ψ=10°, β=550, δ=950, κ_0=8.5×10^{-5}, α=0.95, η_1=1.2, η_2=0.15, a_t=0.0 and a_c=1.0). The results show obviously the different stiffness degradation during compression and tension (that is stronger in tension). A recovery of the compressive stiffness upon crack closure and un-recovery of the tensile stiffness as the load changes between tension and compression is satisfactorily reflected. The evident difference between a pure damage model (without plastic strains) and coupled one (with plastic strains) during one uniaxial load cycle is demonstrated in Fig. 6.15.

The effect of the damage scale factors a_t and a_c on the load-displacement diagram under tension-compression-tension is described in Fig. 6.16 by assuming a_t=0.2 and a_c=0.8. This change of both factors is stronger in compression.

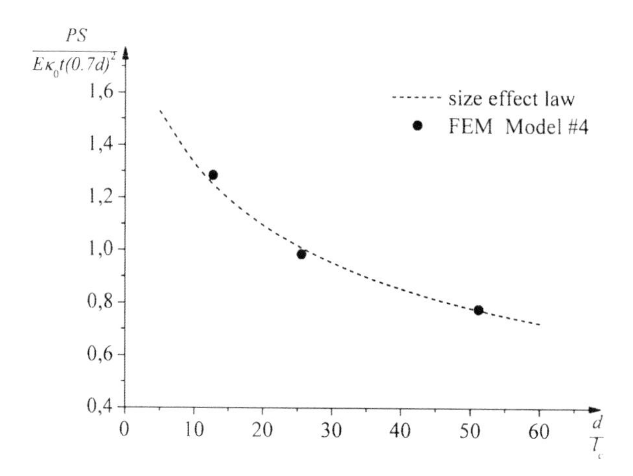

Fig. 6.12 Calculated deterministic size effect for concrete notched beams subjected to quasi-static cyclic three-point bending (using coupled model '4') as compared with size effect law by Bažant (Eq. 5.5) (Bažant 2003) (t – beam thickness, d - beam height, S – beam span)

Finally, Fig. 6.17 demonstrates the 2D FE results with the model '4' for a concrete specimen subjected to uniaxial cyclic compression by taking strain localization into account. All nodes at the lower edge of a rectangular specimen were fixed in a vertical direction. The size of the specimen was arbitrarily chosen: 15 cm (height) and 5 cm (width). To preserve the stability of the specimen, the node in the middle of the lower edge was kept fixed. The deformations were initiated through constant vertical displacement increments prescribed to nodes along the upper edge of the specimen. The lower and upper edges were smooth. The number of triangular finite elements was 896 (the size of elements was not greater than $3\times l_c$). The material constants were: E=30 GPa, ν=0.18, σ_{yc}^{0}=20 MPa, ϕ=25°, ψ=10°, η_1=1.2, η_2=0.7, δ=800, l_c=5 mm, a_t=0.0 and a_c=1.0. To induce strain localization, a weak element was inserted in the middle of height, on edge of the specimen. Due to the lack of the initial experimental data, the calculated stress-strain curve was qualitatively compared with the experimental one by Karsan and Jirsa (1969) (Fig. 6.17).

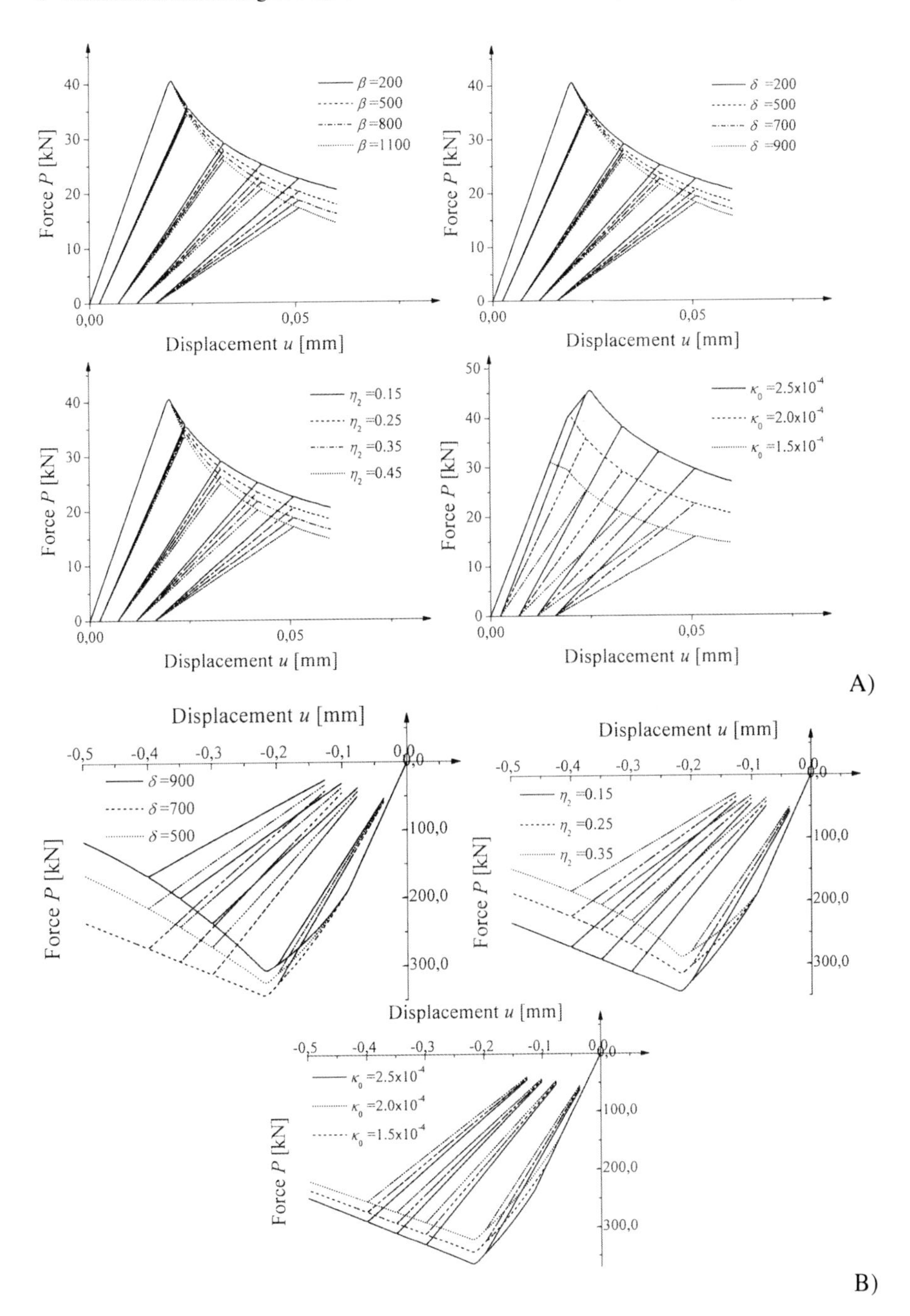

Fig. 6.13 Effect of different material constants on uniaxial response of coupled elasto-plastic-damage model '4' for concrete under: A) cyclic uniaxial tension and B) cyclic uniaxial compression (with damage scale factors a_t=0.0 and a_c=1.0) (Marzec and Tejchman 2010)

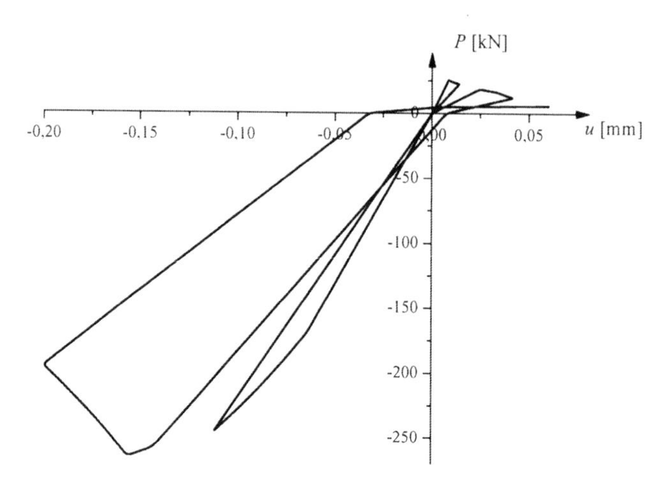

Fig. 6.14 Calculated load-displacement curve with coupled model '4' (with damage scale factors a_t=0.0 and a_c=1.0) during uniaxial tension-compression-tension (Marzec and Tejchman 2010)

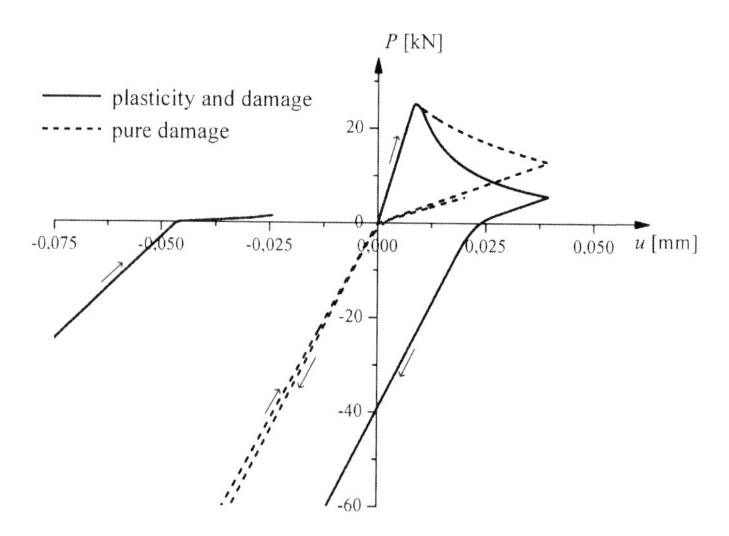

Fig. 6.15 Calculated load-displacement curves with coupled model '4' during uniaxial tension-compression-tension with and without plastic strains (Marzec and Tejchman 2010)

The calculated stress-strain curve (Figs. 6.17c and 6.17d) is qualitatively the same as in a cyclic compressive test by Karsan and Jirsa (1969) (Figs. 6.17c and 6.17e) with respect to material softening and stiffness degradation. The calculated thickness of a localized zone is 3.4 cm (6.8×l_c) and the inclination to the horizontal is about 45° (Fig. 6.17a and 6.17b). These results are very similar to those within elasto-plastic calculations (Bobiński and Tejchman 2004). The shear zone inclination is significantly higher (and more realistic) than that obtained with a simple non-local isotropic damage model (Simone et al. 2002), which was smaller than 35°-40°.

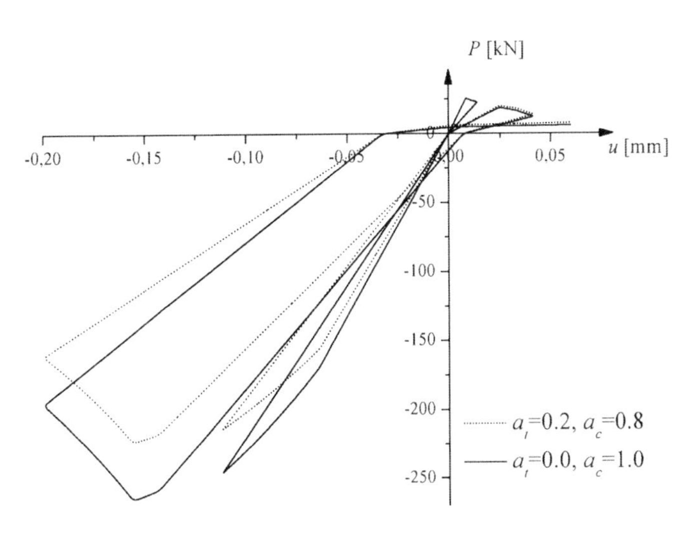

Fig. 6.16 Uniaxial response of coupled elasto-plastic-damage model '4' for concrete under tension-compression-tension for different damage scale factors a_t and a_c (Marzec and Tejchman 2010)

The following conclusions can be derived:

- The FE calculations show that the coupled elasto-plastic damage models used enhanced by a characteristic length of micro-structure in a softening regime can properly reproduce the experimental load-displacement diagrams and strain localization in plain concrete notched beams under tensile loading during quasi-static cyclic bending. All models '1-4' properly capture material softening and the width of a localized zone. The models '1', '3' and '4' are also able to correctly describe the stiffness degradation. The drawback of the model '2' is the lack of possibility to simulate simultaneously both plastic deformation and stiffness degradation during cyclic loading. The model '3' has the smallest number of material constants to be calibrated. The coupled models '3' and '4' indicate the best agreement with cyclic bending experiments under tensile failure. In general, the models 1, 3 and 4 show similar results under tension. The shape and thickness of a localized zone above the notch in concrete beams under tension depends on the coupled formulation.
- A choice of a suitable local state variable for non-local averaging strongly depends on the model used. It should be carefully checked to avoid problems with non-sufficient regularization.

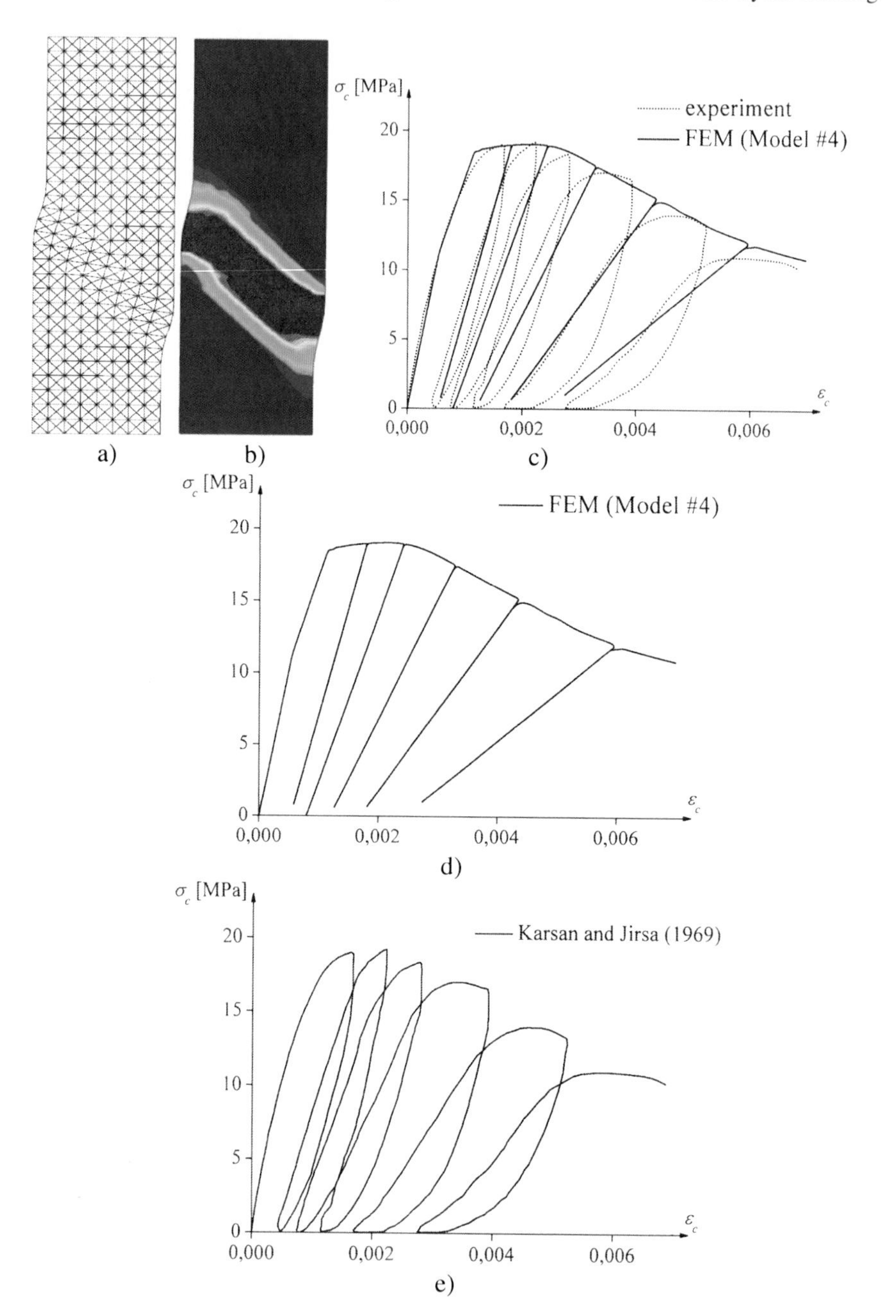

Fig. 6.17 Response of coupled elasto-plastic-damage model '4' for concrete specimen under uniaxial cyclic compression from FE calculations (with damage scale factors a_t=0.0 and a_c=1.0): a) deformed FE mesh, b) contours of calculated non-local parameter, c) calculated and experimental stress-strain curve by Karsan and Jirsa (1969), d) calculated stress-strain stress-strain curve, e) experimental stress-strain curve by Karsan and Jirsa (1969)

• The improved model '4' captures in addition plastic strains and stiffness degradation in both tension and compression, and stiffness recovery effect during cyclic loading by means of a strain equivalence hypothesis (thus the coupling between damage and plasticity is different than in ABAQUS 2004). It is able to properly describe strain localization under both tension and compression due to a presence of a characteristic length of micro-structure. Its drawback is no clear distinction between elastic, plastic and damage strain rates, and a relatively large number of material constants to be calibrated. Most of material constants may be calibrated independently with a monotonic uniaxial compression and tension (bending) test. Standard uniaxial cyclic tests are needed to calibrate damage scale factors.

References

ABAQUS, Theory Manual, Version 5.8, Hibbit. Karlsson & Sorensen Inc. (1998)

Bažant, Z.P., Planas, J.: Fracture and size effect in concrete and other quasibrittle materials. CRC Press LLC (1998)

Bažant, Z.P.: Scaling of Structural Strength. Hermes-Penton, London (2003)

Bobiński, J., Tejchman, J.: Numerical simulations of localization of deformation in quasi brittle materials within non-local softening plasticity. Computers and Concrete 1(4), 433–455 (2004)

Carol, I., Rizzi, E., Willam, K.: On the formulation of anisotropic elastic degradation. Int. J. of Solids and Structures 38(4), 491–518 (2001)

Hansen, N.R., Schreyer, H.L.: A thermodynamically consistent framework for theories of elastoplasticity coupled with damage. International Journal of Solids and Structures 31(3), 359–389 (1994)

Hansen, E., Willam, K.: A two-surface anisotropic damage-plasticity model for plane concrete. In: de Borst, R. (ed.) Procceedings Int. Conf. Fracture Mechanics of Concrete Materials, Paris, Balkema, pp. 549–556 (2001)

Hordijk, D.A.: Local approach to fatigue of concrete. PhD Thesis. Delft University of Technology (1991)

Karsan, D., Jirsa, J.O.: Behaviour of concrete under compressive loadings. Journal of the Structural Division (ASCE) 95(12), 2543–2563 (1969)

Marzec, I.: Application of coupled elasto-plastic-damage models with non-local softening to concrete cyclic behaviour. PhD Thesis, Gdańsk University of Technology (2009)

Marzec, I., Bobiński, J., Tejchman, J.: Simulations of crack spacing in reinforced concrete beams using elastic-plastic and damage with non-local softening. Computers and Concrete 4(5), 377–403 (2007)

Marzec, I., Tejchman, J.: Modeling of concrete behaviour under cyclic loading using different coupled elasto-plastic-damage models with non-local softening. In: Oñate, E., Owen, D.R.J. (eds.) X International Conference on Computational Plasticity-COMPLAS X, pp. 1–4. CIMNE, Barcelona (2009)

Marzec, I., Tejchman, J.: Application of enhanced elasto-plastic damage models to concrete under quasi-static and dynamic cyclic loading. In: Bicanic, N., de Borst, R., Mang, H., Meschke, G. (eds.) Modelling of Concrete Structures, pp. 529–536. Taylor and Francis Group, London (2010)

Marzec, I., Tejchman, J.: Application of coupled elasto-plastic-damage models with non-local softening to cyclic concrete behaviour. Archives of Mechanics (2011) (under review)

Meschke, G., Lackner, R., Mang, H.A.: An anisotropic elastoplastic-damage model for plain concrete. International Journal for Numerical Methods in Engineering 42(4), 702–727 (1998)

Nguyen, G.D.: A thermodynamic approach to constitutive modelling of concrete using damage mechanics and plasticity theory, PhD Thesis, Trinity College, University of Oxford (2005)

Pamin, J., de Borst, R.: Stiffness degradation in gradient-dependent coupled damage-plasticity. Archives of Mechanics 51(3-4), 419–446 (1999)

Peerlings, R.H.J.: Enhanced damage modeling for fracture and fatigue. PhD Thesis, TU Eindhoven, Eindhoven (1999)

Perdikaris, P.C., Romeo, A.: Size effect on fracture energy of concrete and stability issues in three-point bending fracture toughness testing. ACI Material Journal 92(5), 483–496 (1995)

Simone, A., Wells, G.N., Sluys, L.J.: Discontinuous modelling of crack propagation in a gradient enhanced continuum. In: Proc. of the Fifth World Congress on Computational Mechanics, WCCM V, Vienna (2002)

van Vliet, M.R.A., van Mier, J.G.M.: Experimental investigation of size effect in concrete and sandstone under uniaxial tension. Engineering Fracture Mechanics 65(2-3), 165–188 (2000)

Vorechovsky, M.: Interplay of size effects in concrete specimens under tension studied via computational stochastic fracture mechanics. International Journal of Solids and Structures 44(9), 2715–2731 (2007)

Chapter 7
Modelling of Fracture in Reinforced Concrete under Monotonic Loading

Abstract. In this Chapter, the numerical analyses of reinforced concrete bars, beams, columns, corbels and tanks were performed using three enhanced constitutive continuum approaches for concrete: isotropic elasto-plastic model, isotropic damage model and smeared crack model with non-local softening (Chapters 3.1 and 3.3). Attention was paid to strain localization developed in concrete.

7.1 Bars

The 3D elasto-plastic calculations were performed with a horizontal reinforced concrete bar subjected to tension (Małecki et al. 2007, Widuliński et al. 2009) using a large-displacement analysis by Hughes and Winget 1980 (ABAQUS 1998). The element was l=100 cm long, mainly with a cross-section of 100×100 mm^2. Totally, 1250 cube elements were used (Fig. 7.1). The calculations were performed also with elements with a cross-section of 200×200=40000 mm^2 (using 5000 cube elements) and 400×400=160000 mm^2 (using 2500 cube elements). The rod was fixed at the left side. Concrete was modelled with an elasto-plastic constitutive law with non-local softening. A Rankine criterion with non-local isotropic softening and associated flow rule was adopted in a tensile regime of concrete (Eqs. 3.32, 3.40, 3.93 and 3.99). The following material parameters were assumed for concrete: E_c=30.0 GPa, υ_c=0.20 and f_t=2.0 MPa. The tensile strength f_t was assumed according to a Gaussian (normal) distribution around the mean value 2 MPa using a standard deviation s_d=0.05 MPa and a cut-off c_f=±0.1 MPa. To obtain a Gaussian distribution of the concrete strength, a polar form of the so-called Box-Muller transformation (1958) was used. The curves describing the evolution of the tensile yield stress σ_t versus the softening parameter κ_2 are shown in Fig. 7.2A. A linear relationship σ_t=f(κ_2) (Fig. 7.2A) or a non-linear relationship σ_t=f(κ_2) according to Hordijk (1991) (Fig. 7.2B) were assumed. Most of the calculations were carried out with the linear softening curve 'a' of Fig. 7.2A with

J. Tejchman, J. Bobiński: Continuous & Discontinuous Modelling of Fracture, SSGG, pp. 183–296.
springerlink.com

the softening modulus in tension of H_t=0.67 GPa and the parameter κ_u=0.003, wherein κ_u – ultimate value of κ associated with a total loss of the load bearing capacity. The fracture energy for 3 different linear softening functions was approximately: $G_f=g_f \times w_c$=0.075÷0.750 N/mm (g_f – area under the softening function, w_f=25-40 mm - localized zone width for l_c=5 mm). In turn, for non-linear softening, it was equal to G_f=0.90 N/mm. A characteristic length was assumed in the range of l_c=5-15 mm. The non-locality parameter was chosen as m=2. The reinforcement (modelled as 1D-elements) was located in the middle of the concrete bar. An elasto-perfect plastic constitutive law was assumed to model the reinforcement behaviour with E_s=210 GPa, σ_y=440 MPa and υ_s=0.3 (σ_y - yield stress). The calculations were carried out with perfect bond and bond-slip. In the first case, the same displacements along the contact line were assumed for concrete and reinforcement. In the case of bond-slip, the analyses were carried out with an assumption between the bond shear stress τ_b and slip u using two the bond law according to Dörr (1980) (Eqs. 3.114 and 3.115). The results with the bond law by den Uijl and Bigaj (1996) (Chapter 3.4) were similar due to the fact that bond traction values were far from the limiting value (thus, the shape of the law after the peak was unimportant). To consider bond-slip, an interface with a zero thickness was assumed along the contact surface where a relationship between the shear traction and slip was introduced.

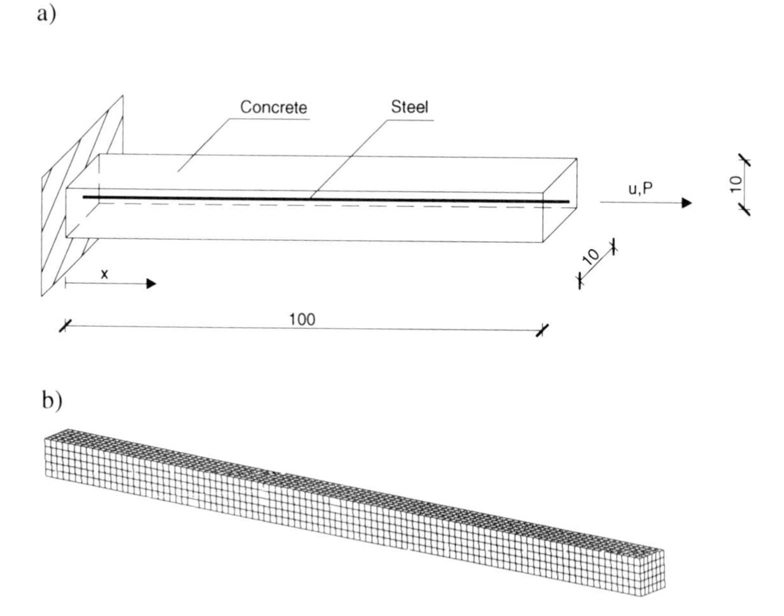

Fig. 7.1 Reinforced concrete bar in pure tension: a) schematically (units in mm), b) FE-mesh (Widuliński et al. 2009)

Effect of characteristic length l_c

Figure 7.3 shows the load-displacement curves for the reinforcement ratio ρ=0.5% using a bond-slip bond by Dörr (1980) with u_o=0.06 mm (Eqs. 3.114 and 3.115) (u_0 - displacement at which perfect slip occurs) at the beginning of a tensile test up to u=4.5 mm assuming the different characteristic lengths l_c=5-15 mm (using a linear softening curve 'a' of Fig. 7.2A and assuming a stochastic distribution of the tensile strength).

The horizontal tensile force increases with increasing characteristic length. The width of localized zones w_c increases with increasing l_c and is about $w_c=5\times l_c$=25 mm for l_c=5 mm, $w_c=4\times l_c$=40 mm for l_c=10 mm and $w_c=3.5\times l_c$=52 mm for l_c=15 mm. In turn, the spacing s_c of localized zones also increases and is $s_c=11\times l_c$=55 mm for l_c=5 mm, $s_c=10\times l_c$=100 mm for l_c=10 mm and $s_c=9\times l_c$=137 mm for l_c=15 mm.

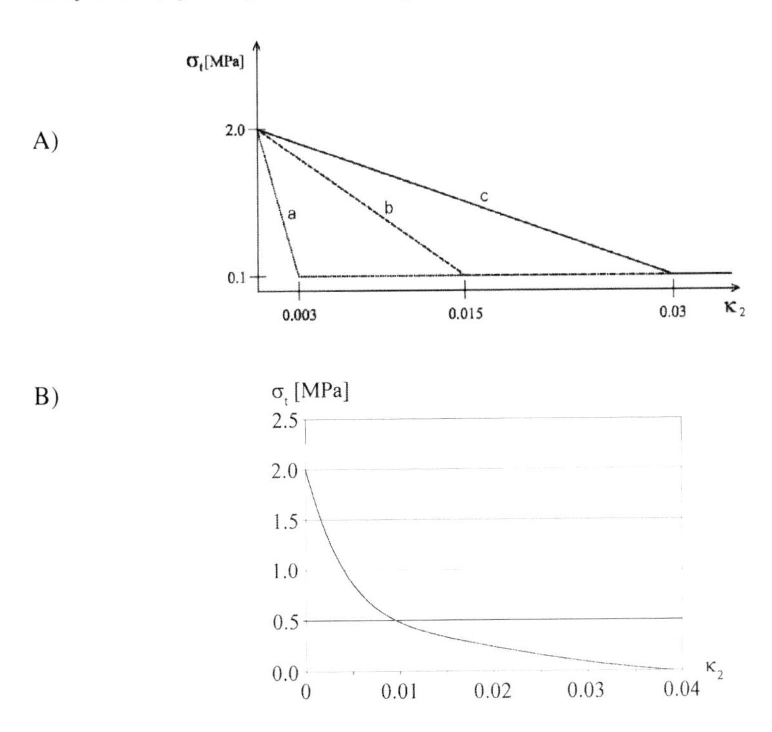

Fig. 7.2 Assumed softening curves $\sigma_t=f(\kappa_2)$ in tensile regime (σ_t – tensile yield stress, κ_2 – softening parameter): A) curves with linear softening, B) curve with non-linear softening by Hordijk (1991) (Widuliński et al, 2009)

The calculated spacing of localized zones s_c was compared with the average crack spacing s according to CEB-FIP Model Code (1991)

$$s_c = \frac{2}{3}\times\frac{\phi_s}{3.6\rho} = \frac{2}{3}\times\frac{8}{3.6\times 0.02} = 74 \text{ mm}, \tag{7.1}$$

Eurocode 2 (1991)

$$s_c = 50 + k_1 k_2 \frac{\phi_s}{4\rho} = 50 + 0.8 \times 1 \frac{8}{4 \times 0.02} = 130 \text{ mm}, \tag{7.2}$$

and the formula by Lorrain et al. (1998)

$$s_c = 1.5c + 0.1 \frac{\phi_s}{\rho} = 1.5 \times 45 + 0.1 \frac{8}{0.02} = 108 \text{ mm}, \tag{7.3}$$

wherein ϕ_s=8 mm is the reinforcing bar diameter, ρ=2% denotes the reinforcement ratio, k_i are the coefficients and c denotes the concrete cover. The calculated spacing of localized zones with l_c=5 mm (s_c=11×l_c=55 mm) is too small and with l_c=10 mm (s_c=10×l_c=100 mm) and l_c=15 mm (s_c=9.5×l_c=137 mm) is too large with the average crack spacing according to Eq. 7.1. In the case of a direct comparison with Eqs. 7.2 and 7.3, the best agreement was achieved with l_c=10-15 mm, respectively.

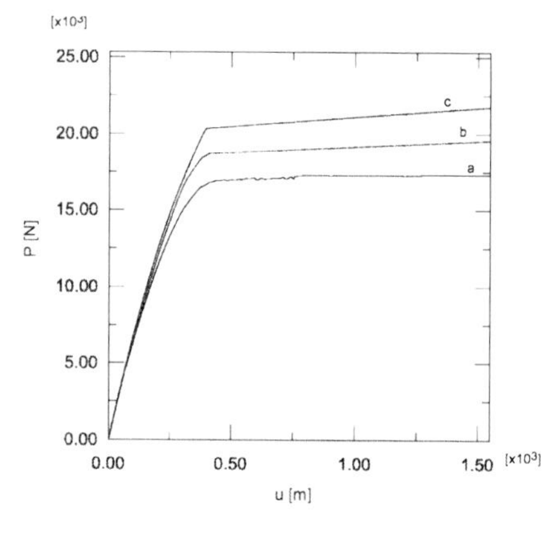

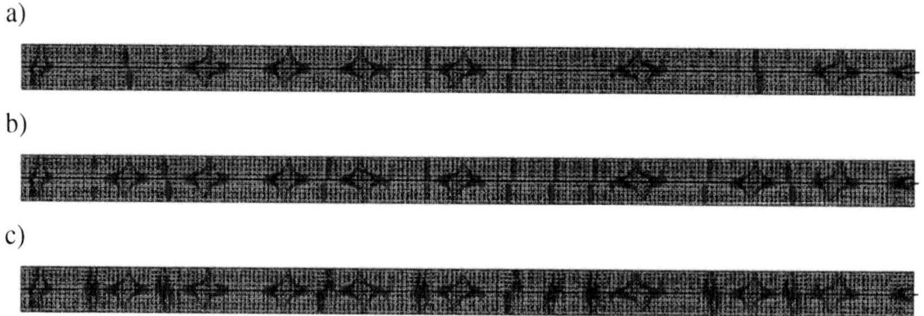

Fig. 7.3 Calculated lad-displacement curves and distributions of nonlocal softening parameter within elasto-plasticity for different characteristic lengths at u=1 mm (ρ=0.5%, u_o=0.06 mm): a) l_c=5 mm, b) l_c=10 mm, c) l_c=15 mm (Widuliński et al, 2009)

Effect of reinforcement ratio

The effect of reinforcement ratio ρ in the range of 0.01%-2.0% on the load-displacement curves and spacing of localized zones is depicted in Figs. 7.4A and 7.5 for l_c=5 mm (softening curve 'a' of Fig. 7.2A, u_o=0.06 mm). An increase of the reinforcement ratio increases the overall horizontal tensile force P. The width of localized zones does not depend on ρ (it is about w_c=5×l_c). In the case of ρ=0.01%, a localized zone occurs only at the right side of the bar. The distance between localized zones slightly increases with decreasing ρ at the beginning of deformation at u=1 mm (it is approximately s_c=11×l_c for ρ=0.5% and s_c=10×l_c for ρ=2.0%).

The resultant horizontal force with the Poisson's ratio for reinforcement υ_s=0.3 was by 10% larger than in the case of υ_s=0 due to the assumption of plane strain (Fig. 7.4B).

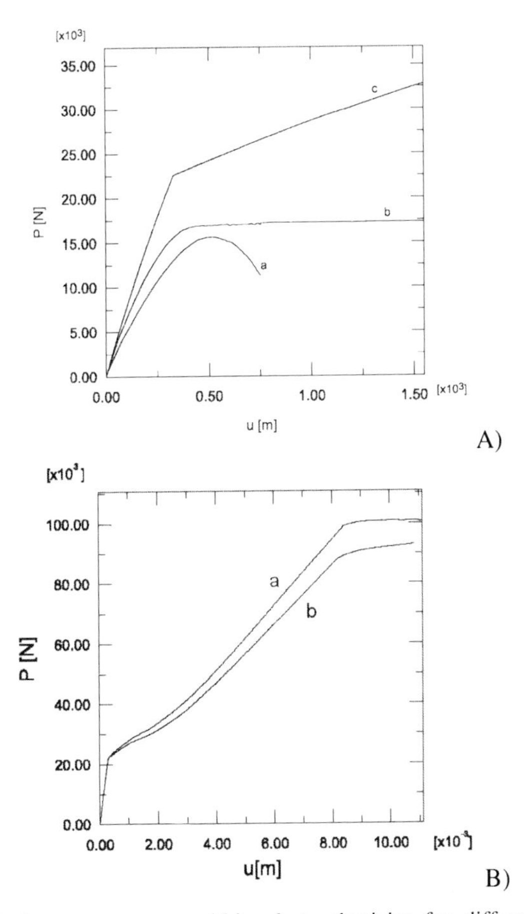

Fig. 7.4 Load-displacement curves within elasto-plasticity for different reinforcement ratios: A) for horizontal displacements u=0-1.5 mm (a) l_c=5 mm, u_o=0.06 mm, ρ=0.01%, b) ρ=0.5%, c) ρ=2.0%), B) for u=0-10 mm (a) (l_c=15 mm, u_o=0.06 mm, ρ=2.0%, υ_s=0.3, υ_s=0) (Małecki et al. 2007, Widuliński et al. 2009)

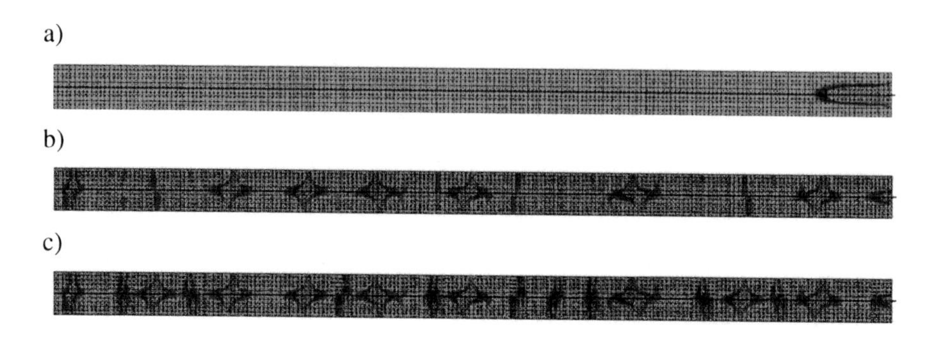

Fig. 7.5 Distributions of the nonlocal softening parameter within elasto-plasticity for different reinforcement ratios at u=1 mm (l_c=5 mm, u_o=0.06 mm): a) ρ=0.01%, b) ρ=0.5%, c) ρ=2.0% (Widuliński et al. 2009)

Effect of cross-section size

The effect of the cross-section on both the horizontal normalized tensile force and spacing of localized zones is insignificant.

Effect of confining compressive pressure

The calculations were performed with ρ=2.0%, l_c=5 mm and u_o=0.06 mm (Fig. 7.6). The effect of confining compressive pressure on both the horizontal tensile force and spacing of localized zones is insignificant.

Effect of non-locality parameter

The calculations were performed with ρ=2.0%, l_c=5 mm (u_o=0.06 mm) and two different non-locality parameters m: m=2 and m=3 (Fig. 7.7). An increase of the parameter m causes an increase of P, w_c (from $w_c=4\times l_c$=20 mm for m=2 up to $w_c=8\times l_c$=40 mm for m=3) and s_c (s_c=33 mm for m=2 and s_c=38 mm for m=3).

Effect of distribution of tensile strength

Figure 7.8 shows the influence of the distribution of the tensile strength on the load displacement curve (l_c=5 mm, ρ =2.0%, m=2, u_o=0.06 mm). The calculations were carried out with a uniform distribution of the tensile strength. The results indicate that the distribution of the tensile strength practically does not affect the overall horizontal tensile force and the width and spacing of localized zones ($w_c=5\times l_c$ and $s_c=11\times l_c$). However, if the distribution of the tensile strength is uniform, the localized zones develop later.

Effect of softening curve shape

The FE-results for two different softening functions: linear (function 'a' of Fig. 7.2A) and non-linear one by Hordijk (1991) (Fig. 7.2B) are demonstrated in Fig. 7.9 (l_c=5 mm, ρ=2.0%, u_o=0.06 mm). In both cases, the area under the softening functions was the same. The overall horizontal tensile force is slightly larger with linear softening. The spacing of localized zones increases from $9\times l_c$ for linear softening 'a' of Fig. 7.2A up to $11\times l_c$ for non-linear softening of Fig. 7.2B. The localized zone width remains similar ($5\times l_c$).

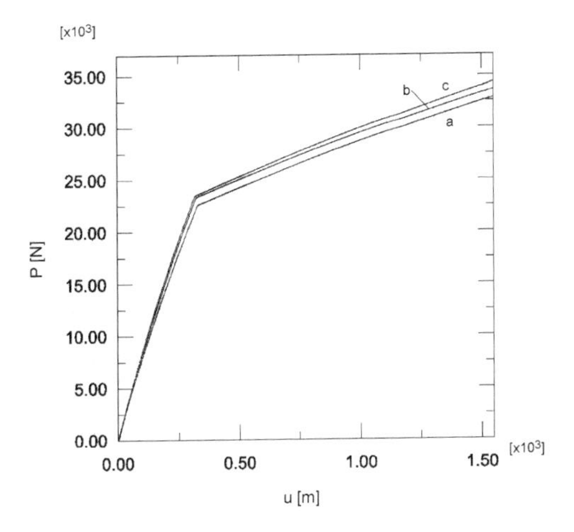

Fig. 7.6 Load-displacement curves for different confining pressures (ρ=2.0%, l_c=5 mm, u_o=0.06 mm): a) 0 MPa, b) 5 MPa, c) 10 MPa (Widuliński et al. 2009)

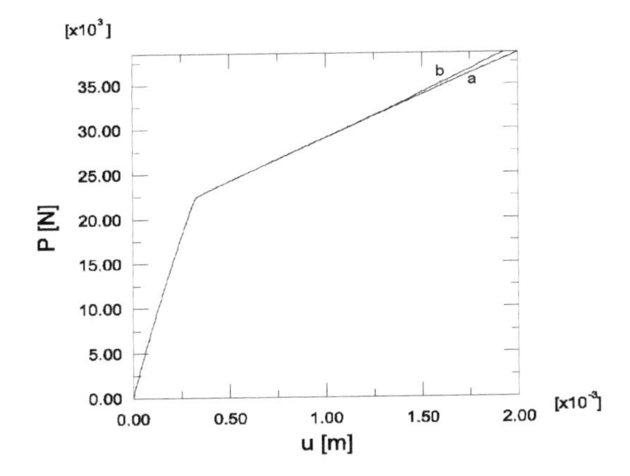

Fig. 7.7 Load-displacement curves for different non-local parameters m (l_c=5 mm, ρ=2.0%, u_o=0.06 mm): a) m=2, b) m=3 (Widuliński et al. 2009)

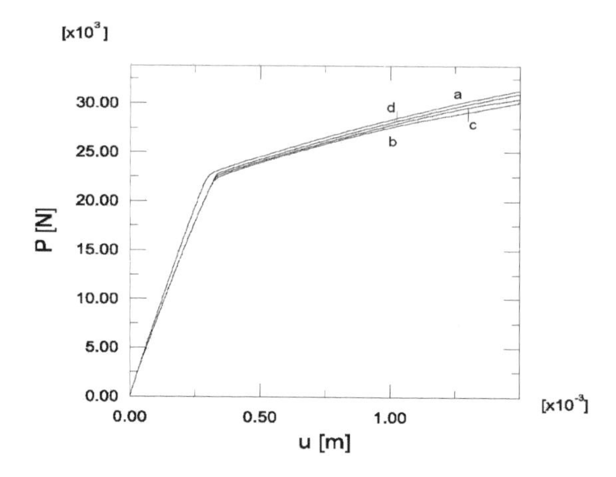

Fig. 7.8 Load-displacement curves for different tensile strength distributions (l_c=5 mm, ρ=2.0%, u_o=0.06 mm): a) uniform, b) stochastic with standard deviation s_d=0.05 MPa, c) stochastic with standard deviation s_d=0.03 MPa, d) stochastic with standard deviation s_d=0.01 MPa (Widuliński et al. 2009)

Effect of stiffness of bond-slip

The effect of the initial bond stiffness in the bond-slip law by Dörr (1980) (Fig. 7.10) for l_c=5 mm and ρ=2.0% (u_o=0.06-0.6 mm) was investigated (Fig. 7.11). Since the bond traction values are far from the limiting value, the cracking process is influenced by the initial bond stiffness only (i.e. by u_o).

The overall horizontal tensile force P increases with increasing initial bond stiffness. The spacing of localized zones increases with decreasing initial bond stiffness (from 10×l_c for the curve 'a' of Fig. 7.10 up to 16×l_c for the curve 'e' of Fig. 7.10). In turn, the width of localized zones increases only for the lowest initial bond stiffness from 5×l_c for the curves 'a-d' of Fig. 7.10 up to 6×l_c for the curve 'e' of Fig. 7.10.

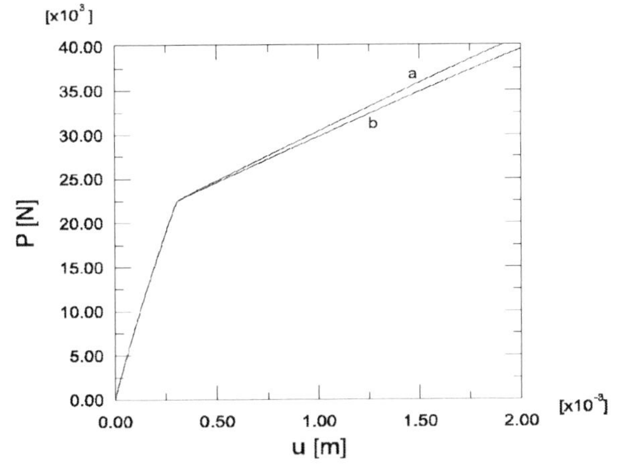

a)

b)

Fig. 7.9 Load-displacement curves and distributions of the nonlocal softening parameter for different softening curves at u=1 mm (l_c=5 mm, ρ=2.0%, u_o=0.06 mm): a) linear softening of Fig. 7.2A with κ_u=0.0015, b) non-linear softening by Hordijk (1991) (Fig. 7.2B) (Widuliński et al. 2009)

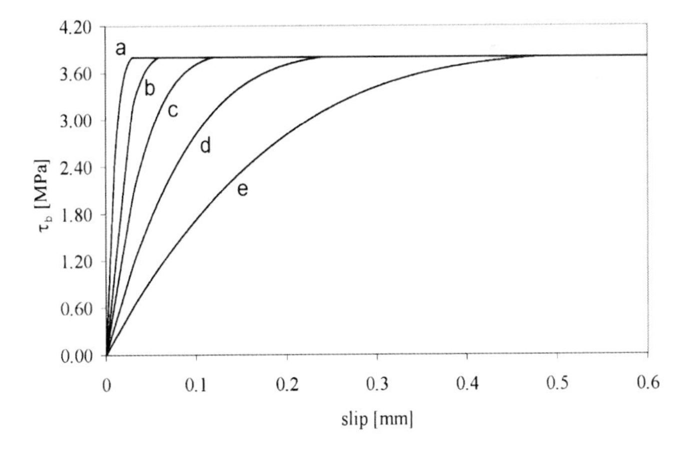

Fig. 7.10 Bond-slip law by Dörr (1980) with different initial bond stiffness (f_t=2.0 MPa): a) u_o=0.03 mm, b) u_o=0.06 mm, c) u_o=0.12 mm, d) u_o=0.24 mm, e) u_o=0.48 mm (u_0 - displacement at which perfect slip occurs) (Widuliński et al. 2009)

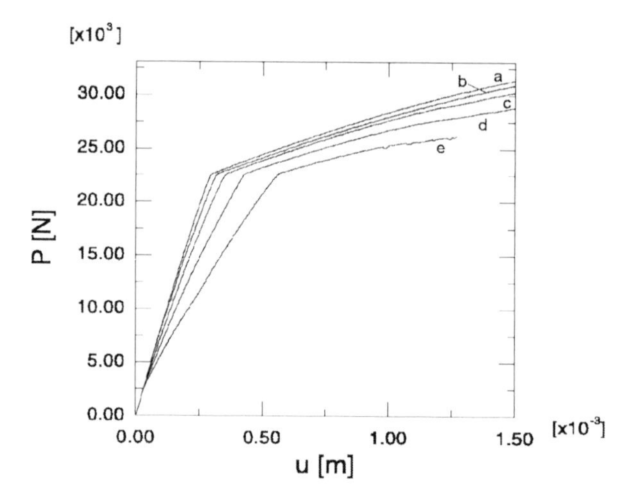

a)

d)

e)

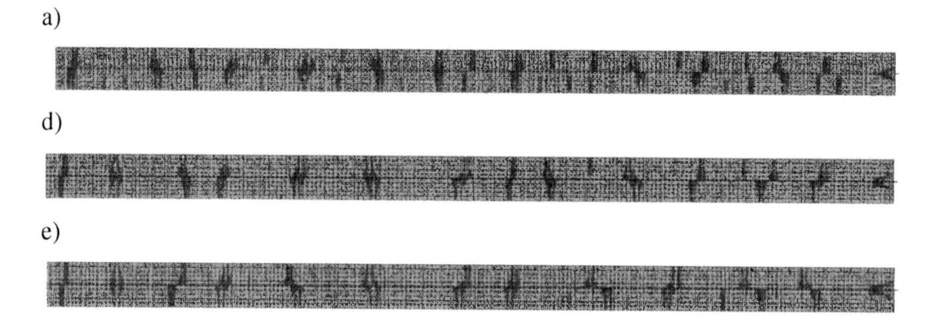

Fig. 7.11 Load-displacement curves and distribution of non-local softening parameter for different initial stiffness of bond slip law by Dörr (1980) at u=1 mm (l_c=5 mm, ρ=2.0%): a) curve 'a' of Fig. 7.10 (u_o=0.03 mm), b) curve 'b' of Fig. 7.10 (u_o=0.06 mm), c) curve 'c' of Fig. 7.10 (u_o=0.12 mm), d) curve 'd' of Fig. 7.10 (u_o=0.24 mm), e) curve 'e' of Fig. 7.10 (u_o=0.48 mm) (Widuliński et al. 2009)

Effect of tensile fracture energy

Figure 7.12 presents the effect of 3 different fracture energies of Fig. 7.2A for l_c=5 mm and ρ=2.0% (u_o=0.06 mm). The higher the fracture energy, the larger is the overall horizontal tensile force and the smaller the spacing of localized zones. The crack spacing decreases from 12×l_c for the curve 'a' of Fig. 7A (G_f=0.075 N/mm) down to 8×l_c for the curve 'c' of Fig. 7.2A (G_f=0.75 N/mm). The width of localized zones remains similar (5×l_c).

Effect of stirrups

Figure 7.13 presents the reinforced concrete bar with stirrups (the spacing was 10-20 cm). The results of Fig. 7.14 show the effect of stirrups on the load-displacement curve and spacing of localized zones (one assumed perfect bond between concrete and stirrups,). The results demonstrate that this effect is negligible.

The obtained FE-results are quantitatively in good agreement with numerical results by Pamin and de Borst (1998) dealing with a similar problem. Our results show that if the tensile strength of concrete is distributed stochastically, the results become similar with perfect bond and bond-slip. Only a significantly larger slip displacement is required for localized zones to evolve in the case of perfect bond. As compared to two-dimensional results under plane strain conditions (Malecki et al. 2007), the differences in FE results are negligible.

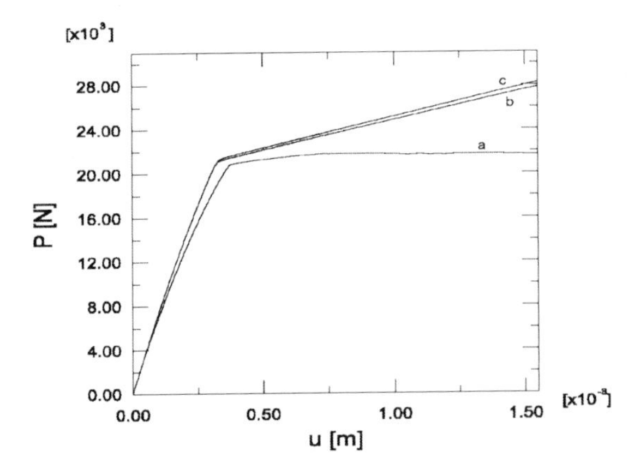

Fig. 7.12 Load-displacement curves for different softening rates at u=1 mm (l_c=5 mm, ρ=2.0%, u_o=0.06 mm): a) curve 'a' of Fig. 7.2A, b) curve 'b' of Fig. 7.2A, c) curve 'c' of Fig. 7.2A (Widuliński et al. 2009)

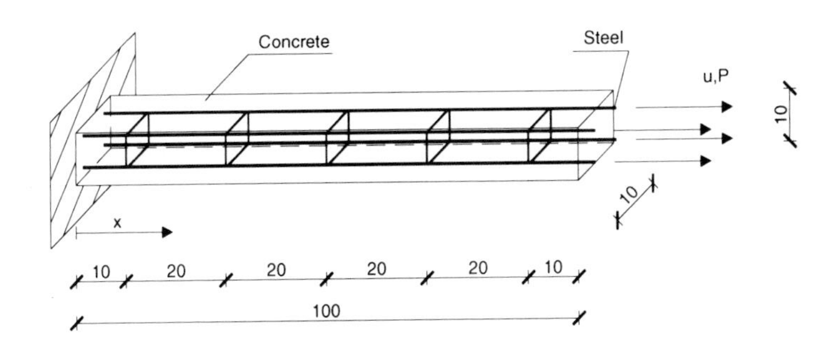

Fig. 7.13 Reinforced concrete bar with stirrups in pure tension (units in mm) (Widuliński et al. 2009)

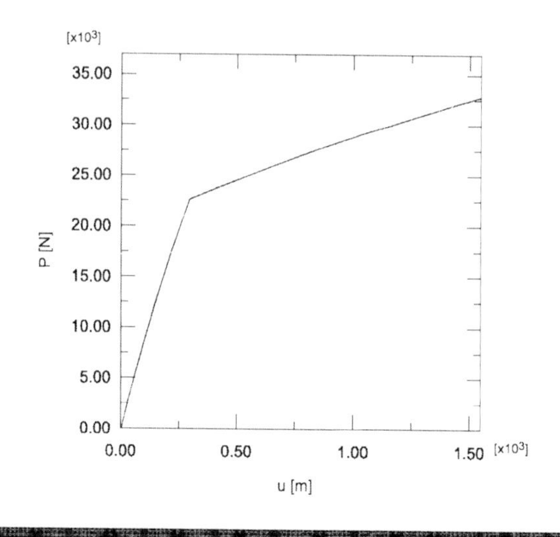

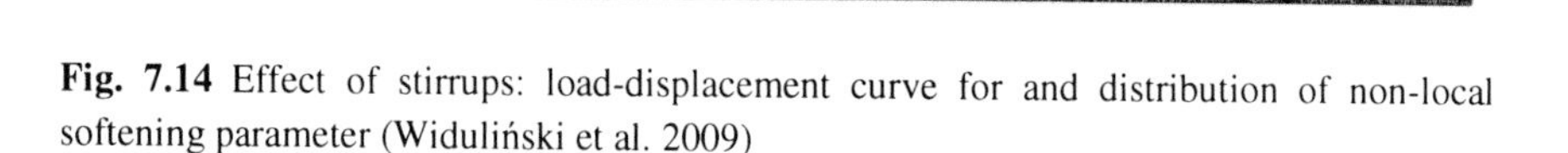

Fig. 7.14 Effect of stirrups: load-displacement curve for and distribution of non-local softening parameter (Widuliński et al. 2009)

The calculations revealed the following points:

- The maximum force increases with increasing reinforcement ratio, characteristic length, initial stiffness of the bond-slip, fracture energy and confining compressive pressure. It is insignificantly influenced by the distribution of the tensile strength in the considered range of the standard deviation and stirrups.
- The width of localized zones increases with increasing characteristic length and non-locality parameter. It increases insignificantly with initial bond stiffness. It does not depend on the reinforcement ratio, shape of the softening curve, distribution of the tensile strength and compressive confining pressure.
- The spacing of localized zones increases with increasing characteristic length, non-locality parameter and softening modulus, and decreasing reinforcement ratio, fracture energy and initial bond stiffness. It does not depend on the distribution of the tensile strength and stirrups.
- The width and spacing of localized zones are similar for perfect bond and usual bond-slip laws if the tensile strength is distributed stochastically. However, the localized zones occur later for perfect bond.
- The spacing of localized zones with l_c=5-10 mm seems to be in satisfactory agreement with code recommendations of CEB-FIP for the crack spacing.

7.2 Slender Beams

Quasi-static plane strain FE simulations of strain localization in long reinforced concrete beams without stirrups were carried out (Marzec et al. 2007). The material was modelled with two different isotropic continuum models: an elasto-plastic and a damage one (Chapters 3.1 and 3.2). A non-local and second-gradient model were used as regularization techniques (Chapter 3.3). The numerical results were compared with the corresponding experimental ones by Walraven (1978).

Several tests were carried out for beams with free ends without shear reinforcement (Walraven 1978, Walraven and Lehwalter 1994) (Chapter 2). The experiments were carried out with 3 different beams with the same width of b=200 mm: h=150 mm, l=2300 mm (small-size beam '1'), h=450 mm, l=4100 mm (medium-size beam '2') and h=750 mm, l=6400 mm (large-size beam '3'). The average cube crushing strength of concrete was 34.2-34.8 MPa. In turn, the average cube splitting strength was 2.49-2.66 MPa. The maximum size of the aggregate in concrete was d_a=16 mm. The concrete cover measured from the bar centre to the concrete surface was 25 mm (beam 1) and 30 mm (beam '2' and '3'), respectively. The effective beam height d was: d=125 mm (beam '1'), d=420 mm (beam '2') and d=720 mm (beam '3'), respectively. The longitudinal reinforcement consisted of uncurtailed bars of deformed cold-drawn steel ratio (with the yielding strength of 440 MPa and reinforcement ratio of 0.79-0.83%): $1\times\phi_s 8$ and $2\times\phi_s 10$ (beam '1'), $1\times\phi_s 20$ and $2\times\phi_s 14$ (beam '2') and $3\times\phi_s 22$ (beam '3'). The beams were incrementally loaded by two symmetric vertical forces at the distance of 1000 mm at the shear span ratio of a/d=3 (a – distance between the vertical forces and beam supports: a=375 mm (beam '1'), a=1250 mm (beam '2') and a=2160 mm (beam '3').

In all beams, the shear-tension type of failure was observed. First, vertical cracks appeared at the beginning of loading. They opened perpendicularly first while later an increasing shear displacement was observed. The arising of an inclined crack leaded to failure. The ultimate vertical forces were: V=29.8 kN (beam '1'), V=70.6 kN (beam '2') and V=100.8 kN (beam '3'), respectively. Thus, a pronounced size effect took place since the normalized shear resistance force $V_n=V/bd$ was decreasing (almost linearly) with increasing effective height d: V_n=1.26 MPa (beam '1'), V_n=0.84 MPa (beam '2') and V_n=0.79 MPa (beam '3'). Due to that, the cracking pattern developed significantly faster in larger beams. In the experiments, main (high) and secondary (low) cracks appeared. The average spacing of main and secondary cracks was: 85 mm and 65 mm (small size beam), 180 mm and 60 mm (medium size beam) and 200 mm and 85 mm (large size beam), respectively.

The plane strain calculations were performed with 3 reinforced concrete beams without stirrups. 3200-22500 quadrilateral elements (composed of four diagonally crossed triangles) were used to avoid volumetric locking (Groen 1997), Fig. 7.15. The maximum element height, 15 mm, and element width, 23 mm, were not greater there than $3\times l_c$ (l_c=10-30 mm) in the region of strain localization in all beams to achieve mesh-objective results.

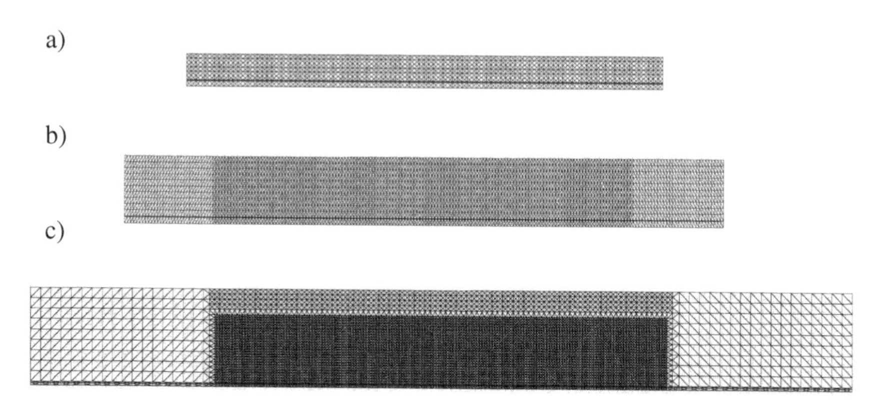

Fig. 7.15 FE-meshes used for calculations: a) small-size beam 1, b) medium-size beam 2, c) large-size beam 3 (Marzec et al. 2007)

FE results within isotropic elasto-plasticity with non-local softening

The following elastic material parameters were assumed for concrete: E_c=28.9 GPa and υ_c=0.20. To simplify calculations, linear relationships between the compressive σ_c and hardening and softening parameter κ_1, and linear softening between tensile stress σ_t versus softening parameter κ_2 were assumed within elasto-plasticity (Fig. 7.16). In the case of a tensile regime, 2 different linear softening curves were assumed with κ_2^u=0.003 and κ_2^u=0.006 (κ_2^u – ultimate value of κ_2 associated with a total loss of the load bearing capacity). The internal friction angle was equal to φ=12° (Eq. 3.27) and the dilatancy angle ψ=8°. (Eq. 3.30). The compressive strength was equal to f_c=34.2 MPa. The tensile strength f_t was taken from a Gaussian (normal) distribution around the mean value 2.49 MPa with a standard deviation 0.05 MPa and a cut-off ±0.1 MPa. To obtain a Gaussian distribution of the concrete strength, a polar form of the so-called Box-Muller transformation (1958) was used. The tensile fracture energy was $G_f=g_f \times w_c$=0.07-0.80 N/mm; g_f – area under the softening tensile function and $w_c \approx 3.5 \times l_c$ - crack zone width with l_c=5-30 mm. A characteristic length l_c was assumed in the range of l_c=5-30 mm. The non-locality parameter was m=2.

The reinforcement was assumed mainly as 2D elements. An elasto-perfectly plastic constitutive law by von Mises was assumed to model the reinforcement behaviour with E_s=210 GPa, ν=0.3, and σ_y=440 MPa (σ_y - yield stress). Thus, the size of these elements ($h_r \times b_r$) was 5×44 mm^2 (beam '1'), 8.8×70 mm^2 (beam '2') and 11×100 mm^2 (beam '3'). The width b_r was equal to the total perimeter of bars divided by 2 due to a contact from the both sides. The calculations were carried out mainly with bond-slip by Dőrr (1980) (Eqs. 3.114 and 3.115). The comparative calculations were also performed with reinforcement modelled as bar elements and as solid 2D-elements assuming the Poisson's ratios as υ_c=0 and υ_s=0 (to approximately induce a plane stress state).

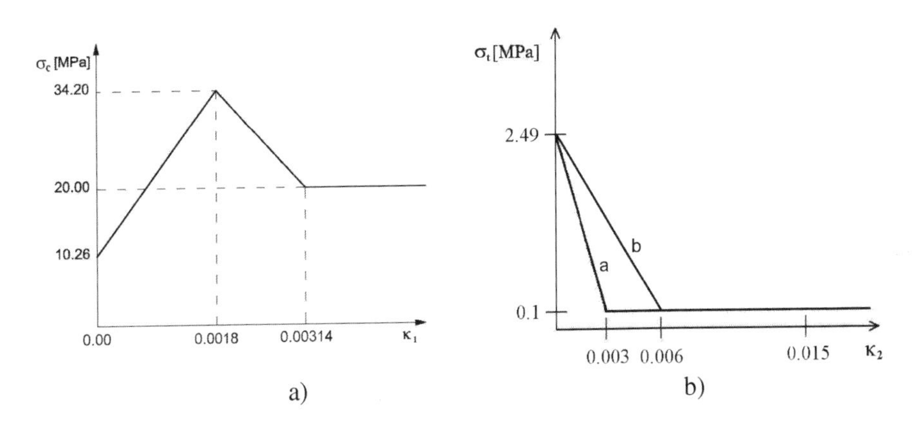

Fig. 7.16 Assumed curves $\sigma_c=f(\kappa_1)$ in compressive (a) and $\sigma_t=f(\kappa_2)$ in tensile regime (b) (σ_c – compressive stress, σ_t – tensile stress, κ_i – hardening-softening parameter) (Marzec et al. 2007)

Effect of characteristic length and fracture energy

Figures 7.17, 7.19 and 7.21 show the load-displacement curves for a medium, small and large size beam using the bond-slip law by Dörr (1980) for different characteristic lengths (l_c=5-30 mm) and two different fracture energies of Fig. 7.16b with κ_2^u=0.003 and κ_2^u=0.006 as compared to the experiments. The reinforcement was assumed as 2D-elements. The distribution of a nonlocal parameter in the beams is depicted in Figs. 7.18, 7.20 and 7.22 as compared to the experimental crack distribution at the ultimate load (Walraven 1978).

The calculated load-displacements curves are in a satisfactory agreement with the experimental ones, in particular for a smaller fracture energy (κ_2^u=0.003, G_f≈0.13 N/mm) and a smaller characteristic length (l_c=10 mm). The calculated ultimate vertical forces are always larger by 5-10% than the experimental ones. The bearing capacity of the beams increases with increasing l_c and fracture energy G_f. The geometry of strain localization is approximately in agreement with experiments (in particular with respect to main localized zones). There exist vertical and inclined localized zones, and high and low localized zones. The width of the localized zones is about w_c=(3-4)×l_c. In turn, the calculated average spacing s of main (high) zones is approximately s=120 mm (12×l_c) for a small-size beam, s=190-210 mm ((7-9)×l_c) for a medium-size beam and s=190-300 mm ((15-19)×l_c) for a large-size beam with κ_2^u=0.003. N/mm. It is s=75-100 mm ((10-15)×l_c) for a small-size beam and s=160-210 mm ((7-16)×l_c) for a medium-size beam at κ_2^u=0.006. Thus, the spacing of localized zones increases with increasing characteristic length and beam height, and decreasing fracture energy. In contrast to experiments, the height of localized zones is in FE-analyses slightly smaller and the number of inclined zones is also smaller.

The FE-results for a small-, medium- and large-size beam with ρ=0.75%, l_c=10 mm, κ_2^u=0.003 and reinforcement assumed as a 1D bar element are given in Figs. 7.23 and 7.24. In this case, the agreement with experimental results is even better. The calculated normalized ultimate shear resistance are: V_n^c=1.29 MPa (small size beam), V_n^c=0.90 MPa (medium size beam) and V_n^c=0.61 (large size beam) (Fig. 7.23). Thus, the ultimate forces differ only by 5% for all beams. Thus, the size effect is satisfactorily reproduced in the FE-analysis. The average spacing of main localized zones (l_c=10 mm, κ_2^u=0.003) is: s=70 mm (7×l_c) (small-size beam), s=170 mm (17×l_c) (medium-size beam) and s=190 mm (19×l_c) for a large-size beam is also close to the experimental outcomes: 85 mm (small-size beam), 160 mm (medium-size beam) and 200 mm (large-size beam). In the case of a large-size beam, except of main localized zones, secondary ones can be observed. The spacing of all (main and secondary) localized zones is about 90 mm (19×l_c) which is also in a good accordance with the experiment (85 mm).

The calculated spacing of localized zones s was also compared with the average crack spacing according to CEB-FIP Model Code (1991)

$$s=\frac{2}{3}\times\frac{\phi_s}{3.6\rho}=\frac{2}{3}\times\frac{9}{3.6\times 0.0075}=223\ \text{mm} \quad \text{(small-size beam)}, \tag{7.4}$$

$$s=\frac{2}{3}\times\frac{\phi_s}{3.6\rho}=\frac{2}{3}\times\frac{16}{3.6\times 0.0075}=395\ \text{mm} \quad \text{(medium-size beam)}, \tag{7.5}$$

$$s=\frac{2}{3}\times\frac{\phi_s}{3.6\rho}=\frac{2}{3}\times\frac{22}{3.6\times 0.0075}=543\ \text{mm} \quad \text{(large-size beam)}, \tag{7.6}$$

and the formula by Lorrain et al. (1998)

$$s=1.5c+0.1\frac{\phi_s}{\rho}=1.5\times 20.5+0.1\frac{9}{0.0075}=150\ \text{mm} \quad \text{(small-size beam)}, \tag{7.7}$$

$$s=1.5c+0.1\frac{\phi_s}{\rho}=1.5\times 22.0+0.1\frac{16}{0.0075}=246\ \text{mm} \quad \text{(medium-size beam)}, \tag{7.8}$$

$$s=1.5c+0.1\frac{\phi_s}{\rho}=1.5\times 19+0.1\frac{22}{0.0075}=322\ \text{mm} \quad \text{(large-size beam)}, \tag{7.9}$$

wherein ϕ_s=9 mm, ϕ_s=16 mm and ϕ_s=22 mm are the mean reinforcing bar diameters in a small, medium and large beam, ρ=0.75% denotes the reinforcement ratio and c denotes the concrete cover. The calculated and experimental spacing of localized zones is significantly smaller than these obtained with different analytical formulas. The effect of a characteristic length and fracture energy on the spacing of localized zones is similar as in the calculations of by Pamin and de Borst (1998) for a medium-size beam using a second gradient-enhanced crack model with a Rankine failure surface approximated by a circular function in the

A)

B)

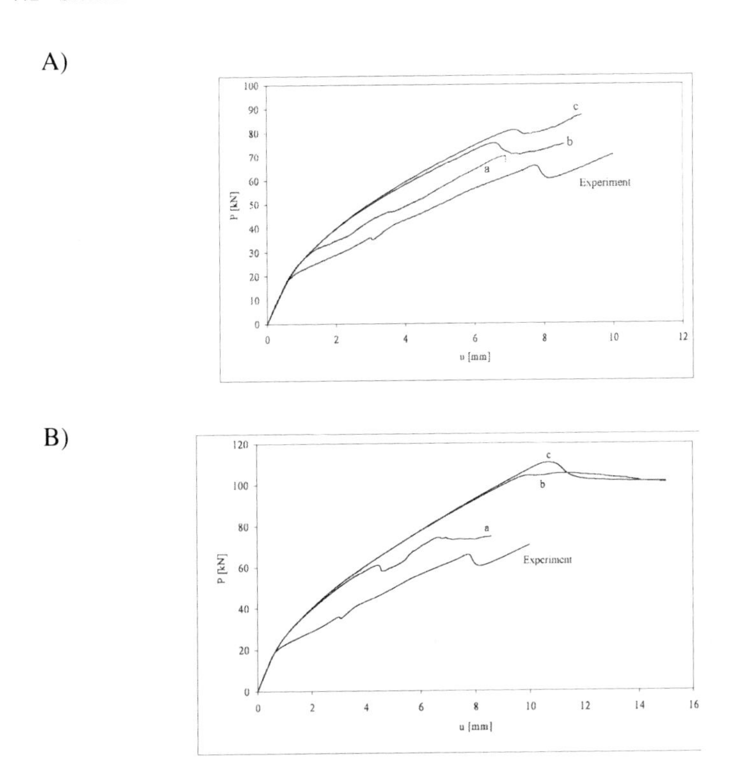

Fig. 7.17 Calculated load-displacement curves for medium-size beam (h=450 mm, bond-slip of by Dörr (1980), ρ=0.75%, a/d=3) as compared to experiments by Walraven (1978) (P – vertical resultant force, u – vertical displacement): A) κ_u=0.003, B) κ_u=0.006, a) l_c=10 mm, b) l_c=20 mm, c) l_c=30 mm (Marzec et al. 2007)

tension-tension regime under plane stress conditions. The geometry of localized zones is also similar.

Effect of tensile strength

The calculations were carried out with the different tensile strength f_t using a linear softening curve 'a' of Fig. 7.16b (bond-slip by Dörr (1980), a/d=3, medium beam, l_c=20 mm, κ_2^u=0.003, ρ=0.75%, reinforcement as 2D elements). The tensile strength was f_t=2.49 MPa, f_t=2.66 MPa and f_t=2.90 MPa, respectively. An increase of the tensile strength obviously causes a linear increase of the ultimate shear force; from V_u=75.5 kN (f_t=2.49 MPa) and V_u=77.5 kN (f_t=2.66 MPa) up to V_u=80.0 kN (f_t=2.9 MPa). Such linear dependency is in accordance with experiments (Kani 1966).

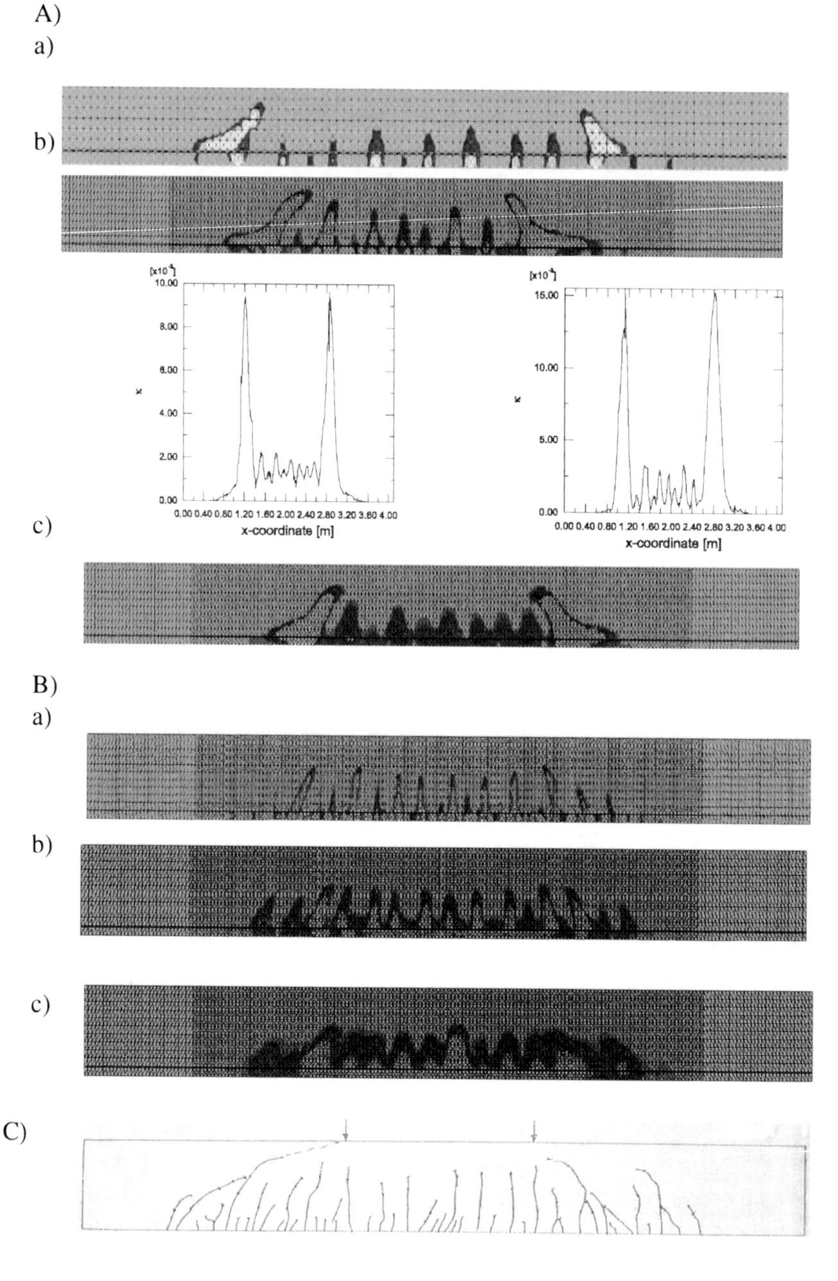

Fig. 7.18 Distribution of non-local softening parameter in medium-size beam at vertical displacement of u=8.5 mm (h=450 mm, bond-slip by Dörr (1980), ρ=0.75%, a/d=3) compared to experiments by Walraven (1978) (C): A) κ_u=0.003, B) κ_u=0.006, a) l_c=10 mm, b) l_c=20 mm, c) l_c=30 mm (Marzec et al. 2007)

Effect of reinforcement ratio

The effect of the reinforcement ratio in a medium beam for l_c=20 mm (using the bond-slip law by Dörr (1980) and curve '*b*' of Fig. 7.16b) was investigated for ρ=1.5% and ρ=2.0%. The spacing of localized zones decreases with increasing ρ (s=190 mm for ρ=0.75%, s=160 mm for ρ=1.5% and s=140 mm for ρ=2.0%). Thus, the effect of ρ on s is significantly stronger than in the bar (Chapter 7.1). The width of localized zones is always the same (3-4)×l_c.

The FE calculations were also performed with υ_s=0 (ρ=0.75%) and υ_c=0 and υ_s=0 (ρ=1.50%). In the first case, the ultimate shear resistance was smaller by 5% (from 70 kN down to 67 kN) and in the second case was reduced by 10% (from 119.2 kN down to 110.1 kN).

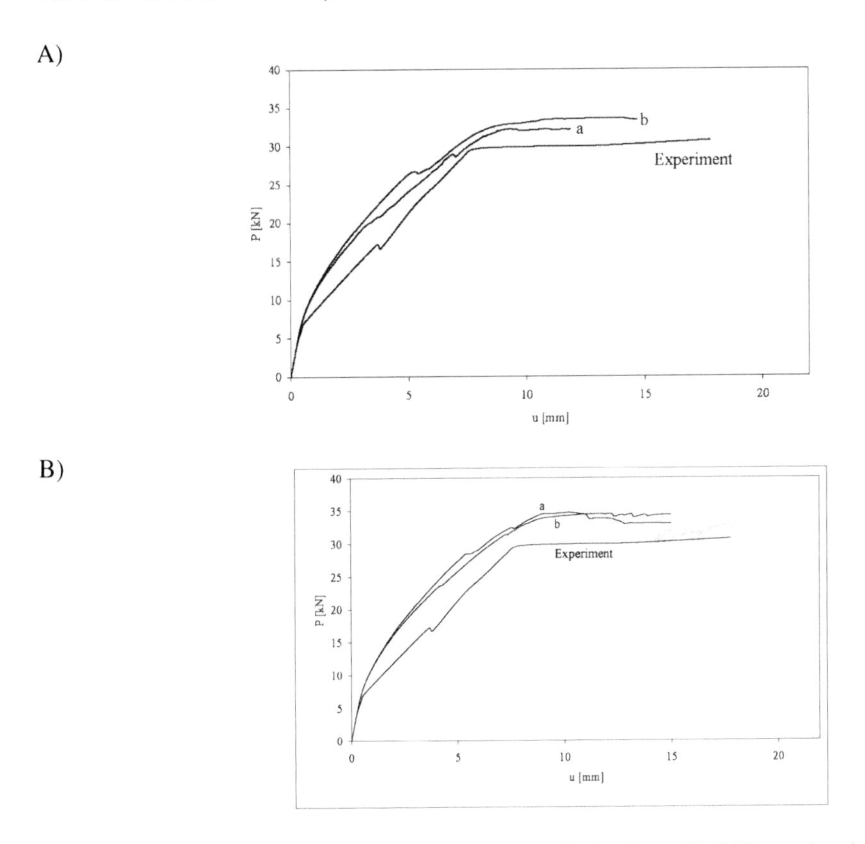

Fig. 7.19 Calculated load-displacement curves for small-size beam (h=150 mm, bond-slip by Dörr (1980), ρ=0.75%, a/d=3) as compared to experiments by Walraven (1978) (P – vertical resultant force, u – vertical displacement, 'e' – experiment by Walraven, 1978): A) κ_u=0.003, B) κ_u=0.006, a) l_c=5 mm, b) l_c=10 mm (Marzec et al. 2007)

Effect of shear span ratio

The effect of the shear span ratio in the range of a/d=2-3.5 on the distribution of localized zones is shown in Fig. 7.25 for a medium beam using a bond-slip by Dörr (1980) (ρ=0.75%, l_c=20 mm, κ_2^u=0.003, reinforcement as 2D elements).

The ultimate shear resistance force V obviously decreases with increasing distance of vertical forces from the supports a/d (from V_u=95.1 kN for a/d=2, V_u=75.5 kN for a/d=3 down to V_u=60 kN for a/d=3.5). The dependency is parabolic what is in accordance with experiments (Kani 1966) and calculations (Jia et al. 2006).

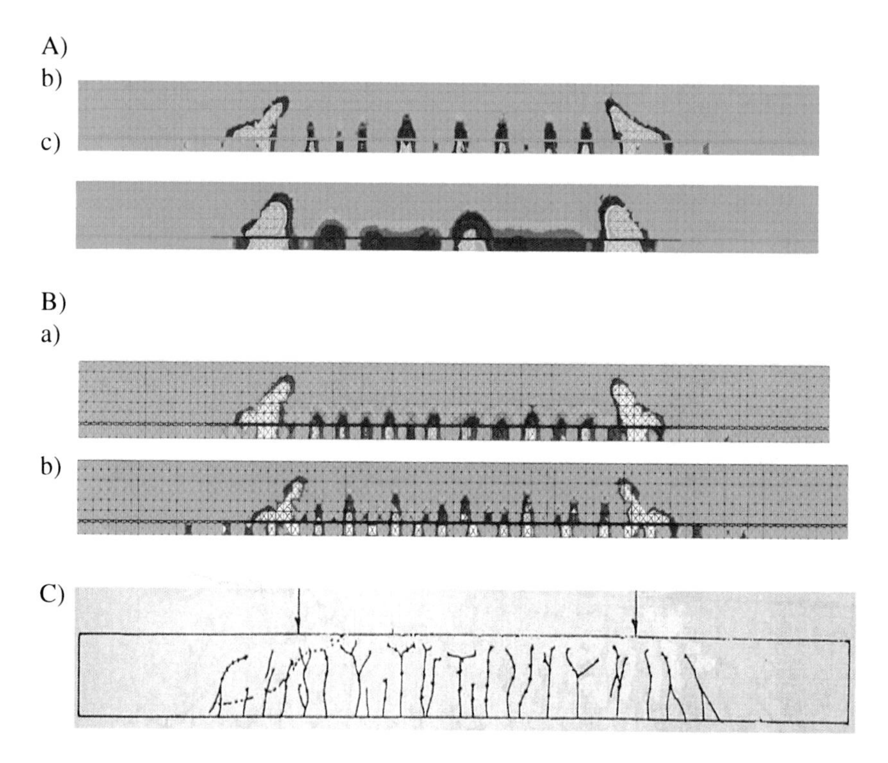

Fig. 7.20 Distribution of non-local softening parameter in small-size beam at vertical displacement of u=8.5 mm (h=150 mm, bond-slip by Dörr, ρ=0.75%, a/d=3) compared to experiments by Walraven (1978) (C): A) κ_u=0.003, B) κ_u=0.006, a) l_c=5 mm, b) l_c=10 mm, c) l_c=20 mm, C) experiment (Marzec et al. 2007)

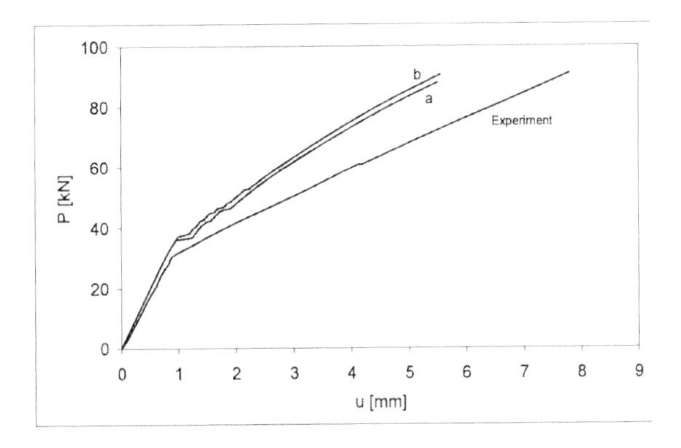

Fig. 7.21 Calculated load-displacement curves for large-size beam (h=750 mm, bond-slip by Dörr (1980), ρ=0.75%, a/d=3, κ_u=0.003) as compared to experiments by Walraven (1978) (P – vertical resultant force, u – vertical displacement, 'e' – experiment by Walraven, 1978): a) l_c=10 mm, b) l_c=20 mm, c) experiment (Marzec et al. 2007)

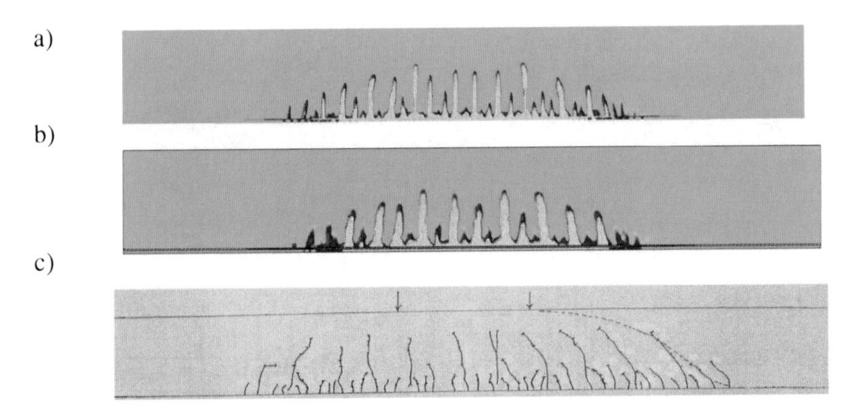

Fig. 7.22 Distribution of non-local softening parameter in large-size beam at vertical displacement of u=8.5 mm (h=750 mm, bond-slip by Dörr (1980), ρ=0.75%, a/d=3, κ_u=0.003) compared to experiments by Walraven (1978): a) l_c=10 mm, b) l_c=20 mm, c) experiment (Marzec et al. 2007)

The width of localized zones slightly increases with increasing a/d (from $w=3\times l_c$ with a/d=2, $w=3.5\times l_c$ with a/d=3, up to $w=4\times l_c$ with a/d=3.5). In turn, the spacing of localized zones slightly increases with increasing a/d (from s=170 mm with a/d=2, s=190 mm with a/d=3-3.5).

Effect of bond-slip

The type of the bond law insignificantly influences the load-displacement curve and width and spacing of localized zones. Since the bond traction values are far from the limiting value, the cracking process is influenced by the initial bond stiffness only. The effect of the stiffness of bond-slip by Dörr (1980) (Eqs. 3.114 and 3.115) (in the range of u_o=0.06-0.24 mm) on the distribution of localized zones is shown in Fig. 7.26 for a medium-size beam (ρ=0.75%, l_c=20 mm, κ_2^u=0.003).

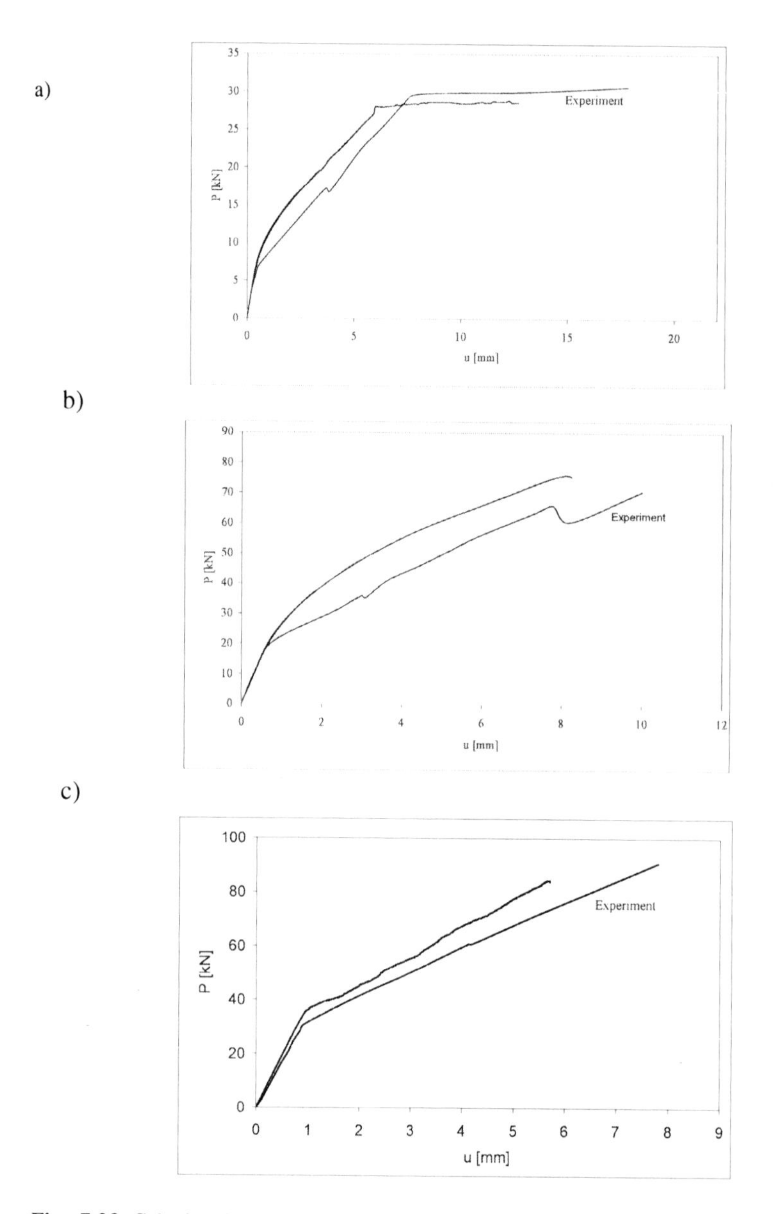

Fig. 7.23 Calculated load-displacement curves (a/d=3, ρ=0.75%, κ_u=0.003, l_c=10 mm, bond-slip by Dörr (1980), reinforcement as 1D element) as compared to experiments by Walraven (1978): a) small beam, b) medium beam, c) large beam (Marzec et al. 2007)

The spacing of localized zones increases with decreasing initial bond stiffness; from s=190 mm (12×l_c) for u_o=0.06 mm, s=340 mm (17×l_c) for u_o=0.12 mm up to s=480 mm (24×l_c) for u_o=0.24 mm. In turn, the width of localized zones decreases with decreasing initial bond stiffness from 3.5×l_c (u_o=0.06 mm) down to 3×l_c (u_o=0.12-0.24 mm).

Figures 7.27 and 7.28 demonstrate the results with perfect bond (medium beam, (ρ=0.75%, l_c=20-30 mm, κ_2^u=0.003). The ultimate vertical force is larger for perfect bond by 5%. The spacing of localized zones is slightly smaller for perfect bond (s=180 mm). Their width is similar.

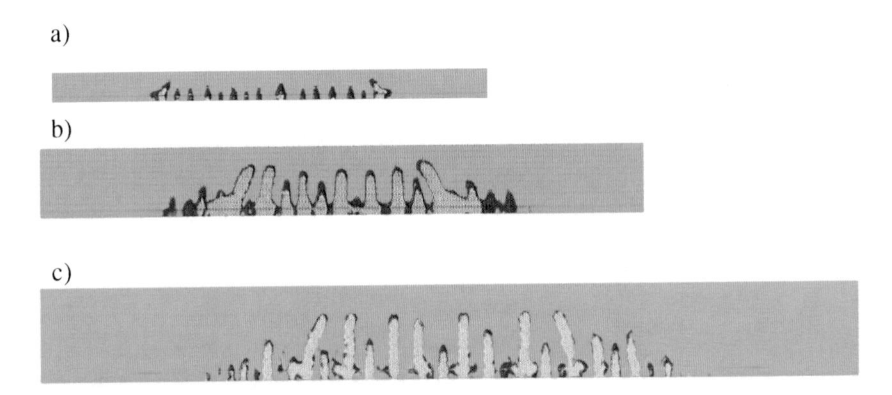

Fig. 7.24 Distribution of non-local softening parameter κ_2 (a/d=3, ρ=0.75%, κ_u=0.003, l_c=10 mm, bond-slip by Dörr (1980), reinforcement as 1D element) in: a) small beam, b) medium beam, c) large beam (Marzec et al. 2007)

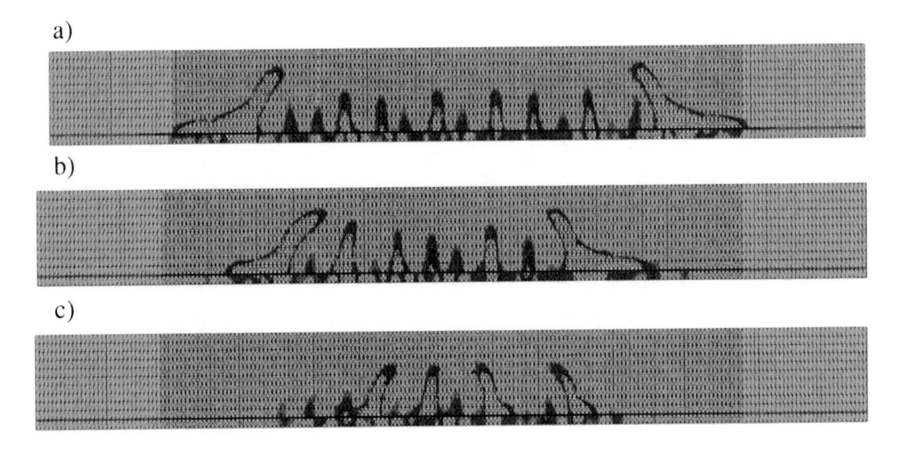

Fig. 7.25 Distribution of the non-local softening parameter κ_2 in medium size beam at vertical displacement of u=8.5 mm (bond-slip of Eqs. 3.114 and 3.115, l_c=20 mm, ρ=0.75%, κ_u=0.003) for different ratios a/d: a) a/d=2, b) a/d=3, c) a/d=3.5 (Marzec et al. 2007)

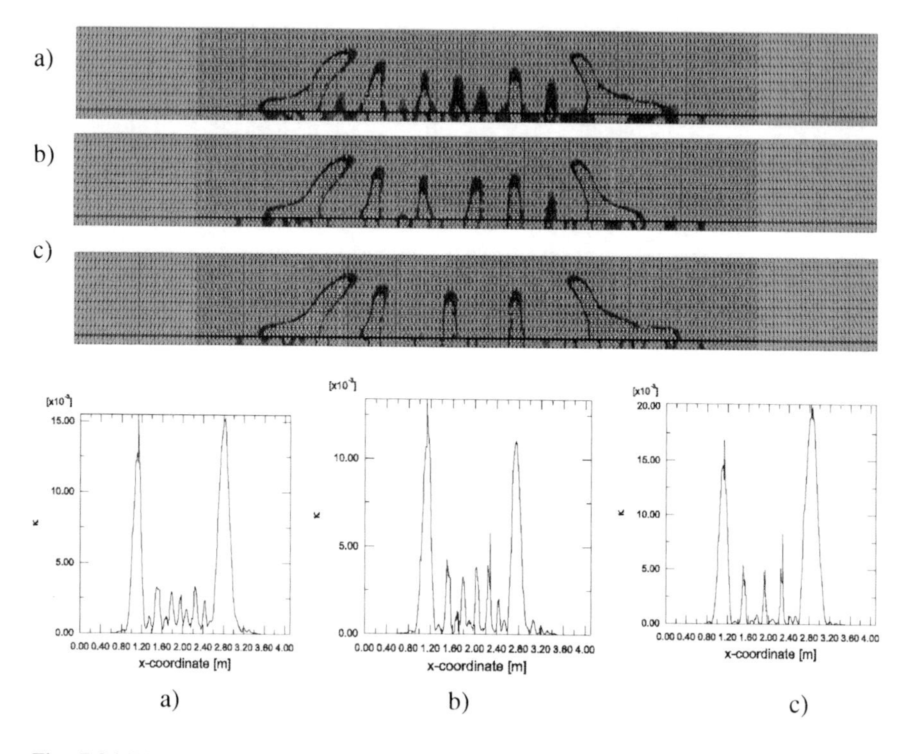

Fig. 7.26 Distribution of non-local softening parameter in medium-size beam at vertical displacement of u=8.5 mm (bond-slip by Eqs. 3.114-3.115), l_c=20 mm, ρ=0.75%, κ_u=0.003): a) u_o=0.06 mm, b) u_o=0.12 mm, c) u_o=0.24 mm (Marzec et al. 2007)

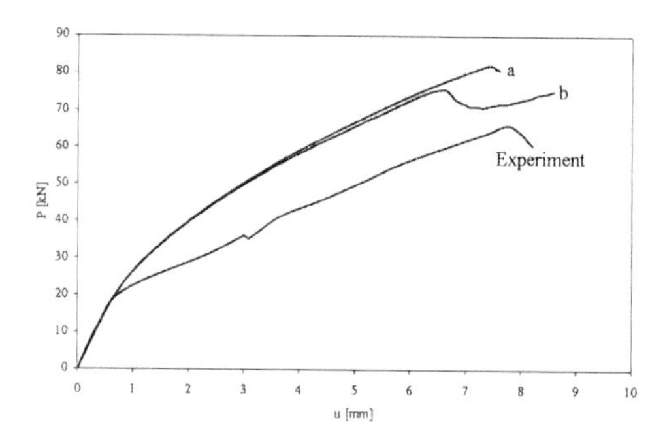

Fig. 7.27 Calculated and experimental load-displacement curves for medium-size beam (a/d=3, ρ=0.75%, κ_u=0.003, l_c=20 mm): a) perfect bond, b) bond-slip (Marzec et al. 2007)

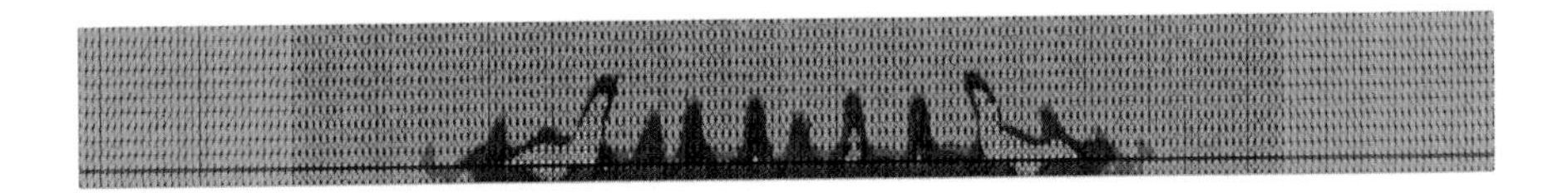

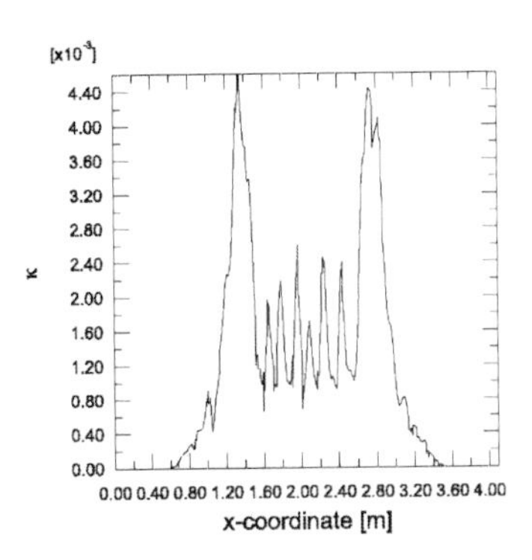

Fig. 7.28 Distribution of non-local softening parameter in medium-size beam at vertical displacement of u=8.5 mm (perfect bond, a/d=3, ρ=0.75%, κ_u=0.003, l_c=20 mm) (Marzec et al. 2007)

FE-results within isotropic damage mechanics with non-local softening

In the case of the isotropic damage model, the following parameters were assumed: κ_0=8.62·10^{-5}, α=0.96, β=200 and l_c=10 mm (Eqs. 3.35, 3.40, 3.93 and 3.99). Figure 7.29 shows the load-displacement curves for a medium, small and large size beam using the bond-slip law by Dörr (1980) (Eqs. 3.114 and 3.115) and a characteristic lengths of l_c=10 mm as compared to the experiments. The reinforcement was assumed as 2D-elements. The distribution of a nonlocal parameter $\bar{\varepsilon}$ in beams is depicted in Fig. 7.30.

The evolution of the vertical force is very similar as in the experiment (Fig. 7.29). The agreement with experiments is even better than within elasto-plasticity since the calculated ultimate vertical forces differ only by 3% from the experimental ones. The calculated normalized ultimate shear forces are: V_u=1.20 MPa (small size beam), V_u=0.86 MPa (medium size beam) and V_u=0.64 (large size beam) respectively.

The calculated average spacing of main localized zones s is approximately s=100 mm (10×l_c) for a small-size beam, s=160 mm (16×l_c) for a medium-size beam and s=240 mm (24×l_c) for a large-size beam (Fig. 7.30). Thus, the spacing of localized zones is similar for a medium beam and larger in the case of a small and large beam as compared to elasto-plastic solutions.

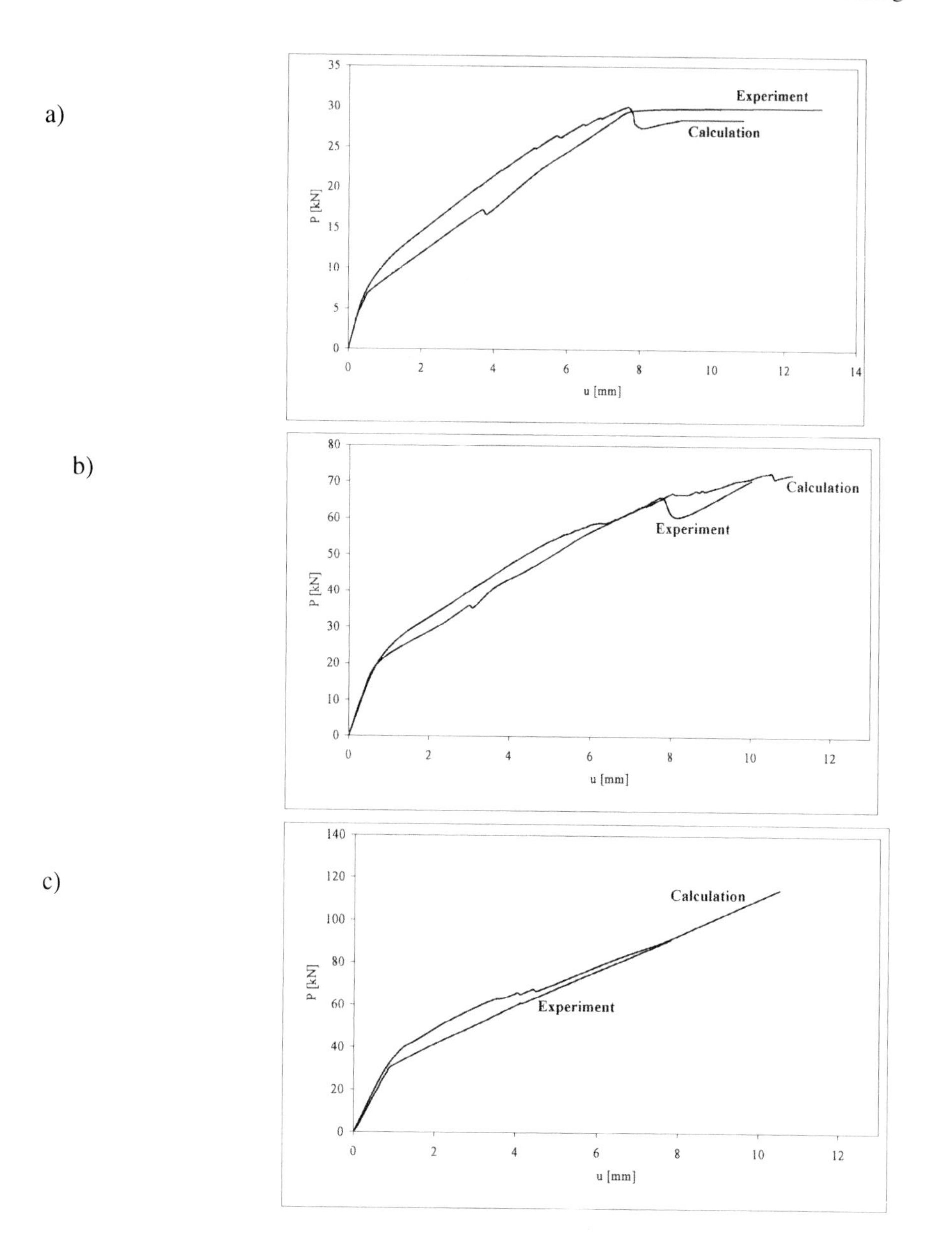

Fig. 7.29 Load-displacement curves (a/d=3, ρ=0.75%, l_c=10 mm, bond-slip, reinforcement as 2D elements) as compared to experiments by Walraven (1978): a) small beam, b) medium beam, c) large beam (damage mechanics) (Marzec et al. 2007)

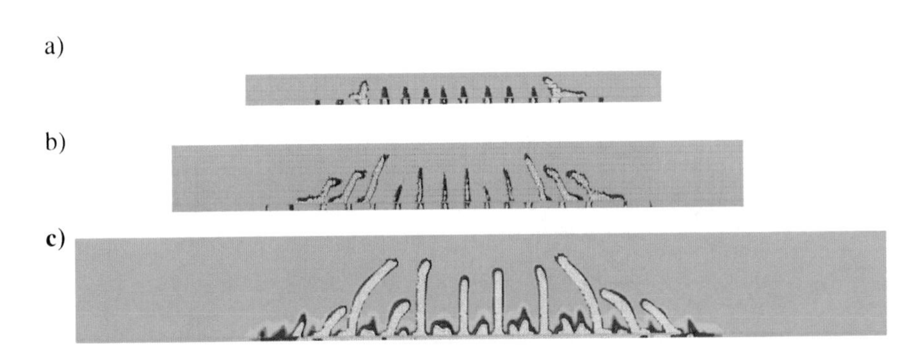

Fig. 7.30 Distribution of the non-local softening parameter (a/d=3, ρ=0.75%, l_c=10 mm, bond-slip, reinforcement as 2D elements) in: a) small beam, b) medium beam, c) large beam (damage mechanics) (Marzec et al. 2007)

FE-results with explicit second-gradient strain isotropic damage approach

The explicit second-gradient strain approach of Chapter 3.3.2 was used. The same material parameters were assumed for concrete when using a damage approach: E=28.9 GPa, ν_c=0.20, κ_0=8.62×10^{-5}, α=0.96, β=200 and k=10. Three different characteristic lengths were assumed: l_c=5 mm, 10 mm and 12.5 mm. An elasto-perfect plastic constitutive law was again assumed to model the reinforcement behaviour. The reinforcement was assumed as 1D bar elements with the cross-section corresponding to the reinforcement area. The bars were fixed at ends. The bond law by Dörr (1980) was used between concrete and reinforcement (Eqs. 3.114 and 3.115).

Figure 7.31 shows the load-displacement curves as compared to the corresponding laboratory tests by Walraven (1978). The evolution of the vertical forces is similar as in the experiment. The calculated normalized shear resistance forces $V_n^c=V/bd$, 1.24 MPa (small beam) and 0.89 MPa (medium beam), match well with the experimental values of 1.26 MPa and 0.84 MPa, respectively. A pronounced size effect obviously takes place since the normalized shear resistance force decreases (almost linearly) with increasing effective height. The calculated average spacing of main localized zones varies from 110 mm up to 130 mm (l_c=5 mm) (Fig. 7.32), which is in satisfactory agreement with experiments and calculations with a non-local model.

Due to its explicit character, the FE calculations were shorter by ca. 30% as compared to the corresponding ones with a non-local approach (Marzec et al. 2007).

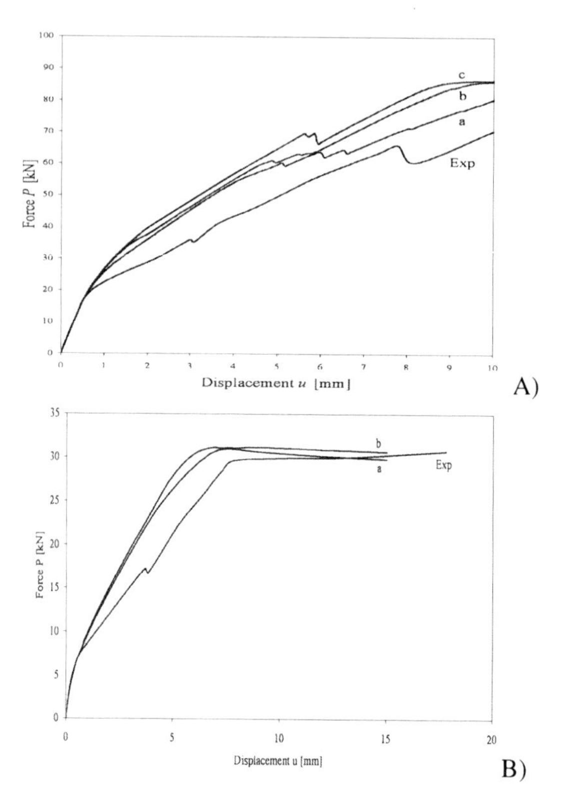

Fig. 7.31 Calculated and experimental load-displacement curve in reinforced concrete beam for small-size beam (A) and medium-size beam (B): l_c=5 mm (a), l_c=10 mm (b), l_c=12.5 mm (c) using explicit second-gradient strain isotropic damage approach (Małecki and Tejchman 2009)

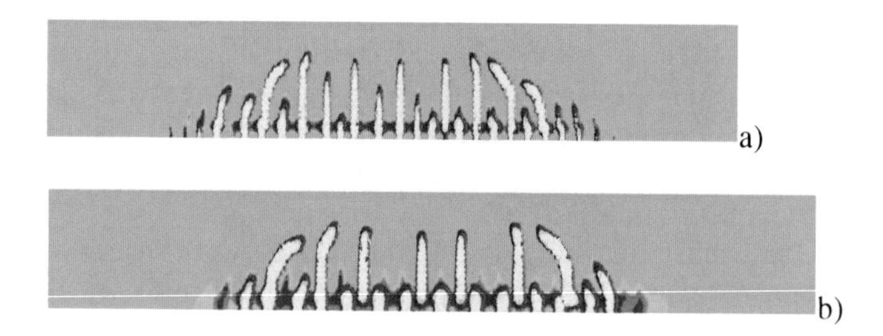

Fig. 7.32 Calculated localized zones in reinforced concrete medium-size beam for different characteristic lengths: l_c=5 mm (a), l_c=10 mm (b) using explicit second-gradient strain isotropic damage approach (Małecki and Tejchman 2009)

The FE-analyses of reinforced concrete slender beams under bending revealed the following points:

• in spite of the simplicity of the used models, the calculated normalized material strength and spacing of main localized zones are in a satisfactory agreement with experiments. The evolution of load-displacement curves is also similar. The differences concern the height and shape of localized zones,
• the FE-results are similar within two different continuum crack models, although a slightly better agreement with experiments was achieved with a damage model,
• the beam strength increases mainly with increasing reinforcement ratio, characteristic length, tensile strength, fracture energy and decreasing beam size and shear span ratio. It is not affected by the type of the bond-slip,
• the calculated ultimate forces differ by about 5 % as compared to experimental ones when l_c=10 mm. To a achieve a better agreement, the characteristic length and fracture energy should be smaller (l_c<10 mm, G_f<0.10 N/mm). However this will be connected to a larger computation time,
• the size effect is realistically captured, the load bearing capacity increases with decreasing element size,
• the width of primary localized zones increases strongly with increasing characteristic length l_c. The width is about (3-4)×l_c,
• the spacing of localized zones increases with increasing characteristic length l_c, tensile softening modulus and decreasing fracture energy, reinforcement ratio and initial bond stiffness. It is not affected by the type of the bond-slip. The spacing is about (7-24)×l_c. In the case of main pronounced cracks, it is in a good agreement with experiments and is smaller than this given by different analytical formulas,
• the reinforcement can be modelled as 2D and 1D elements.
• the results within non-local and second-gradient mechanics are similar. The FE calculations within second-gradient mechanics are shorter by ca. 30% as compared to the non-local approach.

7.3 Short Beams

Chapter presents quasi-static FE-simulations of strain localization in short reinforced concrete beams without shear reinforcement loaded in shear (Skarżyński et al. 2010). Concrete was modelled with 3 different constitutive models. First, an isotropic elasto-plastic model with a Drucker-Prager criterion defined in compression and with a Rankine criterion defined in tension was used (Chapter 3.1.1). Next, an isotropic damage model and an anisotropic smeared crack model were applied (Chapters 3.1.2 and 3.1.3). All models were enhanced in a softening regime by a characteristic length of micro-structure by means of a non-local theory (Chapter 3.3.1). The numerical results were compared with the corresponding laboratory tests by Walraven and Lehwalter (1994).

Laboratory tests were carried out with five different short reinforced concrete beams without shear reinforcement and free at ends (Walraven and Lehwalter 1994) (Chapter 2). The geometry of the specimens is shown in Fig. 7.33. The length varied between 680 mm and 2250 mm and height was between 200 mm and 1000 mm (the beam width b was always 250 mm). The cylinder crushing strength of concrete was about f_c=20 MPa. In turn, the cylinder splitting strength was about f_t=2 MPa. The maximum aggregate size in concrete was d_a^{max}=16 mm. The concrete cover measured from the bar centre to the concrete surface was 40 mm for the smallest beam and 70 mm for the largest one. In the all tests the span-to-depth ratio was 1. The reinforcement ratio of the specimens was 1.1% (to avoid the failure by yielding of bars). The longitudinal reinforcement consisted of uncurtailed bars of deformed cold-drawn steel (with yielding strength of 420 MPa). To obtain a geometrically similar cross-sectional area, various combinations of bar sizes were used (diameters 16, 18, 20 mm). The ratio between the width of the loading plate k and the effective depth d was kept constant (k/d=0.25). The beams were incrementally loaded by a vertical force situated in the middle of beam length. Firstly, at about 40% of the failure load, bending cracks appeared. Afterwards, at about 45-50% of the failure load, the first inclined crack occurred. Contrary to slender beams (Walraven 1978, Walraven and Lehwalter 1994), failure occurred in a gradual gentle way in shear compression by crushing concrete adjacent to the loading plate initiated by the formation of short parallel inclined cracks.

A pronounced size effect took place since the normalized shear resistance force $V_n=V/bd$ was decreasing with increasing effective height d (V – ultimate vertical force): V_n=4.75 MPa (beam V711), V_n=3.20 MPa (beam V022), V_n=2.85 MPa (beam V511) and V_n=2.30 MPa (beam V411). The cracks developed apparently significantly faster in larger beams. The details of the beams, shear cracking loads V_c and shear failure loads V_n are given in Tab. 7.1.

The two-dimensional FE calculations were performed with 4 reinforced concrete beams (h=200-800 mm). The regular meshes with 2720 (h=200 mm) up to 16560 (h=800 mm) quadrilateral elements composed of four diagonally crossed triangles were used to avoid volumetric locking. The maximum finite element height, 15 mm, and finite element width, 10 mm, were not greater than $3\times l_c$ (l_c=5-20 mm) to achieve mesh-objective results. The comparative 3D calculations were performed for the beam of h=200 mm. The mesh with 16320 eight-nodded solid elements was used. The maximum sizes of finite elements were again not greater than $3\times l_c$ (l_c=10-20 mm). The following elastic material parameters were assumed for concrete: E=28.9 GPa (modulus of elasticity) and υ=0.20 (Poisson's ratio). The cylinder compressive strength was f_c=20 MPa and the tensile strength was f_t=2 MPa. The deformation was induced by prescribing a vertical displacement at the mid-point of the beam top.

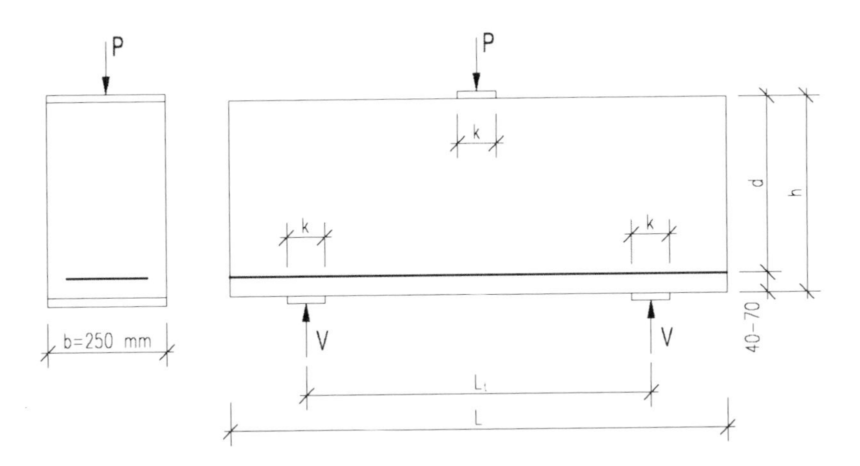

Fig. 7.33 Geometry of reinforced concrete beams used in laboratory tests (Walraven and Lehwalter 1994)

Table 7.1 Specimen properties and failure loads (Skarżyński et al. 2010)

beam specimen	h [mm]	d [mm]	L [mm]	L_t [mm]	A_{sl} [mm^2]	reinforcement bars	f_c [N/mm^2]	V_c [kN]	V_n [kN]
V711	200	160	320	680	606	3ø16	18.1	70	165
V022	400	360	720	1030	1020	4ø18	19.9	125	270
V511	600	560	1120	1380	1570	5ø20	19.8	150	350
V411	800	740	1480	1780	2040	2 (4ø18)	19.4	225	365
V211	1000	930	2250	1860	2510	2 (4ø18)	20.0	240	505

The calculations were carried out with bond-slip using a relationship between the bond shear stress τ_b and slip u according to Dörr (1980) (Eqs. 3.114 and 3.115) due to the fact that bond traction values were far from the limiting value (thus, the shape of the bond law after the peak was unimportant). To investigate the effect of the bond stiffness, several numerical tests were carried out with a different value of u_0 changing from 0.06 mm (Dörr 1980) up to 1.0 mm (Haskett et al. 2008).

FE results with enhanced elasto-plastic model

Preliminary FE calculations have shown a certain effect of a characteristic length of micro-structure, compressive fracture energy, tensile fracture energy, softening rate in tension and compression, softening type (linear and non-linear) and stiffness of end-slip on both the nominal beam strength, width and spacing of localized zones (Tabs.7.2 and 7.3). The beam load bearing capacity increased with increasing characteristic length, tensile fracture energy and compressive fracture energy. In turn, the spacing of localized zones increased with increasing characteristic length and softening rate, and decreasing tensile fracture energy, compressive fracture energy and bond stiffness. The calculated width of localized tensile and compressive zones increased with increasing characteristic length l_c and was equal approximately to (1.5-4)$\times l_c$ with l_c=5-20 mm. The ultimate vertical force P was smaller for the 3D model by 5% only.

On the basis of our preliminary calculations, the further analyses were performed with a 2D model, using a characteristic length of l_c=5 mm, a non-local parameter m=2, linear hardening and softening in compression and linear softening in tension (Fig. 7.34). The tensile fracture energy was G_f=50 N/m and compressive fracture energy was G_c=1500 N/m. The tensile fracture energy was calculated as $G_f=g_f\times w_f$; g_f – area under the entire softening function (with $w_f\approx 4\times l_c$ – width of tensile localized zones, l_c=5 mm). In turn, the compressive fracture energy was calculated as $G_c=g_c\times w_c$ (g_c– area under the entire softening/hardening function up to κ_I=0.006, $w_c\approx 4\times l_c$ – width of compressive localized zones, l_c=5 mm). The internal friction angle was $\varphi=14^\circ$ and the dilatancy angle was chosen as $\psi=8^\circ$. The displacement u_o at which perfect slip occurred was assumed as 0.24 mm (Eqs. 3.114 and 3.115). The distribution of material parameters was uniform in all beams.

Figure 7.35 shows the calculated force-displacement curves (P – vertical force at the mid-point of the beam top, u – vertical displacement of this mid-point) for the beams of h=200-800 mm. The distribution of a non-local tensile and compressive softening parameter is depicted in Figs. 7.36 and 7.37 at the beam failure. In addition, the distribution of non-local tensile softening parameter is shown at the normalized vertical force of $V/(bdf_c)$=0.10 as compared to the experimental crack pattern (Fig. 7.38).

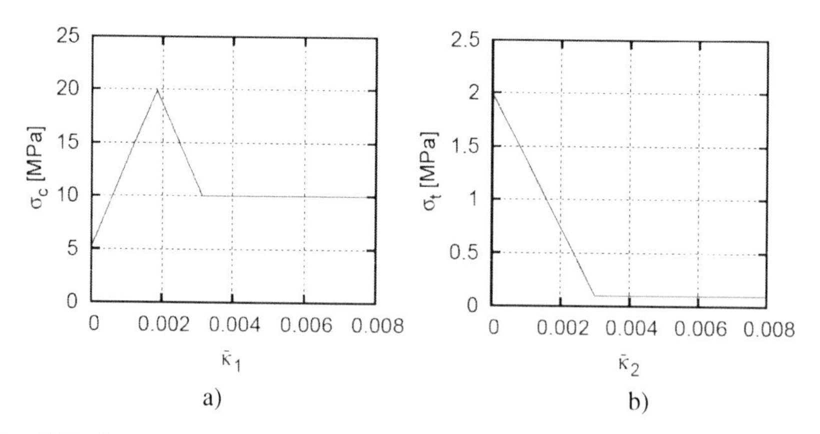

Fig. 7.34 Assumed hardening/softening functions for FE-calculations: a) $\sigma_c=f(\kappa_1)$ in compression, b) $\sigma_t=f(\kappa_2)$ in tension (σ_t – tensile stress, σ_c – compressive stress, κ_i – hardening/softening parameter) (Skarżyński et al. 2010)

The calculated failure forces are in a satisfactory agreement with the experimental ones (Tab. 7.3), but are always larger by 10%–20% (the differences increase with increasing beam size). The geometry of localized zones matches well the experimental crack pattern (Fig. 7.38). The vertical and inclined long and short localized zones were numerically obtained. The experimental crack pattern was obviously non-symmetric. The widths of calculated tensile and compressive localized zones were about $w_f=w_c=4\times l_c$ (Figs. 7.36 – 7.38). In turn, the calculated average spacing s of main localized tensile zones was: s=80 mm (h=200 mm), s=90 mm (h=400 mm), s=170 mm (h=600 mm) and s=150 mm (h=800 mm), respectively.

The calculated spacing of localized zones s was again compared with the average crack spacing according to CEB-FIP Model Code (1991):

$$s=\frac{2}{3}\times\frac{\phi_s}{3.6\rho}=\frac{2}{3}\times\frac{16}{3.6\times0.011}=270\quad \text{mm}\quad (h=200\text{ mm}), \tag{7.10}$$

$$s=\frac{2}{3}\times\frac{\phi_s}{3.6\rho}=\frac{2}{3}\times\frac{18}{3.6\times0.011}=303\quad \text{mm}\quad (h=400\text{ mm}), \tag{7.11}$$

$$s=\frac{2}{3}\times\frac{\phi_s}{3.6\rho}=\frac{2}{3}\times\frac{20}{3.6\times0.011}=337\quad \text{mm}\quad (h=600\text{ mm}), \tag{7.12}$$

$$s=\frac{2}{3}\times\frac{\phi_s}{3.6\rho}=\frac{2}{3}\times\frac{18}{3.6\times0.011}=303\quad \text{mm}\quad (h=800\text{ mm}), \tag{7.13}$$

and the formula by Lorrain et al. (1998):

$$s=1.5c+0.1\frac{\phi_s}{\rho}=1.5\times32+0.1\frac{16}{0.011}=193\quad \text{mm}\quad (h=200\text{ mm}), \tag{7.14}$$

$$s=1.5c+0.1\frac{\phi_s}{\rho}=1.5\times31+0.1\frac{18}{0.011}=210\quad \text{mm}\quad (h=400\text{ mm}), \tag{7.15}$$

$$s=1.5c+0.1\frac{\phi_s}{\rho}=1.5\times30+0.1\frac{20}{0.011}=227\quad \text{mm}\quad (h=600\text{ mm}), \tag{7.16}$$

$$s=1.5c+0.1\frac{\phi_s}{\rho}=1.5\times51+0.1\frac{18}{0.011}=240\quad \text{mm}\quad (h=800\text{ mm}), \tag{7.17}$$

wherein ϕ_s is the mean bar diameter, ρ=1.1% denotes the reinforcement ratio and c denotes the concrete cover. The spacing of localized zones obtained from numerical simulations is significantly smaller than these obtained with analytical formulas.

Table 7.2 Summary of FE-input data (Skarżyński et al. 2010)

FE simulation Nr.	beam height h [mm]	width of tensile localized zones w_f [mm]	tensile fracture energy G_f [N/m]	width of compressive localized zones w_c [m]	compressive fracture energy G_c [N/m]	charact. length l_c [mm]	bond model
1a		15	50	20	1500	5	
1b	200	15	50	20	1500	10	bs
1c		35	50	25	1750	20	
2a		15	100	20	1500	5	
2b	200	20	100	20	1500	10	bs
2c		40	100	25	1750	20	
3a		15	200	20	1500	5	
3b	200	35	200	25	1750	10	bs
3c		60	200	25	1750	20	
4a	400	15	50	15	1500	5	bs
4b		35	50	25	1750	10	
5a	400	20	100	15	1500	5	bs
5b		40	100	25	1750	10	
6a	600	15	50	15	1500	5	bs
6b		15	100	25	1750		
7a	800	15	50	15	1500	5	bs
7b		15	100	25	1750		
8a		15	50	20	1500		bs
8b	200	15	50	20	1500	5	bs (u_0=0.12 mm)
8c		15	50	20	1500		bs (u_0=0.24 mm)
8d		15	50	20	1500		bs (u_0=1 mm)
9a		40	100	25	1750		bs
9b	400	40	100	25	1750	10	bs (u_0=0.12 mm)
9c		40	100	25	1750		bs (u_0=0.24 mm)
9d		40	100	25	1750		bs (u_0=1 mm)
10a	200	15	50	20	1500	10	pb
10b		15	50	20	1500		bs

Table 7.2 (*continued*)

11a	400	40	100	25	1750	10	pb
11b		40	100	25	1750		bs
12a		15	50	15	900		
12b	200	15	50	20	1500	5	bs
12c		15	50	20	1800		
13a		15	50	20	900		
13b	400	35	50	25	1750	10	bs
13c		35	50	25	2250		
14a	200	35	50	25	1750	20	3D, bs
14b	200	20	100	20	1400	10	3D, bs
15	200	35	50	25	1750	20	3D bs (u_0=1 mm)

'*bs*' – bond slip (u_0=0.06 mm), '*pb*' – perfect bond.

Table 7.3 Data summary of experiments, FE-results and analytical formulae (crack spacing) (Skarżyński et al. 2010)

Nr. of FE-simulation (Tab. 7.2)	beam height h [mm]	failure vertical force (experiments) [kN]	failure vertical force (FEM) [kN]	spacing of localized tensile zones from FEM s [mm]	crack spacing by CEB-FIP model (1991) [mm]	crack spacing by Lorrain et al. (1998) [mm]
1a			182	105		
1b	200	165	185	105	270	193
1c			187	160		
2a			186	80		
2b	200	165	190	80	270	193
2c			193	160		
3a			190	105		
3b	200	165	192	105	270	193
3c			197	160		
4a	400	270	285	180	303	210
4b			287	145		
5a	400	270	291	60	303	210
5b			295	90		

Table 7.3 (*continued*)

6a	600	350	400	110	337	227
6b			405	170		
7a	800	365	425	85	303	240
7b			435	150		
8a	200	165	182	105	270	193
8b			178	105		
8c			187	105		
8d			175	105		
9a	400	270	295	145	303	210
9b			296	145		
9c			297	145		
9d			275	230		
10a	200	165	195	80	270	193
10b			190	80		
11a	400	270	305	145	303	210
11b			295	180		
12a	200	165	170	80	270	193
12b			182	105		
12c			185	80		
13a	400	270	275	360	303	210
13b			295	145		
13c			297	180		
14a	200	165	195	160	270	193
14b			220	80	270	193
15	200	165	175	160	270	193

FE results with enhanced smeared crack model and enhanced damage model
The following parameters were assumed in both models: E=28.9 GPa, υ=0.2, $\kappa_0=10^{-4}$, α=0.95 and β=500 (isotropic damage approach Eqs. 3.35 and 3.40), E=28.9 GPa, υ=0.2, $\kappa_0=10^{-4}$, α=0.95, β=500 and k=10 (isotropic damage approach Eqs. 3.38 and 3.40), E=28.9 GPa, υ=0.2, $\kappa_0=10^{-4}$, α=0.95, β=500, α_1=0.1, α_2=1.16, α_3=2.0 and γ=0.2 (isotropic damage approach Eq. 3.39 and 3.40) and E=28.9 GPa, υ=0.2, p=4.0, b_1=3.0, b_2=6.93, f_t=2.0 MPa, ε_{su}=0.006 and ε_{nu}=0.006 (smeared crack approach, Chapter 3.1.3). The results are shown in Figs. 7.39 and 7.40 with a smeared rotating crack model and in Figs. 7.41 and 7.42 with damage models (the results using Eq. 3.38 were similar to those by Eq. 3.35).

The force-displacement curves are very similar as those obtained with an elasto-plastic model. The calculated forces at failure are usually larger by 2%–20% than the experimental ones.

The calculated geometry of localized zones within a smeared crack approach is similar as this within elasto-plasticity except of beams with h>400 mm where the localized zones are more diffuse. The effect of crack type assumed in the model (fixed or rotating crack model) was insignificant. In turn, large discrepancies occur in the distribution of localized zones when using the damage model. The inclined localized zones were not obtained in FE analyses (only one vertical at bottom mid-point of a tensile type).

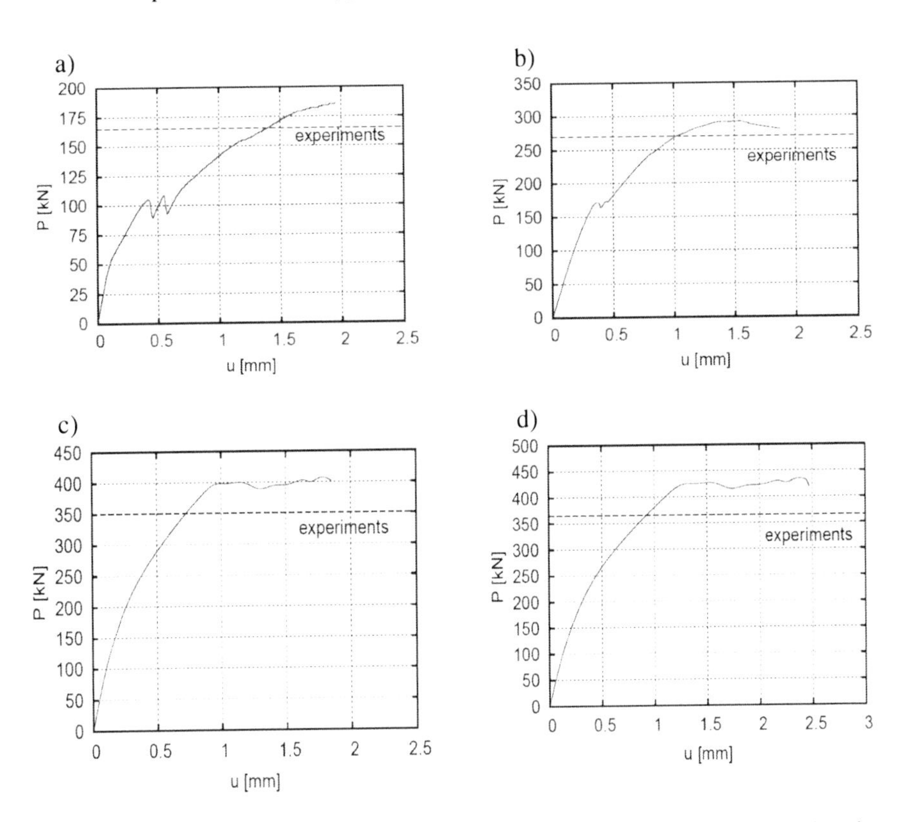

Fig. 7.35 Calculated force-displacement curves within elasto-plasticity (as compared to the experimental maximum vertical force) for different beams: a) h=200 mm, b) h=400 mm, c) h=600 mm, d) h=800 mm (P – resultant vertical force, u – vertical displacement) (Skarżyński et al. 2010)

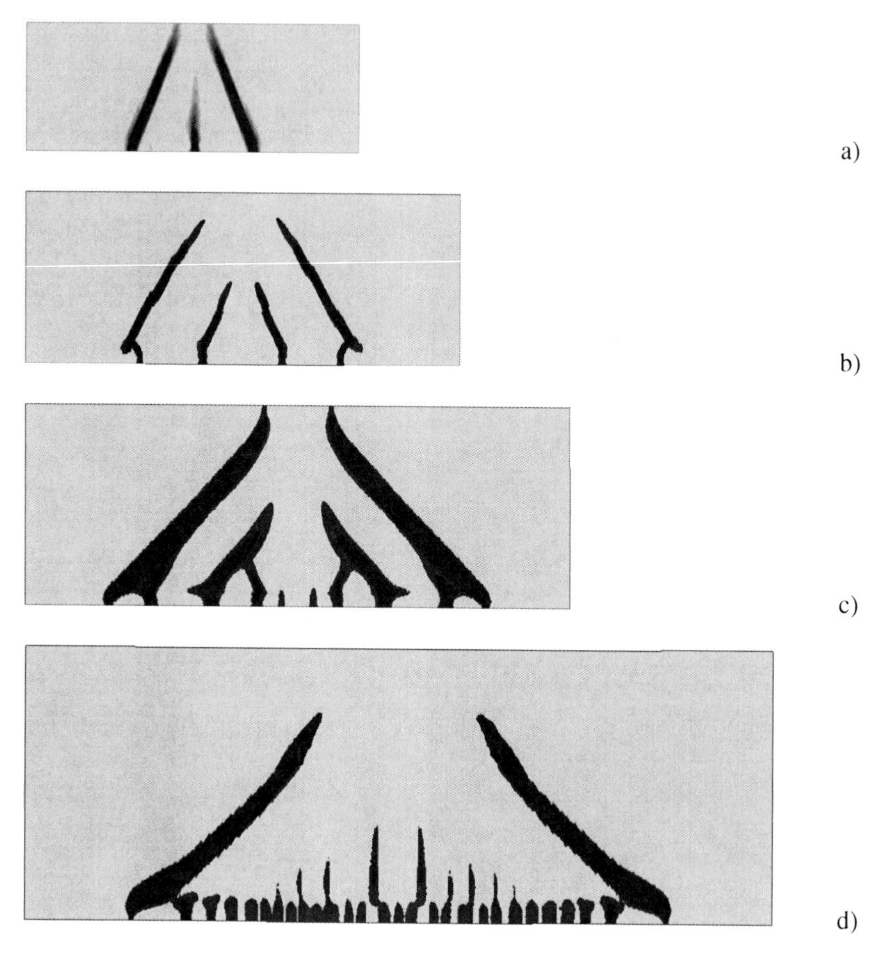

Fig. 7.36 Distribution of calculated non-local tensile softening parameter $\overline{\kappa}_2$ within elasto-plasticity in beams at failure for different beams: a) h=200 mm, b) h=400 mm, c) h=600 mm, d) h=800 mm (the beams are not proportionally scaled) (Skarżyński et al. 2010)

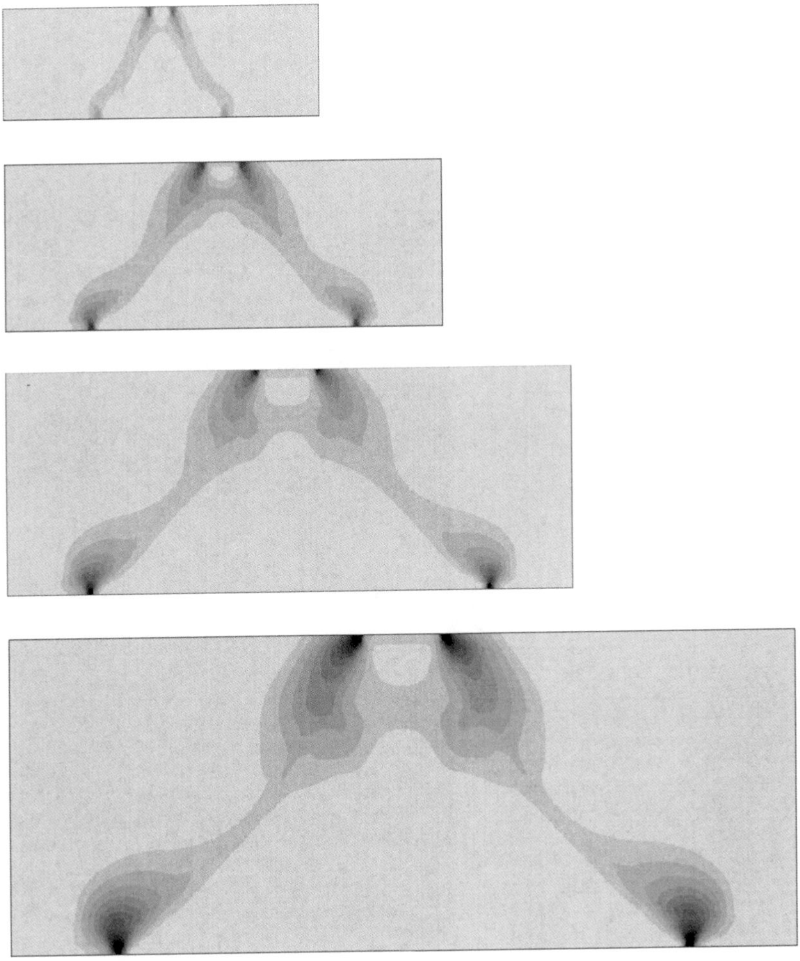

a)

b)

c)

d)

Fig. 7.37 Distribution of the non-local compressive softening parameter $\overline{\kappa}_1$ within elasto-plasticity for different beams at vertical displacement of u=10 mm: a) h=200 mm, b) h=400 mm, c) h=600 mm, d) h=800 mm (the beams are not appropriately scaled) (Skarżyński et al. 2010)

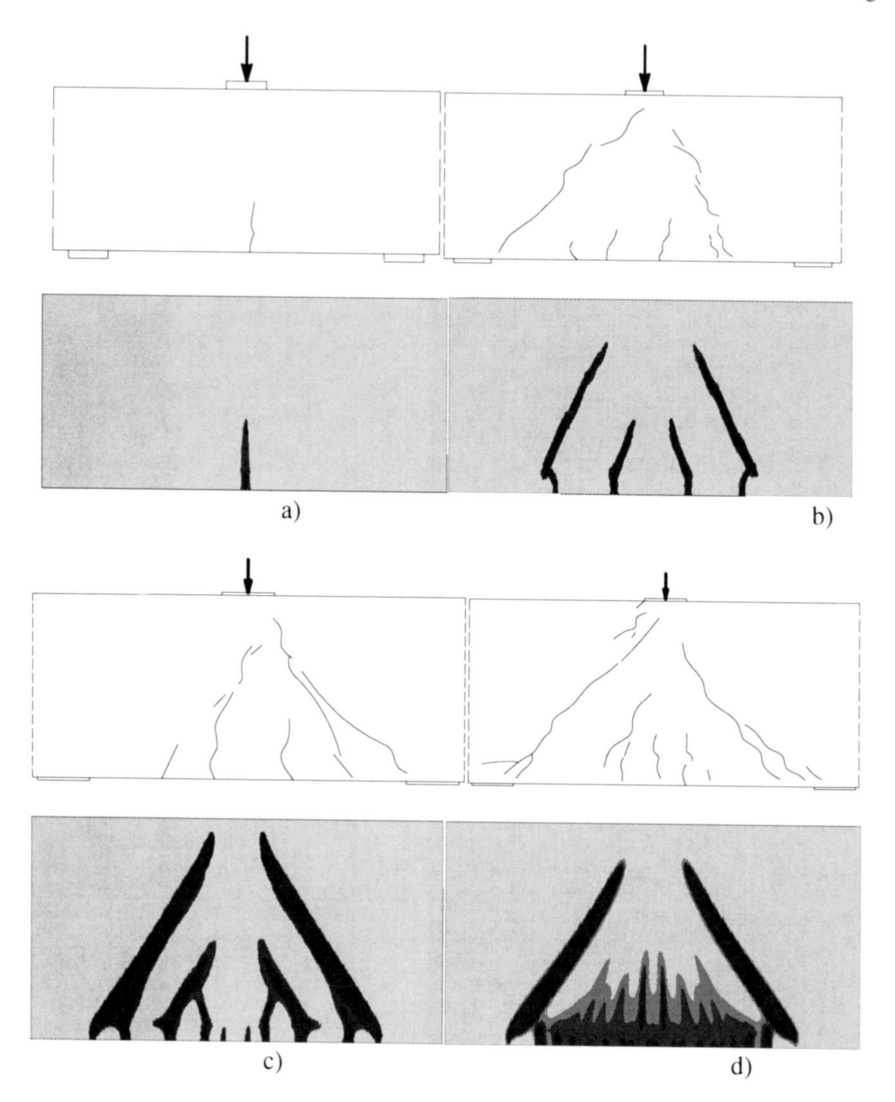

Fig. 7.38 Comparison of distribution of non-local tensile softening parameter $\overline{\kappa}_2$ within elasto-plasticity in short reinforced concrete beams at the normalized vertical force $V/(bdf_c)$=0.10 with experimental crack patterns for different beams: a) h=200 mm, b) h=400 mm, c) h=600 mm, d) h=800 mm (the beams are not proportionally scaled) (Skarżyński et al. 2010)

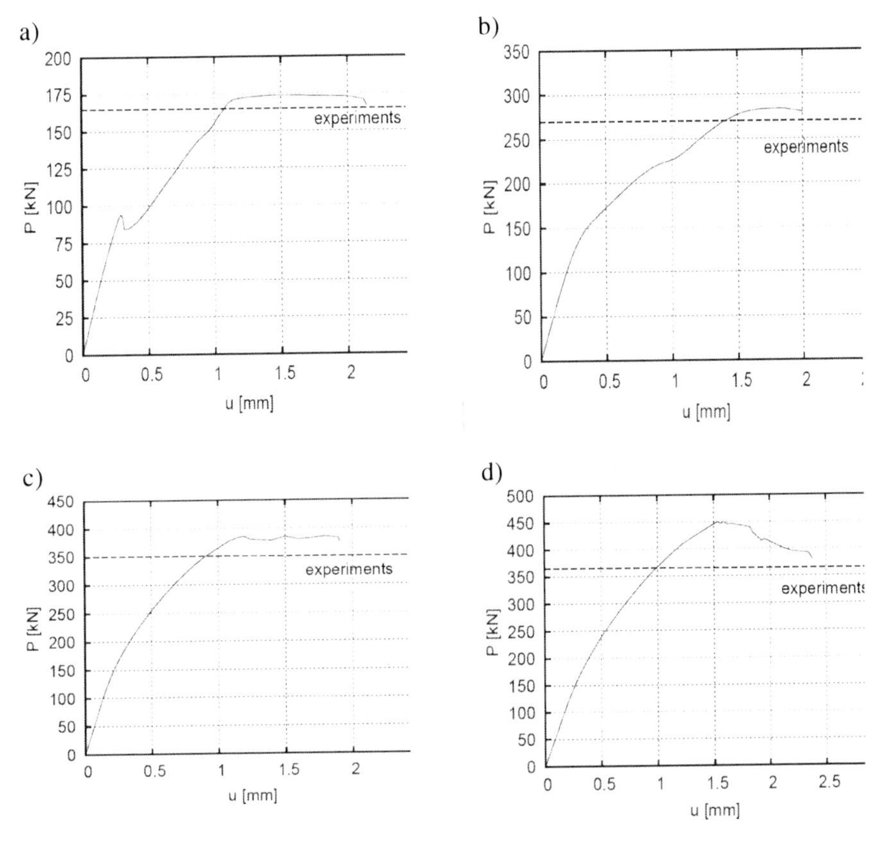

Fig. 7.39 Calculated force-displacement curves with smeared crack approach (Chapter 3.1.3) (as compared to the experimental maximum vertical force) for different beams: a) h=200 mm, b) h=400 mm, c) h=600 mm, d) h=800 mm (P – resultant vertical force, u – vertical displacement) (Skarżyński et al. 2010)

Figure 7.43 shows a comparison between the calculated (with 3 continuum models) and experimental size effect: the relative shear stress $V/(bdf_c)$ at failure as a function of the effective beam depth. In addition, the size effect law by Bažant (Bažant and Planas 1998) (being valid for structures with large cracks) is enclosed (Eq. 5.5). The experimental and theoretical beam strength shows a strong size dependence. The experimental and theoretical results are close to the size effect law by Bažant (Bažant and Planas 1998).

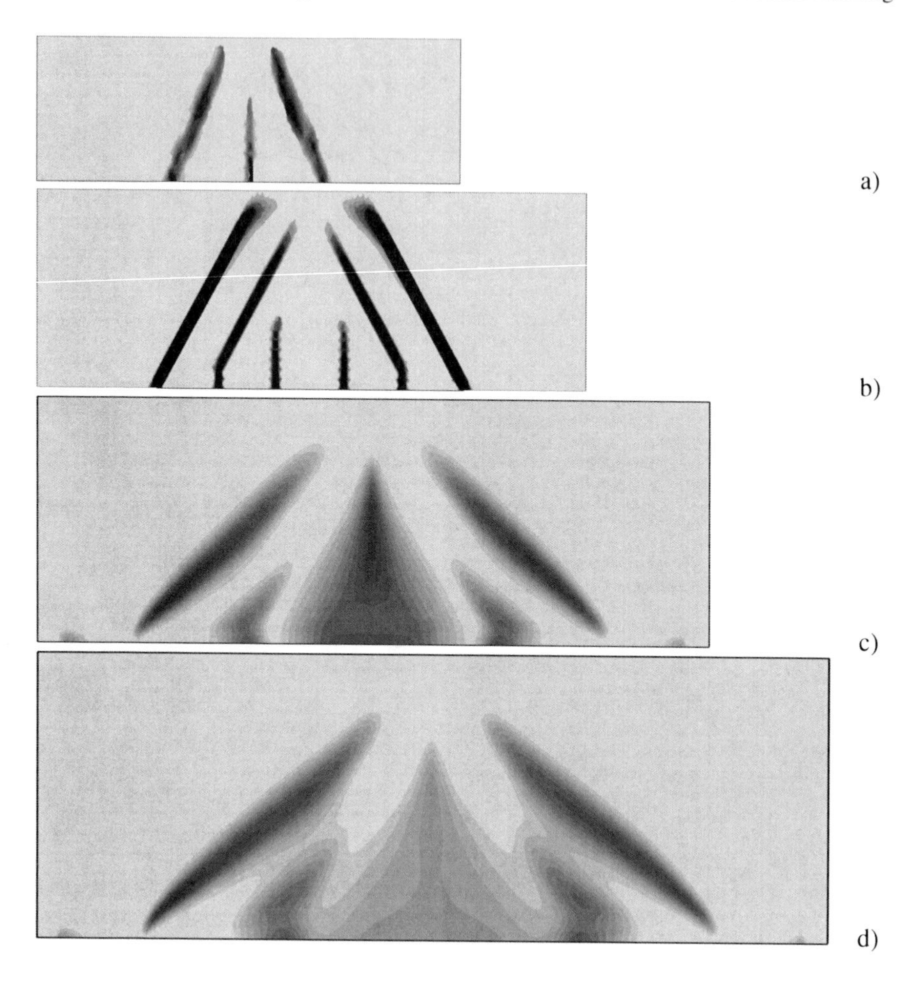

Fig. 7.40 Distribution of calculated non-local strain within smeared crack approach (Chapter 3.1.3) in different beams at failure: a) h=200 mm, b) h=400 mm, c) h=600 mm, d) h=800 mm (the beams are not proportionally scaled) (Skarżyński et al. 2010)

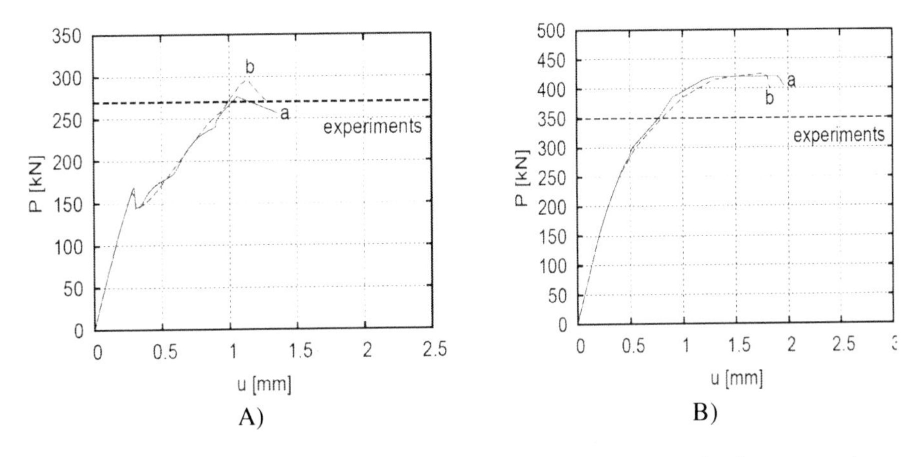

Fig. 7.41 Calculated force-displacement curves within damage mechanics (as compared to the experimental maximum vertical force) for 2 beams: A) h=400 mm, B) h=600 mm, a) equivalent strain measure by Eq. 3.35, b) equivalent strain measure by Eq. 3.39 (P – resultant vertical force, u – vertical displacement) (Skarżyński et al. 2010)

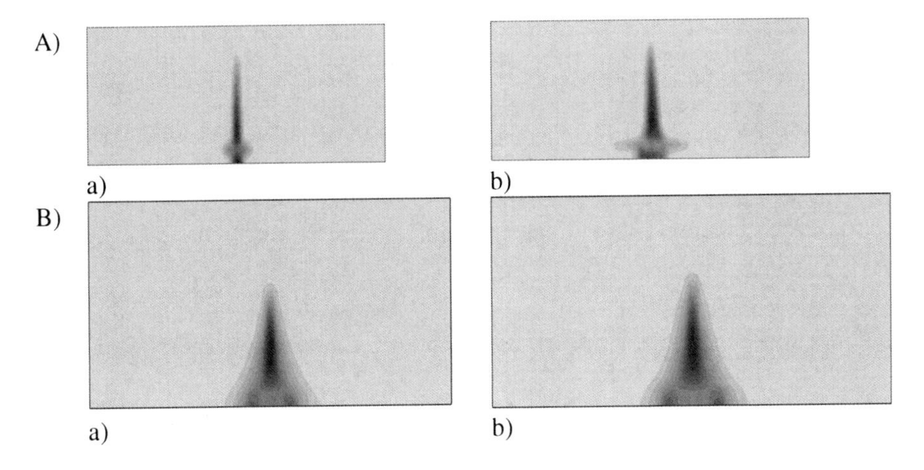

Fig. 7.42 Distribution of calculated non-local equivalent strain measure within damage mechanics in two beams at failure: A) h=400 mm, B) h=600 mm, a) equivalent strain measure by Eq. 3.35, b) equivalent strain measure by Eq. 3.39 (Skarżyński et al. 2010)

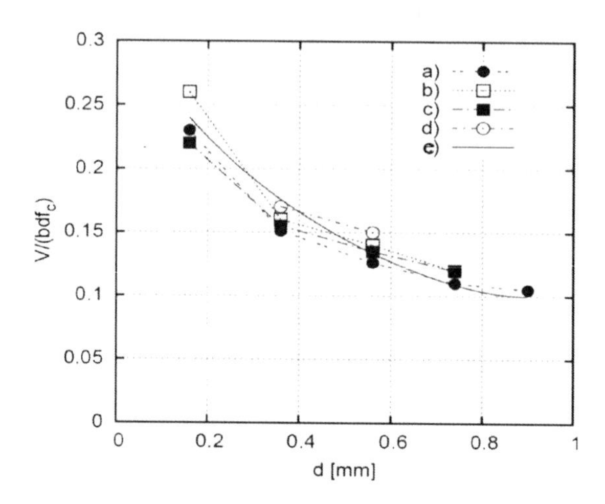

Fig. 7.43 Calculated size effect in reinforced concrete beams from FE-analyses compared to experiments (Walraven and Lehwalter 1994) and to the size effect law by Bažant (Bažant and Planas 1998) (*b* – beam width, *d* – effective beam height, f_c – compressive strength of concrete, V_u – ultimate vertical force): a) experiments, b) FE-calculations (elasto-plasticity), c) FE-calculations (smeared crack model, Chapter 3.1.3), d) FE-calculations (damage mechanics, Eqs. 3.35 and 3.40), e) size effect law by Bažant (Skarżyński et al. 2010)

The FE-simulations have shown that three different simple continuum crack models enhanced by non-local softening are able to capture a deterministic size effect in short reinforced concrete beams without shear reinforcement subjected to shear-tension failure. From the obtained results the following conclusions can be drawn:

- the calculated material strength in reinforced concrete beams of a different size was always higher by 2%-20% as compared to experimental ones. Thus, the models need an improvement,
- the geometry of localized zones was in a good agreement within elasto-plasticity, in a medium agreement within a smeared crack approach and a completely false within damage mechanics. Thus, the isotropic damage model needs improvements to describe localized shear zones in reinforced concrete elements under shear-tension failure,
- the calculated spacing of localized tensile zones increased with increasing characteristic length, softening rate and beam height and decreasing fracture energy and bond stiffness. The calculated and experimental spacing was significantly smaller than this from analytical formulae.

7.4 Columns

Reinforced concrete columns supporting slabs and beams and subject to eccentric compression belong to the most important structure elements. Their role still grows due to the increasing size of multi-story high buildings. In particular, nowadays the use of high-strength concrete columns and concrete-filled steel tubular columns (Kim and Yang 1995, Kilpatrick and Rangan 1999) is becoming very popular worldwide. The buckling behaviour of reinforced concrete columns depends upon many different factors as: column slenderness, load eccentricity, boundary conditions at ends, area and shape of the cross-section of concrete, area and spacing of the vertical and horizontal reinforcement, reinforcement ratio, compressive and tensile strength of concrete, strength of reinforcement, type and character of load (short or long–term, monotonic or cycling), concrete shrinkage and creep. To calculate the optimum concrete and reinforcement area, and thus to decrease building costs and increase net floor space, a realistic prediction of the effect of these factors on stresses in the entire column element is needed. However, the buckling behaviour of reinforced concrete columns is a complex phenomenon due to cracks in concrete.

There have been many experimental studies on reinforced concrete columns. The effect of the load eccentricity was investigated on columns by Makovi (1969), Gruber and Menn (1978), Kiedroń (1980), Billinger and Symons (1995) and Lloyd and Rangan (1996). In turn, Billinger and Symons (1995), and Kim and Yang (1995) studied the effect of the slenderness of columns. Szuchnicki (1973) analysed the influence of the cross-section area. In turn, the effect of creep was investigated by Kordina and Warner (1975), the effect of lateral pre-stressing by Gardner et al. (1992), the effect of the vertical reinforcement by Lloyd and Rangan (1996), Saenz and Martin (1963), Martin and Olivieri (1966) and Kim and Yang (1995), and the effect the horizontal reinforcement by Oleszkiewicz et al. (1973), Korzeniowski (1997) and Nemecek and Bittnar (2004). In turn, the influence of the concrete strength was shown in tests by Kiedroń (1980), Billinger and Symons (1995), Lloyd and Rangan (1996), Saenz and Martin (1963), and Kim and Yang (1995). A deterministic size effect was investigated by Bažant and Kwon (1994). The results of experiments have evidently shown that bearing capacity of columns decreases with increasing load eccentricity, slenderness, ratio of the end fixing and creep. The increase of concrete strength influences significantly the bearing capacity for small eccentricities. The lateral pre-stressing and horizontal reinforcement increase the bearing capacity of cylindrical elements. The failure load exhibits a strong size effect (the bearing capacity decreases as the column size increases).

There exist several theoretical models to calculate the bearing capacity of elements under eccentric compression on the basis of analytical assumptions and FEM (Kim and Yang 1995, Bromst and Viest 1958, Pfrang and Siess 1964, Bažant et al. 1991, El-Metwally and Chen 1989, Xie et al. 1997, Fragomeni and Mendis 1997, Baglin and Scott 2000, Mendis 2000).

The intention of calculations is to numerically predict the behaviour of reinforced concrete columns subject to eccentric compression with consideration of cracks (Majewski et al. 2008, 2009). The analysis was carried out with a finite element method based on elasto-plasticity with non-local softening (Eqs. 3.27-3.32,

3.93 and 3.99). The effect of the load eccentricity, column slenderness, area of the vertical longitudinal reinforcement, bond between concrete and reinforcement, fracture energy, distribution of the tensile strength and characteristic length of micro-structure on the failure strength of columns was studied. Two-dimensional plane strain and three dimensional FE-simulations were carried out. The theoretical results of failure forces were mainly compared with corresponding comprehensive experiments carried out by Kim and Yang (1995). In addition, they were compared with experiments by Kim and Lee (2000), Hsu (1988), Lloyd and Rangan (1996) and Bažant and Kwon (1994).

Experiments on columns by Kim and Yang (1995)

A series of laboratory tests was carried out for 30 tied reinforced columns with a square cross-section of $b \times h$=80×80 mm^2 and 3 slenderness ratios λ of 10, 60 and 100 (Fig. 7.44). The corresponding column heights were l=0.24 m, l=1.44 m and l=2.40 m, respectively. Three different concrete compressive strengths of 25.5 MPa, 63.5 MPa and 86.2 MPa and 2 different longitudinal steel ratios of 1.98% (4ϕ6) and 3.95% (8ϕ6) using a symmetric reinforcement were applied. The splitting tensile strength of concrete was 3.4 MPa, 5.5 MPa and 6.2 MPa, respectively. The maximum size of the aggregate in concrete was d_{max}=13 mm. The concrete cover measured from the bar centre to the concrete surface was 15 mm. The ties made of 3 mm plain steel bars had a spacing of 60 mm (reduced to 30 mm at ends). The boundary conditions at the ends were both hinged. The steel end plates were fixed at the column ends with bolts. Each end plate had a groove (the initial load eccentricity was always e=24 mm). The rate of loading was controlled by a constant increment rate of the vertical displacement.

Most of columns with the slenderness ratio λ=10 failed at the mid-height of columns by increased compressive strain in concrete. The failure of the columns with λ=60 and λ=100 occurred due to increased tensile steel strain near the mid-height. The increment of the ultimate load decreased with a growth of the slenderness ratio. An increase of the longitudinal steel ratio was more effective in slender columns with a higher compressive strength than in short columns with a low compressive strength.

The FE analyses were mainly performed for the slenderness ratio of λ=100, compressive strength of concrete of f_c=25.5 MPa, reinforcement ratio of ρ=1.98% and eccentricity of e=24 mm (Kim and Yang 1995). One assumed mainly that l_c=h/20=4.0 mm. For 2D-calculations, quadrilateral elements (composed of four diagonally crossed triangles) were always used. Totally, 2900 (for l=0.24 m), 7200 (for l=1.44 m) and 15000 (for l=2.40 m) triangular elements were used. The mesh was refined in the middle of the columns for l=1.44-2.40 m. Thus, the maximum element size, 8 mm, was not greater there than 3×l_c in all columns to get mesh-objective results. In the case of 3D computations (only for λ=100), about 18000 8-noded solid elements were used (8×5×442). The maximum element size was also 8 mm along the height. The column was hinged at the both ends. The deformation was induced by increments of the vertical displacement prescribed to both ends through steel plates with a large stiffness. In 2D plane strain calculations, the vertical reinforcement was assumed mainly in the form of 2D 3-node triangular elements. The width of each steel column composed of 2D elements was assumed to be 2.0 mm and their length in the direction perpendicular to the deformation plane was 32 mm for ρ=1.98% and 64 mm for ρ=3.95%. In 2D calculations, one simulation was also performed with vertical 2-node truss elements (describing reinforcement) with a diameter of 9.0 mm. In turn, in 3D calculations, one used always vertical 2-node truss elements with a circular cross-section and diameter of 6 mm. The vertical reinforcement was fixed to the horizontal edge steel plates. The stirrups were not taken into account since the FE-study concerned mainly the column behaviour up to peak (their influence is very important in the post-peak regime). A bond-slip law by Dörr (1980) was assumed (Eqs. 3.114 and 3.115).

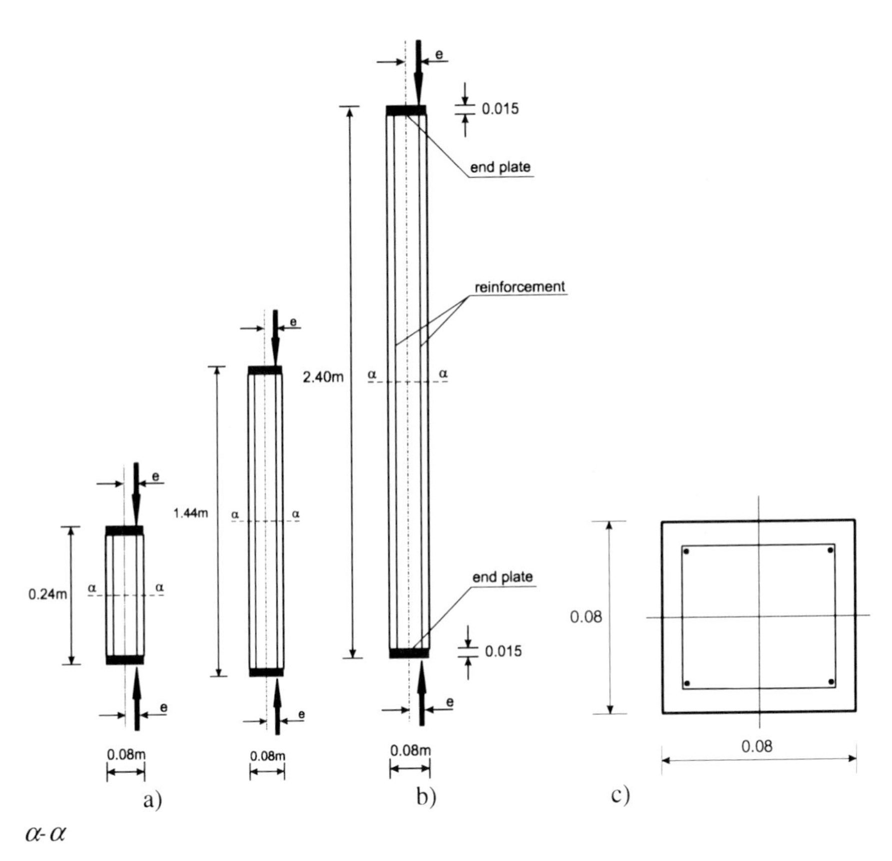

Fig. 7.44 Geometry of reinforced concrete columns in experiments by Kim and Yang (1995): a) λ=10, b) λ=60, c) λ=100

The assumed diagram describing the evolution of uniaxial compressive yield stress σ_c versus a hardening-softening parameter κ_1 is shown in Fig. 7.45a. The evolution of the curve up to the peak is according to the Polish Standard (2002). The modulus of elasticity was equal to E=23.665 GPa and Poisson's ratio was υ=0.20. The internal friction angle was assumed to be φ=14° and the dilatancy angle was taken as ψ=8°. The assumed diagram describing the evolution of the tensile yield stress σ_t versus the softening parameter κ_2 (assumed in most of our calculations) is shown in Fig. 7.45b (it follows the proposition of Hordijk (1991) Eq. 3.55). One assumed that the tensile strength was equal to f_t=0.1×f_c=2.295 MPa. The plastic strain associated with a total loss of load-carrying capacity of the material in tension was about κ_2^u=0.0035. The fracture energy was, thus, approximately G_f=g_f×w_c≈0.02 N/mm (g_f – area under the softening function, w_c=12 mm - width of a localized zone). The non-local parameter was chosen m=2. An elasto-perfect plastic constitutive law by von Mises was assumed to model the reinforcement behaviour with E_s=210 GPa, υ_c=0.3 and f_{yd}=387 MPa (f_{yd} - yield stress). The volume weight of the concrete column was not taken into account.

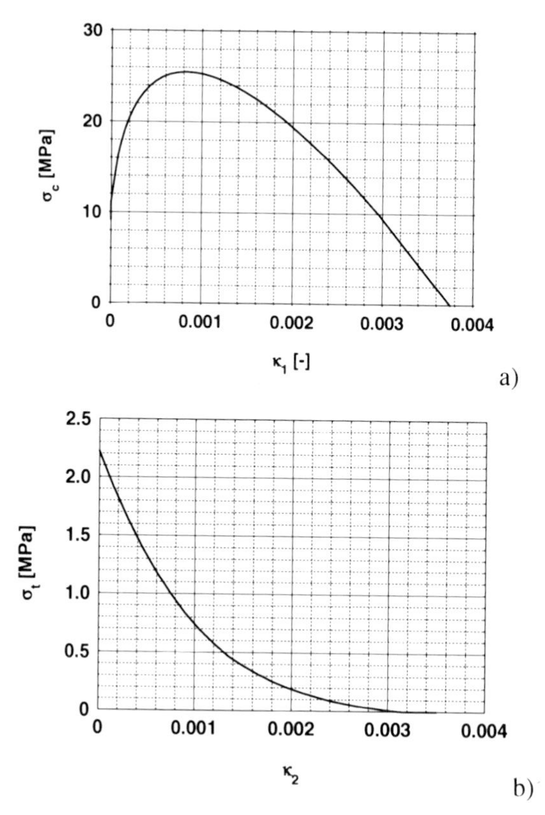

Fig. 7.45 Assumed curves: a) σ_c=f_1(κ_1) in the compressive regime and b) curve σ_t=f_2(κ_2) in the tensile regime (σ_c – compressive yield stress, σ_t – tensile yield stress, κ_i – hardening (softening) parameter, i=1, 2) (Majewski et al. 2008)

Effect of slenderness ratio λ

Figure 7.46 shows the calculated evolution of the vertical force versus the vertical displacement of the top edge u for 3 different slenderness ratios of λ=10, λ=60 and λ=100 using a characteristic length l_c=4.0 mm, perfect bond between concrete and reinforcement, reinforcement ratio ρ=1.98%, eccentricity e=24 mm and uniform distribution of the tensile strength f_t. In turn, Fig. 7.47 compares the calculated curves with experimental ones (the sign '+' for ε_{11} denotes a tensile strain and the sign'-'for ε_{11} stands for a compressive strain). A comparison between the calculated and measured ultimate load and lateral deflection at the column mid-height is given in Tab. 7.4. The evolution of vertical normal stresses in the concrete and reinforcement along the cross-section at the height of $H=l$, $H=0.75l$ and $H=0.5l$ (measured from the bottom) versus the vertical displacement is demonstrated in Fig. 7.48 (for λ=100). The distribution of the non-local parameter $\bar{\kappa}_2$ along the column length at the tensile reinforcement is depicted in Fig. 7.49.

The ultimate load obviously decreases with increasing slenderness ratio. A comparison between the numerical predictions and the experimental results shows a fair agreement in the ultimate load and lateral deflection (Tab. 7.4). The maximum differences in the ultimate load P are about 10%. The height of the tensile zone at mid-height is approximately 4 cm. The symmetric reinforcement is not optimum since the normal stress in the reinforcement of the tensile zone is far below the yield stress (Fig. 7.48). The width of the localized zones w_c is about w_c=5×l_c=20 mm (λ=10) and w_c=3×l_c=12 mm (λ=60-100). In turn, the spacing s_c (on the basis of the non-local parameter $\bar{\kappa}_2$ by counting the peaks) is approximately s_c=8×l_c=32 mm (λ=10) and s_c=6.5×l_c=26 mm (λ=60 and λ=100) (Fig. 7.49).

In the computations, the columns failed similarly as in experiments. The column with λ=10 failed by increased compressive strain in concrete. The failure of the columns with λ=60 and λ=100 took place due to increased tensile strain in reinforcement near the mid-height.

Effect of reinforcement ratio ρ

The effect of the reinforcement ratio in the range 0.6%-5.26% is shown in Figs. 7.50 and 7.51 and Tab. 7.5 using perfect bond, uniform distribution of f_t, l_c=4 mm, e=24 mm and λ=100. The ultimate load increases obviously with increasing reinforcement ratio. The width of the localized zones, w_c=3×l_c=12 mm, does not depend on ρ. In turn, the distance between localized zones at mid-region decreases with increasing ρ up to ρ=3.95%; it is approximately s_c=32 mm (8×l_c) for ρ=0.65%, s_c=26 mm (6.5×l_c) for ρ=1.98%, and s_c=24 mm (6×l_c) for both ρ=3.95% and ρ=5.26%. The difference between calculations and experiments increases with increasing ρ. For ρ=3.95%, it is rather large, namely 30%. This difference is mainly due to an assumption of plane strain conditions; the maximum compressive normal stress in concrete (for ρ=3.95%), 38 MPa, is larger than its uniaxial compressive strength f_c=25.5 MPa. Thus, the ultimate load is larger. An extreme maximum of $\bar{\kappa}_2$ in Figs. 7.51c and 7.51d was obtained at the place of the change of the element size.

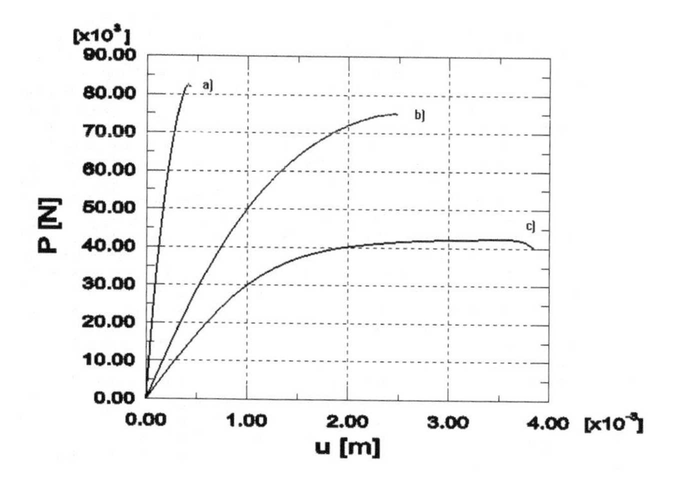

Fig. 7.46 Load-displacement curves for different slenderness: a) λ=10, b) λ=60, c) λ=100 (ρ=1.98%, e=24 mm, l_c=4 mm, P – load, u – vertical displacement of the top edge) (Majewski et al. 2008)

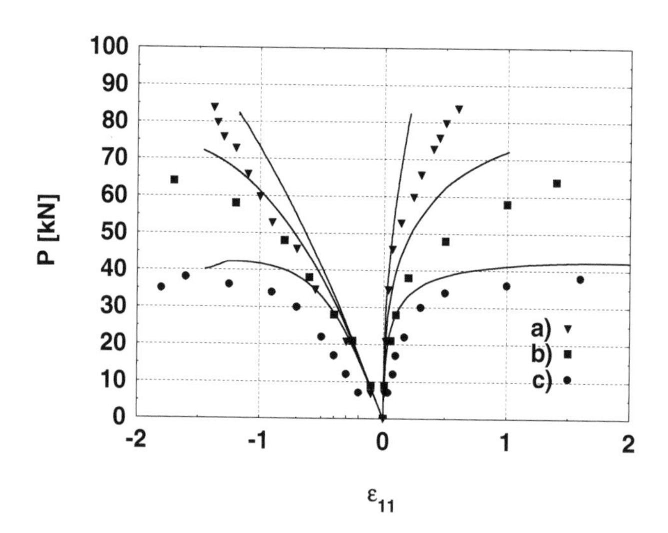

Fig. 7.47 Calculated load-strain curves (solid lines) versus experimental results by Kim and Yang (1995) for a different slenderness: a) λ=10, b) λ=60, c) λ=100 (ρ=1.98%, e=24 mm, l_c=4 mm, P – vertical load, ε_{11} – vertical strain in reinforcement at the mid-height) (Majewski et al. 2008)

Effect of characteristic length l_c

Figure 7.52 demonstrates the effect of a characteristic length on strain localization (using perfect bond, uniform distribution of f_t, λ=100, e=24 mm, l_c=4-24 mm and ρ=1.98%). The ultimate force slightly increases from 42.28 kN (l_c=4 mm) up to 44.40 kN (l_c=24 mm). Several localized zones occur in the tensile zone only for l_c<8 mm. If l_c≥8 mm, one large localized zone appears in the tensile zone at the column mid-height; e.g. for l_c=14 mm, its width is about 200 mm (8.5×l_c) (Fig. 7.52d). In all cases, one localized zone occurs in the compression zone (Figs. 7.52a and 7.52c).

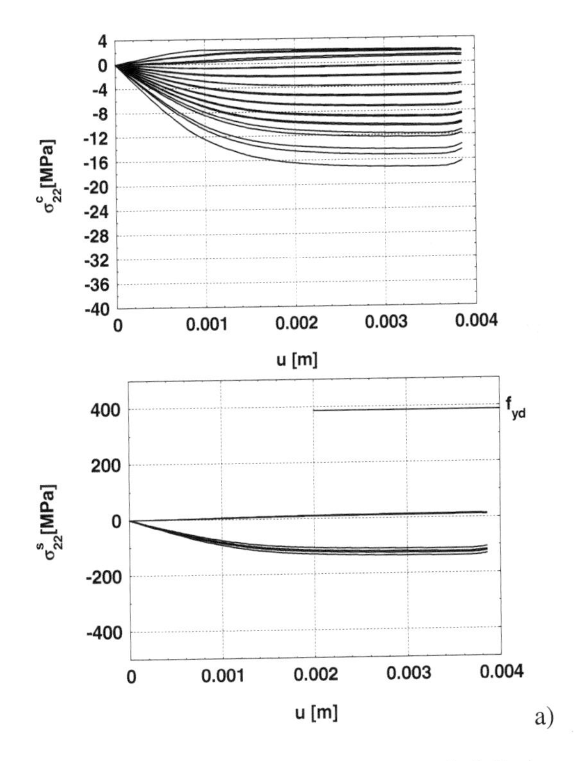

Fig. 7.48 Evolution of vertical normal stresses σ_{22} against vertical displacement in concrete (index 'c') and steel (index 's') in horizontal section at the height $H=l$ (a), $H=0.75l$ (b), and $H=0.5l$ (c) for λ=100 (ρ=1.98%, e=24 mm, l_c=4 mm, f_{yd}=387 MPa – yield stress) (Majewski et al. 2008)

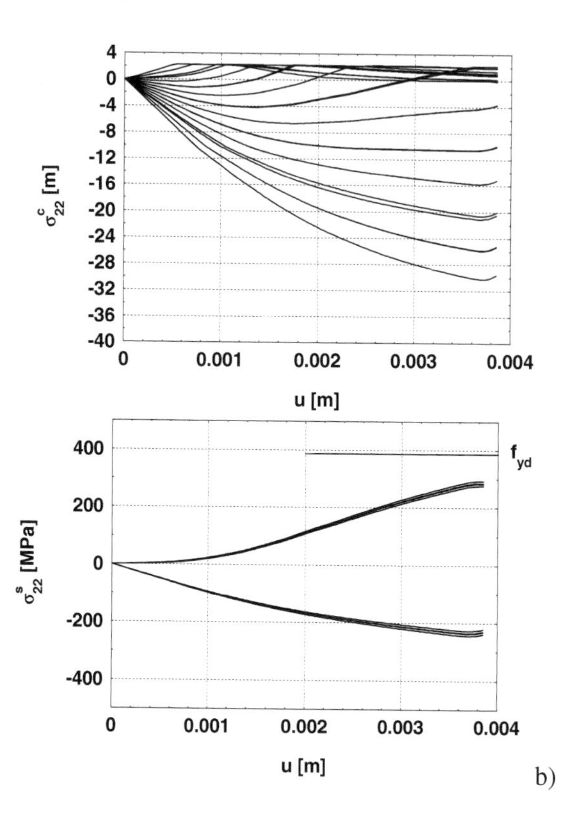

Fig. 7.48 (*continued*)

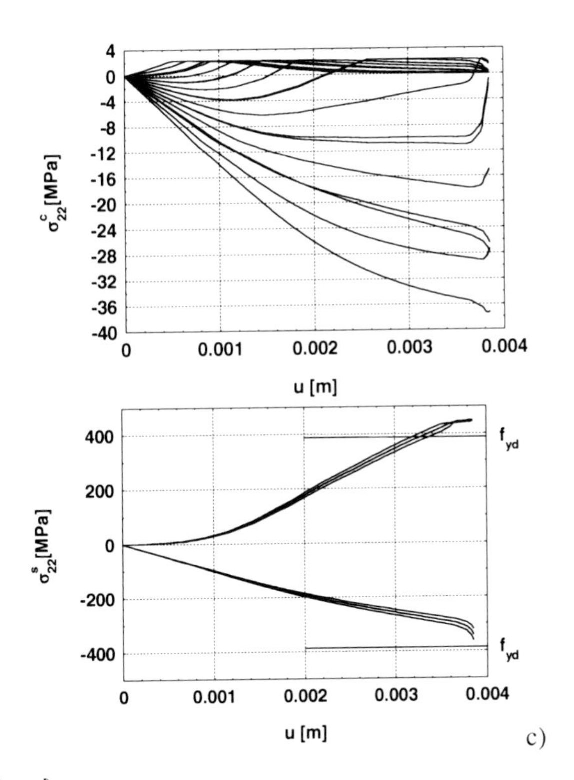

Fig. 7.48 *(continued)*

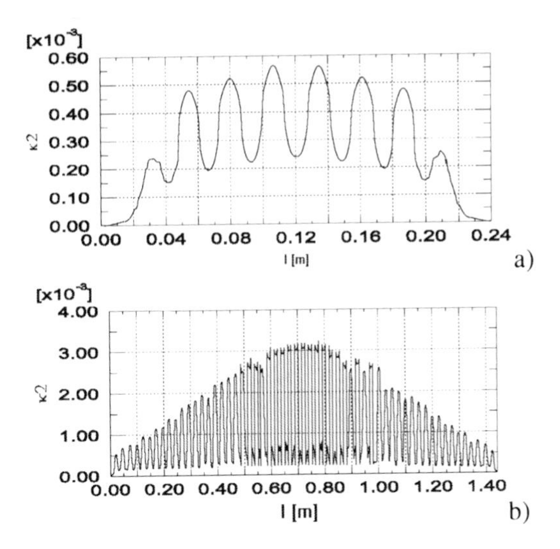

Fig. 7.49 Distribution of non-local parameter $\bar{\kappa}_2$ along column length l for ultimate load with different slenderness: a) λ=10, b) λ=60, c) λ=100 (ρ=1.98%, e=24 mm, l_c=4 mm) (Majewski et al. 2008)

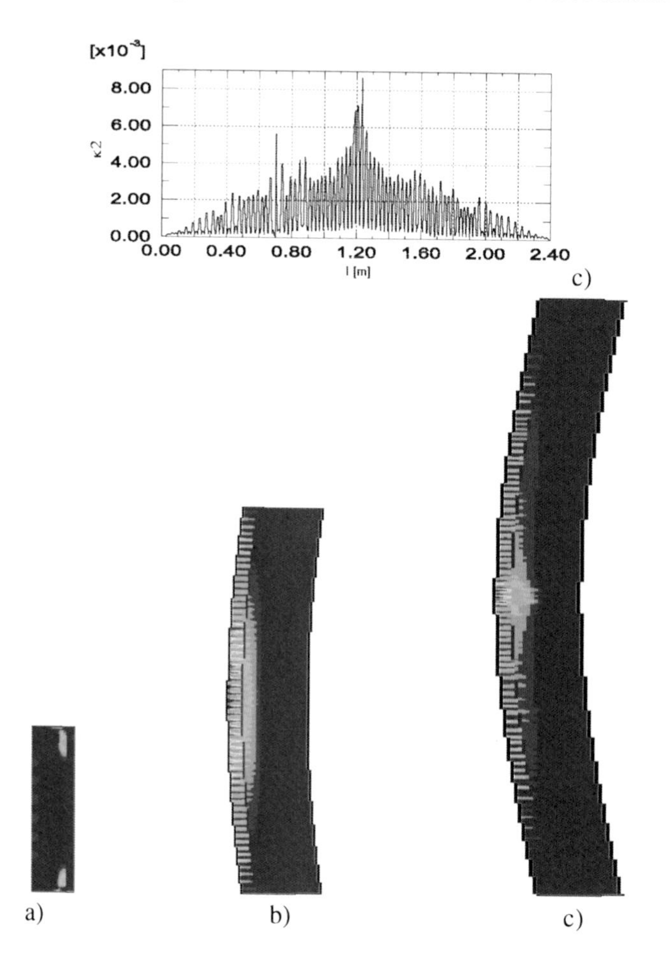

Fig. 7.49 (*continued*)

Table 7.4 Ultimate loads and lateral deflections at mid-height for different slenderness ratios (ρ=1.98%) from experiments and 2D-simulations (Majewski et al. 2008)

Slenderness λ	Ultimate load N [kN] Experiments	Ultimate load N [kN] 2D-calculations	Lateral deflection [mm] Experiments	Lateral deflection [mm] 2D-calculations
10	83.1	82.1	0.4	0.4
60	63.7-65.7	75.3	14.9-16.2	14.9
100	35.0-38.2	42.3	29.8-32.7	40.7

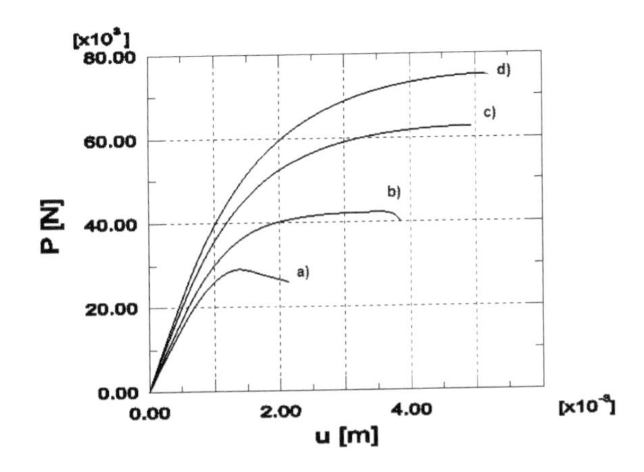

Fig. 7.50 Calculated load-displacement curves for different reinforcement ratios: a) ρ=0.65%, b) ρ=1.98%, c) ρ=3.95%, d) ρ=5.26% (e=24 mm, l_c=4 mm, λ=100, P – load, u – vertical displacement of the top edge) (Majewski et al. 2008)

Effect of eccentricity *e*

The effect of eccentricity on FE-results is depicted in Fig. 7.53 (using perfect bond, uniform distribution of f_t, λ=100, ρ=1.98% and l_c=4 mm). The ultimate load decreases obviously with increasing eccentricity. The width of localized fracture zones is similar, 12 mm ($4\times l_c$). The spacing increases with increasing e; from 26 mm ($6.5\times l_c$) at e=12 mm up to 30 mm ($7.5\times l_c$) at e=40 mm.

Effect of distribution of tensile strength

The calculations were carried out with a stochastic distribution of the tensile strength using perfect bond (λ=100, e=24 mm, ρ=1.98% and l_c=4 mm). The tensile strength was taken from a normal (Gaussian) distribution around the mean value 2.25 MPa with a standard deviation s_d=0.05 MPa and a cut-off c_t=±0.1 MPa. To obtain a Gaussian distribution of the concrete strength, a polar form of the so-called Box-Muller transformation (1958) was used.

The results indicate that the distribution of the tensile strength does not affect the ultimate load and width of localized zones ($w_c=3\times l_c$) for the assumed stochastic parameters s_d and c_t. The spacing of localized zones is slightly larger; about 32 mm ($8\times l_c$) against 26 mm ($6.5\times l_c$) in the case of the uniform distribution of f_t.

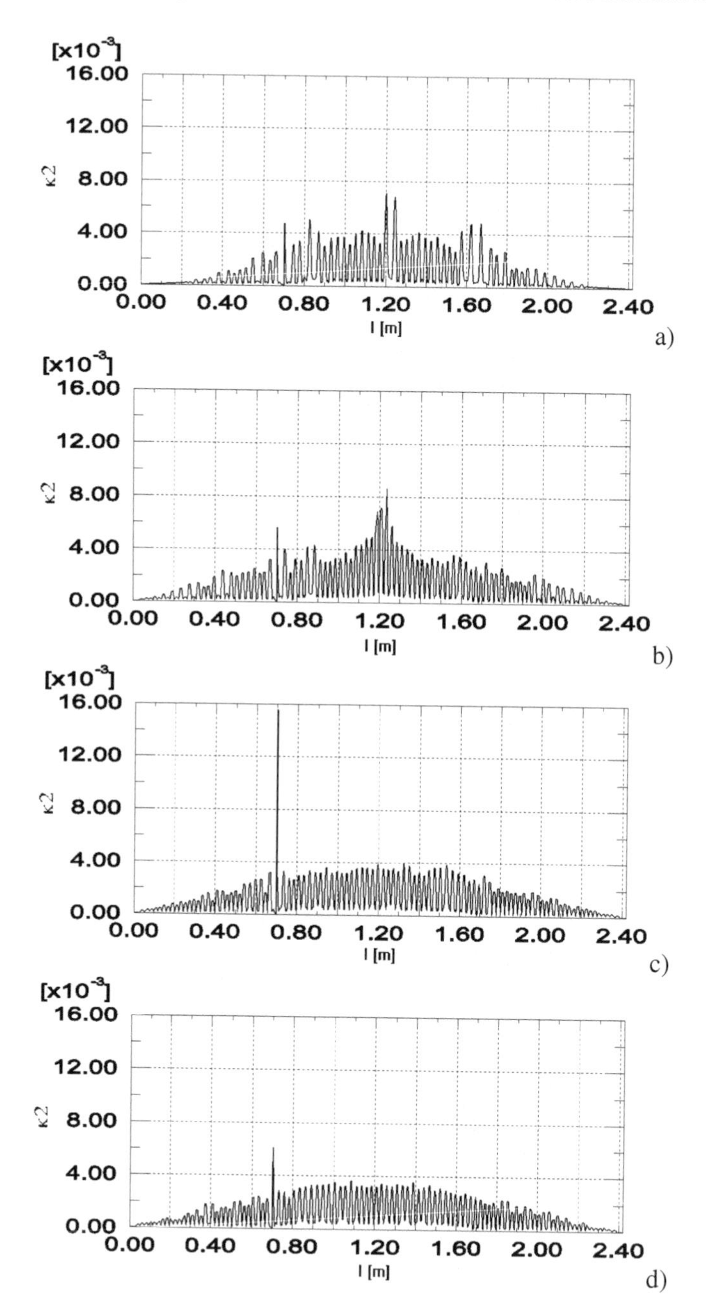

Fig. 7.51 Distribution of non-local parameter $\bar{\kappa}_2$ along the column length l for ultimate load at different reinforcement ratios: a) ρ=0.65%, b) ρ=1.98%, c) ρ=3.95%, d) ρ=5.26% (λ=100, e=24 mm, l_c=4 mm) (Majewski et al. 2008)

Fig. 7.51 (*continued*)

Table 7.5 Ultimate loads and lateral deflections at mid-height for different reinforcement ratios (λ=100) from experiments and 2D-simulations (Majewski et al. 2008)

Reinforcement ratio ρ	Ultimate load N [kN] Experiments	Ultimate load N [kN] 2D-calculations	Lateral deflection [mm] Experiments	Lateral deflection [mm] 2D-calculations
0.6	(-)	29.0	(-)	28.8
1.32	(-)	34.8	(-)	37.6
1.98	35.0-38.2	42.23	29.84-32.72	40.7
2.63	(-)	49.4	(-)	42.7
3.95	47.0-49.0	62.6	36.2-38.2	49.0
5.26	(-)	75.0	(-)	43.4

(-) - experiments were not performed.

Effect of bond-slip

The influence of the bond-slip stiffness by Dörr (1980) (Eqs. 3.114 and 3.115) on the results is shown in Figs. 7.54 and 7.55 for λ=100, ρ=1.98%, e=24 mm l_c=4 mm and uniform distribution of f_t. The simulations were carried out with u_o=0.06 mm and u_o=0.3 mm (Eqs. 3.93 and 3.94). The ultimate load, about 40-41 kN, is smaller in both cases by about 5% as compared to perfect bond. The crack width is similar (3×l_c=12 mm). The cracks spacing is larger for bond-slip than for perfect bond and grows with decreasing bond stiffness. It increases from s_c=26 mm (6.5×l_c) for perfect bond up to s_c=54 mm (13.5×l_c) for bond-slip with u_o=0.06 mm and up to s_c=250 mm (62×l_c) for bond-slip with u_o=0.3 mm.

Effect of fracture energy

The calculations were carried out with a larger fracture energy, namely G_f=0.05 N/mm assuming λ=100, ρ=1.98%, e=24 mm, l_c=4 mm, perfect bond and uniform distribution of f_t.

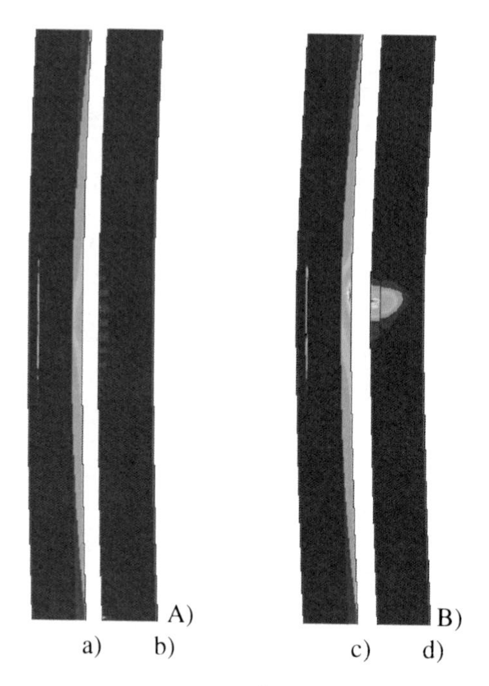

Fig. 7.52 Distribution of non-local parameters $\bar{\kappa}_1$ (a) and $\bar{\kappa}_2$ (b) along column length l for ultimate load with different characteristic lengths: A) l_c=4 mm, B) l_c=14 mm (ρ=1.98%, e=24 mm, λ=100) (Majewski et al. 2008)

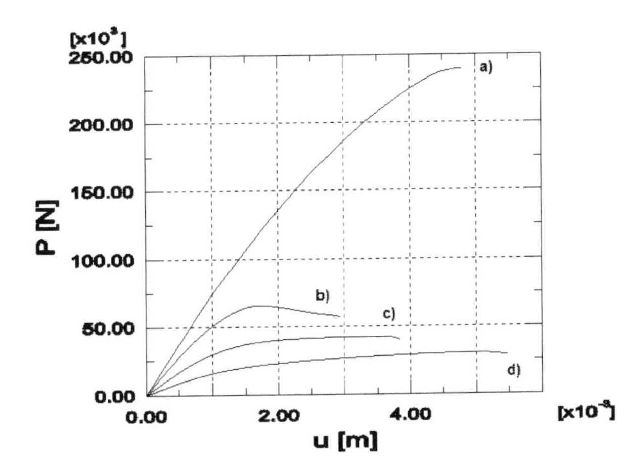

Fig. 7.53 Load-displacement curves for different eccentricities: a) e=0 mm, b) e=12 mm, c) e=24 mm, d) e=40 mm (ρ=1.98%, l_c=4 mm, λ=100, P – load, u – vertical displacement of the top edge) (Majewski et al. 2008)

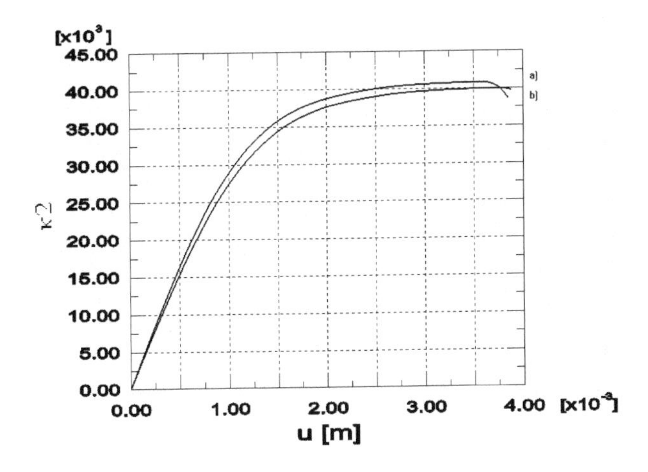

Fig. 7.54 Load-displacement curves: using bond-slip by Dörr (1980) (Eqs. 3.114 and 3.115): a) u_o=0.06 mm, b) u_o=0.3 mm (ρ=1.98%, l_c=4 mm, λ=100, e=24 mm, P – load, u – vertical displacement of the top edge) (Majewski et al. 2008)

The larger fracture energy (smaller softening rate), the larger the ultimate load by ca. 6%. In turn, the crack spacing becomes slightly smaller, i.e. 22-24 mm ($6\times l_c$) (Fig. 7.56a).

Effect of reinforcement type

Figure 7.57 presents the results for $\bar{\kappa}_2$ with reinforcement assumed as 1D elements for λ=100, ρ=1.98%, e=24 mm, l_c=4 mm, perfect bond and uniform distribution of f_t.

The ultimate vertical force is the same as for the 2D reinforcement model. However, the spacing of localized zones increases from s_c=26 mm (6.5×l_c) up to s_c=38 mm (9.5×l_c). It is caused by a different distribution of stresses in reinforcement elements and concrete ones near the reinforcement.

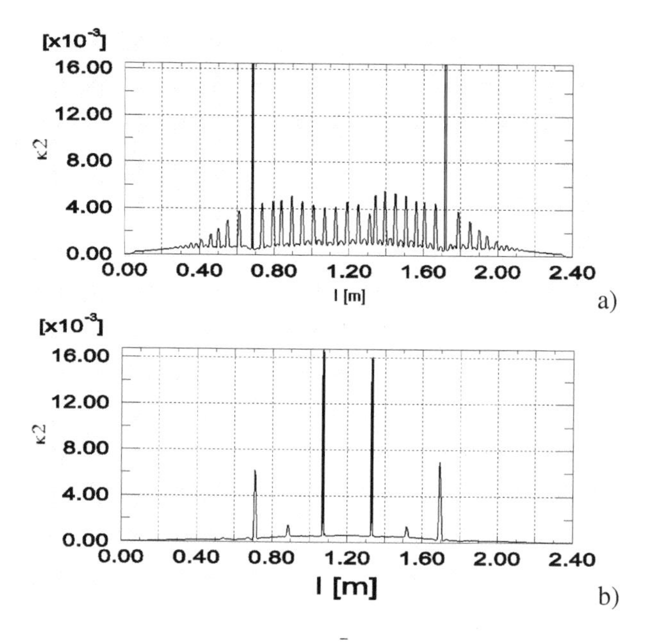

Fig. 7.55 Distribution of non-local parameter $\bar{\kappa}_2$ along column length l for ultimate load using bond-slip by Dörr (1980) a) u_o=0.06 mm, b) u_o=0.3 mm (ρ=1.98%, l_c=4 mm, λ=100, e=24 mm) (Majewski et al. 2008)

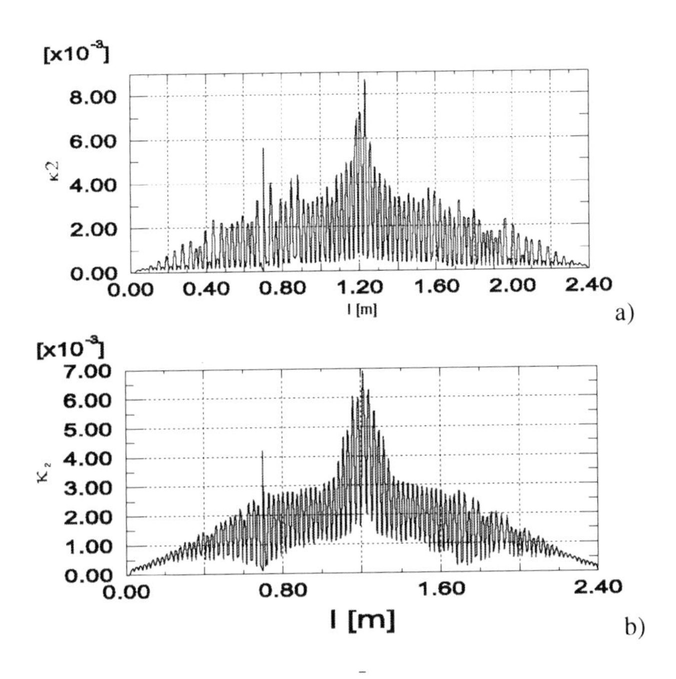

Fig. 7.56 Distribution of non-local parameter $\bar{\kappa}_2$ along column length l for ultimate load using: a) fracture energy with non-linear softening (G_f=0.05 N/mm), b) fracture energy with non-linear softening (G_f=0.02 N/mm), (ρ=1.98%, l_c=4 mm, λ=100, e=24 mm, P – load, u – vertical displacement of the top edge) (Majewski et al. 2008)

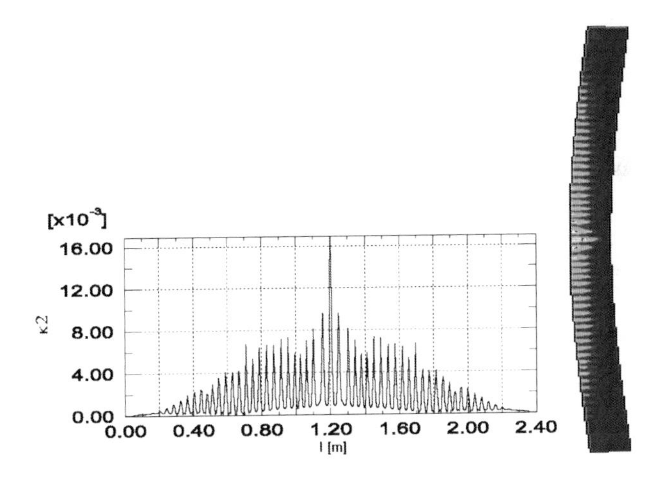

Fig. 7.57 Distribution of non-local parameter $\bar{\kappa}_2$ along column length l for ultimate load with reinforcement assumed as 1D elements (ρ=1.98%, l_c=4 mm, λ=100, e=24 mm, perfect bond, κ_2^u=0.035) (Majewski et al. 2008)

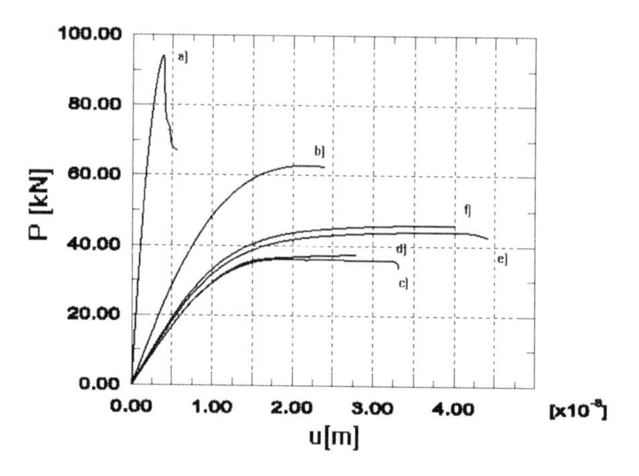

Fig. 7.58 Load-displacement curves from 3D calculations: a) λ=10, perfect bond, ρ=1.98%, b) λ=60, perfect bond, ρ=1.98%, c) λ=100, perfect bond, ρ=1.98%, d) λ=100, bond-slip, ρ=1.98%, e) λ=100, perfect bond, ρ=3.95%, f) λ=100, bond-slip, ρ=3.95% (e=24 mm, l_c=4 mm, P – load, u – vertical displacement of the top edge) (Majewski et al. 2008)

Effect of 3D-calculations

The 3D calculations were carried out for e=24 mm, l_c=4 mm, ρ=1.98% (λ=10-100) and ρ=3.95% (λ=100) assuming perfect bond and an uniform distribution of f_t. In addition, two calculations were carried out with bond-slip by Dörr (1980) for u_o=0.06 mm (λ=100, ρ=1.98% and 3.95%). Figure 7.58 shows the calculated load-displacement curves. In turn, the distribution of the non-local parameter κ_2 along the column length for the ultimate load is demonstrated in Fig. 7.59. Table 7.5 compares the numerical results with experimental ones.

All calculated buckling loads differ by only 10% from experimental ones. The buckling load for bond-slip is slightly larger than perfect bond. For perfect bond, the crack spacing is larger in 3D-simulations than in 2D-simulations; it is about 40 mm (10×l_c) for λ=10 and λ=60, and 42 mm (10.5×l_c) for λ=100 at ρ=1.98%, and 35 mm (8.8×l_c) for λ=100 at ρ=3.95%. For bond-slip and λ=100, it is 44 mm (11×l_c) for ρ=1.98% and ρ=3.95%. By going from 2D to 3D calculations, the approximation of the ultimate load improves for a medium and slender column but gets worse for a stocky one due to slightly different failure mechanisms in stocky columns (Figs. 7.59a and 7.59a). In the slender columns, the failure mechanism is similar (Figs. 7.59bc and 7.59b-f). The incorporation of bond slip leads to a decrease of the ultimate load in 2D simulations and to its increase in 3D analyses. It is caused by a different model of reinforcement assumed in calculations.

The calculated spacing of fracture process zones from 3D calculations for perfect bond and bond-slip at ρ=1.98%, s_c=40-42 mm, was compared with the average crack spacing s_c according to 3 different formulas: a) by CEB-FIP Model Code (1991)

$$s_c = \frac{2}{3} \times \frac{\phi_s}{3.6\rho} = \frac{2}{3} \times \frac{6}{3.6 \times 0.0198} = 56 \text{ mm}, \tag{7.18}$$

b) by Eurocode 2 (1991)

$$s_c = 50 + k_1 k_2 \frac{\phi_s}{4\rho} = 50 + 0.8 \times 1 \frac{6}{4 \times 0.0198} = 111 \text{ mm}, \tag{7.19}$$

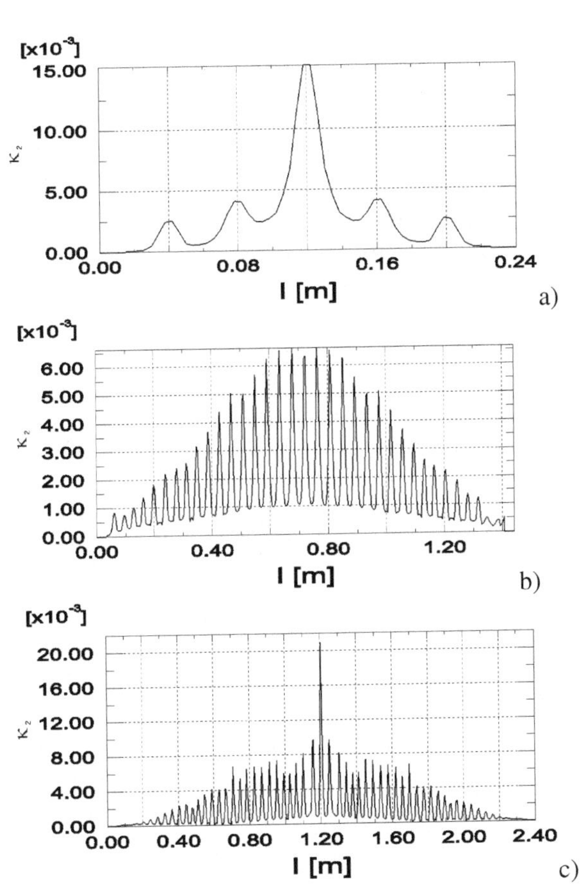

Fig. 7.59 Distribution of non-local tensile parameter along the column length for the ultimate load from 3D calculations: a) λ=10, perfect bond, ρ=1.98%, b) λ=60, perfect bond, ρ=1.98%, c) λ=100, perfect bond, ρ=1.98%, d) λ=100, bond-slip, ρ=1.98%, e) λ=100, perfect bond, ρ=3.95%, f) λ=100, bond-slip, ρ=3.95% (e=24 mm, l_c=4 mm) (Majewski et al. 2008)

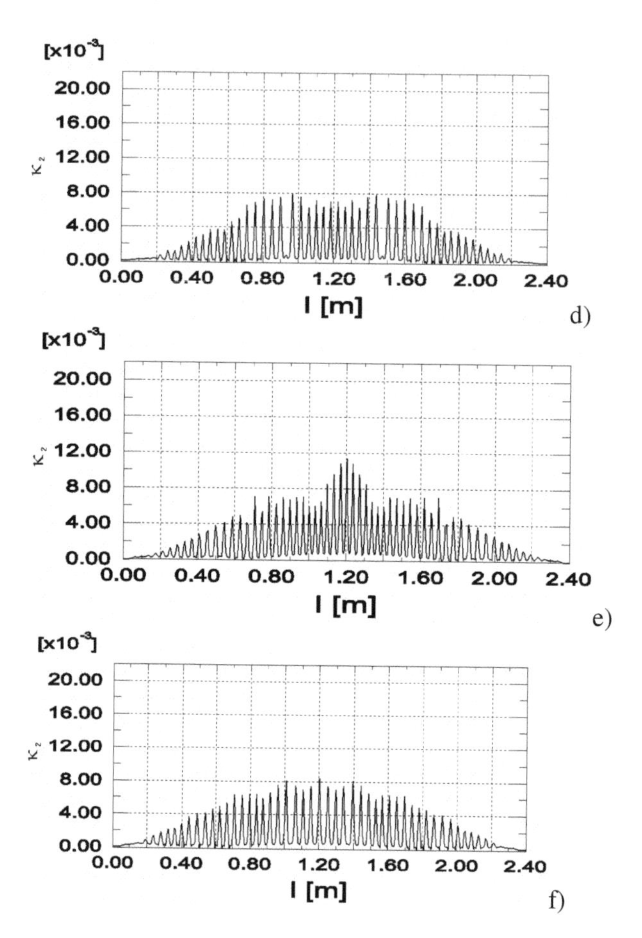

Fig. 7.59 *(continued)*

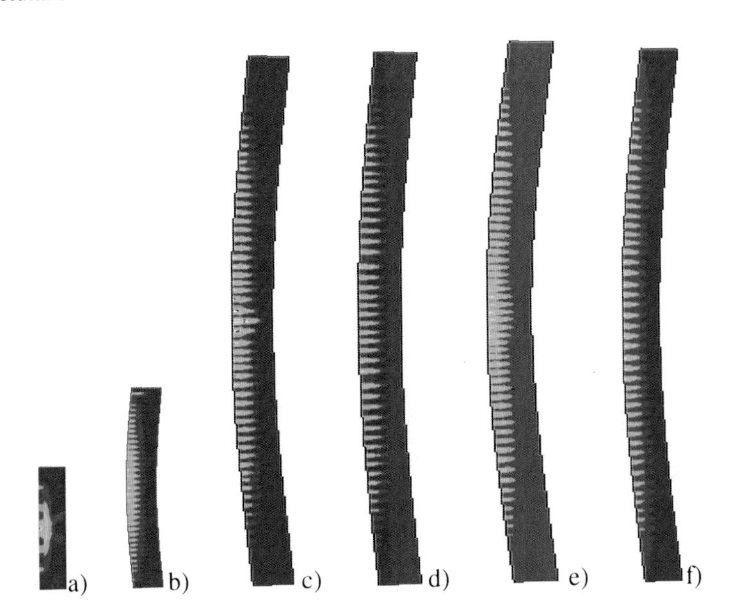

Fig. 7.59 (*continued*)

c) by Lorrain et al. (1998)

$$s_c = 1.5c + 0.1\frac{\phi_s}{\rho} = 1.5\times 15 + 0.1\frac{6}{0.0198} = 53 \text{ mm}, \tag{7.20}$$

wherein ϕ_s=6 mm is the reinforcing bar diameter, ρ=1.98% denotes the reinforcement ratio, k_i are the coefficients and c denotes the concrete cover. Thus, the calculated spacing of localized zones with l_c=4 mm (s_c=42 mm) is in good agreement with the average crack spacing according to CEB-FIP Model Code (1991) and Lorrain et al. (1998). In the case of ρ=3.95% and bond-slip, the match between the calculations (s_c=44 mm) and analytical formula (s_c=38 mm) is even in a better agreement.

Table 7.6 Ultimate loads and lateral deflections at mid-height for different slenderness ratios from experiments and 3D-simulations (Majewski et al. 2008)

Slenderness λ	Ultimate load N [kN] Experiments	Ultimate load N [kN] 3D-calculations	Lateral deflection [mm] Experiments	Lateral deflection [mm] 3D- calculations
10 (ρ=1.98%) (perfect bond)	83.1	94.6	0.403	0.55
60 (ρ=1.98%) (perfect bond)	63.7-65.7	62.9	14.9-16.2	16.2
100 (ρ=1.98%) (perfect bond)	35.0-38.2	36.0	29.8-32.7	36.0
100 (ρ=3.95%) (perfect bond)	47.0-49.0	44.0	36.2-38.2	43.7
100 (ρ=1.98%) (bond-slip)	35.0-38.2	37.4	29.8-32.7	30.0
100 (ρ=3.95%) (bond-slip)	47.0-49.0	46.0	36.2-38.2	41.7

Table 7.7 summarizes some numerical and analytical results with respect to the spacing of localized zones.

Table 7.7 Spacing of localized zones from FE-calculations and analytical formulae (Majewski et al. 2008)

Slender-ness λ	Reinforce-ment ratio ρ	Eccentri-city e [mm]	Bond	CEB-FIB Model Code (1991) [mm]	Eurocode (1991) [mm]	Lorrain et al. (1998) [mm]	FEM 2D [mm]	FEM 3D [mm]
10	1.98	24	*pb*	56	111	53	32	40
60	1.98	24	*pb*	56	111	53	26	40
100	1.98	24	*pb*	56	111	53	26	42
100	0.65	24	*pb*	171	235	115	32	-
100	3.95	24	*pb*	28	80	38	24	35
100	5.26	24	*pb*	21	73	34	24	-
100	1.98	12	*pb*	56	111	53	26	-
100	1.98	40	*pb*	56	111	53	30	-
100	1.98	24	*bs-1*	56	111	53	54	44
100	1.98	24	*bs-2*	56	111	53	250	-

pb – perfect bond, *bs* – bond slip, *bs-1* - u_o=0.06 mm, *bs-2* - u_o=0.30 mm.

FE analysis of experimental tests with columns by Kim and Lee (2000), Hsu (1988), Lloyd and Rangan (1996) and Bažant and Kwon (1994)

The geometries of selected experimental reinforced concrete columns subjected to numerical analyses are given in Fig. 7.60 and Tab. 7.8. The maximum size of finite elements in all columns was not greater there than $3\times l_c$ to get mesh-objective results (l_c=4 mm). All columns were hinged at the both ends. The deformation was induced by imposing increments of a vertical displacement prescribed to both ends through steel plates with a large stiffness. The assumed diagrams describing the uniaxial compressive yield stress σ_c versus the vertical total strain ε and tensile yield stress σ_t versus a softening parameter κ_2 are shown in Fig. 7.61. The internal friction angle was always assumed to be φ=14° and the dilatancy angle was taken as ψ=10° (Eq. 3.27). In the case of the lack of information on the tensile strength in experiments (Hsu 1988), one assumed the tensile strength equal to $f_t\approx 1.73$ MPa ($=0.08\times f_c$). The fracture energy in tension was mainly $G_f=g_f\times w_c\approx 100$ N/mm (g_f – area under the softening function, $w_c=3\times l_c$ mm – width of localized zones). The non-local parameter was equal to m=2. The vertical and horizontal reinforcement were assumed in the form of 2-node truss elements. The vertical reinforcement was composed of 4 or 6 bars which were fixed at ends (as in experiments). The volume weight of the concrete column was not taken into account. In the case of calculations including stirrups, a perfect bond was assumed between concrete and ties. The Poisson's ratio of concrete was 0.20. The modulus of elasticity of reinforcement was E_s=200 GPa. The stirrups were taken in into account only in the calculations of a size effect for reinforced columns by Bažant and Kwon (1994).

The numerical results compared to the experimental ones have been summarized in Tab. 7.9. In addition the values of the buckling forces according to the Polish Standard (2002) were enclosed.

Experiments on columns by Kim and Lee (2000)

The FE-analyses were performed for one square column 100×100 mm^2 with the slenderness ratio of λ=45 undergoing uniaxial bending (denoted as SS0) and biaxial bending (denoted as SS45). The eccentricity was 40 mm (in one direction during uniaxial bending or in two directions during biaxial bending). The maximum size of the aggregate in concrete was d_a^{max}=8 mm. The ties made of 4.7 mm plain steel bars were at a distance of 100 mm (reduced to 50 mm at ends). About 19500 8-noded solid elements were used (8×8×304).

The evolution of the calculated vertical force versus the lateral deflection as compared to experiment is shown in Fig. 7.62. The calculated failure force of 124.9 kN for uniaxial bending is comparable to the experimental values of 119-126 kN (Tab. 7.9). In the case of biaxial bending, the calculated failure force of 114.4 kN is larger than the experimental forces of 103-106 kN by 10% (Tab. 7.9).

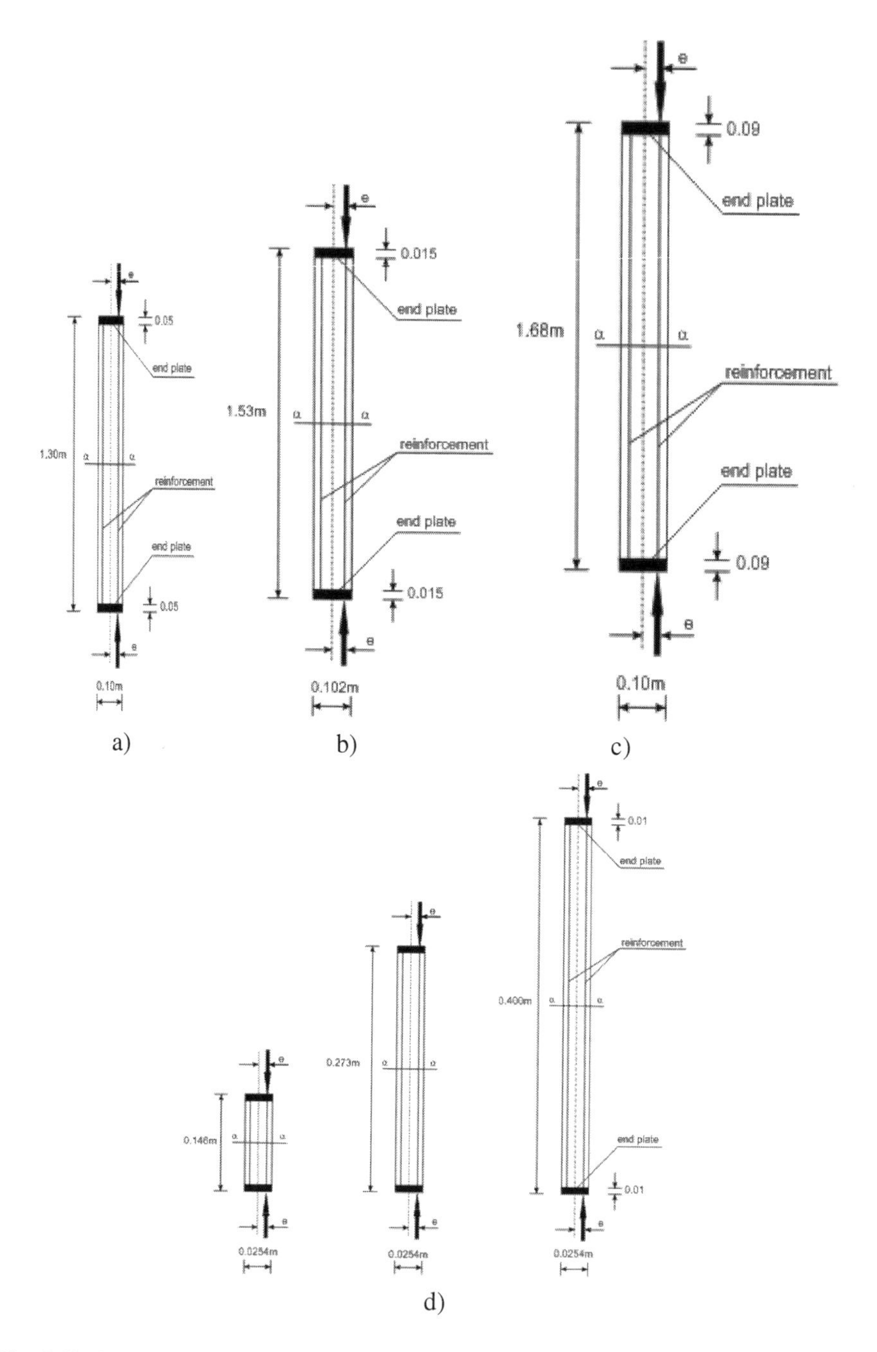

Fig. 7.60 Geometry of reinforced concrete columns in experiments by: a) Kim and Lee (2000), b) Hsu (1988), c) Lloyd and Rangan (1996) and d) Bažant and Kwon (1994) (Majewski et al. 2009)

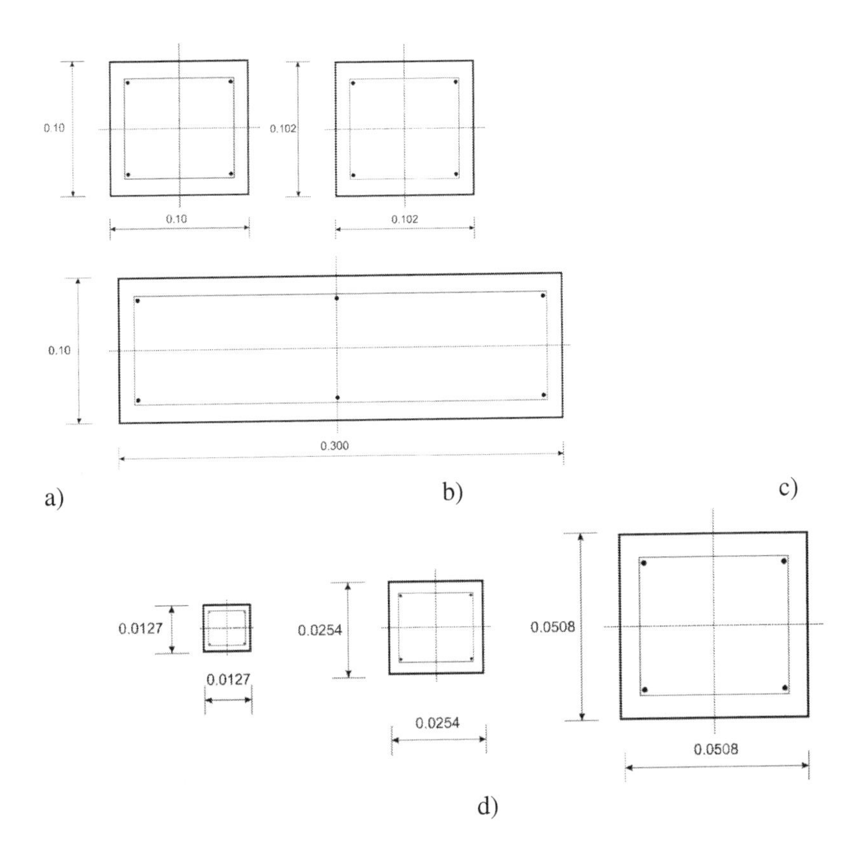

Fig. 7.60 (*continued*)

Moreover, the calculated shape of the load-displacement curve after peak is different, viz. the calculated behaviour of columns is too ductile. The distribution of a non-local parameter $\bar{\kappa}_2$ along the column length at the tensile reinforcement is depicted in Fig. 7.63. The calculated mean spacing of localized zones is 40 mm (uniaxial bending) and 36 mm (biaxial bending), and is smaller than the average crack spacing according to: CEB-FIP Model Code (1991), 62 mm, Eurocode 2 (1991), 127 mm, and Lorrain et al. (1998), 73 mm. The calculated deflections at ultimate load (mid-height), 15 mm (uniaxial bending) and 13 mm (biaxial bending), are comparable to the measured value of 16-18 mm for uniaxial bending and 8.5-9.5 mm (biaxial bending), respectively (Tab. 7.9).

In the computations, the columns failed similarly as in experiments. The failure took place in the compressive part of concrete at mid-height.

The effect of a smaller fracture energy in tension (G_f=50 N/mm) on FE-results was insignificant. The calculated failure forces were similar. In turn, the calculated mean spacing of localized zones was slightly larger (by 10%).

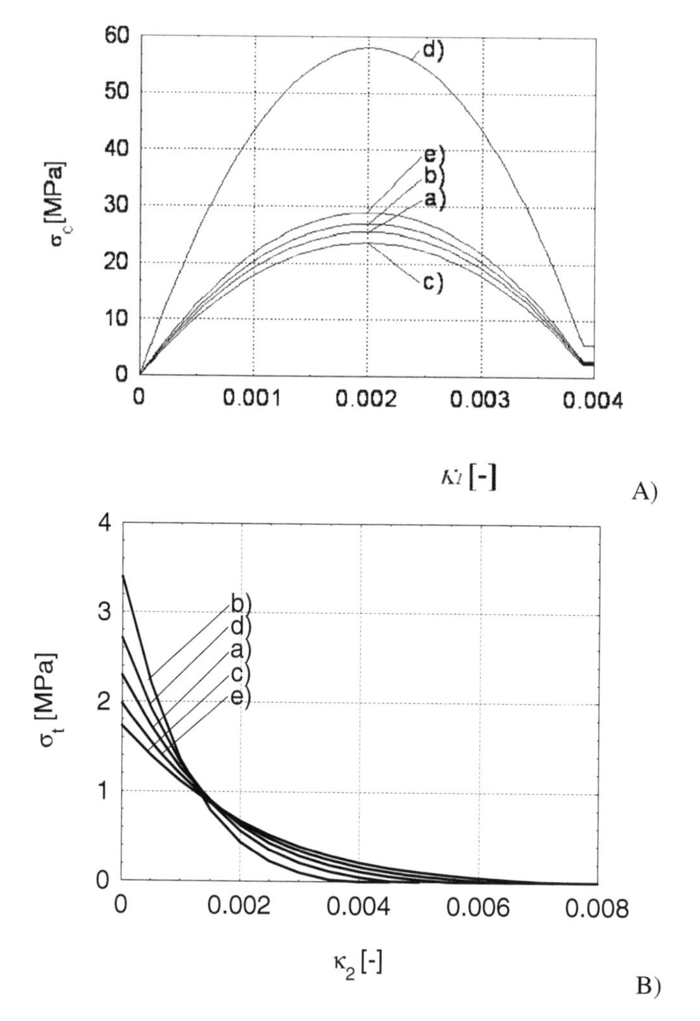

Fig. 7.61 Assumed curve $\sigma_c=f_1(\varepsilon)$ in the compressive regime (A) and tensile regime (B) for experiments by: a) Kim and Yang (1995), b) Kim and Lee (2000), c) Hsu (1988), d) Lloyd and Rangan (1996) and e) Bažant and Kwon (1994) (Majewski et al. 2009)

Table 7.8 Details of test columns by Kim and Lee (2000), b) Hsu (1988), c) Lloyd and Rangan (1996) and d) Bažant and Kwon (1994) (Majewski et al. 2009)

Tests	Cross-section area $b\times d$ [mm^2]	Length l [m]	Slender-ness	Eccentricity [mm]	Reinforce-ment (ratio [%]/number and diameter [mm])
1a, 1b [1]	100×100	1.30	45	40	2.14 (4ϕ 8.3)
2a [2]	102×102	1.53	52	127	2.75 (4ϕ 9.5)
2b [2]	102×102	1.53	52	76.2	2.75 (4ϕ 9.5)
3a [3]	102×305	1.68	57	30	1.44 (6ϕ 12)
3b [3]	102×305	1.68	57	40	1.44 (6ϕ 12)
4a [4]	25.4×25.4	0.146	19.9	6.35	4.91 (4ϕ 3.2)
4b [4]	25.4×25.4	0.273	37.2	6.35	4.91 (4ϕ 3.2)
4c [4]	25.4×25.4	0.400	54.6	6.35	4.91 (4ϕ 3.2)
Tests	Concrete cover [mm]	Concrete compres-sive strength [MPa]	Concrete Tensile strength [MPa]	Modulus of elasticity of concrete [GPa]	Yield stress in vertical reinforce-ment [MPa]
1a, 1b [1]	23	27.0	3.4	24.30	436
2a [2]	24	23.6	1.73	30.04	307
2b [2]	24	23.6	1.73	30.04	307
3a [3]	15	58.	2.71	38.392	430
3b [3]	15	58.	2.71	38.392	430
4a [4]	5.1	29	1.98	32.5	552
4b [4]	5.1	29	1.98	32.5	552
4c [4]	5.1	29	1.98	32.5	552

[1] Kim and Lee (2000), [2] Hsu (1988), [3] Lloyd and Rangan (1996), [4] Bažant and Kwon (1994).

Experiments on columns by Hsu (1988)

The FE-analyses were performed for 2 square columns 102×102 mm^2 with the slenderness ratio of λ=52 (denoted by HS). The eccentricities were: e=76.2 mm and e=127 mm. About 5000 8-noded solid elements were used (5×5×200).

The evolution of the calculated vertical force versus the lateral deflection as compared to experiment is shown in Fig. 7.64. The calculated failure loads of 29.4 kN (e=76.2 mm) and 53.4 kN (e=127 mm) agree with the corresponding experimental values 28.7 kN and 53 kN, respectively (Tab. 7.9). The distribution of a non-local parameter $\bar{\kappa}_2$ along the column length in concrete (at the tensile reinforcement) is depicted in Fig. 7.65. The calculated spacing of localized zones (s_c=58 mm) is in good agreement with the average crack spacing according to CEB-FIP Model Code (1991), 64 mm, and Lorrain et al. (1998), 70 mm.

In the computations, the columns failed similarly as in experiments. The failure took place due to increased tensile strain in reinforcement near the mid-height.

Table 7.9 Data summary of tests, FE-analysis and standard values (forces, mid-height deflections, spacing of localized zones); experiments by Kim and Lee (2000), b) Hsu (1988), c) Lloyd and Rangan (1996) and d) Bažant and Kwon (1994) (Majewski et al. 2009)

Tests	Failure peak force (experiment) [kN]	Failure peak force (FEM) [kN]	Failure peak force (Polish Standard 2002) [kN]	Deflection at maximum load (experiment) [mm]	Deflection at maximum load (FEM) [mm]
1a [1]	119-126	124.9	103.6	16.0-18.1	15
1b [1]	103-106	114.4	77.0	8.4-9.5	13
2a [2]	28.7	29.9	25.9	-	12
2b [2]	54	53.4	41.7	-	11
3a [3]	471	528.9	342.9	32.1	15
3b [3]	422	354.3	281.2	27.6	15
4a [4]	10.3-13.0	13.32	9.55	0.97-1.24	1.67
4b [4]	11.7-12.6	10.8	8.9	2.01-2.41	2.19
4c [4]	10.5-11.1	8.6	8.0	3.76-4.78	4.2
Tests	Spacing of localized zones (FEM) [mm]	Crack spacing (standard CEB-FIP 1991)	Crack spacing (standard Eurocode 1991)	Crack spacing (Lorrain et al. 1998)	
1a [1]	40	62	127	73	
1b [1]	36	62	127	73	
2a [2]	58	64	119	70	
2b [2]	58	84	119	70	
3a [3]	47	152	214	105	
3b [3]	50	152	214	105	
4a [4]	*	12	63	14	
4b [4]	*	12	63	14	
4c [4]	*	12	63	14	

[1] Kim and Lee (2000), [2] Hsu (1988), [3] Lloyd and Rangan (1996), [4] Bažant and Kwon (1994).
(*) – one localized zone.

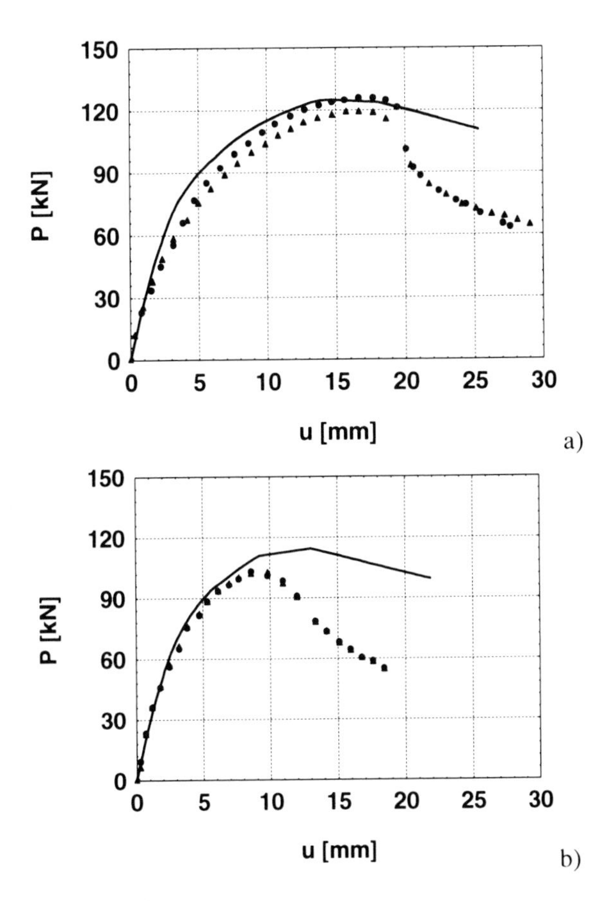

Fig. 7.62 Calculated load-deflection curves (solid lines) versus experimental results by Kim and Lee (2000) (λ=45, ρ=2.85%, e=40 mm, l_c=4 mm): a) uniaxial bending, b) biaxial bending (Majewski et al. 2009)

Experiments on columns by Lloyd and Rangan (1996)

The FE-analyses were performed for 2 selected rectangular columns 102×305 mm^2 (denoted as IVA and IVB) with the slenderness ratio of λ=57 using the eccentricities e=30 mm and e=40 mm. The maximum size of the aggregate in concrete was d_{max}=7-10 mm. The ties made of 4 mm plain steel bars were at a distance of 60 mm. About 15000 8-noded solid elements were used (8×10×190).

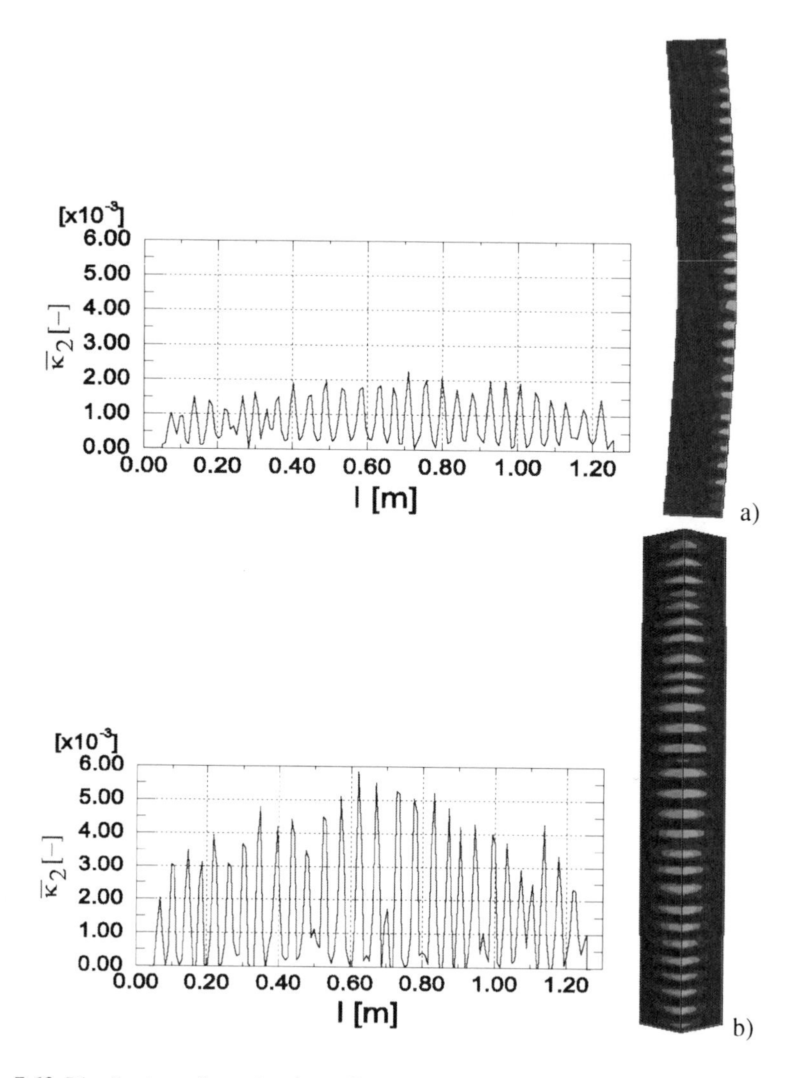

Fig. 7.63 Distribution of non-local tensile parameter along column length l for ultimate load (λ=45, ρ=2.85%, e=40 mm, l_c=4 mm) in experiments by Kim and Lee (2000): a) uniaxial bending, b) biaxial bending (2 perpendicular side surfaces were shown) (Majewski et al. 2009)

The evolution of the calculated vertical force versus the lateral deflection as compared to experiment is shown in Fig. 7.66. The calculated failure loads of 529 kN (e=30 mm) and 354 kN (e=40 mm) significantly differ from the corresponding experimental values of 471 kN and 422 kN, respectively. The distribution of a non-local parameter $\bar{\kappa}_2$ along the column length in concrete (at the tensile reinforcement) is depicted in Fig. 7.67. The mean spacing of localized zones is 47 mm (*e*=30 mm) and 50 mm (*e*=40 mm), respectively, and is 3 times smaller than the average crack spacing according to CEB-FIP Model Code (1991), 154 mm. The calculated mid-height deflection at ultimate load, 15 mm, is slightly smaller than the measured value of 27.6-32.1 mm (Tab. 7.9).

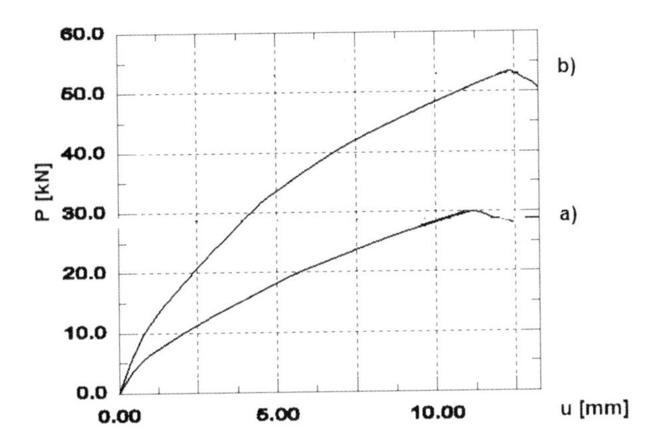

Fig. 7.64 Calculated load-deflection curves for experiments by Hsu (1988) (λ=52, ρ=2.95%, l_c=5 mm): a) *e*=127 mm, b) *e*=76.2 mm; experimental peak values: 28.7 kN (*e*=127 mm) and 54 kN (*e*=76.2 mm) (Majewski et al. 2009)

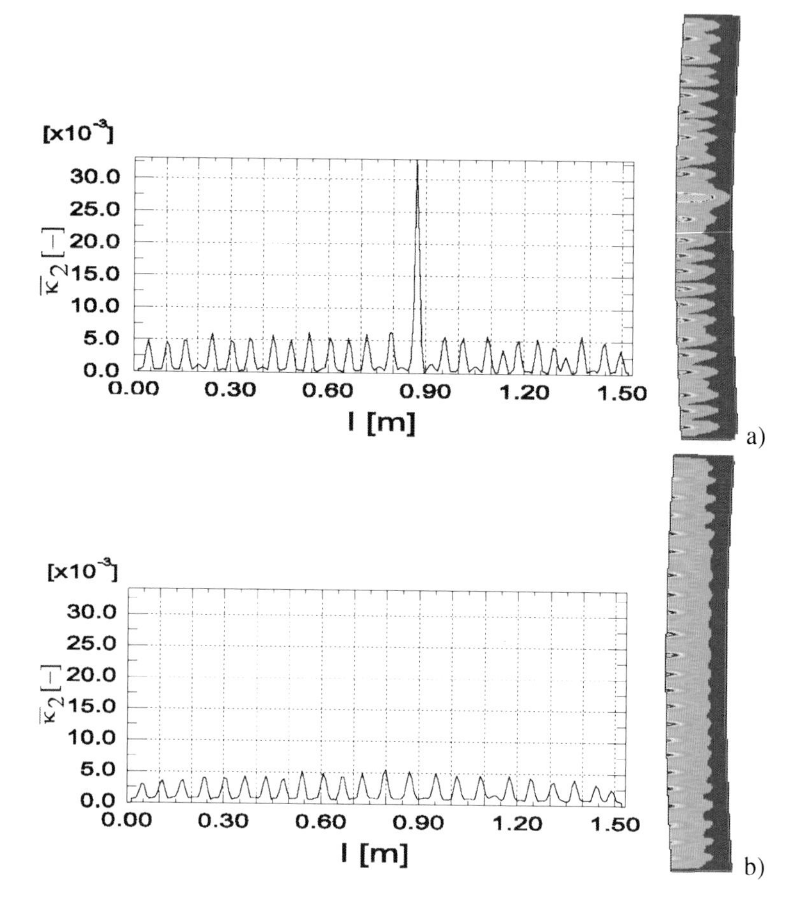

Fig. 7.65 Distribution of non-local tensile parameter along column length l for ultimate load (λ=52, ρ=2.95%, l_c=5 mm) in tests by Hsu (1988): a) e=76.2 mm, b) e=127 mm (Majewski et al. 2009)

In the computations, the columns failed similarly as in experiments; they broke at mid-length in the compressive zone.

Experiments on columns by Bažant and Kwon (1994)

The FE-analyses were performed for 3 square columns 25.4×25.4 mm^2 (denoted by M1, M2 and M3) with the slenderness ratio of λ=19.2, 35.8 and 52.5, respectively. The maximum size of the aggregate in concrete was d_{max}=3.35 mm. The ties made of 1.59 mm plain steel bars were at a distance of 15.2 mm (reduced to 10.2 mm at ends). In the experiment, the columns broke right at mid-length. In the calculations, the stirrups were taken into account. About 15200 8-noded solid elements were used (5×5×200).

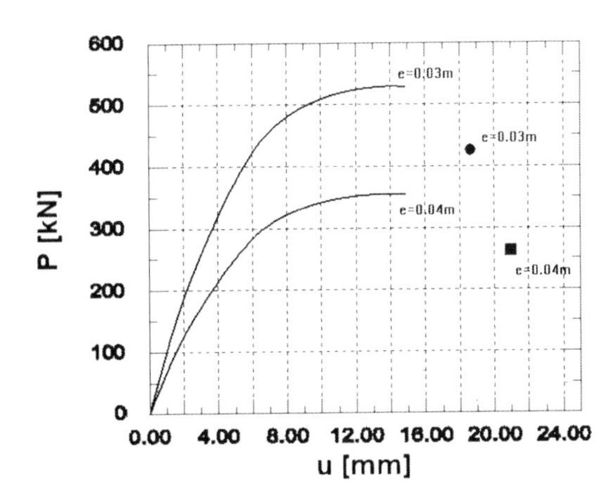

Fig. 7.66 Calculated load-deflection curves for experiments by Lloyd and Rangan (1996) as compared to experimental ultimate loads (λ=58, ρ=1.44%, l_c=5 mm): a) e=30 mm, b) e=40 mm

The evolution of the calculated vertical forces versus the lateral deflection as compared to experiment is shown in Fig. 7.68. The calculated failure forces of 13.3 kN (λ=19.2), 10.8 kN (λ=35.8) and 8.6 kN (λ=52.5) differ only by 10%-20% from the corresponding experimental values 10.3 kN - 13.3 kN (λ=19.2), 11.7 kN - 12.6 kN (λ=35.8) and 10.5-11.1 kN (λ=52.5), respectively (Tab. 7.9). Thus, a deterministic size effect was properly captured. The distribution of the non-local parameter $\bar{\kappa}_2$ along the column length at the tensile reinforcement is depicted in Fig. 7.69. Only one diffuse localized zone was obtained. The calculated spacing of fracture process according to CEB-FIP Model Code (1991) is 12 mm. The calculated deflections at ultimate load, 1.67 mm (λ=19.2 mm), 2.19 mm (λ=35.8 mm) and 4.2 mm (λ=52.5 mm) compare well with the measured value of 0.97-1.24 mm (λ=19.2 mm), 2.01-2.41 mm (λ=35.8 mm) and 3.76-4.78 mm (λ=52.5 mm) (Tab. 7.9).

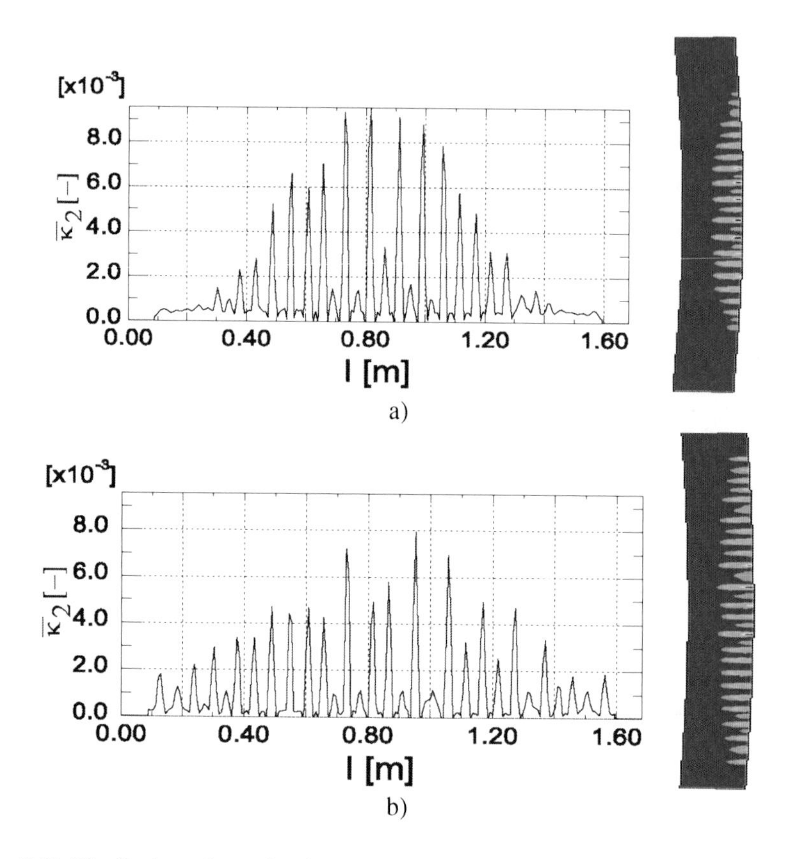

Fig. 7.67 Distribution of non-local parameter along column length l for ultimate load (λ=58, ρ=1.4%, l_c=5 mm) in tests by Lloyd and Rangan (1996): a) e=30 mm, b) e=40 mm (Majewski et. 2009)

In the computations, the columns failed similarly as in experiments; they broke at mid-length in the compressive zone.

The difference between the calculated and standard failure force by PN-B-03264 (2002) was higher by 25%-30% (Tab. 7.9).

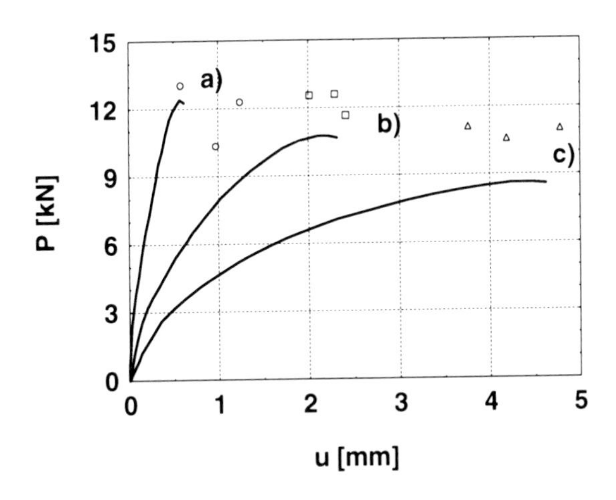

Fig. 7.68 Calculated load-deflection curves for experiments by Bažant and Kwon (1994) (e=6.35 mm, ρ=4.91%, l_c=5 mm): a) λ=19.2, b) λ=35.8, c) λ=52.5 (Majewski et al. 2009)

The following conclusions can be drawn from the FE-analysis of the behaviour of reinforced concrete elements under eccentric compression:

- the numerical results of buckling loads and lateral deflections are in satisfactory accordance with almost all experiments. Only in the case of tests by Lloyd and Rangan, the differences were significant (about 17% for the force and 50% for the deflection),
- the standard buckling forces are always smaller than the test results,
- the numerical results of buckling loads and lateral deflections are in a satisfactory accordance with experimental results, in particular for three-dimensional simulations. The 2D results of the buckling strength are larger than experimental values for a larger reinforcement ratio due to the assumption of plane strain conditions in reinforcement and concrete,
- the buckling strength of columns increases mainly with decreasing slenderness, eccentricity and increasing reinforcement area. In turn, the effect of a characteristic length l_c is smaller than 5% (the strength increases with increasing characteristic length of micro-structure), In the case of perfect bond, the strength can be slightly larger (2D-analysis) or slightly smaller (3D-analysis) as compared to bond-slip,
- the width of localized zones at ultimate load in a tensile zone increases with increasing characteristic length and decreasing slenderness ratio. The width of zones lies in the range of (3.0-8.5)$\times l_c$,

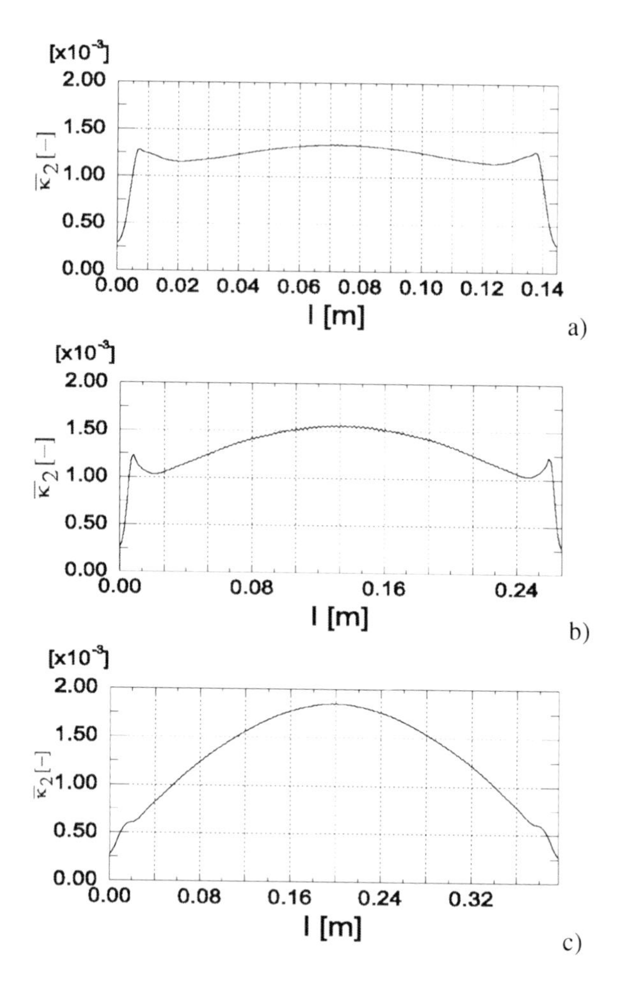

Fig. 7.69 Distribution of non-local parameter along column length l for ultimate load (e=6.35 mm, ρ=4.91%, l_c=5 mm) in experiments by Bažant and Kwon (1994): a) λ=19.2, b) λ=35.8, c) λ=52.5 (Majewski et al. 2009)

- the spacing of localized zones in a tensile zone is (8-12)$\times l_c$. The spacing of localized zones in a tensile zone increases with increasing characteristic length and eccentricity, and decreasing reinforcement ratio, fracture energy and initial bond stiffness. It is larger for the reinforcement modelled as 1D elements and a stochastic distribution of the tensile strength. The spacing decreases with increasing (2D-analyses) and decreasing slenderness ratio (3D – analyses) when perfect bond is assumed. The spacing is about (6-14)$\times l_c$,

- the spacing of localized zones in 3D-simulations is larger than in 2D-simulations for perfect bond and smaller for bond-slip,
- the average crack spacing according to CEB-FIP Model Code (1991) and the formula by Lorrain et al. (1998) is well reproduced by numerical analyses,
- the effect of stirrups and fracture energy in tension and compression on the ultimate load is negligible.

7.5 Corbels

Corbels are structural members widely used in practical engineering. They occur as reinforced concrete elements additionally strengthened by steel fibers or FRP sheets. Although they are very popular structural elements, the detailed information about their complex behaviour is still lacking. Codes contain only simplified relationships which are not able to describe strain localization connected closely to both ductile flexural and shear brittle failure mode.

Extensive experimental and theoretical studies on the behaviour of corbels were performed in the last decade. The earliest analytical concept for reinforced concrete corbels was proposed by Niedenhoff (1961) using a so-called strut-and-tie model (called also a 'truss analogy' approach). Later this model was improved among others by Mehmel and Backer (1965) who proposed a statistically indeterminate five-bar truss and by Hagberg (1983) who allowed in the truss for the depth variability of a compression zone. In turn, the experiments were carried out by e.g. Kriz and Raths (1965), Robinson (1969), Mattock et al. (1976), Yong and Balaguru (1994), Foster et al. (1996), Nagrodzka-Godycka (2001), Campione et al. (2005) and Souza (2010), which significantly contributed to an identification of different failure mechanisms occurring in corbels. In the case of the FEM, the behaviour of reinforced concrete corbels was investigated among others by Will et al. (1972), van Mier (1987), Renuka Prasad et al. (1993), Strauss et al. (2006) and Manzoli et al. (2008). Van Mier (1987) used a smeared crack approach by Rots et al. (1985) assuming an elasto-plastic Mohr-Coulomb yield criterion in compression. Renuka Prasad et al. (1993) analysed corbels with a smeared-fixed crack model using a constitutive model by Channakeshava and Iyengar (1988). In turn, Strauss et al. (2006) modelled the experiments by Kriz and Raths (1965) and by Fattuhi (1990) by taking into account the advantage of a smeared crack approach with a yield surface by Menétrey and Willam in compression available in the programme ATENA (Červenka1985). A simplified regularization technique in the form of a crack-band model was introduced. It allowed one to obtain mesh-insensitive force-displacements curve only, but it did not preseve the well-posedness of the boundary value problem. In turn, the three-dimensional RC corbels in tests by Mehmel and Freitag (1967) were modelled by Manzoli et al. (2008) by means of a strong discontinuity approach to obtain mesh-independent results. The constitutive concrete behaviour was described with an isotropic damage model. The best agreement between numerical and experimental results with respect to the load bearing capacity of corbels was achieved by Strauss et al. (2006) for 2D calculations (the difference between the experimental and numerical failure force was 7%), and by Manzoli et al. (2008) for 3D calculations where the numerical ultimate load was higher by 11% than the experimental one. Concerning the crack pattern, the best agreement with experiments was achieved again by Manzoli et al. (2008).

The aim of the present research is to properly describe a load-displacement curve and strain localization in reinforced concrete corbels using relatively simple continuum models for concrete (with respect to implementation and calibration) enhanced by a characteristic length of micro-structure by means of a non-local theory. The initial 2D calculations were performed within enhanced elasto-plasticity to model tests by Fattuhi (1990) and by Mehmel and Freitag (1967) in order to compare our numerical results with those by Strauss et al. (2006) and by Manzoli et al. (2008). Next, the numerical 2D and 3D results were comprehensively compared with the corresponding tests by Campione et al. (2005) using 3 different enhanced continuum models: an isotropic elasto-plastic, an isotropic damage and an anisotropic smeared crack model. Such comprehensive comparative analysis of the mixed tensile-shear failure mode has not been performed yet.

Tests by Mehmel and Freitag (1967)

The numerical calculations were carried out with a corbel with horizontal stirrups of Fig. 7.70a. The shear span-depth ratio was a/h=0.77 and the reinforcement ratio ρ=0.46%. The main longitudinal reinforcement consisted of two 14 mm and four 16 mm bars. The reinforcement had the ultimate strength of 430 MPa. The concrete parameters were as follows: compressive strength f_c=22.6 MPa and tensile strength f_t=2.26 MPa.

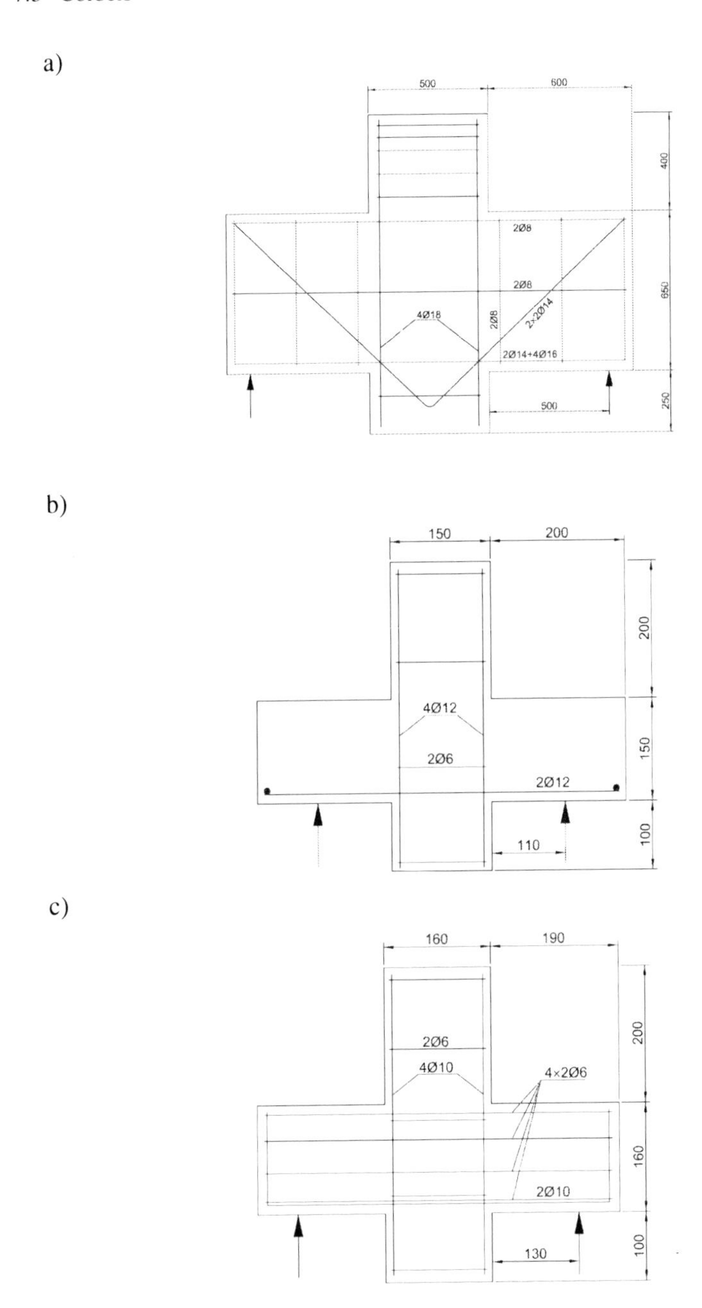

Fig. 7.70 Geometry of reinforced concrete corbels in different tests: a) by Mehmel and Freitag (1967), b) by Fattuhi (1990), c) by Campione et al. (2005)

Tests by Fattuhi (1990)

The corbel denoted as 'F25' without shear reinforcement was subjected to numerical calculations (Fig. 7.70b). The shear span-depth ratio was equal to $a/h=0.74$ and the reinforcement ratio was $\rho=0.99\%$. The two 12 mm bars had the yield strength of 452 MPa and ultimate strength of 684 MPa. The compressive and tensile concrete strengths were $f_c=36.1$ MPa and $f_t=3.02$ MPa, respectively.

Tests by Campione (2005)

Laboratory tests were carried out among others with short corbels with shear reinforcement (Fig. 7.70c). The shear span-effective depth ratio was $a/d=0.93$. The maximum compressive strength of concrete was $f'_c=48.5$ MPa and maximum splitting tensile strength was $f'_t=4.09$ MPa. The reinforcement ratio was $\rho=0.7\%$. The 10 mm bars had the yield strength of 488 MPa.

Analytical truss model

Ultimate failure forces in RC corbels were also analytically calculated on the basis of a strut-and-tie model frequently used in the engineering practice. This model (Hagberg 1983) (Fig. 7.71) takes into account the variability depth of a compression zone and the shear span-effective depth ratio (a/d)

$$F_v = \frac{F_s}{\tan\beta_1} + \frac{F_{sw}}{\tan\beta_2} = \frac{F_x}{\tan\beta}, \tag{7.21}$$

where F_v – ultimate vertical force, $F_x=F_s+F_{sw}$ - reinforcement force ($F_s = A_s f_{ys}$ - force in bars and $F_{sw} = A_s f_{y,sw}$ - force in stirrups). The value of $\tan\beta$ was calculated as

$$\left[1 - \frac{2f_c db}{F_x}\right]\tan^2\beta + \frac{2f_c ba}{F_x}\tan\beta + 1 = 0, \tag{7.22}$$

where b is the corbel width. In turn, the angles β_1 and β_2 denote the inclination of compressive members.

FE input data

Tests by Mehmel and Freitag (1967)

The 2D calculations were carried out with the elasto-plastic model only (Eqs. 3.27-3.32, 3.93 and 3.99). A regular mesh with 1194 quadrilateral elements composed of four diagonally crossed triangles was used by taking into account the symmetry of the system. The 3-node constant strain triangles were used. Concrete was assumed as a uniform material. The finite element height and width were equal to 25 mm and were not greater than $3\times l_c$ ($l_c=25$ mm) to achieve mesh-objective results (Bobiński and Tejchman 2004). The elastic constants for concrete were: E=21.9 GPa and ν_c=0.20. The compressive and tensile strengths were f_c=22.6 MPa and f_t=2.26 MPa, respectively. The internal friction angle was φ=20° and the dilatancy angle was ψ=10°. In the case of tension, the parameter κ_2^u=0.0017 was assumed (Fig. 7.72Aa). Thus, the tensile fracture energy was G_f =100 N/m. It was calculated as G_f=g_f×w_f; g_f – area under the entire softening function (with $w_f \approx 2l_c$ – width of tensile localized zones with l_c=5-10 mm). In the case of compression (Fig. 7.72Ab), we assumed κ_1^u=0.0085. Thus, the compressive fracture energy G_c was 5000 N/m. The non-locality parameter was chosen as m=2 on the basis of other calculations (Bobiński and Tejchman 2004).

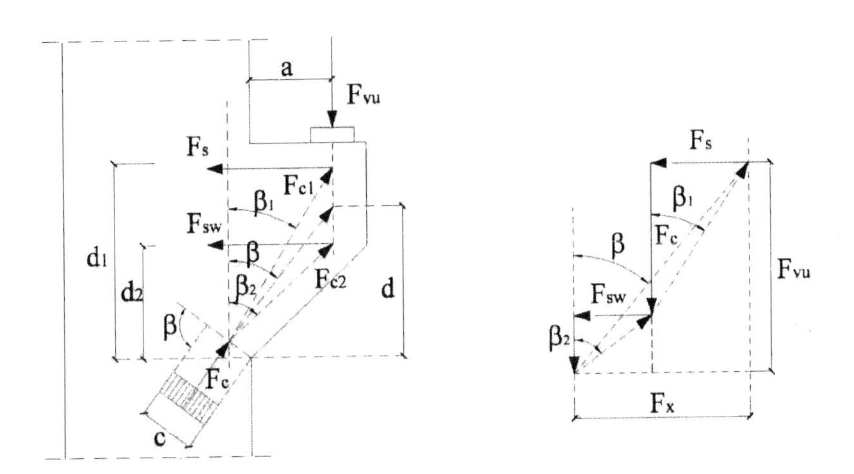

Fig. 7.71 Forces in corbel with horizontal stirrups using strut-and tie model by Hagberg (1983)

The bars and stirrups were assumed as 1D truss elements. The perfect bond was assumed between concrete and bars and between concrete and stirrups. The reinforcement behaviour was described by an elasto-plastic model according to Gonzales-Vidosa et al. (1988) (Fig. 7.72c) with E_s=206 GPa, σ_y=430 MPa for bars and strirrups (σ_y – yield stress). The deformation was induced by a vertical displacement applied in the middle of the bearing plate situated on the column.

Tests by Fattuhi (1990)

The 2D calculations were carried out with the elasto-plastic model only (Eqs. 3.27-3.32). A regular mesh with four-noded elements composed of a cross-diagonal patch of four three-nodal constant strain triangles was used (in total 1860 elements). The symmetry of the system was taken into account. The finite element height and width were equal to 5 mm and were not greater than 3×l_c (l_c=5 mm).

The following material constants were assumed for concrete: E_c=33.5 GPa and ν_c=0.20, f_c=36.1 MPa, f_t=3.02 MPa, φ=20°, ψ=10°, G_f =100 N/m, G_c=5000 N/m, l_c=10 mm and m=2 with compressive and tensile yield relationships shown in Fig. 7.72B. The reinforcement consisting of bars only was assumed as 1D elements fixed at ends. The reinforcement behaviour was described by the elasto-plastic idealization by Gonzales-Vidosa et al. (1988) (Fig. 7.72c) with E_s=200 GPa and σ_y=452 MPa. The perfect bond was assumed.

Tests by Campione (2005)

In the case of 2D calculations, regular meshes with 1860 quadrilateral elements composed of four diagonally crossed triangles were used. In turn, a 3D model included 5376 cube-shaped elements (the 8-node elements with 8 integration points and a selectively reduced-integration technique were applied (Abaqus 1998). A symmetry of the system was assumed. The finite element height and width were equal to 5 mm and were not greater than 3×l_c. The material constants were for concrete: E_c=34.7 GPa, ν_c=0.20, f_c=48.5 MPa, f_t=3.7 MPa (calculated as f_t=0.9×f_t'), φ=20°, ψ=10°, l_c=5-10 mm and m=2. The calculations were carried out with various hardening-softening parameters. In the case of tension, the following parameters were assumed: κ_2^u=0.0018 and κ_2^u=0.0036 (Fig. 7.72Ca). Thus, the tensile fracture energy was G_f =50 N/m or G_f =100 N/m. In the case of compression (Fig. 7.72Cb), we assumed κ_1^u=0.0135 or κ_1^u=0.027. Thus, the compressive fracture energy was G_c=5000 N/m or G_c=10000 N/m, respectively. The input data for FE-analyses with the elasto-plastic concrete model are included in Tab. 7.10.

In the case of the isotropic damage model, we used the following material constants: E=34.7 GPa, ν_c=0.20, α=0.96, β=200, and κ_0=1.06×10^{-4} (Eqs. 3.35, 3.93 and 3.99), and E=34.7 GPa, ν=0.2, κ_0=1.06×10^{-4}, α=0.96, β=200, α_1=0.1, α_2=1.16, α_3=2.0 and γ=0.2 (Eqs. 3.39, 3.93 and 3.99).

When the smeared crack approach was used, the following material constants were assumed: E=34.7 GPa, ν=0.2, p=2.0, b_1=3.0, b_2=6.93, f_t=2.0 MPa, ε_{su}=0.1 and ε_{nu}=0.00182 (Eqs. 3.52-3.60, 3.93 and 3.101).

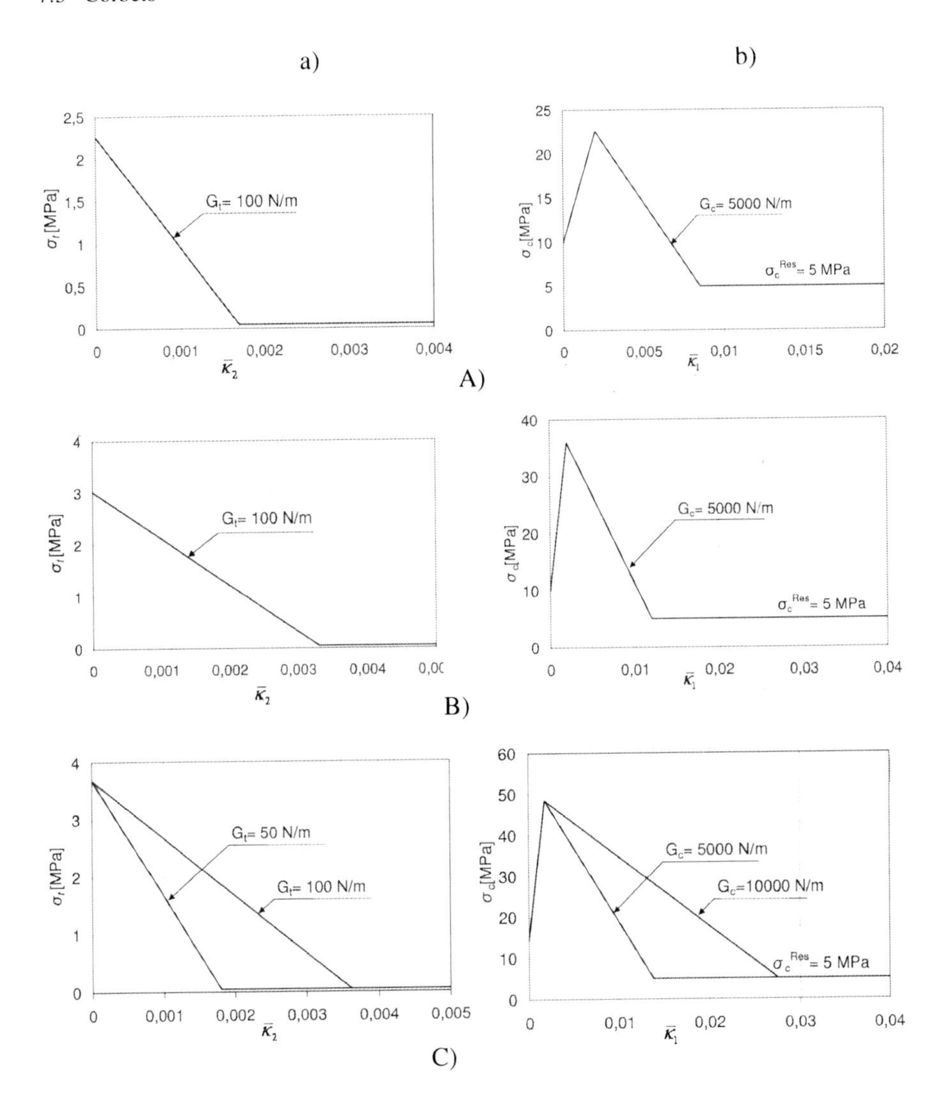

Fig. 7.72 Assumed curves in FE-calculations: a) tensile stress versus non-local parameter $\sigma_t = f(\bar{\kappa}_2)$ for concrete, b) compressive stress versus non-local parameter $\sigma_c = f(\bar{\kappa}_2)$ for concrete, c) yield stress σ_y versus strain ε for reinforcement according to Gonzales-Vidosa et al. for tests by Mehmel and Freitag (1967) (A), Fattuhi (1990) (B) and Campione et al. (2005) (C)

c)

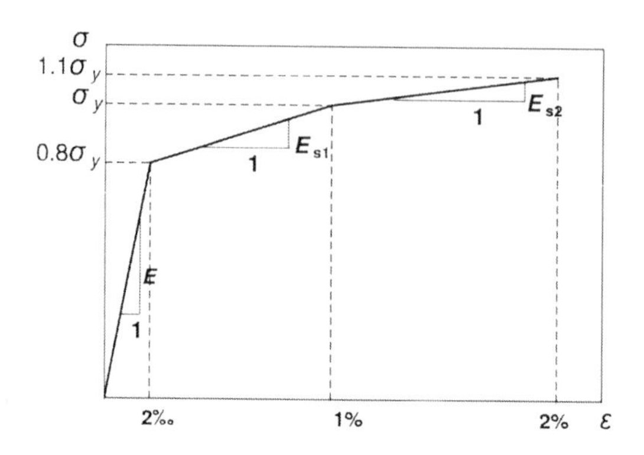

Fig. 7.72 *(continued)*

The horizontal bars and stirrups were assumed as 1D elements. The steel behaviour was described by the elasto-plastic law following Gonzales-Vidosa et al. (1988) (Fig. 7.72c) with E_s=200 GPa, σ_y=488 MPa for bars and σ_y=445 MPa for stirrups. Either perfect bond or bond-slip according to CEB-FIP (1990) or Dörr (1980) (Eq. 3.114-3.119) was assumed between bars and concrete. In calculations with slip-bond, the same displacements were enforced between reinforcement (truss) and concrete (solid) elements in nodes at the specimen surfaces.

Initial FE results within elasto-plasticity with non-local softening for corbels by Mehmel and Freitag (1967) and by Fattuhi (1990)

The numerical results show a satisfactory agreement with experimental results for reinforced concrete corbels by Mehmel and Freitag (1967) (Figs. 7.73 and 7.74) and Fattuhi (1990) (Figs. 7.75 and 7.76) with respect to the geometry of localized zones and vertical failure forces.

The calculated failure force in our model was even better reproduced than the calculated one by Manzoli et al (2008) as compared to the experiment by Mehmel and Freitag (1967) (Fig. 7.73). In turn, the calculated failure force, initial stiffness and softening rate with reference to the corbel by Fattuhi (1990) were also more precisely captured by our model than by the model by Strauss et al. (2005) (Fig. 7.75).

Table 7.10 The summary of the input data for FE-analyses with elasto-plastic concrete model in 2D simulations (tests by Campione et al. 2005)

FE simulation	Reinforcement	l_c [mm]	G_f [N/m]	G_c [N/m]	Bond model
1	2∅10+4∅6	10	100	10000	CEB-FIP (1990)
2	2∅10+4∅6	5	100	10000	CEB-FIP (1990)
3	2∅10+4∅6	5	50	5000	CEB-FIP (1990)
4	2∅10+4∅6	5	100	5000	CEB-FIP (1990)
5	2∅10+4∅6	5	100	10000	Dörr (1980)
6	2∅10+4∅6	5	100	10000	perfect bond
7	2∅10	5	100	5000	CEB-FIP (1990)
8	-	5	50	5000	-

The main localized zones were better captured in our calculations than by other authors (Figs. 7.74 and 7.76) as compared to experiments (in particular to tests by Mehmel and Freitag (1967)). First, a vertical localized zone appeared due to bending and next several inclined ones developed due to shear from the bottom tensile edge to the upper compressive corner. In the case of tests by Fattuhi (1990), the experimental inclined localized zone is more curved, but the general strain distribution was properly reproduced. The calculated vertical localized zone was shorter than the experimental one.

The difference between numerical and experimental vertical failure forces for test by Mehmel and Freitag (1967) and test by Fattuhi (1990) was in the range of 1%. As compared to the theoretical value by a strut-and-tie model (Eqs. 7.21 and 7.22), the experimental maximum vertical force in the test by Mehmel and Freitag (1967) was lower by 8%, while in the test by Fattuhi (1990) was higher by 6% (Tab. 7.11).

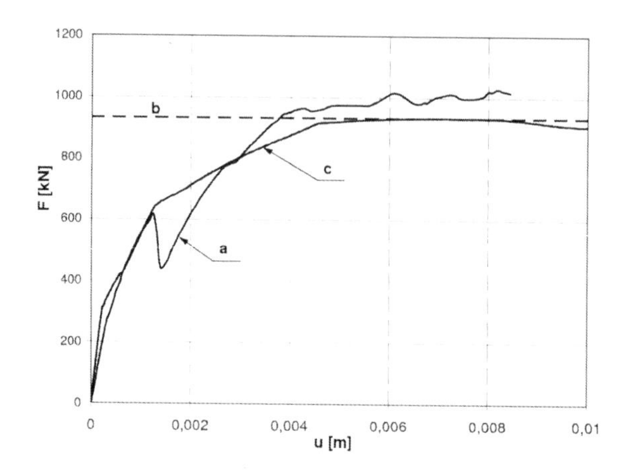

Fig. 7.73 Load-displacement curves from 2D calculations using elasto-plastic model (a) as compared with experimental result by Mehmel and Freitag (1967) (b) and numerical result by Manzoli (2008) (c) (Syroka et al. 2011)

FE results within elasto-plasticity with non-local softening for corbels by Campione et al. (2005)

Effect of characteristic length and fracture energy

Figure 7.77 shows the calculated 2D load-displacement curves (F - vertical force at the support, u - vertical displacement at the support) for reinforced concrete corbels by Campione et al. (2005) possessing shear reinforcement with the different characteristic length of micro-structure (l_c=5 mm or l_c=10 mm), various tensile fracture energy (G_f=50 N/m or G_f=100 N/m) and compressive fracture energy (G_c=5000 N/m or G_c=10000 N/m) and the bond-slip law for bars of Fig. 3.13 (note that by varying l_c, the softening curve can be scaled in order to keep the fracture energy G_f unchanged). The experimental curve was also attached. The distribution of a non-local tensile softening parameter $\bar{\kappa}_2$ during 3 subsequent phases (for the vertical force equal to 50 kN, 100 kN and at the peak, respectively) is depicted in Fig. 7.78 (l_c=5 mm and l_c=10 mm, G_f=50 N/m, G_c=10000 N/m).

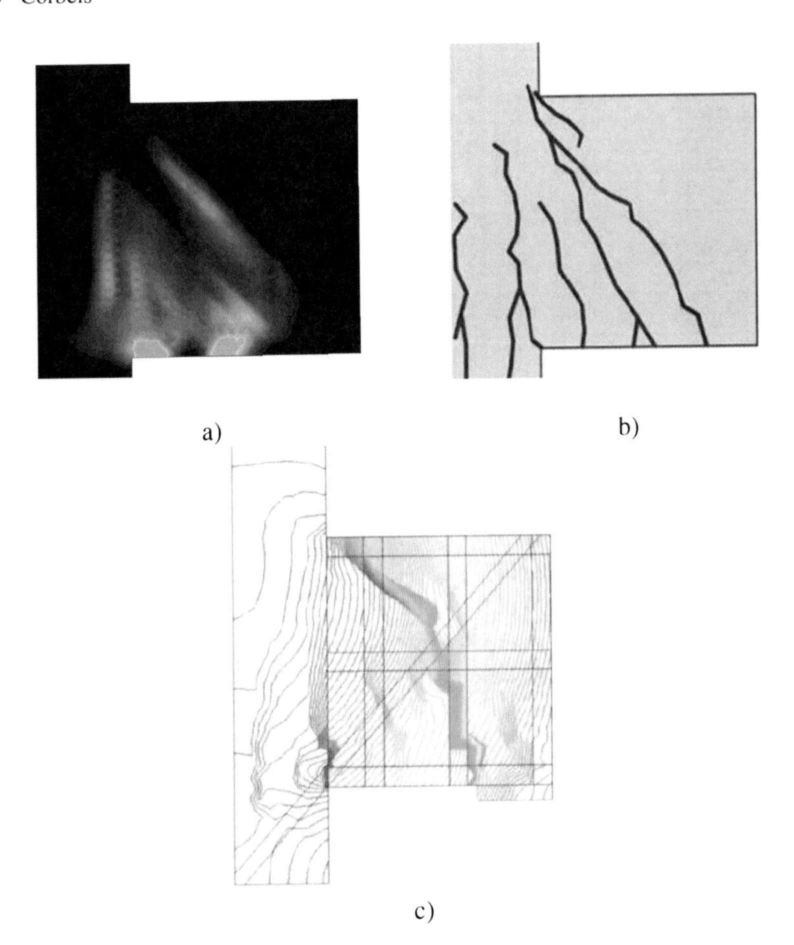

Fig. 7.74 Calculated pattern of localized zones from our 2D calculations with elasto-plastic model (a) as compared to experimental result by Mehmel and Freitag (1967) (b) and numerical result of displacement contours by Manzoli (2008) (c) (Syroka et al. 2011)

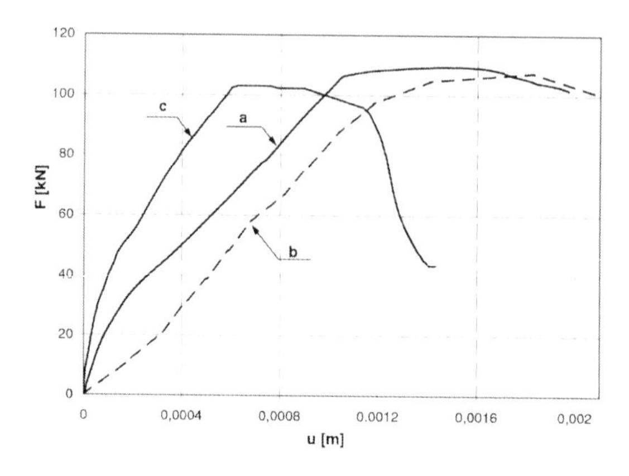

Fig. 7.75 Load-displacement curves from our 2D calculations using elasto-plastic model (a) compared with experimental result by Fattuhi (1990) (b) and numerical results by Strauss et al. (2006) (c) (Syroka et al. 2011)

The calculated 2D geometry of localized zones is in a satisfactory agreement with the experimental one. There exist vertical and inclined localized zones as in the experiment. The sequence of the formation of localized tensile zones during loading is close to laboratory results (Fig. 7.79). An increase of the characteristic length causes a growth of the width of localized zones w by 30% (w=20 mm with l_c=5 mm). However, a strong discrepancy between FE analyses and experiments with respect to the evolution of the load-displacement curve is evident. Firstly, the calculated maximum vertical force is higher by 20%. Secondly, the calculated initial stiffness of the specimen is significantly larger than the experimental one. This difference is probably caused by an inaccurate measurement of the vertical displacement during initial loading in the test (the elastic modulus of concrete was assumed in FE analyses the same as in the experiment).

The influence of the tensile fracture energy on the load bearing capacity is negligible. The lower tensile fracture energy causes a small increase of the width of localized zones. The load bearing capacity of corbels increases with increasing compressive fracture energy. For G_c=10000 N/m, the ultimate force is higher by 5% than for G_c=5000 N/m. The pattern of localized zones is not affected by the variation of the tensile and compressive fracture energy. An inclined localized zone propagating from the support zone reaches the upper corner earlier for the lower compressive fracture energy.

The experimental vertical forces were also compared to the theoretical one calculated on the basis of a strut-and-tie model (Eqs. 7.21 and 7.22). The theoretical value was higher (similarly as the numerical one) than the experimental one by 16% (Tab. 7.11).

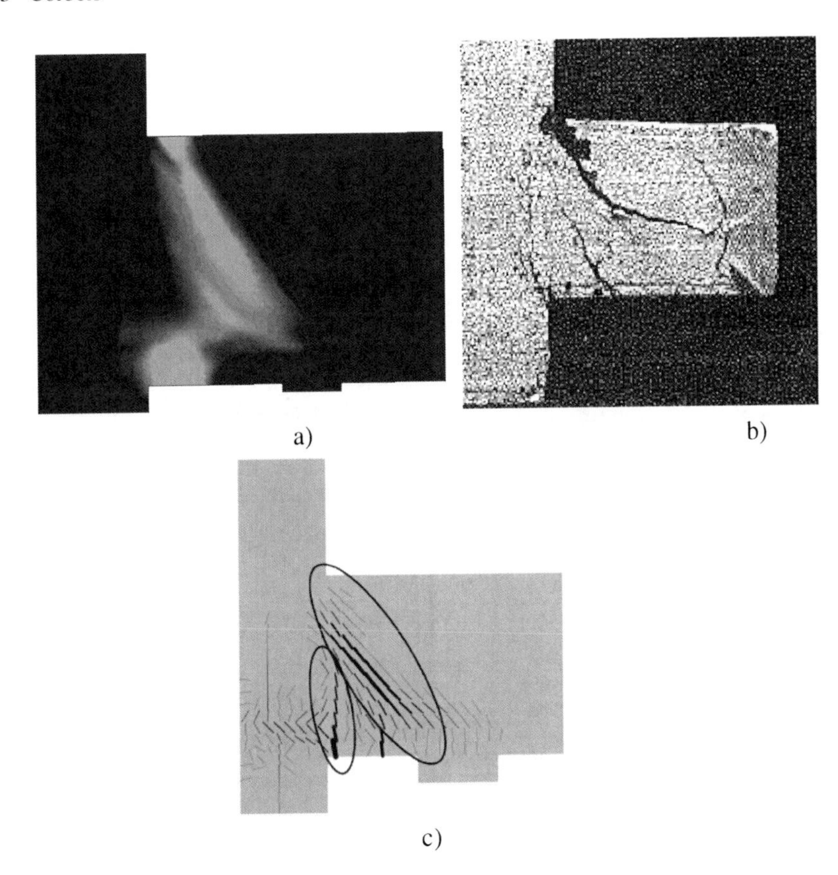

Fig. 7.76 Calculated pattern of localized zones from our 2D calculations with elasto-plastic model (a) compared to experimental results by Fattuhi (1990) (b) and numerical results by Strauss et al. (2006) (c) (Syroka et al. 2011)

Afterwards, FE results were compared for 3 different bond definitions between concrete and reinforcement: perfect bond, bond-slip by Dörr (1980) (Eqs. 3.114 and 3.115) or by CEB-FIP (1990) (Eqs. 3.116-3.119). The effect of the bond-slip law on the pattern of localized zones and a load-displacement diagram was found to be negligible in contrast to FE calculations with reinforced concrete beams (Marzec et al. 2007), where the spacing of localized zones increased with decreasing initial bond stiffness.

Table 7.11 Data summary of failure force from experiments, 2D FE results with elasto-plastic concrete model and analytical formula using strut-and-tie model by Hagberg (1983)

FE simulation of Tab. 7.8	Maximum vertical force from FE analyses F_N [kN]	Maximum vertical force from experiments F_E [kN]	F_N/F_E	Maximum vertical force from truss model F_T [kN]	F_N/F_T
		Campione et al (2007)			
1	129.51	98.83	1.31	114.54	1.13
2	128.50	98.83	1.30	114.54	1.12
3	121.95	98.83	1.23	114.54	1.06
4	122.02	98.83	1.23	114.54	1.07
5	128.43	98.83	1.3	114.54	1.12
6	128.35	98.83	1.3	114.54	1.12
7	79.68	77.60	1.03	76.59	1.04
8	14.01	13.35	1.05	-	-
		Mehmel and Freitag (1967)			
1	933.29	933	1.00	1020.39	0.92
		Fattuhi (1990)			
1	109.55	108.5	1.01	103.52	1.06

Effect of horizontal bars and stirrups

Figure 7.80 presents the numerical and experimental 2D results for concrete corbels and for reinforced concrete with and without shear reinforcement (l_c=5 mm, G_f=100 N/m, G_c=5000 N/m). The presence of shear reinforcement improves material ductility and increases the load bearing capacity of reinforced concrete corbels by 50% (Tab. 7.11). In turn, the presence of bending reinforcement increases about six times the load bearing capacity of concrete corbels (Tab. 7.11). The calculated ultimate force for the reinforced concrete corbel without shear reinforcement and for the concrete corbel is nearly the same as the experimental one (Tab. 7.11). However, the calculated initial stiffness is too large. The reinforced concrete corbel without stirrups failed due to the appearance of a vertical (flexural) and an inclined (shear) localized zone (Fig. 80Bb) as in the experiment. The concrete corbel failed due to the appearance of a vertical localized zone as in the experiment (Fig. 80Bd).

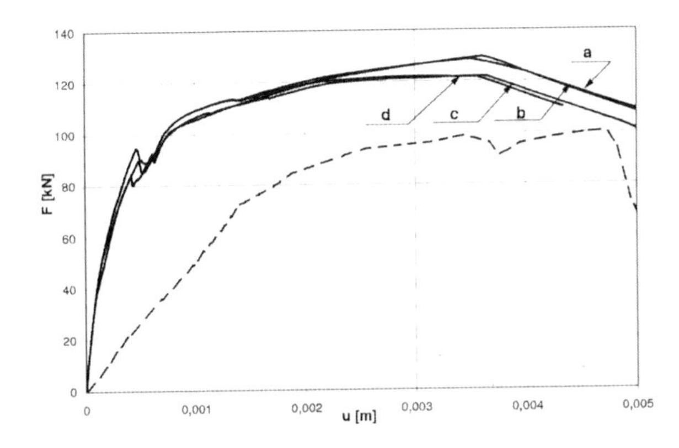

Fig. 7.77 Load-displacement curves for different characteristic lengths and various compressive and tensile fracture energy from 2D calculations (solid lines) compared to the experiments by Campione et al. (2005): a) l_c=5 mm, G_f=100 N/m, G_c=10000 N/m, b) l_c=10 mm, G_f=100 N/m, G_c=10000 N/m, c) l_c=5 mm, G_f=100 N/m, G_c=5000 N/m, d) l_c=5 mm, G_f=50 N/m, G_c=5000 N/m (bond-slip law by CEB-FIP 1990) (Syroka et al. 2011)

Effect of 3D conditions

The load bearing capacity (Fig. 7.81) and geometry of localized zones (Fig. 7.82) in 3D analyses are similar as in 2D simulations. The width of localized zones is also similar. A decrease of compressive fracture energy reduces the load bearing capacity of reinforced concrete corbels. The calculated slope of softening is in good agreement with the experiment.

FE results within damage mechanics and smeared crack approach with non-local softening for corbels by Campione et al. (2005)

Figures 7.83-7.85 demonstrate the FE results for two remaining constitutive models (Chapters 3.1.2 and 3.1.3) (l_c=5 mm, perfect bond) for reinforced concrete corbels with shear reinforcement by Campione et al. (2005) (the results with a rotating and a fixed crack approach were similar).

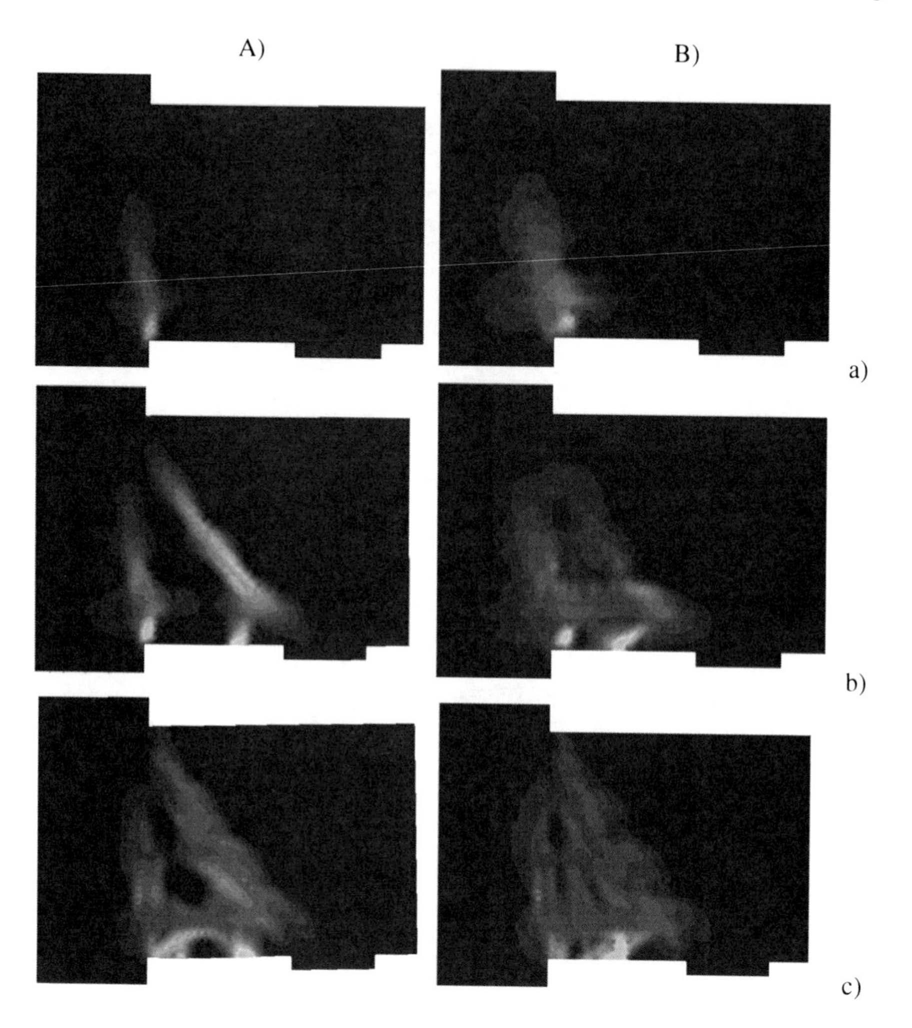

Fig. 7.78 Distribution of non-local softening tensile parameter $\bar{\kappa}_2$ for different characteristic lengths from 2D calculations for tests by Campione et al. (2005): A) l_c=5 mm, B) l_c=10 mm (G_f=100 N/m, G_c=10000 N/m, bond-slip law by CEB-FIP (1990), a) F=50 kN, b) F=100 kN, c) at peak (Syroka et al. 2011)

In the case of the isotropic damage model, the agreement with experiments is satisfactory, although the vertical localized zone continuously develops and the inclined zone is too steep (Fig. 7.84). In the case of the modified von Mises definition of the equivalent strain measure $\tilde{\varepsilon}$ (Eq. 3.38), the vertical localized zone first appears and next separately the inclined one. In turn, with the definition of the equivalent strain measure $\tilde{\varepsilon}$ by Haüsler-Combe (Eq. 3.39), the inclined zone is

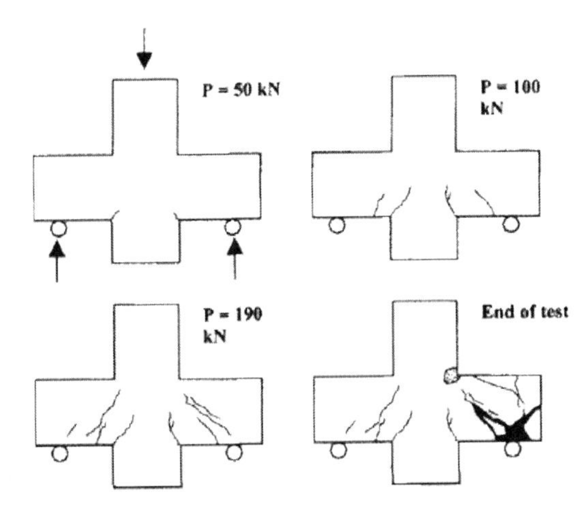

Fig. 7.79 Experimental crack pattern for reinforced concrete corbels with shear reinforcement (Campione et al. 2005)

connected with the vertical one. The width of the vertical strain localization in the first stage is similar in both cases and larger by 10% than for the elasto-plastic constitutive model. The calculated load bearing capacities are 20% higher than in the experiment. Both curves exhibit similar softening as in the experiment.

The smeared crack approach is also able to properly capture the propagation of localized zones (Fig. 7.85). The sequence of localized zones (first, the vertical and next, the inclined one) is similar to the one observed in the experiment. As compared with the elasto-plastic model, the width of vertical zone is much larger for the smeared crack approach, moreover it has a different slope. The calculated maximum vertical force (Fig. 7.85) is significantly too high as compared to the experiment (by 45%). A similar outcome in the form of a high calculated vertical failure force with the smeared crack model was also obtained by Souza (2010).

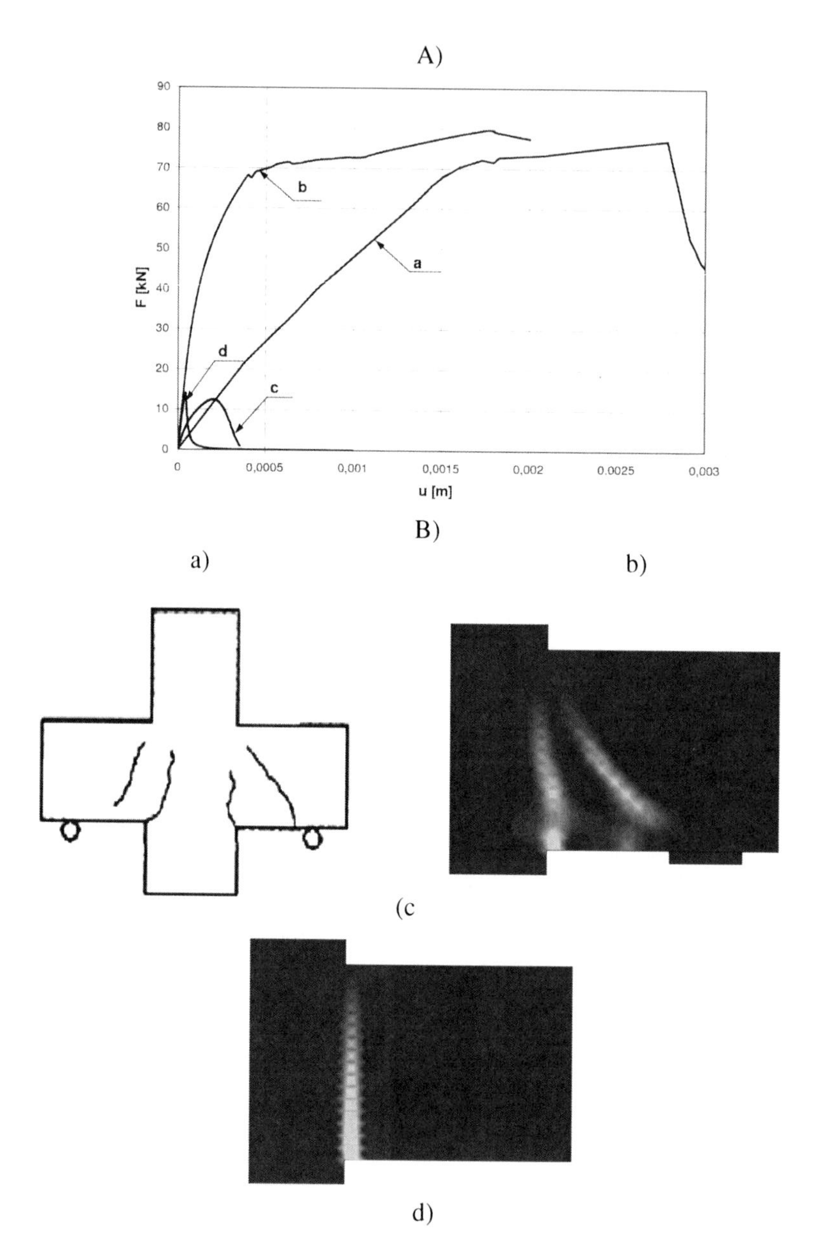

Fig. 7.80 The calculated 2D load-displacement curves (A) and distribution of non-local softening tensile parameter $\bar{\kappa}_2$ (B) for corbels as compared to experiments (Campione et al. 2005): a) reinforced concrete corbel without shear reinforcement (experiment), b) reinforced concrete corbel without shear reinforcement (FEM), c) concrete corbel (experiment), d) plain concrete corbel (FEM) (Syroka et al. 2011)

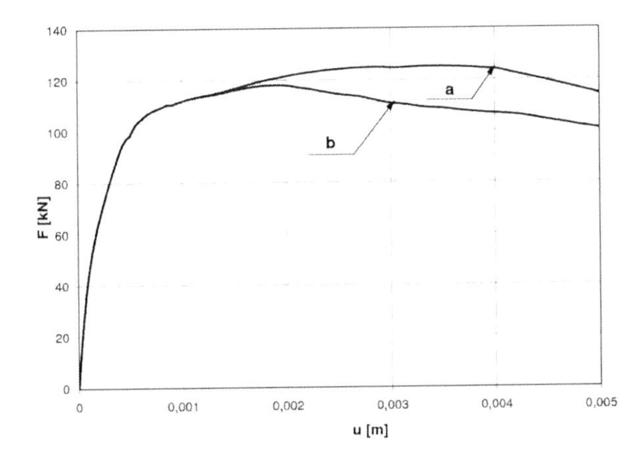

Fig. 7.81 Load-displacement curves from 3D FE calculations with different compressive fracture energies: a) G_c=10000 N/m, b) G_c=5000 N/m (G_f=50 N/m, l_c=5 mm, perfect bond) for tests of Campione et al. (2005) (Syroka et al. 2011)

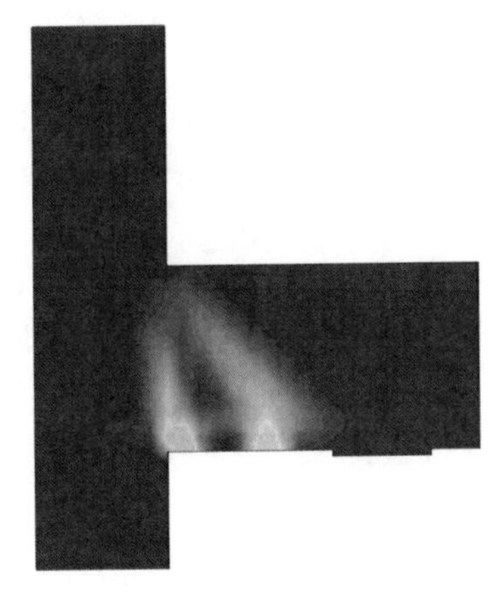

Fig. 7.82 Distribution of non-local softening parameter $\bar{\kappa}_2$ from 3D simulations with elasto-plastic model with non-local softening (G_c=5000 N/m, G_f=50 N/m l_c=5 mm, perfect bond) for tests of Campione et al. (2005) (Syroka et al. 2011)

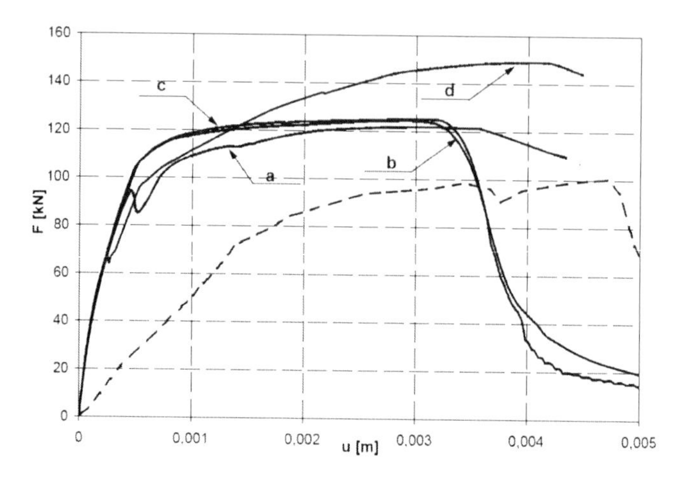

Fig. 7.83 Load-displacement curves from 2D calculations for various concrete models (solid lines) with non-local softening compared to the experimental curve (dashed line): a) elasto-plastic model, b) damage model with modified von Misses equivalent strain measure (Eq. 3.38), c) damage mechanics with equivalent strain measure by Häussler-Combe (Eq. 3.39), d) smeared fixed crack approach (for tests of Campione et al. (2005) (Syroka et al. 2011)

The following conclusions can be derived from the comprehensive FE analysis of reinforced concrete corbels under a mixed tensile-shear failure mode:

• The best agreement with experimental results was obtained using the elasto-plastic model with non-local softening (with the compressive fracture energy G_c=5000 N/m and characteristic length of l_c=5 mm) with respect to a load-displacement diagram and a geometry of localized zones. A perfect agreement in the case of a vertical failure force was achieved with the elasto-plastic model for experiments by Mehmel and Freitag (1967) and by Fattuhi (1990). In the case of tests by Campione et al. (2005), the calculated ultimate force was higher by 20% within both elasto-plasticty and isotropic damage mechanics and by 45% with the anisotropic smeared crack model.

• The effect of tensile fracture energy on the ultimate vertical force was insignificant. A decrease of the compressive fracture energy slightly reduced the ultimate vertical force.

• An increase of a characteristic length slightly increased the ultimate vertical force and significantly width of localized zones. The characteristic length did not influence the pattern of localized zones.

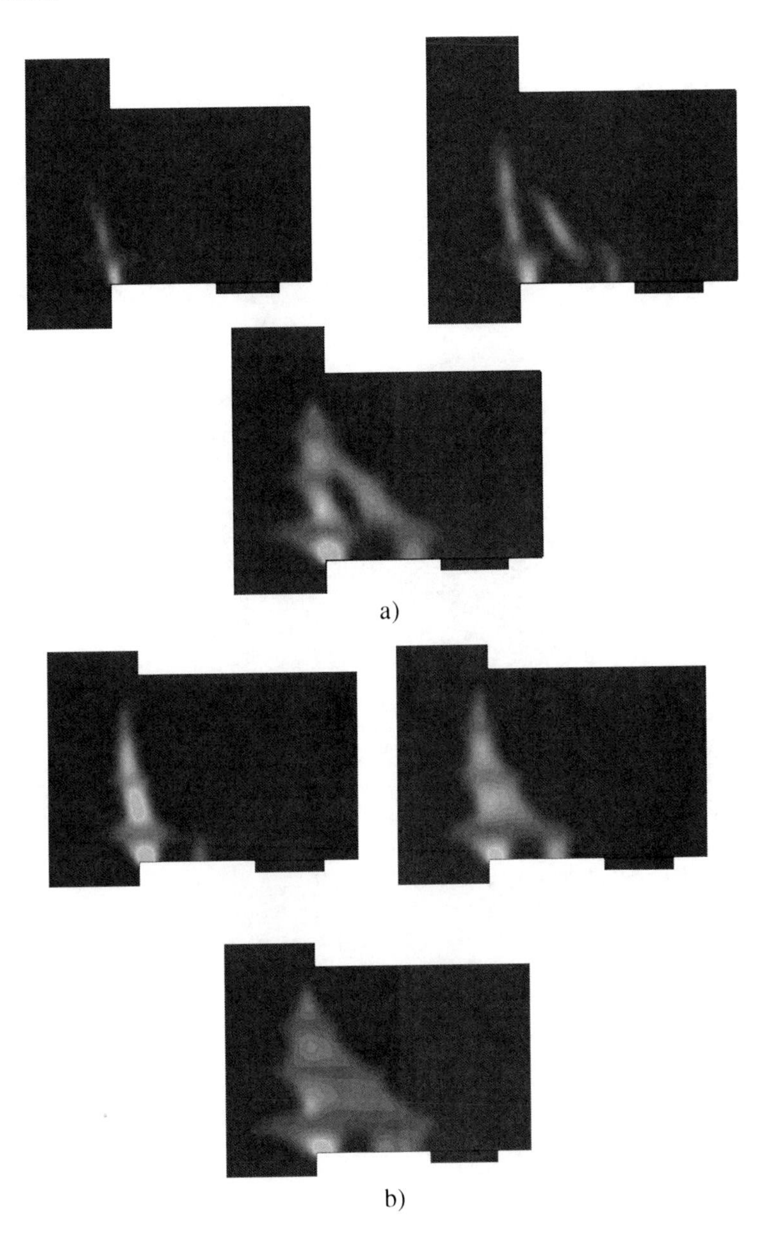

Fig. 7.84 Calculated pattern of localized zones with continuum damage model assuming: a) modified von Misses equivalent strain measure (Eq. 3.38) and b) equivalent strain measure by Häussler-Combe (Eq. 3.39) (for tests of Campione et al. 2005) (Syroka et al. 2011)

Fig. 7.85 Calculated pattern of localized zones with smeared fixed crack approach (for tests of Campione et al. 2005) (Syroka et al. 2011)

- The presence of horizontal stirrups increased the vertical failure force (by 50%) and ductility of reinforced concrete corbels.
- The calculated load bearing capacity was similar in 2D and 3D calculations.
- Concerning the simulated geometry of localized zones, the most satisfactory agreement was achieved with the elasto-plastic model, next with the damage model and finally with the smeared fixed crack model.
- The theoretical failure force by a strut-and-tie model was larger by 7% as compared to the experimental results by Campione et al. (2005), by 8% lower as compared to tests by Mehmel and Freitag (1967) and by 6% higher as compared to tests by Fattuhi (1990).

7.6 Tanks

Frame or wall corners belong to very important structural elements. Their dimensioning is not straightforward due to a complex stress caused by possible failure modes (e.g. reinforcement yielding, concrete splitting, concrete failure in compression and anchorage failure). Numerous experiments on reinforced concrete frame corners (Nilsson 1973, Stroband and Kolpa 1981, 1983, Kordina 1984, Akkermann 2000) show that the stress state and related crack distribution depend strongly on the direction of a loading bending moment: negative (closing) or positive (opening). The reinforcement detail in corners usually chosen for negative moments is unsafe for positive ones. There still exist standards (e.g. Polish Standard 2002) which do not distinguish the type of reinforcement between these two different moments.

Chapter shows an example of failure of a wall corner in full-scale reinforced concrete sewage water tanks subjected to a positive bending moment. Initially, the wall corner was dimensioned within linear elasticity without distinguishing a sign of a bending moment (negative or positive). The calculations were carried out an isotropic elasto-plastic model by Rankine (Eq. 3.32) with non-local softening (Eqs. 3.93 and 3.99). Numerical results were compared with measured wall displacements and deformation of concrete samples taken from the tank corner. Finally, an effective repairing method of failed tanks was proposed.

Three rectangular reinforced concrete tanks were constructed in a sewage treatment plant. Their length was 120.45 m, width 42 m and height 6.48 m. Each tank was 5.28 m sunken in the soil which was strengthened with gravel columns. The tanks were composed of 4 separate technological cells located on a foundation slab of 40 cm thick. Each tank was divided into 5 segments. The wall thickness varied between 35 cm (top) - 50 cm (bottom). The used material was: concrete C30/37 and steel bars AII and AIIIN. During a design phase, the internal forces in the tanks were calculated using FEM for a full 3D model within linear elasticity. Fig. 7.86 shows the reinforcement detail of the wall corner constructed. The reinforcement was assumed according to the Polish Standard (2002), except of the fact that the bents of vertical bars in vertical walls were inserted into a foundation slab in an opposite direction to simplify reinforcement works.

During the virgin water filling of one tank (at this moment, the tank walls were not covered with soil from the outside), an excessive horizontal displacement of vertical walls occurred. The maximum displacement of the exterior wall at the top was 75 mm at the water height of 5.7 m. The shape of the displacement curve indicated a wall rotation against the bottom. The maximum settlement of tanks was about 30 mm. The filling process was stopped and the tank was emptied. After tank emptying, a maximum permanent horizontal wall displacement at the top reduced to 20 mm.

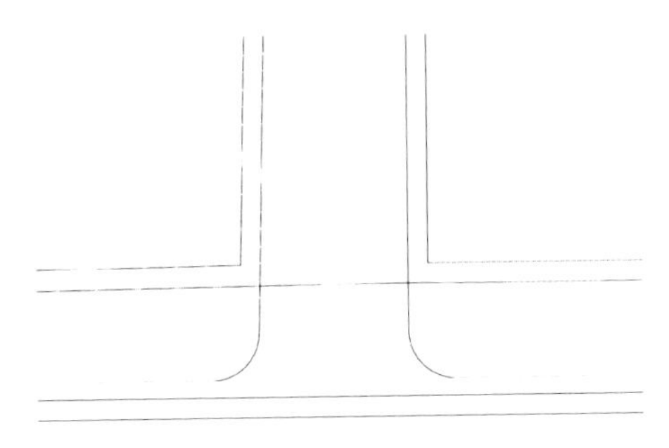

Fig. 7.86 Reinforcement detail of wall bottom corner (Bobiński and Tejchman 2009)

No surface cracks were found on the walls and slabs. The soil investigations by a CPT probe confirmed its proper densification.

The plane strain calculations were performed with a tank fragment embracing the wall corner shown in Fig. 7.87. The tank was located on an elastic space. Totally, 23'000 triangular elements were used: 15'000 for concrete, 2'000 for reinforcement and 6'000 for soil. The number of finite elements in the corner was 10'000 (Fig. 7.88). The maximum finite element size was not greater than $3\times l_c$ (l_c=10 mm) to achieve mesh-objective results. The following elastic material parameters were assumed for concrete: E_c=28.9 GPa and ν_c=0.20. The tensile strength f_t was 2 MPa and the tensile fracture energy was G_f=50 N/m-200 N/m as the plastic parameters (Eq. 3.32). The reinforcement bars were assumed as 1D-elements. For the reinforcement, an elasto-perfect plastic constitutive law was assumed with E_s=210 GPa and σ_y=420 MPa (σ_y – yield stress). Our calculations were carried out with bond-slip. The analyses were carried out with a relation between the bond shear stress τ_b and slip u using a simple bond law according to Dörr (1980) (Eqs. 3.114 and 3.115). The elastic constants of springs simulating the sub-soil stiffness were so adjusted to obtain the measured soil settlement during the virgin water filling.

The behaviour of the structure was analyzed during initial tank filling for 4 different variants of reinforcement in the wall corner subjected to a positive (opening) bending moment (Fig. 7.89). The bearing capacity of the structure and wall displacements were calculated. Next, water loading was increased in the calculations up to the structure failure (assumed as a begin of material softening). Depending upon the reinforcement variant, a different bearing capacity was obtained (Tab. 7.12).

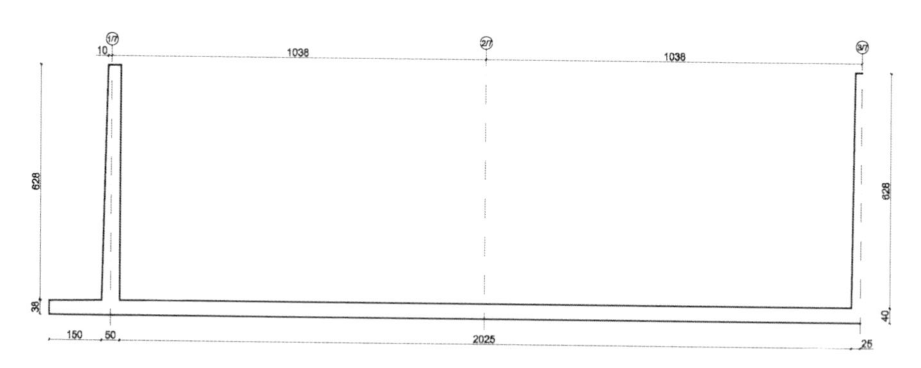

Fig. 7.87 Tank fragment assumed for plane strain FE calculations (Bobiński and Tejchman 2009)

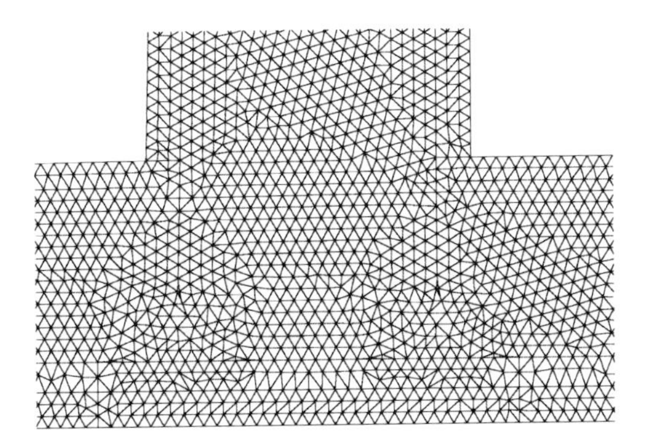

Fig. 7.88 FE mesh for wall corner (Bobiński and Tejchman 2009)

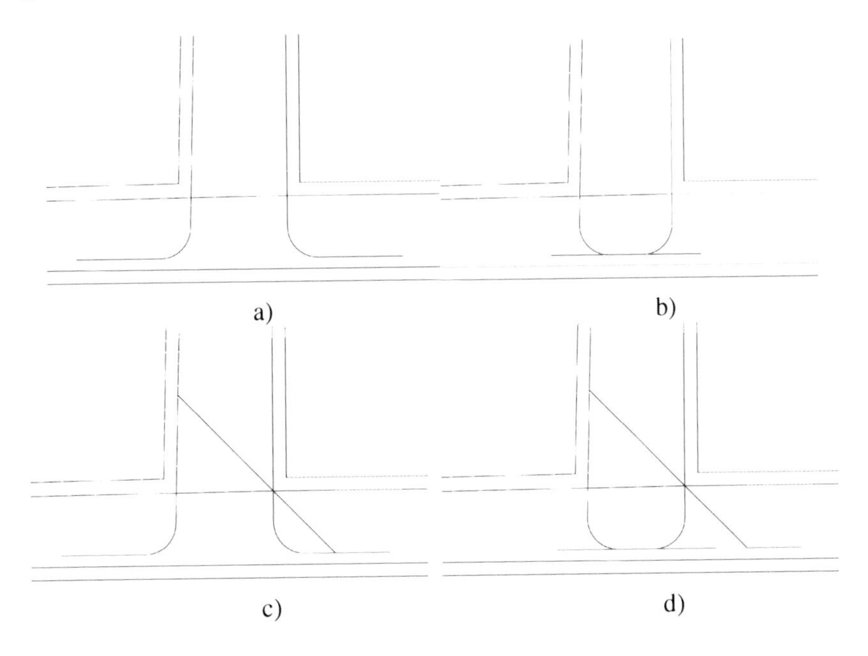

Fig. 7.89 4 different variants of reinforcement of wall corner assumed in FE calculations (Bobiński and Tejchman 2009)

In all cases, a diagonal interior crack occurred (Fig. 7.90) what contributed to a successive wall rotation. It occurred perpendicularly to the angle bisector between the vertical wall and horizontal slab, at the distance of 20 cm from the interior corner. The distribution of cracks in concrete samples taken from the wall corner (Fig. 7.91) confirmed the numerical results with respect to strain localization. The reversal bars (case c) or diagonal bars (case d) improved the bearing capacity by 15% only. The calculated bending moment carried by the failed corner with strain localization was about 250 kNm (on the basis of the distribution of normal stresses).

Table 7.12 Numerical results for different reinforcement variants of tank wall corner (Bobiński and Tejchman 2009)

Reinforcement variant of Fig. 7.89	Horizontal displacement of wall top with 5.7 m water	Strength increase during water loading	Horizontal displacement of wall top for increased during water loading (strength limit)
	[mm]	[%]	[mm]
a)	8	117	40
b)	7	135	60
c)	7	132	60
d)	7	207	80

To repair the tanks, they were strengthened with 3 horizontal beams fixed at the wall top at the height of 5.5 m (Fig. 7.92). The cracked structure was numerically analyzed during repeated filling after strengthening (plane strain). Figure 7.93 shows the calculated deformation. The existing diagonal localized zone did not grow. No additional localized zones occurred. The bearing capacity of the tank increased by 50%.

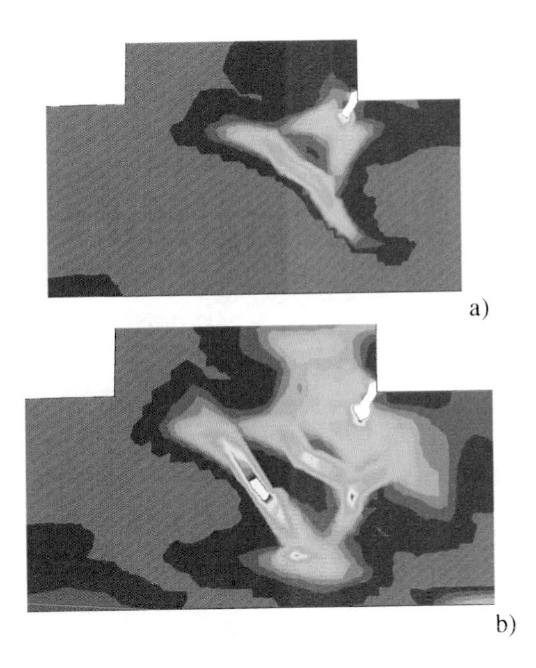

Fig. 7.90 Formation of strain localization in the wall corner during strength limit for two reinforcement variants of Fig. 7.89: a) variant 'a', b) variant 'd' (Bobiński and Tejchman 2009)

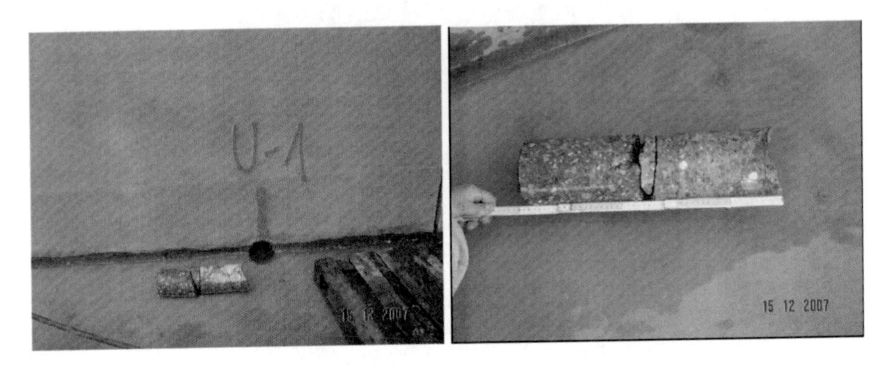

Fig. 7.91 Concrete samples taken from wall corner (Bobiński and Tejchman 2009)

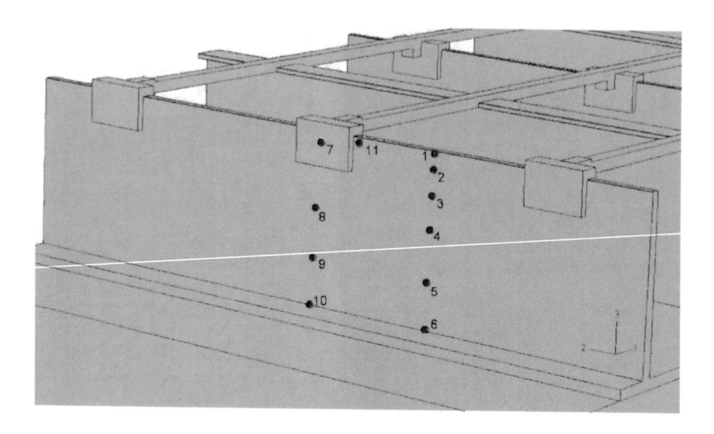

Fig. 7.92 Method for tank strengthening (Bobiński and Tejchman 2009)

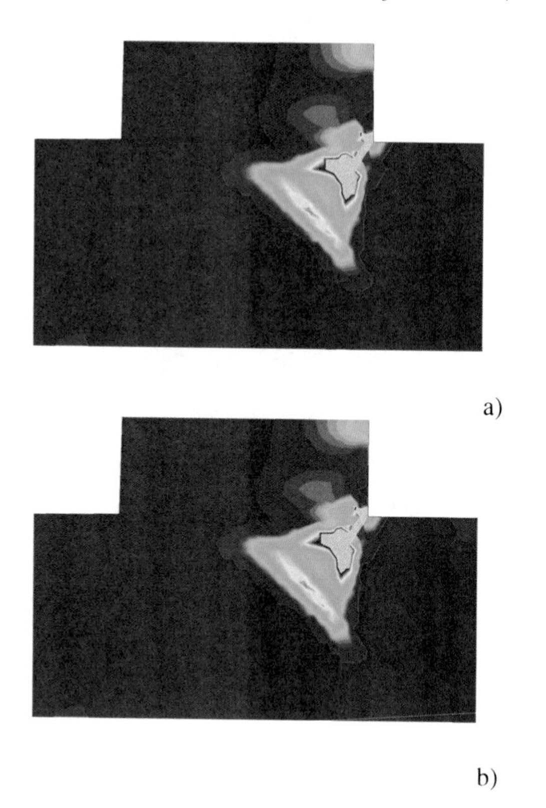

Fig. 7.93 Formation of strain localization in the wall corner during: a) initial water filling, b) repeated water filling, c) strength limit after repeated filling (Bobiński and Tejchman 2009)

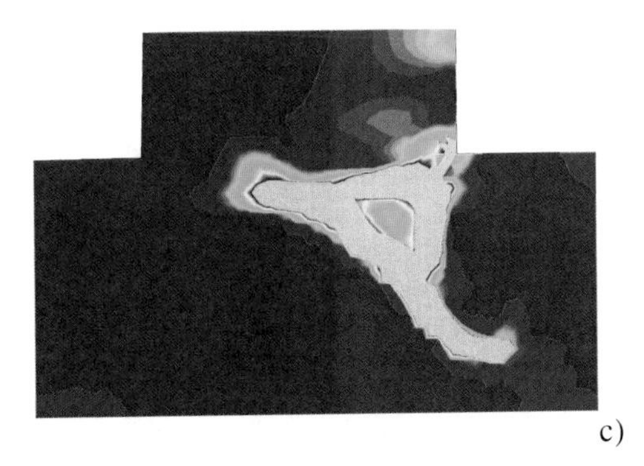

c)

Fig. 7.93 (*continued*)

Next, the 3D analysis was carried out with the tank after strengthening within linear elasticity (500'000 elements were used). The corner stiffness was decreased 20-40 times. The numerical results showed that: a) the crack width was smaller than 0.1 mm, b) the maximum wall displacement was smaller than 1.0 cm, c) bending moments in the corner were smaller than the permissible bending moment in the cracked corner estimated to be 250 kNm, d) the bending moments (which changed in the wall due the wall fixing by horizontal beams at the top) were carried by the existing vertical and horizontal wall reinforcement.

During tank filling after its strengthening, strains and displacement were measured along the wall. The measured maximum horizontal displacement of the wall top was 1.4 cm. The measured maximum tensile stresses in concrete, 2.4 MPa, were smaller than the mean tensile strength of concrete 2.9 MPa.

The numerical analysis of a reinforced concrete sewage tank wherein corners were subjected to a positive bending moment showed:

• linear elastic analysis is not always suitable for a proper engineering calculation of reinforced concrete structures. In contrast, a non-linear elasto-plastic analysis is capable to realistically capture strain localization,

• dimensioning of wall corners loaded by a negative bending moment can significantly differ from this with a positive bending moment.

• independently of the reinforcement detail, a diagonal interior crack always occurs in corners subjected to a negative moment,

• a diagonal crack in the interior of the corner has to be absolutely covered by reinforcement.

The results on strain localization in reinforced concrete elements and structures (Chapter 7) show that the most realistic numerical results are provided by an elasto-plastic approach with non-local softening.

References

ABAQUS, Theory Manual. Version 5.8, Hibbit, Karlsson & Sorensen Inc. (1998)

Akkermann, J.: Rotationsverhalten von Stahlbeton-Rahmenecken. PhD Thesis. Univesity of Karlsruhe (2000)

Baglin, P.S., Scott, R.H.: Finite element modelling of reinforced concrete beam-column connections. ACI Structural Journal 97(6), 886–894 (2000)

Bažant, Z.P., Cedolin, L., Tabbara, M.R.: New method of analysis for slender columns. ACI Structural Journal 88(4), 391–401 (1991)

Bažant, Z.P., Kwon, Y.W.: Failure of slender and stocky reinforced concrete columns: tests of size effect. Materials and Structures 27(2), 79–90 (1994)

Bažant, Z., Planas, J.: Fracture and size effect in concrete and other quasi-brittle materials. CRC Press LLC (1998)

Billinger, M., Symons, M.: Slender reinforced concrete columns produced from high-strength concrete. In: Proceedings of CIA/FIP Conference, Brisbane, pp. 223–232 (1995)

Bobiński, J., Tejchman, J.: Numerical simulations of localization of deformation in quasi-brittle materials within non-local softening plasticity. Computers and Concrete 1(4), 433–455 (2004)

Bobiński, J., Tejchman, J.: Dimensioning of tank wall corner subjected to positive bending moments. Report. Gdańsk University of Technology (2009)

Box, G.E.P., Muller, M.E.: A note of the generation of random normal deviates. Annals of Mathematical Statistics 29(2), 610–611 (1958)

Broms, B., Viest, I.M.: Ultimate strength analysis of long restrained reinforced concrete columns. Journal of the Structural Division (ASCE) 84(ST3), 1635-1–1635-30 (1958)

Campione, G., La Mendola, L., Papia, M.: Flexural behaviour of concrete corbels containing steel fibers or wrapped with FRP sheets. Materials and Structures 38(6), 617–625 (2005)

CEB-FIP, Commité Euro-International du Béton-Fédération International de la Précontrainte. Model Code 1990, bulletin d'information Lusanne, Shwitzerland (1990)

Červenka, V.: Constitutive model for cracked reinforced concrete. ACI Journal 82(6), 877–882 (1985)

Channakeshava, C., Sundara Raja Iyengar, K.T.: Elasto-plastic-cracking analysis of reinforced concrete structures. Journal of the Structural Division (ASCE) 114(11), 2421–2438 (1988)

den Uijl, J.A., Bigaj, A.: A bond model for ribbed bars based on concrete confinement. Heron 41(3), 201–226 (1996)

Dörr, K.: Ein Beitrag zur Berechnung von Stahlbetonscheiben unter besonderer Berücksichtigung des Verbundverhaltens. PhD thesis. Darmstadt University (1980)

El-Metwally, S.E., Chen, W.F.: Nonlinear behaviour of R/C frames. Computers and Structures 32(6), 1203–1209 (1989)

Eurocode 2 "Design of Concrete Structures, part 1-1" (1991)

Fattuhi, N.I.: Strength of SFRC corbels subjected to vertical load. Journal of Structural Engineering ASCE 116(3), 701–718 (1990)

Foster, S.J., Powell, R.E., Selim, H.S.: Performance of high strength concrete corbels. ACI Structural Journal 93(5), 555–563 (1996)

Fragomeni, S., Mendis, P.A.: Instability analysis of normal and high-strength reinforced concrete walls. Journal of Structural Engineering ASCE 123(5), 680–684 (1997)

Gardner, N.J., Ramakrishna, M.G., Tak-Fong, W.: Laterally prestressed eccentrically loaded slender columns. ACI Structural Journal 89(5), 547–554 (1992)

Gitman, I.M., Askes, H., Sluys, L.J.: Coupled-volume multi-scale modelling of quasi-brittle material. European Journal of Mechanics A/Solids 27(3), 302–327 (2008)

Gonzalez-Vidosa, F., Kotsovos, M.D., Pavlovic, M.N.: Symmetrical punching of reinforced concrete slabs: An analytical investigation based on non-linear finite element modeling. ACI Structural Journal 85(3), 241–250 (1988)

Groen, A.E.: Three-dimensional elasto-plastic analysis of soils. PhD Thesis. Delft University (1997)

Gruber, L., Menn, C.: Berechnung und Bemessung schlanker Stahlbetonstützen. Bericht 84. Institut für Baustatik ETH Zürich (1978)

Hagberg, T.: Design of concrete brackets: On the application of the truss analogy. ACI Journal 80(1-2), 3–12 (1983)

Haskett, M., Pehlers, D.J., Mohamed Ali, M.S.: Local and global bond characteristics of steel reinforcing bars. Engineering Structures 30(2), 376–383 (2008)

Häußler-Combe, U., Pröchtel, P.: Ein dreiaxiale Stoffgesetz fur Betone mit normalen und hoher Festigkeit. Beton- und Stahlbetonbau 100(1), 52–62 (2005)

Hordijk, D.A.: Local approach to fatigue of concrete. PhD Thesis. Delft University of Technology (1991)

Hsu, C.T.T.: Analysis and design of square and rectangular columns by equation of failure surface. ACI Structural Journal 85(2), 167–179 (1988)

Hughes, T.J.R., Winget, J.: Finite Rotation Effects in Numerical Integration of Rate Constitutive Equations Arising in Large Deformation Analysis. International Journal for Numerical Methods in Engineering 15(12), 1862–1867 (1980)

Jia, X., Song, W., Yuan, Y.: A smeared crack model for reinforced concrete beam via continuum damage mechanics. In: Meschke, G., de Borst, R., Mang, H., Bicanic, N. (eds.) Computational Modelling of Concrete Structures, EURO-C 2006, pp. 363–369. Taylor and Francis (2006)

Kani, G.N.J.: Basic facts concerning shear failure. ACI-Journal Proceedings 63(6), 675–692 (1996)

Kiedroń, K.: Method for calculations of critical force of columns subject to eccentric compression. PhD Thesis. Technical University of Wrocław (1980) (in Polish)

Kilpatrick, A.E., Rangan, B.V.: Tests on high-strength concrete-filled steel tabular columns. ACI Structural Journal 96(2), 268–274 (1999)

Kim, J., Yang, J.: Buckling behaviour of slender high-strength concrete columns. Engineering Structures 17(1), 39–51 (1995)

Kim, J.K., Lee, S.S.: The behaviour of reinforced concrete columns subjected to axial force and biaxial bending. Engineering Structures 22(11), 1518–1526 (2000)

Kordina, K., Warner, R.F.: Über den Einfluß des Kriechens auf die Ausbiegung schlanker Stahlbetonstützen. In: Deutscher Ausschuß für Stahlbeton, Berlin, Heft 250 (1975)

Kordina, K.: Bewehrungsführung in Ecken und Rahmenendknoten. In: DAfStb Heft 354, Berlin (1984)

Korzeniowski, P.: Effectiveness of increasing load bearing capacity of RC columns by raising the strength of concrete and amount of reinforcement. Archives of Civil Engineering XLIII(2), 149–164 (1997)

Kriz, L.B., Raths, C.H.: Connections in precast concrete – strength of corbels. Journal Prestressed Concrete Institute 10(1), 16–61 (1965)

Lloyd, N.A., Rangan, B.V.: Studies on high-strength concrete columns under eccentric compression. ACI Structural Journal 93(6), 631–638 (1996)

Lorrain, M., Maurel, O., Seffo, M.: Cracking behaviour of reinforced high-strength concrete tension ties. ACI Structural Journal 95(5), 626–635 (1998)

Majewski, T., Bobiński, J., Tejchman, J.: FE-analysis of failure behaviour of reinforced concrete columns under eccentric compression. Engineering Structures 30(2), 300–317 (2008)

Majewski, T., Bobiński, J., Tejchman, J.: FE-analysis of failure behaviour of reinforced concrete columns under eccentric compression. Report. Gdańsk University of Technology (2009)

Makovi, J.: Über den Einfluβ der Hysteresis in the Arbeitslinie des Betons auf das Verformungs- und Tragverhalten exzentrisch belasteter Stahlbetondruckglieder. PhD Thesis. Darmstadt University (1969)

Malecki, T., Marzec, I., Bobiński, J., Tejchman, J.: Effect of a characteristic length on crack spacing in a reinforced concrete bar under tension. Mechanics Research Communications 34(5-6), 460–465 (2007)

Małecki, T., Tejchman, J.: Analysis of strain localization in reinforced concrete elements with explicit second-gradient strain damage approach. In: Proceedings of International Conference Computer Methods in Mechanics, CMM (2009)

Manzoli, O.L., Oliver, J., Diaz, G., Huespe, A.E.: Three-dimensional analysis of reinforced concrete members via embedded discontinuity finite elements. Structures and Materials Journal 1(1), 58–83 (2008)

Martin, I., Olivieri, E.: Tests of slender reinforced concrete columns bent in double curvature. In: Symp. Reinforced Concrete Columns, SP-13, pp. 55–74. ACI, Detroit (1966)

Marzec, I., Bobiński, J., Tejchman, J.: Simulations of crack spacing in reinforced concrete beams using elastic-plasticity and damage with non-local softening. Computers and Concrete 4(5), 377–403 (2007)

Mattock, A.H., Chen, K.C., Soongswang, K.: The behaviour of reinforced concrete corbels. Journal Prestressed Concrete Institute 21(2), 52–77 (1976)

Mehmel, A., Becker, G.: Zur Schubbemessung des kurzen Kragamers. Der Bauingenieur 40(6), 224–231 (1965)

Mehmel, A., Freitag, W.: Tragfahigkeitsversuche an Stahlbetonkonsolen. Bauingenieur 42(10), 362–369 (1967)

Mendis, P.A.: Behaviour of slender high-strength concrete columns. ACI Structural Journal 97(6), 895–901 (2000)

Menétrey, P., Willam, K.J.: Triaxial failure criterion for concrete and its generalization. ACI Structural Journal 92(3), 311–318 (1995)

Nagrodzka-Godycka, K.: Reinforced concrete corbels. Experimental research, theory and design. Monograph, vol. 21. Wydawnictwo Politechniki Gdańskiej, Gdańsk (2001) (in Polish)

Nemecek, J., Bittnar, Z.: Behaviour of normal and high strength concrete columns: experiments and simulation. In: Zingoni, A. (ed.) Proc. Int. Conf. Progress in Structural Engineering, Mechanics and Computation, pp. 1551–1555. Taylor and Francis Group, London (2004)

Niedenhoff, H.: Untersuchungen über das Tragverhalten von Konsolen und kurzen Kragarmen, Dissertation, T. H. Karlsruhe, 115 (1961)

Nilsson, H.E.: Reinforced concrete corners and joints subjected to bending moment. In: National Swedish Building Research, Document D7:1973, Stockholm (1973)

Oleszkiewicz, S., Ruppert, J., Najib, S.: Influence of horizontal reinforcement on the bearing capacity of columns. In: Proc. Polish Conference of Civil Engineers. PZITB, Krynica (1973) (in Polish)

Pamin, J., de Borst, R.: Simulation of crack spacing using a reinforced concrete model with an internal length parameter. Archive of Applied Mechanics 68(9), 613–625 (1998)

Pfrang, E.O., Siess, C.P.: Behaviour of restrained reinforced concrete columns. Journal of the Structural Division (ASCE) 90(ST5), 113–135 (1964)

PN-B-03264:2002 standard: Concrete, reinforced concrete and prestressed structures. Static calculations and design (2002) (in Polish)

Renuka Prasad, H.N., Channakeshava, C., Raghu Prasa, B.K., Sundara Raja Iyengar, K.T.: Nonlinear finite element analysis of reinforced concrete corbel. Computers & Structures 46(2), 343–354 (1993)

Robinson, J.R.: L'Armature des consoles courtes. In: Festschrift Franz. Ernst und Sohn Verlag, Berlin (1969)

Rots, J.G., Nauta, P., Kusters, G.M.A., Blauwaauwendraad, J.: Smeared crack approach and fracture localization in concrete. Heron 30(1), 1–48 (1985)

Saenz, I., Martin, I.: Tests of slender reinforced concrete columns with high slenderness ratios. ACI Journal Proceedings 60(5), 589–615 (1963)

Skarżyński, L., Bobiński, J., Tejchman, J.: FE investigations of a deterministic size effect in reinforced concrete beams under mixed shear-tension failure (2010) (under preparation)

Souza, R.A.: Experimental and numerical analysis of reinforced concrete corbels strengthened with fiber reinforced polymers. In: Bicanic, N., de Borst, R., Mang, H., Meschke, G. (eds.) Computational Modelling of Concrete Structures, pp. 711–718. Taylor and Francis Group, London (2010)

Strauss, A., Mordini, A., Bergmeister, K.: Nonlinear finite element analysis of reinforced concrete corbels at both deterministic and probabilistic levels. Computers and Concrete 3(2-3), 123–144 (2006)

Stroband, J., Kolpa, J.J.: The behavipour of reinforced concrete column-to beam-joints, part 1, Corners subjected to negative moments, Report 5-83-9. University Delft (1983)

Stroband, J., Kolpa, J.J.: The behaviour of reinforced concrete column-to beam-joints, part 2, Corners subjected to positive moments, Report 5-81-5. University Delft (1981)

Szuchnicki, W.: Bearing capacity of columns with a variable cross-section. In: Proceedings of Polish Conference of Civil Engineers, PZITB, Krynica (1973) (in Polish)

Syroka, E., Bobiński, J., Tejchman, J.: FE analysis of reinforced concrete corbels with enhanced continuum models. Finite Element Methods in Analysis and Design 47(9), 1066–1078 (2011)

Walraven, J.C.: The influence of depth on the shear strength of lightweight concrete beams without shear reinforcement. TU-Delft Report 5-78-4. Delft University (1978)

Walvaren, J., Lehwalter, N.: Size effects in short beams loaded in shear. ACI Structural Journal 91(5), 585–593 (1994)

Widuliński, L., Bobiński, J., Tejchman, J.: FE-analysis of spacing of localized zones in reinforced concrete bars under tension using elasto-plasticity with non-local softening. Archives of Civil Engineering 55(2), 257–281 (2009)

van Mier, J.G.M.: Examples of nonlinear analysis of reinforced concrete structures with DIANA. Heron 32(3), 1–147 (1987)
Will, G.T., Uzamerii, S.M., Sihna, S.K.: Application of finite element method to the analysis of reinforced concrete beam-column joint. In: Proc. of Conference on Finite Element Method in Civil Engineering, CSCE, EIC, Canada, pp. 745–766 (1972)
Xie, J., Elwi, A.E., MacGregor, J.: Performance of high-strength concrete columns – a parametric study. ACI Structural Journal 94(2), 91–102 (1997)
Yong, Y.K., Balaguru, P.: Behaviour of reinforced high – strength concrete corbels. Journal of Structural Engineering ASCE 120(4), 1182–1201 (1994)

Chapter 8
Deterministic and Statistical Size Effect in Plain Concrete

Abstract. The numerical FE investigations of a deterministic and stochastic size effect in concrete beams of a similar geometry under three point bending were performed within an elasto-plasticity with a non-local softening. The FE analyses were carried out with four different sizes of notched and unnotched beams. Deterministic calculations were performed with a uniform distribution of the tensile strength. In turn, in stochastic calculations, the tensile strength took the form of random correlated spatial fields described by a truncated Gaussian distribution. In order to reduce the number of stochastic realizations without losing the calculation accuracy, Latin hypercube sampling was applied. The numerical outcomes were compared with the size effect law by Bažant and by Carpinteri.

A size effect phenomenon (nominal strength varies with the size of structure) is an inherent property of the behaviour of many engineering materials. In the case of concrete materials, both the nominal strength and material brittleness (ratio between the energy consumed during the loading process after and before the stress-strain peak) always decrease with increasing element size under tension (Bažant 1984, Carpinteri 1989, Bažant and Planas 1998). Thus, concrete becomes perfectly brittle on a sufficiently large scale. The results from laboratory tests which are scaled versions of the actual structures cannot be directly transferred to them. The physical understanding of size effects is of major importance for civil engineers who try to extrapolate experimental outcomes at laboratory scale to results which can be used in big scale situations. Since large structures are beyond the range of testing in laboratories, their design has to rely on a realistic extrapolation of testing results with smaller element sizes.

Two size effects are of a major importance in quasi-brittle and brittle materials: deterministic and statistical one (the remaining size effects are: boundary layer effect, diffusion phenomena, hydration heat or phenomena associated with chemical reactions and fractal nature of crack surfaces) (Bažant and Planas 1998). Currently there exist two different theories of size effect in quasi-brittle structures:

J. Tejchman, J. Bobiński: Continuous & Discontinuous Modelling of Fracture, SSGG, pp. 297–341.
springerlink.com

the energetic-statistical theory (Bažant and Planas 1998, Bažant 2004) and fractal theory (Carpinteri et al. 1994, 1995).

According to Bažant and Planas (1998) and Bažant (2004) the deterministic size effect is caused by the formation of a region of intense strain localization with a certain volume (micro-crack region - called also fracture process zone FPZ) which precedes macro-cracks and cannot be appropriately scaled in laboratory tests. Strain localization volume is not negligible to the cross-section dimensions and is large enough to cause significant stress redistribution in the structure and associated energy release. The specimen strength increases with increasing ratio l_c/D (l_c – characteristic length of the micro-structure influencing both the size and spacing of localized zones, D – characteristic structure size). In turn, a statistical (stochastic) effect is caused by the spatial variability/randomness of the local material strength. The first statistical theory was introduced by Weibull (1951) (called also the weakest link theory) which postulates that a structure is as strong as its weakest component. The structure fails when its strength is exceeded in the weakest spot, since stress redistribution is not considered. The Weibull's size effect model is a power law and is of particular important for large structures that fail as soon as a macroscopic fracture initiates in one small material element. It is not however able to account for a spatial correlation between local material properties, does not include any characteristic length of micro-structure (i.e. it ignores a deterministic size effect) and it underestimates the experimental size effect. Combining the energetic theory with the Weibull statistical theory led to a general energetic-statistical theory (Bažant and Planas 1998). The deterministic size effect was obtained for not too large structures and the Weibull statistical size effect was obtained as the asymptotic limit for very large structures. In turn, according to Carpinteri et al. (1994, 1995, 2007), the size effect is caused by the multi-fractality of a fracture surface only which increases with a spreading disorder of the material in large structures (stress redistribution and energy release during strain localization and cracking are not considered).

Two size effects laws proposed by Bažant (Bažant and Planas 1998, Bažant 2004) (called Size Effect Laws SEL) for geometrically similar structures allow to take into account a size difference by determining the tensile strength of structures without notches and pre-existing large cracks (the so-called type 1 size effect law) and of notched structures or structures with pre-existing cracks (the so-called type 2 size effect law) (Fig. 8.1). In the first type structures, the maximum load is reached as soon as a macroscopic crack initiates from the fully formed localized region of non-negligible size developed at a smooth surface. In the second type structures, cracks grow in a stable manner prior to the maximum load. Only the first type of structures is significantly affected by material randomness causing a pronounced statistical size effect. The material strength is bound for small sizes by a plasticity limit whereas for large sizes the material follows linear elastic fracture mechanics.

The following analytical formulae for a deterministic size effect predicted by asymptotic matching were proposed by Bažant (Bažant and Planas 1998)

$$\sigma_N(D) = f_r^{\infty}(1+\frac{rD_b}{D})^{\frac{1}{r}} \text{ (type 1 size effect law SEL)} \tag{8.1}$$

and

$$\sigma_N(D) = \frac{Bf_t}{\sqrt{1+\dfrac{D}{D_o}}} \text{ (type 2 size effect law SEL),} \tag{8.2}$$

where σ_N is the nominal strength, D is the characteristic structure size, f_r^{∞}, D_b and r denote the positive constant representing unknown empirical parameters to be determined; f_r^{∞} represents the solution of the elastic-brittle strength reached as the nominal strength for large structures, r controls the curvature and shape of the law and D_b is the deterministic characteristic length having the meaning of the thickness of the cracked layer (if D_b=0, the behaviour is elastic-brittle, Eq. 8.1). In turn, in Eq. 8.2, f_t denotes the tensile strength, B is the dimensionless geometry-dependent parameter (depending on the geometry of the structure and crack) and D_o denotes the size-dependent parameter called transitional size (both unknown parameters to be determined).

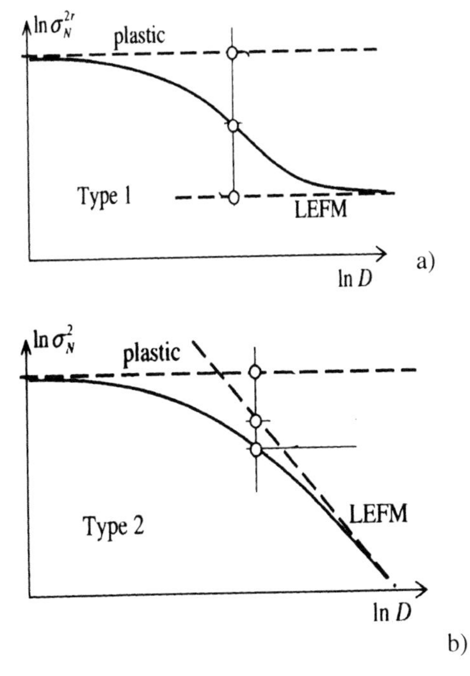

Fig. 8.1 Size effect models SEL by Bažant (2004) in logarithmic scale with σ_N - nominal strength, D – element size: a) type 1 (structures without notches and pre-existing large cracks), b) type 2 (notched structures) (material strength is bound for small sizes by plasticity limit whereas for large sizes, the material follows Linear Elastic Fracture Mechanics LEFM)

Another approach to the size effect was proposed by Carpinteri et al. (1994, 1995, 2007) (called Multi-Fractal Scaling Law MFSL) (Fig. 8.2). In this fractal approach, the nominal strength σ_N under tension decreases in a hyperbolic form with increasing characteristic structure size D

$$\sigma_N(D) = \sqrt{A_I + (A_2 / D)}, \tag{8.3}$$

where A_I and A_2 are the empirical constants. The approach does not distinguish between a deterministic or statistical size effect. The MFSL behaviour in the bilogarithmic plane $\ln\sigma_N$ versus $\ln D$ is non-linear and shows two asymptotes with slope -1/2 for small structures and slope zero for the largest ones, respectively. It predicts a transition from a disordered regime at the smallest scales to an ordered regime at the largest scales. According to Bažant and Yavari (2005, 2007c), the cause of a size effect is certainly energetic-statistical not fractal and the multi-fractal scaling law is a purely empirical formula and good enough only for the type 1 size effect (at crack initiation) and only for sizes not so large that the Weibull statistical size effect would intervene (MFSL does not capture a transition to the Weibull size effect for very large sizes). The disadvantage of both size effect laws is that they do not explicitly present the empirical constants to calculate the size effect in advance. In addition, a transition between two size effect types by Bažant remains still to be challenge.

The fits of the size effect law by Bažant (2004) and the multi-fractal scaling law by Carpintieri et al. (1994) to experimental data for concrete elements (van Vliet 2000) and reinforced concrete beams failing by shear (Bažant and Yavari 2007 c) show that both laws are only similar for experiments at laboratory scale but can significantly differ when the structure is very small or very large (Fig. 8.3) that can have serious consequences in the second case.

In spite of many experiments exhibiting the noticed size effect in concrete and reinforced concrete elements under different loading types (Walraven and Leihwalter 1994, Wittmann et al.1990, Elices et al. 1992, Bažant and Chen 1997,

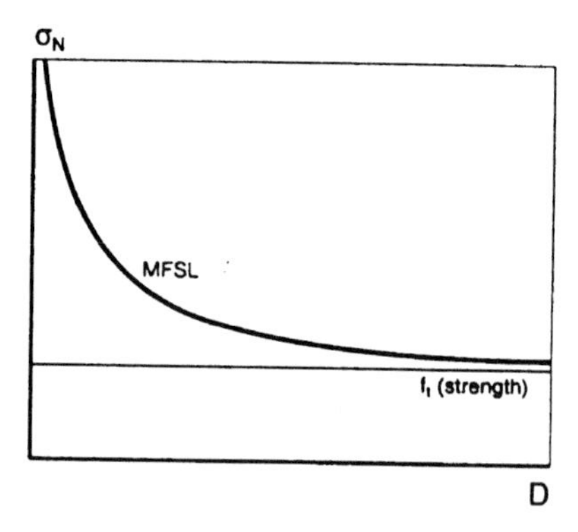

Fig. 8.2 Size effect by Carpintieri et al (1994): nominal strength σ_N versus specimen size D

Bažant and Planas 1998, Koide et al. 1998, van Vliet 2000, Chen et al. 2001, Le Bellego et al. 2003, van Mier and van Vliet 2003, Bažant 2004, Bažant and Yavari 2005, Vorechovsky 2007, Yu 2007), the scientifically (physically) based size effect is not taken into account in a practical design of engineering structures, that may contribute to their failure (Bažant and Planas 1998, Yu 2007). Instead, a purely empirical approach is sometimes considered in building codes which is doomed to yield an incorrect formula since physical foundations are lacking.

The goal of the numerical simulations is to investigate a deterministic and statistical size effect mainly in flexural resistance of notched and un-notched beam elements of a similar geometry under quasi-static three-point bending by considering the influence of strain localization. A finite element method with an elasto-plastic constitutive model using a Rankine'a criterion with non-local softening (Eqs. 3.32, 3.93 and 3.97) was used. Two-dimensional calculations were performed with four different concrete beam sizes of a similar geometry. Deterministic calculations were performed assuming constant values of tensile strength. In turn, statistical analyses were carried out with spatially correlated homogeneous distributions of tensile strength which were assumed to be random. Truncated Gaussian random tensile strength fields were generated using a conditional rejection method (Walukiewicz et al. 1997) for correlated random fields. The approximated results were obtained using a Latin hypercube sampling method (McKay et al. 1979, Bažant and Lin 1985, Florian 1992, Huntington and Lyrintzis 1998) belonging to a group of variance reduced Monte Carlo methods (Hurtado and Barbat 1998). This approach enables one a significant reduction of the sample number without losing the accuracy of calculations. The numerical results of load-displacements diagrams with notched beams were compared with corresponding laboratory tests performed by Le Bellego et al. (2003). The effect of the correlation length was also investigated. The FE results were compared with the size effect law SEL by Bažant and MFSL by Carpinteri.

The combined statistical and deterministic size effects were simulated by Carmeliet and Hens (1994), Frantziskonis (1998), Gutierrez and de Borst (1998), Gutierrez (2006), Vorechovsky (2007), Bažant et al. (2007a, 2007b), Yang and Xu (2008) and Bobiński et al. (2009). The most comprehensive combined calculations were performed by Vorechovsky (2007) for unnotched concrete specimens under uniaxial tension with a micro-plane material model and crack band model using Latin hypercube sampling. A squared exponential autocorrelation function with a correlation length of 80 mm was used. His results showed that the strength of many specimens, which parameters were obtained from random sampling, could be larger than a deterministic one in small specimens in contrast to large specimens which rather obeyed the weakest link model. The difference between a deterministic material strength and a mean statistical strength grew with increasing size. The structural strength exhibited a gradual transition from Gaussian distribution to Weibull distribution at increasing size. As the ratio of autocorrelation length and specimen size decreased, the ratio of spatial fluctuation of random field realizations grew. In the work by Yang and Xu (2008), a heterogeneous cohesive crack model to predict macroscopic strength of materials based on meso-scale random fields of fracture properties was proposed. A concrete notched beam subjected to mixed-mode fracture was modeled. Effects of various important parameters on the crack paths, peak loads, macroscopic ductility and overall reliability (including the variance of random fields, the correlation length, and the shear fracture resistance) were investigated and discussed.

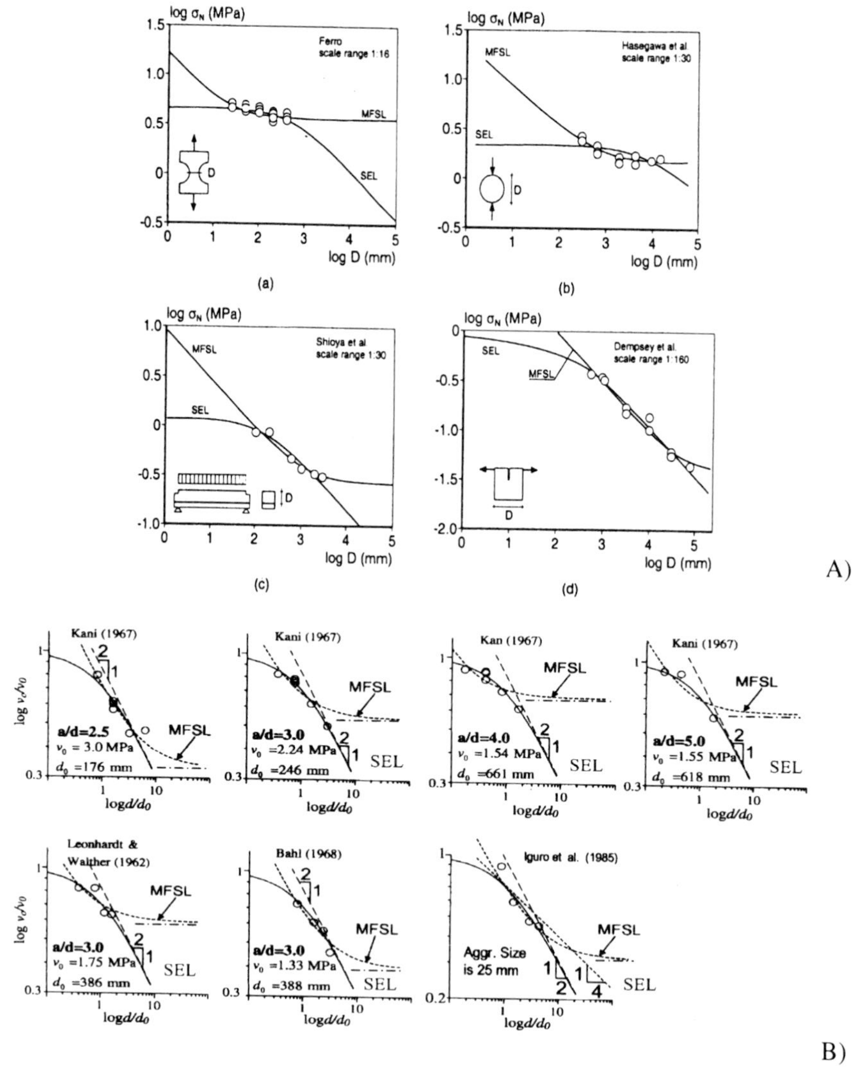

Fig. 8.3 Fits of the Size Effect Law by Bažant (2003) (SEL) and the Multi Fractal Scaling Law by Carpintier et al (1994) (MFSL) to experimental data: A) for concrete elements (van Vliet 2000) and B) for reinforced concrete beams failing by shear (Bažant and Yavari 2007c)

Our calculations with beams follow the research presented by Vorechovsky (2004, 2007) by using an alternative stochastic approach. In contrast to his simulations: a) free-supported concrete beams under bending were analyzed, b) a more sophisticated regularization technique was used in the softening regime, namely non-local theory, which ensured entirely mesh–independent results with respect to load-displacement diagrams and widths of localized zones (in contrast to the crack band model which provides only mesh-independent load-displacement diagrams), c) an original method of the random field generation with a different homogeneous correlation function was used.

In addition, a deterministic effect was examined in concrete during uniaxial compression using a Drucker-Prager's criterion with non-local softening (Eqs. 3.27-3.30, 3.93 and 3.97).

8.1 Notched Beams

Deterministic and statistical calculations

The two-dimensional FE-calculations (Bobiński et al. 2009) of free supported notched beams with free ends under bending (assuming constant values of tensile strength f_t) were performed with 4 different beam sizes of a similar geometry $h\times L_t$: 8×32 cm^2 (called small-size beam), 16×64 cm^2 (called medium-size beam), 32×128 cm^2 (called large-size beam) and 192×768 cm^2 (called very large-size beam) (h – beam height, L_t – total beam length). The span length L was equal to $3h$ for all beams (Fig. 8.4). The size of the first 3 beams was similar as in corresponding experiments carried out by Le Bellego et al. (2003). The quadrilateral elements divided into triangular elements were used to avoid volumetric locking. 7628 triangular (small-size beam), 14476 (medium-size beam), 28092 (large size beam) and 104310 (large-size beam) triangular elements were used, respectively. The mesh was particularly very fine in the region of a notch (Fig. 8.5) to properly capture strain localization in concrete (where the finite element size was equal to $1/3\times l_c$, l_c=5 mm). The ratio between the width of this fine region and beam length was always the same.

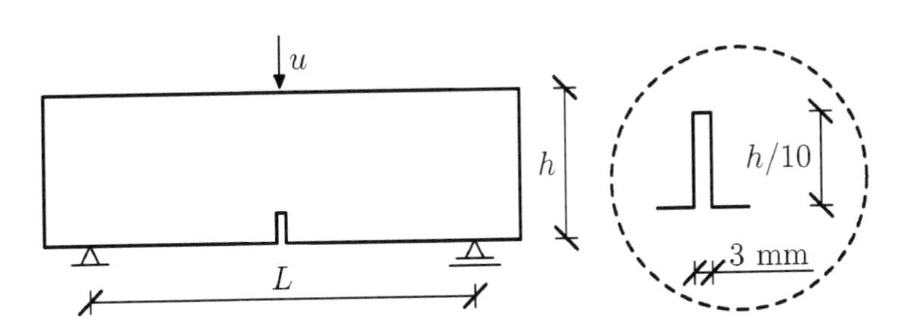

Fig. 8.4 Notched concrete beams used for calculations (L=3×h) (Bobiński et al. 2009)

To describe the behaviour of concrete under tension during three-point bending, a Rankine criterion was used with a yield function with isotropic softening (Eq. 3.32). To model the concrete softening under tension, the exponential curve by Hordijk (1991) with the tensile strength of the concrete of f_t=3.6 MPa was assumed (κ_u=0.005 b_1=3.0, b_2=6.93) (Eq. 3.55). The modulus of elasticity was assumed to be E=38.5 GPa and the Poisson ratio was υ=0.24 (Le Bellego et al. 2003). The calculations were performed under plane strain conditions (the differences between the results obtained within Rankine plasticity under plane stress and plane strain conditions are insignificant). A large-displacement analysis available in the ABAQUS finite element code (1998) was used (although the influence of such analysis was negligible). In this method, the current configuration of the body was taken into account. The Cauchy stress was taken as the stress measure. The conjugate strain rate was the rate of deformation. The rotation of the stress and strain tensor was calculated with the Hughes-Winget method (1980). The non-local averaging was performed in the current configuration.

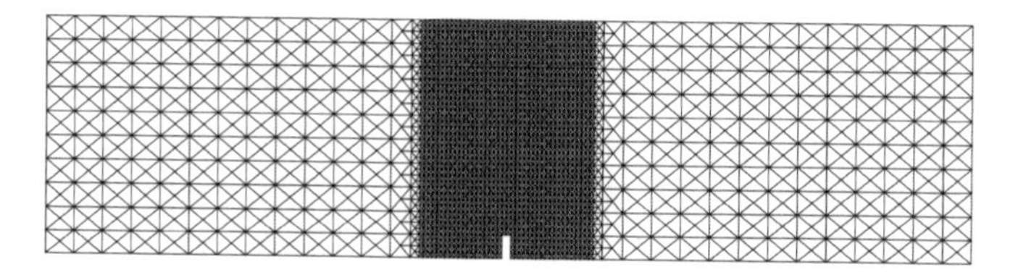

Fig. 8.5 FE mesh in the case of a medium-size beam (Bobiński et al. 2009)

A quasi-static deformation of a small, medium and large beam was imposed through a constant vertical displacement increment $\Delta \boldsymbol{u}$ prescribed at the upper mid-point of the beam top. To capture a snap-back behaviour in a very large size beam, the so-called arc-length technique was used. The actual load vector $\boldsymbol{P}$ was defined as $\lambda \boldsymbol{P}_{\max}$ where λ – multiplier and $\boldsymbol{P}_{\max}$ – maximum constant load vector. In general, the determination of the length of the arc the $\boldsymbol{P}$–$\boldsymbol{u}$ space ($\boldsymbol{u}$ – displacement vector) involves the displacements of all nodes (as e.g. the Riks method available in ABAQUS Standard 1998). However, for problems involving strain localization, it is more suitable to use an indirect displacement control method, where only selected nodal displacements are considered to formulate an additional condition in the $\boldsymbol{P}$–$\boldsymbol{u}$ space. The horizontal distance between two nodes lying on the opposite sides of the notch was chosen as a control variable *CMOD* (crack mouth open displacement). The indirect displacement algorithm was implemented with the aid of two identical and independent FE-meshes and some additional node elements to exchange the information about the displacements between these meshes.

The Monte Carlo method was used in statistical calculations. Application of the method in stochastic problems of mechanics requires the following steps: simulation of random variables or fields describing the problem under consideration (variability of material parameters, initial imperfections in structure geometrics and others), solution of the problem for each simulated realization, creation of a set of results and its statistical description. Contrary to stochastic finite element codes, the Monte Carlo method does not impose any restriction to the solved random problems. Its only limitation is the time of calculations. For example, to reproduce exactly the input random data of initial geometric imperfection of a shell structure problem, at least 2000 random samples should be used (Bielewicz and Górski 2002). Any nonlinear calculations for such number of initial data are, however, impossible due to excessive computation times. To determine a minimal, but sufficient number of samples (which allows one to estimate the results with a specified accuracy), a convergence analysis of the outcomes was proposed (Górski 2006). It was estimated that in case of various engineering problems only ca. 50 realizations had to be considered. For example in the shell structure limit load analysis (Górski 2006), the change of the error of limit load mean values between 50 and 150 samples equaled 2% and the standard deviations error was 12%. A further decrease of sample numbers can be obtained using Monte Carlo variance reduction methods.

In the papers by Tejchman and Górski (2007, 2008), two methods: a stratified and a Latin sampling method were considered. It should be pointed out that these methods were not used for the generation of two-dimensional random fields as, for example, in the paper by Vorechovsky (2007), but for their classification. For that reason, the single realization was generated according to the initial data, i.e. the theoretical mean value and the covariance matrix was exactly reproduced. The statistical calculations according to the proposed version of the Latin sampling method were performed in two steps (Tejchman and Górski 2007, 2008). First, an initial set of random samples was generated in the same way as in the case of a direct Monte Carlo method. Next, the generated samples were classified and arranged in increasing order according to the chosen parameters (i.e. their mean values and the gap between the lowest and the highest values of the fields). From each subset defined in this way, only one sample was chosen for the analysis. The selection was performed in agreement with the theoretical background of the Latin sampling method. The numerical calculations were performed only for these samples. It was proved that using the Latin sampling variance reduction method the results can be properly estimated by several realizations only (e.g. 12-15) (Tejchman and Górski 2007, 2008).

To generate the random field, the original conditional-rejection method described by Walukiewicz et al. (1997), Bielewicz and Górski (2002), Górski (2006), Tejchman and Górski (2007), and Tejchman and Górski (2008) was used. The method makes it possible to simulate any homogeneous or non-homogeneous truncated Gaussian random field described on regular or irregular spatial meshes. An important role in the calculations was played by the propagation base scheme covering sequentially the mesh points and the random field envelope which allowed one to fulfill the geometric and boundary conditions of the structure of the model. Random fields of practically unlimited sizes could be generated.

Various properties of concrete may be considered as randomly distributed. In the present work, only fluctuations of its tensile strength were taken into account. Two parameters described the random field should be chosen, i.e. the distribution of the random variable in a single point of the field and a function defining the correlation between these points. The distribution of a single random variable took the form of a truncated Gaussian function with the mean concrete tensile strength of $\bar{f}_t = 3.6$ MPa. Additionally, it was assumed that the concrete tensile strength values changed between f_t=1.6 MPa and f_t=5.6 MPa ($f_t = 3.6 \pm 2.0$ MPa). To fulfil this condition, the standard deviation $s_{f_t} = 0.424$ MPa was used in the calculations. The coefficient of variations describing the field scattering was $\mathrm{cov} = s_{f_t} / \bar{f}_t = 0.118$ ($\bar{f}_t$=3.6 MPa - the mean tensile strength). Since $5 s_{f_t} = 5 \times 0.424 = 2.12$ MPa, the cut of variables did not change the theoretical Gauss distribution distinctly (Fig. 8.6). The Irvin's characteristic length $(EG_f)/f_t^2$ (G_f - tensile fracture energy) which controls the length of the fracture process zone (Bažant and Planas 1998) varied between 0.100 m and 0.351 m.

Randomness of tensile strength f_t has to be described by a correlation function. For lack of the appropriate data, the correlation function is usually chosen arbitrarily. It is evident that the fluctuation of any material parameters should be described by a homogeneous function, which confirms that the correlation between random material variables vanishes when the random point distance increases. Any non-homogeneous correlation function, for example Wiener or Brown, defines strong correlation between every point of the field, and such a definition of material parameters is unrealistic. The simplest choice is a standard first order correlation function $K(x_1, x_2) = e^{-\lambda_{x1} \Delta x_1} e^{-\lambda_{x2} \Delta x_2}$. Here, the following (more general) second order and homogeneous correlation function was adopted (Bielewicz and Gorski 2002)

$$K(\Delta x_1, \Delta x_2) = s_{f_t}^2 \times e^{-\lambda_{x1} \Delta x_1} (1 + \lambda_{x_1} \Delta x_1) e^{-\lambda_{x2} \Delta x_2} (1 + \lambda_{x_2} \Delta x_2), \tag{8.4}$$

where Δx_1 and Δx_2 is are the distances between two field points along the horizontal axis x_1 and vertical axis x_2, λ_{x1} and λ_{x2} are the decay coefficients (damping parameters) characterizing a spatial variability of the specimen properties (i.e. describe the correlation between the random field points). The second order homogeneous function (Eq. 8.4) was proved to be very useful in engineering calculations (Knabe et al. 1998).

In finite element methods, continuous correlation function (Eq. 8.4) has to be represented by the appropriate covariance matrix. For this purpose, the procedure of local averages of the random fields proposed by Vanmarcke (1983) was adopted. After an appropriate integration of the function (Eq. 8.4), the following expressions describing the variances D_w and covariances K_w were obtained (Knabe et al. 1998)

$$D_w(\Delta x_1, \Delta x_2) = \frac{2}{\lambda_{x_1}\Delta x_1} s_{f_t}^2 \left[2 + e^{-\lambda_{x_1}\Delta x_1} - \frac{3}{\lambda_{x_1}\Delta x_1}\left(1 - e^{-\lambda_{x_1}\Delta x_1}\right)\right] \times$$
$$\times \frac{2}{\lambda_{x_2}\Delta x_2} s_{f_t}^2 \left[2 + e^{-\lambda_{x_2}\Delta x_2} - \frac{3}{\lambda_{x_2}\Delta x_2}\left(1 - e^{-\lambda_{x_2}\Delta x_2}\right)\right], \quad (8.5)$$

$$K_w(\Delta x_1, \Delta x_2) = \frac{e^{\lambda_{x_1}\Delta x_1}}{\left(\lambda_{x_1}\Delta x_1\right)^2} s_{f_t}^2 \left\{\left[\cos\left(\lambda_{x_1}\Delta x_1\right) - \sin\left(\lambda_{x_1}\Delta x_1\right)\right] + 2\lambda_{x_1}\Delta x_1 - 1\right\} \times$$
$$\times \frac{e^{\lambda_{x_2}\Delta x_2}}{\left(\lambda_{x_2}\Delta x_2\right)^2} s_{f_t}^2 \left\{\left[\cos\left(\lambda_{x_2}\Delta x_2\right) - \sin\left(\lambda_{x_2}\Delta x_2\right)\right] + 2\lambda_{x_2}\Delta x_2 - 1\right\} \quad . \quad (8.6)$$

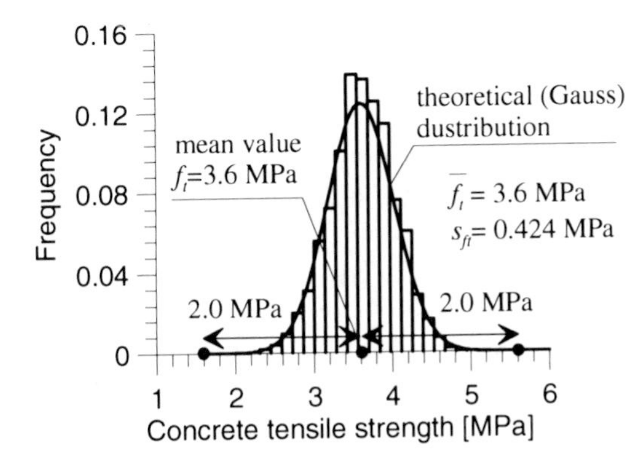

Fig. 8.6 Distribution of the concrete strength values for a single point of the mesh (Bobiński et al. 2009)

We took mainly into account a strong correlation of the tensile strength f_t in a horizontal direction λ_{x1}=1 1/m and a weaker in a vertical direction λ_{x2}=3 1/m in Eq. 8.4 (due to the way of specimen's preparation). In this way, the layers forming during the concrete placing were modeled. The range of significant correlation was approximately 80 mm in the horizontal direction and 30 mm in the vertical direction (Fig. 8.7). The smaller the λ parameter, the shorter is the correlation range. The dimension of the random field was identical as the finite element mesh. The same random values were assumed in 4 neighboring triangular elements.

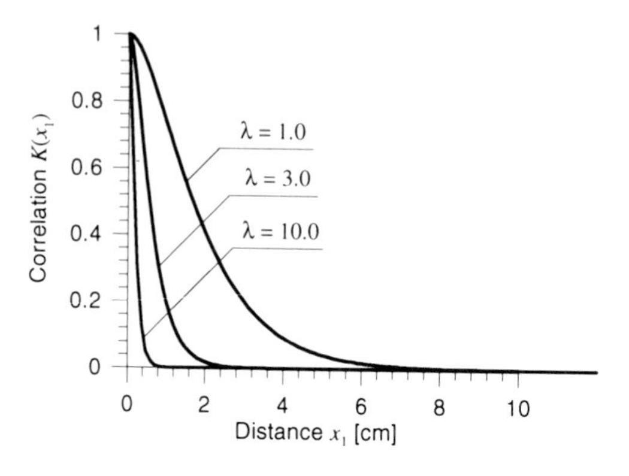

Fig. 8.7 The correlation distances for different coefficients λ [1/m] (Bobiński et al. 2009)

Using the conditional-rejection method, 2000 field realizations of the tensile strength were generated. Next, the generated fields were classified according to two parameters: the mean value of the tensile strength and the gap between the lowest and the highest value of the tensile strength. The joint probability distribution (so-called "ant hill") is presented in Fig. 8.8. One dot represents one random vector described by its mean value and the difference between its extreme values. The two variable domains were divided in 12 intervals of equal probabilities (see vertical and horizontal lines in Fig. 8.8). Next, according to the Latin hypercube sampling assumptions, 12 random numbers in the range 1-12 were generated (one number appeared only once) using the uniform distribution. The generated numbers formed the following 12 pairs: 1 – 4, 2 – 7, 3 – 3, 4 – 11, 5 – 5, 6 – 8, 7 – 1, 8 – 6, 9 – 2, 10 – 9, 11 – 10 and 12 – 12. According to these pairs, the appropriate areas (subfields) were selected (they are presented as rectangles in Fig. 8.9). From each subfield only one realization was chosen and used as the input data for FEM calculations. In this way, the results of 12 realizations were analyzed. Figure 8.9 shows a stochastic distribution of the tensile strength in one arbitrary concrete beam in the area close to the notch.

FE results of deterministic size effect

Figure 8.10 shows the evolution of the calculated normalized vertical force $PL/tf_t(0.9h)^2$ versus the normalized vertical beam displacement u/h for four different beam heights h: 8 cm, 16 cm, 32 cm and 192 cm with constant values of the tensile strength of f_t=3.6 MPa. The thickness of the specimen was equal to t=4 cm (as in laboratory experiments). A distribution of the non-local softening parameter is shown close to the notch (Fig. 8.11). Moreover, the numerical results of a deterministic size effect compared to the size effect model SEL 2 by Bažant for notched concrete specimens (Bažant and Planas 1998) (Eq. 8.2) are shown (Fig. 8.12).

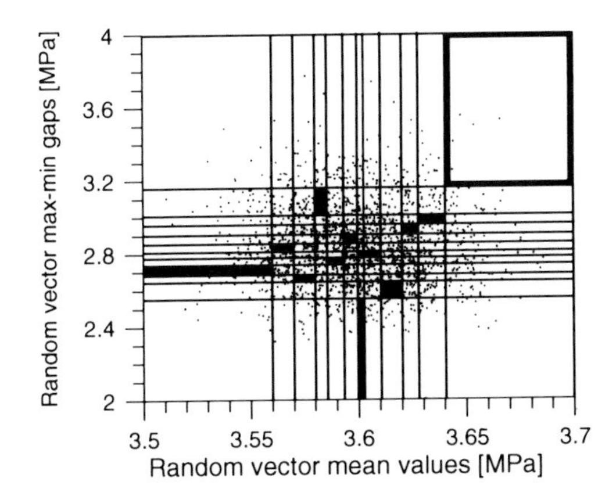

Fig. 8.8 Selection of 12 pairs of random samples using Latin hypercube sampling: 1 – 4, 2 – 7, 3 – 3, 4 – 11, 5 – 5, 6 – 8, 7 – 1, 8 – 6, 9 – 2, 10 – 9, 11 – 10 and 12 – 12 (Bobiński et al. 2009)

The beam strength and beam brittleness obviously increased with increasing beam size. This pronounced deterministic size effect is in agreement with the size effect model by Bažant of Fig. 8.1b (Bažant and Planas 1998). For a very large size beam, a so-called snap-back behaviour occurred (decrease of strength with decreasing deformation). The mean width of a localized zone above the notch was 15.08 mm (h=8 cm), 15.10 mm (h=16 cm), 18.02 mm (h=32 cm) and 18.05 mm (h=192 cm) at u/h=1.000‰, 0.494‰, 0.234‰ and 0.105‰, respectively.

The calculated vertical forces for a small, medium and large beam are in good accordance with the experiments by la Bellego et al. 2003 (Fig. 8.13). The calculated width of the localized zone is similar as in experiments, i.e. about 20 mm (on the basis of acoustic emission, Pijaudier-Cabot et al. 2004).

FE results of statistical size effect

12 selected random samples using Latin hypercube sampling are shown in Fig. 8.8 (λ_{x1}=1 1/m, λ_{x2}=3 1/m, $s_{f_t} = 0.424$). The 12 different evolutions of the vertical normalized force versus the vertical normalized displacement are shown in Fig. 8.14 for 3 different beam heights h: 8 cm (small beam), 32 cm (large beam) and 192 cm (very large beam), respectively. Figure 8.15 demonstrates the calculated width of a localized zone. In turn, 5 arbitrary deformed FE-meshes for a small-size beam are shown in Fig. 8.16. The size effect is shown in Fig. 8.17.

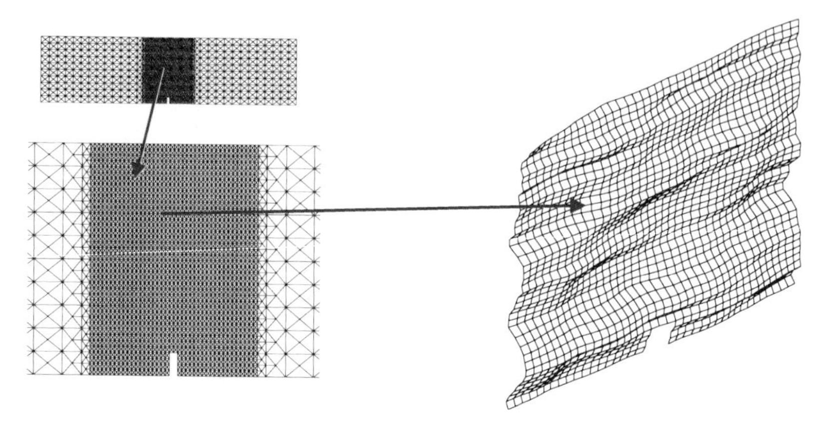

Fig. 8.9 Stochastic distribution of tensile strength f_t close to the notch in small-size beam (strong correlation, small standard deviation) (Bobiński et al. 2009)

The normalized maximum vertical force decreases with decreasing beam height h (Fig. 8.14). For h=8 cm, it changes between 2.92-3.38 kN. The mean stochastic P_{max}=3.08 kN (with the standard deviation of 0.126 kN) is practically the same as the deterministic value P_{max}=3.13 kN (it is smaller by only 2%). If the beam height is h=32 cm, the maximum vertical force varies between 7.73-8.85 and the mean stochastic force P_{max}=8.30 kN (with the standard deviation of 0.334 kN) is smaller by only 0.6% than the deterministic value (P_{max}=8.35 kN). For the beam height of h=192 cm, the maximum vertical force varies between 26.05-28.72 kN and the mean stochastic force P_{max}=27.56 kN is again smaller by only 0.6% than the deterministic value of P_{max}=27.72 kN (the standard deviation equals 0.692 kN).

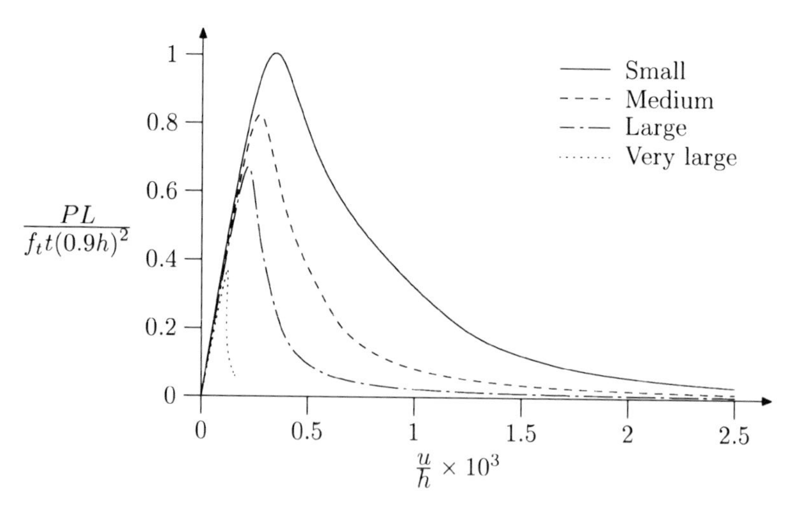

Fig. 8.10 Normalized force-displacement curves with constant values of tensile strength for 4 notched beams under three-point bending (Bobiński et al. 2009)

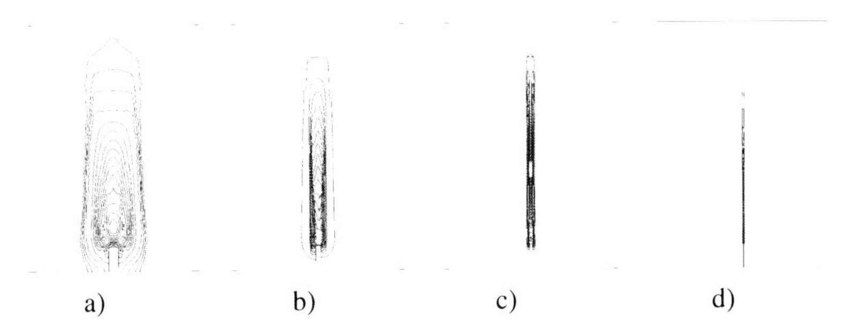

Fig. 8.11 Calculated contours of non-local softening parameter $\overline{\kappa}$ above the notch for three-point bending of small (a), medium (b) large (c) and very large (d) notched concrete beam with constant values of tensile strength (Bobiński et al. 2009)

The stochastic size effect in notched concrete beams is very small; the difference between the deterministic material strength and mean statistical strength is practically negligible.

The load-displacement curves for a very large beam are not smooth in softening regime when the tensile strength is distributed stochastically. The scatter of the maximum vertical force around its mean value is similar for all beam sizes (Fig. 8.17). The deformation field above the notch is strongly non-symmetric (Fig. 8.16). The mean width of the localized zone above the notch is slightly higher than the deterministic value, namely: w=16.56 mm (h=8 cm), w=18.88 mm (h=32 cm) and w=19.67 mm (h=192 cm), Fig. 8.15.

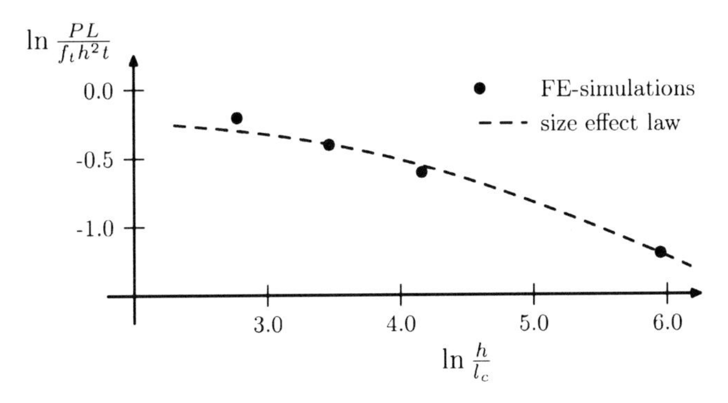

Fig. 8.12 Relationship between calculated normalized concrete strength $\ln\sigma=\ln[PL/(f_t h^2 t)]$ and ratio $\ln(h/l_c)$ compared to size effect law by Bažant of Fig. 8.1b (Bažant and Planas 1998) for constant values of tensile strength (h- beam height, l_c - characteristic length) (Bobiński et al. 2009)

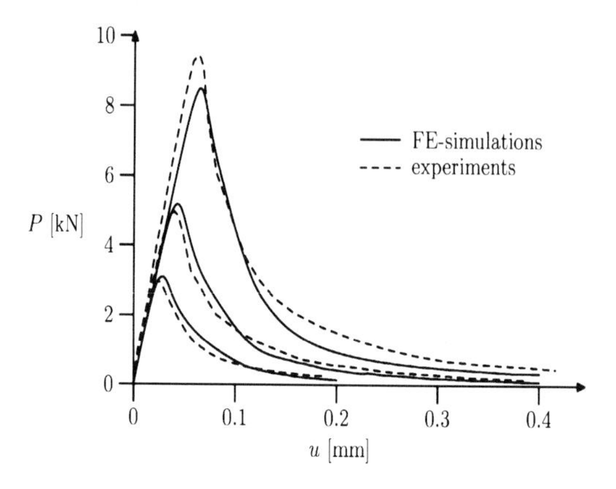

Fig. 8.13 The load-displacement curves from FE-calculations with constant values of tensile strength compared to the experiments by Le Bellego et al. (2003) with 3 notched concrete beams: h=8 cm (lower curves), h=16 cm (medium curves) and h=32 cm (upper curves) (Bobiński et al. 2009)

Our results are close to those given by Vorechovsky (2007). However, in contrast to his results, the difference between stochastic and deterministic values and the scatter of stochastic values in our calculations are similar independently of the beam size. In contrast to simulations by Yang and Xu (2008), which were performed with one notched beam only, the strong tortuousness of crack trajectories was not obtained for a small beam. Beside this fact, the evolution of stochastic load-displacement curves was similar.

Effect of sample number

The calculations were carried out with a small size beam using a direct Monte Carlo method with 30 samples (Fig. 8.18) (λ_{x1}=1 1/m, λ_{x2}=3 1/m, s_{f_t} =0.424 MPa). Almost similar results (mean P_{max}=3.07 kN with s_{f_t} =0.138 MPa) were obtained as in the case of Latin hypercube sampling with 12 samples (mean P_{max}=3.06 kN).

Effect of correlation range

In addition, the calculations were carried out with a small-size beam assuming a very small correlation length of 10 mm (Fig. 8.7) by assuming λ_{x1}=10 1/m, λ_{x2}=10 1/m and s_{f_t} = 0.424 MPa in Eq. 8.3. The results (Figs. 8.19 and 8.20) show that the mean stochastic vertical force, P_{max}=3.08 kN, and mean width of the localized zone, w=16.56 mm, are similar as the results with λ_{x1}=1 1/m and λ_{x2}=3 1/m. However, the scatter of forces is significantly smaller.

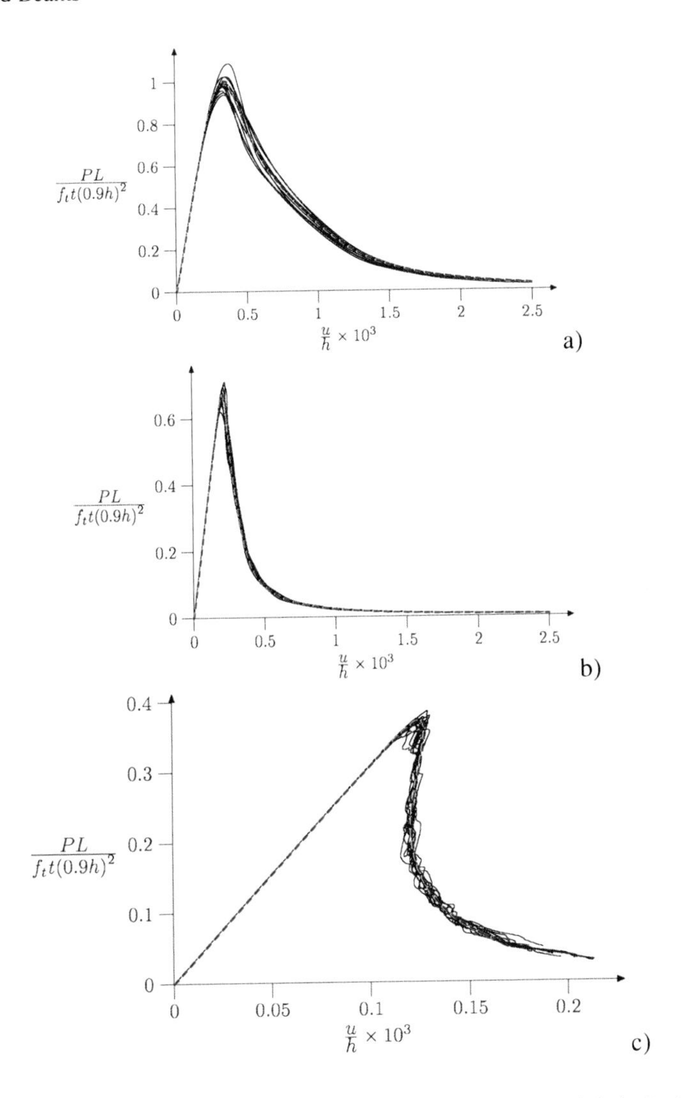

Fig. 8.14 Normalized force-displacement curves in the case of deterministic (red dashed lines) and random calculation (solid lines) for 3 notched beams under three-point bending: a) small-size beam (h=8 cm), b) large-size beam (h=32 cm), c) very large-size beam (h=192 cm) (λ_{x1}=1 1/m, λ_{x2}=3 1/m, s_{ft}=0.424 MPa) (Bobiński et al. 2009)

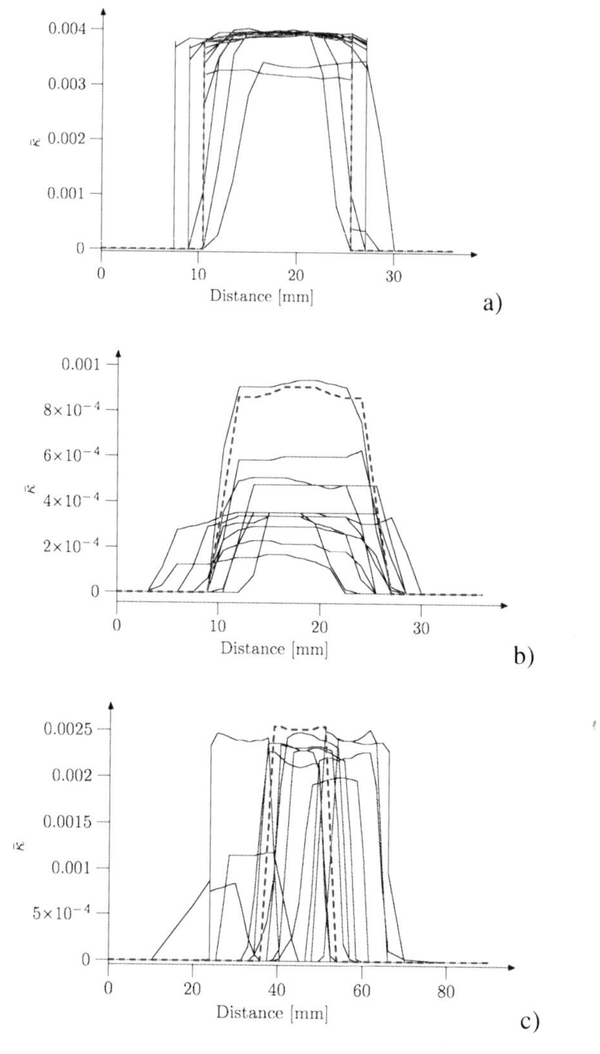

Fig. 8.15 Distribution of non-local softening parameter above the notch in the case of deterministic (red dashed lines) and random calculation (solid lines) for 3 notched beams under three-point bending: a) small-size beam (h=8 cm), b) large-size beam (h=32 cm), c) very large-size beam (h=192 cm) (λ_{x1}=1 1/m, λ_{x2}=3 1/m, s_{ft}=0.424 MPa) (Bobiński et al. 2009)

8.2 Unnotched Beams

Very similar deterministic and stochastic calculations were carried out with concrete beams of Chapter 8.1 without notch using the similar input and material data (Syroka et al. 2011). The two-dimensional FE-analysis of free-supported

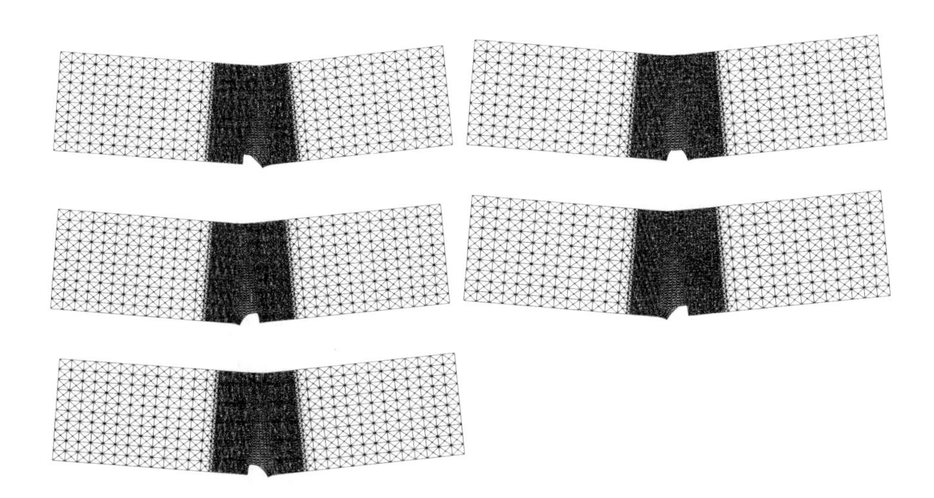

Fig. 8.16 Five arbitrary deformed FE meshes for a small-size beam (h=8 cm, u/h=0.25%) with random distribution of tensile strength (λ_{x1}=1 1/m, λ_{x2}=3 1/m, s_{ft}=0.424 MPa) (Bobiński et al. 2009)

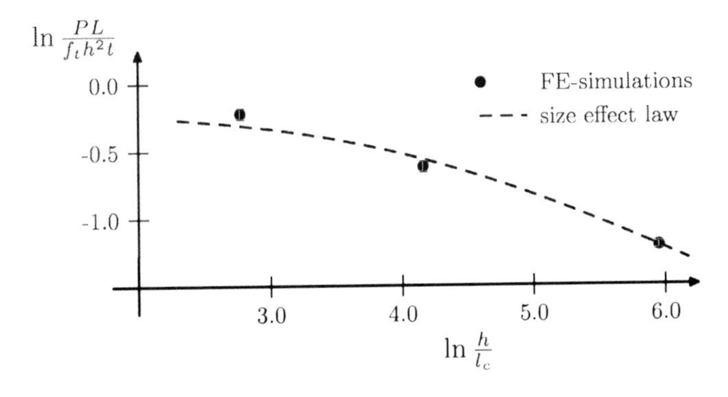

Fig. 8.17 Relationship between calculated normalized concrete strength ln σ=ln $[PL/(f_t h^2 t)]$ and ratio ln (h/l_c) compared to the size effect law by Bažant (Bažant and Planas 1998) for stochastic values of tensile strength (Bobiński et al. 2009)

unnotched beams was mainly performed with 4 different beam sizes of a similar geometry $D\times L_t$: 8×32 cm^2 (called small-size beam), 16×64 cm^2 (called medium-size beam), 32×128 cm^2 (called large-size beam), 192×768 cm^2 (called very large-size beam) (D – beam height, L_t – beam length), Fig. 8.21. The span length L was equal to 3D for all beams. The depth of the specimens was t=4 cm. The size $D\times L_t\times t$ of the first 3 beams was similar as in the corresponding experiments carried out by Le Bellego et al. (2003) and Skarżyński et al. (2009). The quadrilateral elements divided into triangular elements were used to avoid volumetric locking. Totally, 13’820 (small-size beam), 39’900 (medium-size beam), 104’780 (large-size beam) and 521’276 (very large-size beam) triangular elements were used, respectively The computation time varied between 3 hours (small-size beam) and 3 days (very large beam) using PC 3.2 MHz.

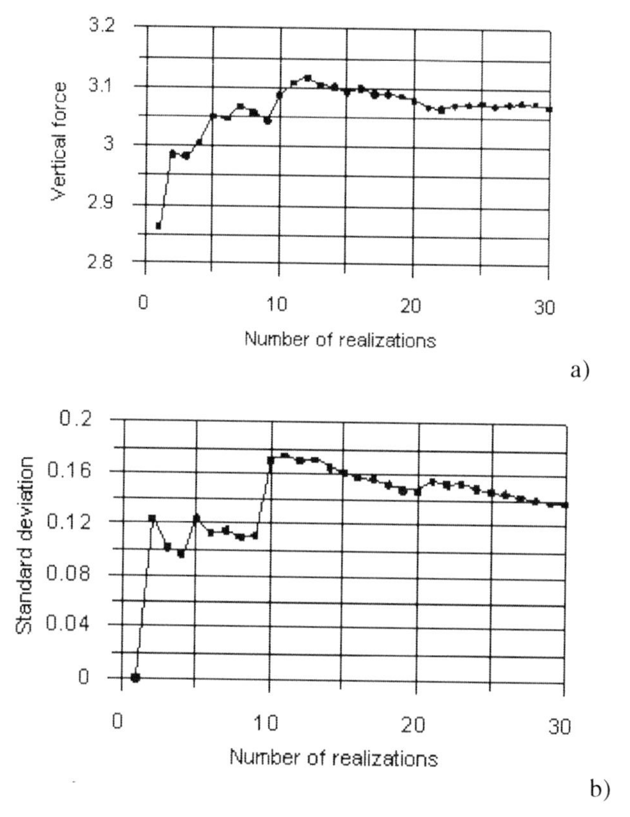

Fig. 8.18 Small size beam with random distribution of tensile strength (h=8 cm) using a direct Monte Carlo method with 30 samples: maximum vertical force with expected values (a) and standard deviation (b) (λ_{x1}=1 1/m, λ_{x2}=3 1/m, s_{ft}=0.424 MPa) (Bobiński et al. 2009)

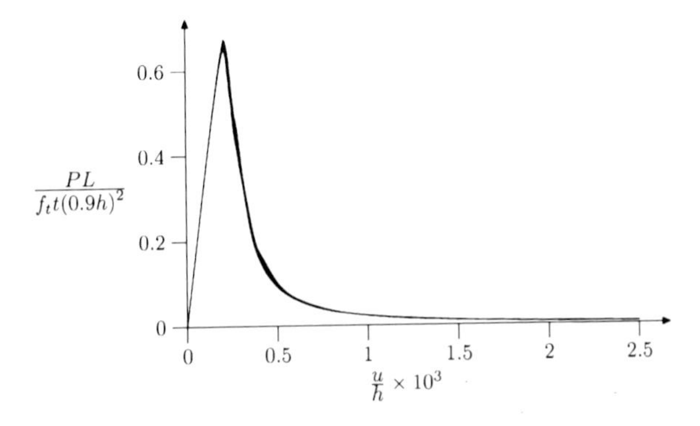

Fig. 8.19 Normalized force–displacement curves with and random distribution of tensile strength for notched small beam under three-point bending (h=8 cm) for smaller correlation length (λ_{x1}=10 1/m, λ_{x2}=10 1/m, s_{f_t} = 0.424 MPa) (Bobiński et al. 2009)

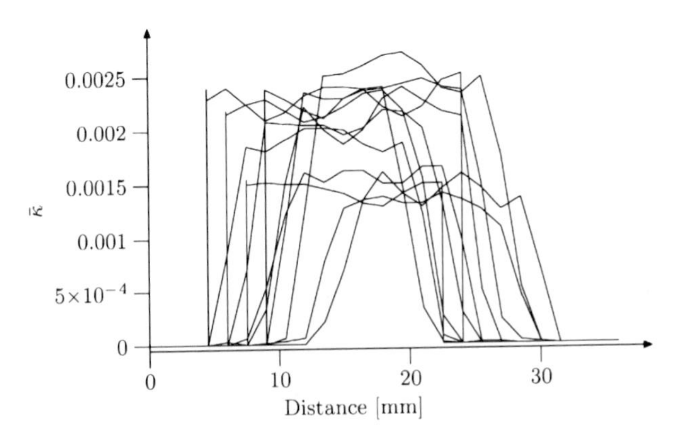

Fig. 8.20 Distribution of non-local softening parameter random distribution of tensile strength for notched small beam under three-point bending (h=8 cm) for small correlation length (λ_{x1}=10 1/m, λ_{x2}=10 1/m, s_{f_t} = 0.424 MPa) (Bobiński et al. 2009)

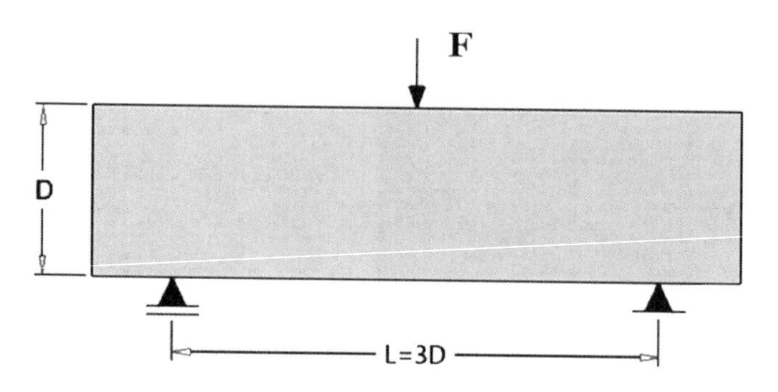

Fig. 8.21 Geometry of free-supported unnotched concrete beams subjected to three-point bending (F – vertical force) (Syroka et al. 2011)

All specimens had again the constant uniformly distributed tensile strength f_t=3.6 MPa. In order to properly capture strain localization in concrete, the mesh was very fine in the mid-part of the beam (Fig. 8.21) (where the element size was not greater than $3 \times l_c$). The width of this region s of Fig. 8.22 was determined with preliminary calculations: s=12 cm (D=8 cm), s=18 cm (D=16 cm), s=24 cm (D=32 cm) and s=192 cm (D=192 cm). A quasi-static deformation of a small and medium beam was imposed through a constant vertical displacement increment Δu prescribed at the upper mid-point of the beam top.

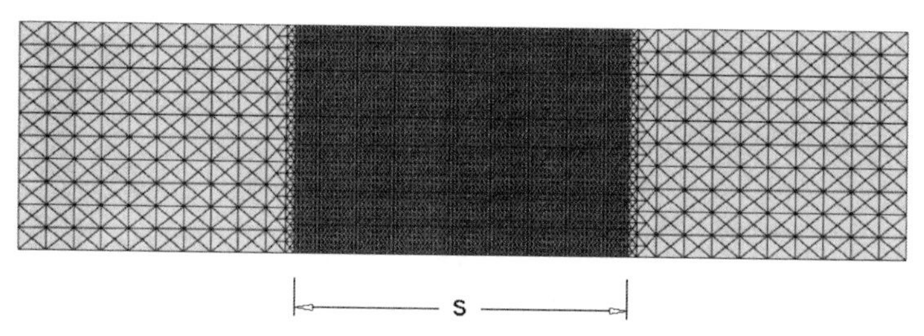

Fig. 8.22 Assumed FE mesh in small-size beam (s – width of region with finer mesh) (Syroka et al. 2011)

Correlated random fields describing a fluctuation of the tensile strength were used to capture a stochastic size effect. The distribution of this single random variable f_t took the form of a truncated Gaussian function with the mean concrete tensile strength of 3.6 MPa (as in calculations with notched beams, Fig. 8.6). The concrete tensile strength values again changed between 1.6 MPa and 5.6 MPa ($f_t = 3.6 \pm 2.0$ MPa). The homogeneous correlation function by Eq. 8.4 was adopted (Bielewicz and Górski 2002). We took again into account a stronger correlation of the tensile strength f_t in a horizontal direction λ_{x1}=1.0 1/m and a weaker correlation in a vertical directions λ_{x2}=3.0 1/m in Eq. 8.4 (due to the way of the specimen preparation by means of layer-by-layer from the same concrete block). The dimension of the random field was identical as the finite element mesh. The same random values were assumed in four neighbouring triangular elements. To generate the random fields, the conditional-rejection method was again used. The selection was performed by the Latin sampling method (Fig. 8.23). The generated numbers formed the following 12 pairs: 1 – 4, 2 – 7, 3 – 3, 4 – 11, 5 – 5, 6 – 8, 7 – 1, 8 – 6, 9 – 2, 10 – 9, 11 – 10 and 12 – 12 (Fig. 8.23). Figure 8.24 shows the distribution of the concrete tensile strength in a small-size (Fig. 8.24a) and very large-size concrete beam (Fig. 8.24b).

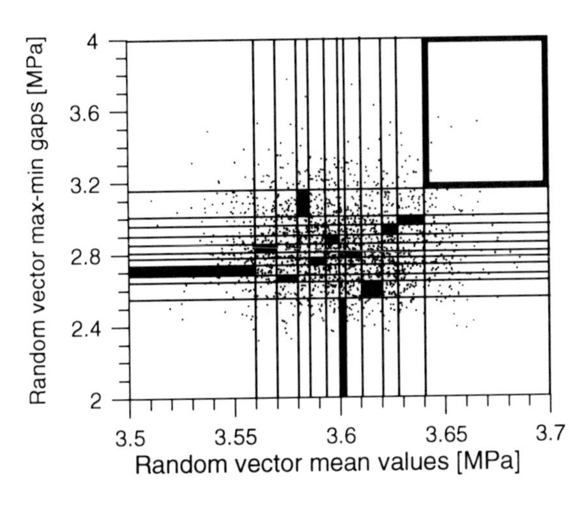

Fig. 8.23 Selection of 12 pairs of random samples using Latin hypercube sampling: 1 – 4, 2 – 7, 3 – 3, 4 – 11, 5 – 5, 6 – 8, 7 – 1, 8 – 6, 9 – 2, 10 – 9, 11 – 10 and 12 – 12 (Syroka et al. 2011)

Deterministic size effect

The evolution of the normalized vertical force $1.5FL/(f_tD^2t)$ versus the normalized deflection u/D for four different beam sizes with the constant values of the tensile strength f_t is shown in Fig. 8.25. The distribution of non-local softening parameter $\bar{\kappa}$ in the mid-region of beams is demonstrated in Fig. 8.26. In Fig. 8.27, our FE results were confronted with FE results for similar notched beams of Chapter 8.1 (Bobiński et al. 2009).

The maximum deterministic vertical forces were: F_{max}=3.83 kN (D=8 cm), F_{max}=6.75 kN (D=16 cm), F_{max}=12.57 kN (D=32 cm) and F_{max}=66.18 kN (D=192 cm), respectively. The strength and ductility strongly increased with decreasing beam height. The normalized nominal (flexural) strength $\sigma_N/f_t=1.5F_{max}L/(D^2tf_t)$ varied between 1.1 (D=192 cm) and 1.5 (D=8 cm). For the large and very large-size beam, the snap-back behaviour occurred. (the strength's decrease with decreasing deformation). It was in particular very strong for the very large-size beam. Note that the snap-back behaviour happened in notched very large concrete beams only (Bobiński et al. 2009).

The width of a localized zone for all beam sizes was about w=1.5 cm (at the same normalized flexural stress of 1.0, Fig. 8.25). In turn, the height of the localized zone h measured at the peak load increased non-linearly with increasing beam height D, i.e.: 24 mm, 34 mm, 40 mm, and 48 mm for the small (D = 80 mm), medium (D = 160 mm), large (D = 320 mm) and very large beam (D=1920 mm), respectively. The larger the beam, the lower was the ratio of the localized zone height to the beam height h/D: 0.3 (D=80 mm), 0.212 (D=160 mm), 0.125 (D=320 mm) and 0.025 (D=1920 mm).

A pronounced deterministic size effect took place in computations (Fig. 8.27). The deterministic size effect is significantly stronger than in notched concrete beams.

When comparing the numerical results with the size effect model by Bažant (Eq. 8.1), the best fit was achieved with a high parameter r=4 (with f_r^{∞}=3.55 MPa and D_b=112 mm). However, based on the recent results by Bažant et al. (2007a, 2007b), the parameter r (which controls both the curvature and slope of the size effect curve) seems to be close to 1. Therefore, a second deterministic characteristic length l_p was introduced (Bažant et al. 2007a, 2007b) to better describe the size effect law by taking into account a perfect plastic rage for extremely small structure sizes D

$$\sigma_N(D) = f_r^{\infty}\left(1+\frac{rD_b}{l_p+D}\right)^{\frac{1}{r}}. \tag{8.7}$$

This formula represents the full size range transition from the perfectly plastic behaviour (for $D\to 0$, $D \trianglelefteq l_p$) to the elastic brittle behaviour (for $D\to\infty$, $D>>D_b$) through the quasi-brittle one. The second deterministic characteristic length l_p governs the transition to plasticity for small sizes D. The case $l_p \neq 0$ shows the plastic limit for vanishing size D. This case is asymptotically equivalent to the case of l_p=0 for large D.

The asymptotic prediction for small and large sizes leads to

$$\lim_{D\to 0}\sigma_N(D) = f_r^{\infty}(1+rD_b/l_p) \quad \text{and} \quad \lim_{D\to 0}\sigma_N(D) = f_r^{\infty}. \tag{8.8}$$

The parameter l_p equals

$$l_p = \frac{rD_b}{\eta_p - 1} \tag{8.9}$$

with η_p - the ratio between the maximum plastic and elastic strength.

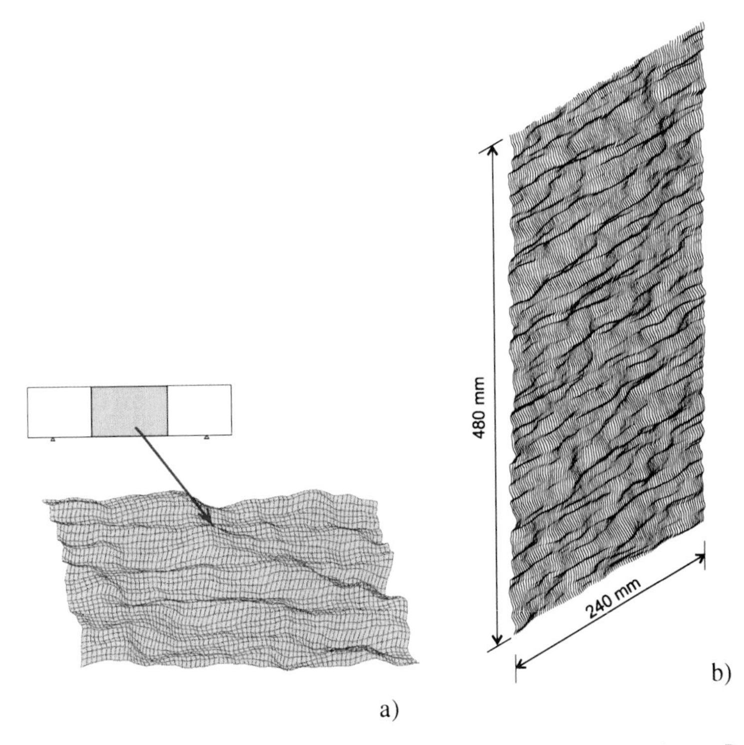

Fig. 8.24 Distribution of concrete tensile strength in small-size concrete beam D=8 cm (region 8×12 cm^2) (a) and in very large-size concrete beam D=192 cm (region 24×48 cm^2) (b) (Syroka et al. 2011)

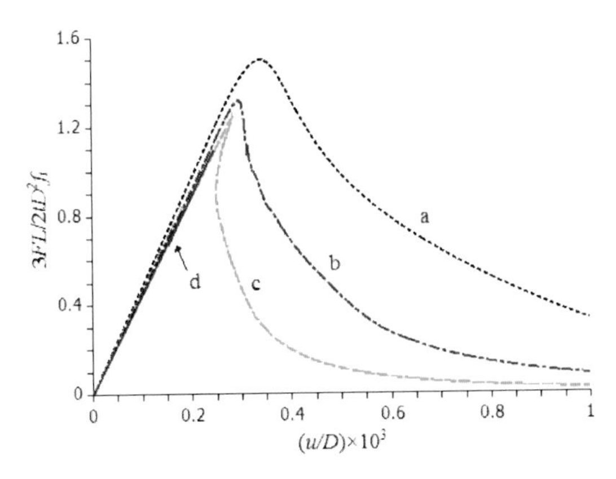

Fig. 8.25 Normalized horizontal normal (flexural) stress-deflection curves $1.5FL/(f_tD^2t)$=f(u/D) under 3-point bending with constant values of tensile strength for 4 different concrete beam heights: small D=8 cm (dashed line 'a'), medium D=16 cm (dotted-dashed line 'b'), large D=32 cm (dotted line 'c'), very large D=192 cm (solid line 'd') (F – vertical force, L – beam length, D – beam height, t- beam thickness, f_t - tensile strength) (Syroka et al. 2011)

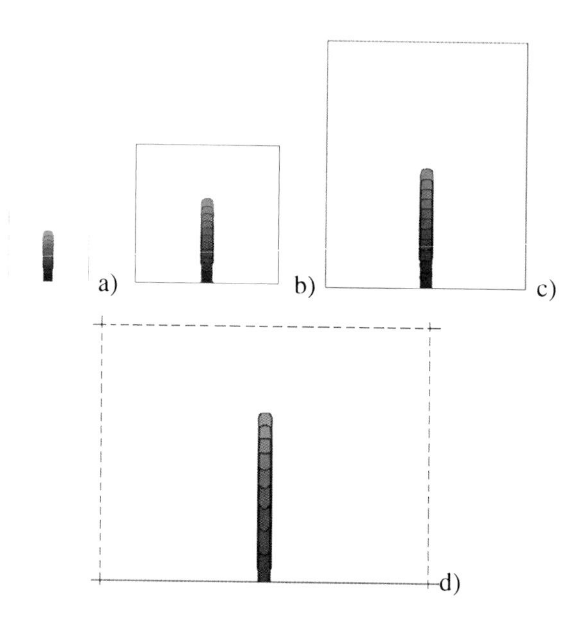

Fig. 8.26 Distributions of non-local softening parameter $\bar{\kappa}$ in concrete beams (mid-region) from deterministic calculations sizes at σ_N=0.45 MPa in: a) small D=8 cm, b) medium D=16 cm, c) large D=32 cm and d) very large-size beam D=192 cm (figure 'd' is not appropriately scaled) (Syroka et al. 2011)

The parameter l_p was determined with additional FE calculations for $D \rightarrow 0$ and $D \rightarrow \infty$. Thus, four additional geometrically similar concrete elements were numerically analyzed by us with D=0.2 cm, D=2 cm, D=4 cm and D=384 cm. On the basis of the nonlinear regression method by Leveneberg-Marquardt, the following parameters were found to fit Eq. 8.7: f_r^{∞}=3.782 MPa, D_b=40 mm, l_p= 13.6 mm, r=1.0. The agreement of our FE results for 8 elements with Eq. 8.7 is almost perfect (Fig. 8.28).

Statistical size effect

The 12 different evolutions of the normalized vertical force $FL/(f_t D^2 t)$ versus the normalized vertical deflection u/D from stochastic calculations are shown in Fig. 8.29 (the deterministic curve is also attached).

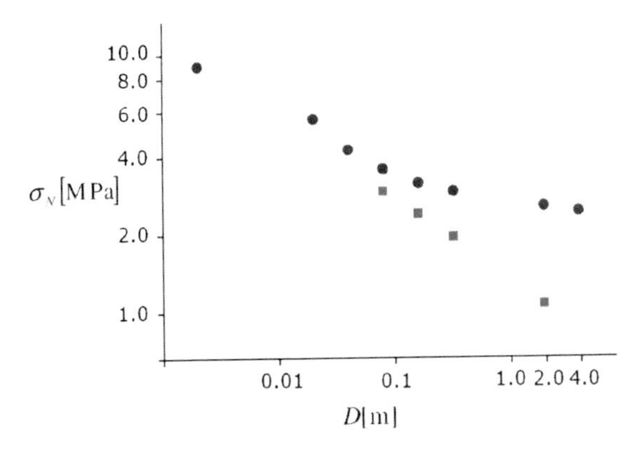

Fig. 8.27 Calculated nominal strength versus beam height from deterministic FE calculations for notched concrete beams (Bobiński et al. 2009, Chapter 8.1) (green diamonds) and unnotched beams (red circles) (Syroka et al. 2011)

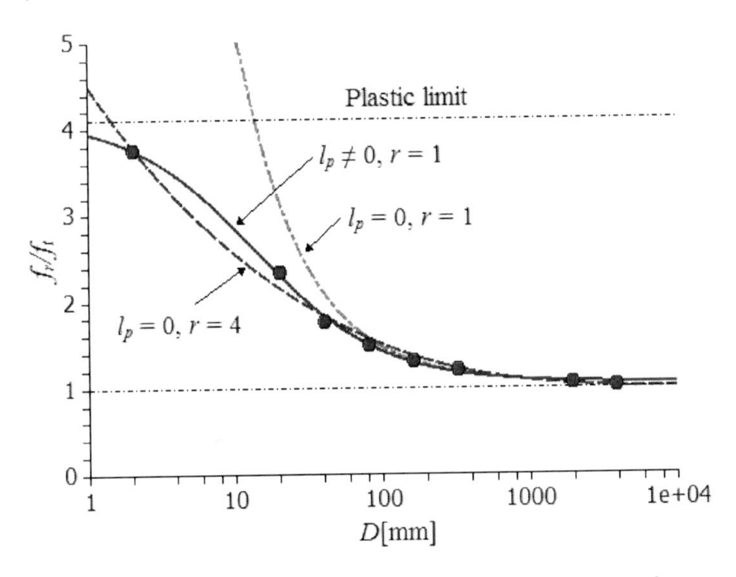

Fig. 8.28 Calculated normalized flexural tensile strength $f_r/f_t=1.5F_{max}L/(f_tD^2t)$ versus beam height D in unnotched concrete beams from deterministic FE calculations (red circles) versus beam height D compared with the deterministic size effect model by Bažant (blue dashed line by Eq. 8.1 with r=1, green dotted line by Eq. 8.1 with r=4 and red solid line by Eq. 8.7 with r=1) (Syroka et al. 2011)

The deterministic normalized vertical force is located in the range of stochastic values for a small and medium-size beam or is the maximum values for a large and very large-size beam. For the height of D=8 cm, the maximum vertical force changes between 3.267-4.08 kN, and the mean value F_{mean}=3.72 kN is by 3%

smaller than the deterministic value F=3.83 kN (the coefficient of variation cov=0.063). If the beam height is D=16 cm, the maximum vertical force varies between 5.61-6.82 kN and the mean stochastic force F_{mean}=6.25 kN (with the coefficient of variation cov=0.057) is smaller by 7% than the deterministic value (F=6.75 kN). For the both beams, the single maximum stochastic vertical force can be higher than the deterministic one. If the beam height is D=32 cm, the maximum vertical force changes between 10.31-12.25 kN, and the mean stochastic F_{mean}=11.07 kN (with the coefficient of variation cov=0.053) is smaller by 12% than the deterministic value of F=12.57 kN. Finally, in the case of the very large-size beam D=192 cm, the maximum vertical force changes between 54.32-59.18 kN and the mean stochastic F_{mean}=57.14 kN (the variation coefficient equals cov=0.027) is smaller by 14% than the deterministic value of F=66.18 kN. Thus, both the mean stochastic nominal strength and coefficient of variation always decrease with increasing size D and the influence of the random distribution of the tensile strength on the nominal strength is stronger for larger structures (Fig. 8.30). In addition, the calculations were carried out with a small-size beam, assuming a correlation length lower than the dimension of a single finite element. A scatter of the vertical force was small (the coefficient of variation strongly depends on the correlation range of correlation).

Figures 8.31-8.33 show some results for a localized zone from stochastic analyses (concerning the propagation way through finite elements with the different tensile strength – Fig. 8.31, zone height – Fig. 8.33 and zone width - Fig. 8.32). The random fields of f_t do not affect the mean width of a localized zone, which is again about 1.5 cm for all beam sizes (Fig. 8.32). A localized zone can be strongly non-symmetric and curved (Figs. 8.31). It occurs at the mean distance of about 2.0 cm (small-size beam) and of about 40 cm (very large-size beam) from the beam-centre (Fig. 8.31). The mean height of localized zones h at peak was closed to the deterministic outcomes.

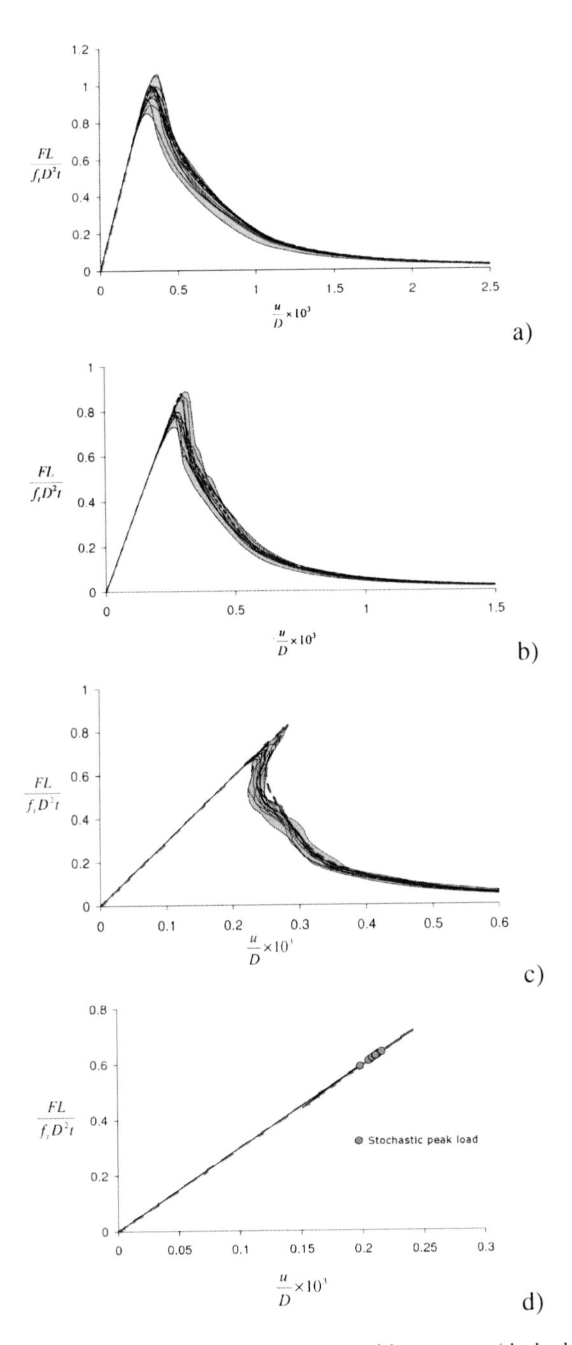

Fig. 8.29 Normalized vertical force-deflection curves with constant (dashed red line) and random (solid lines) value of tensile strength for 4 different beam heights: a) small D=8 cm, b) medium D=16 cm, c) large D=32 cm, d) very large-size beam D=192 cm (Syroka et al. 2011)

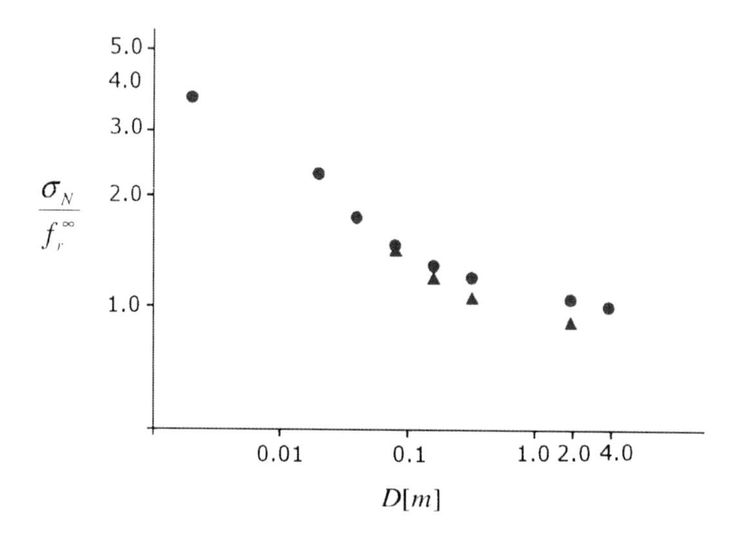

Fig. 8.30 Calculated normalized nominal strength $\sigma_N(D)/f_r^\infty$ versus beam height D from deterministic (circles) and stochastic (triangles) FE calculations for unnotched concrete beams (Syroka et al. 2011)

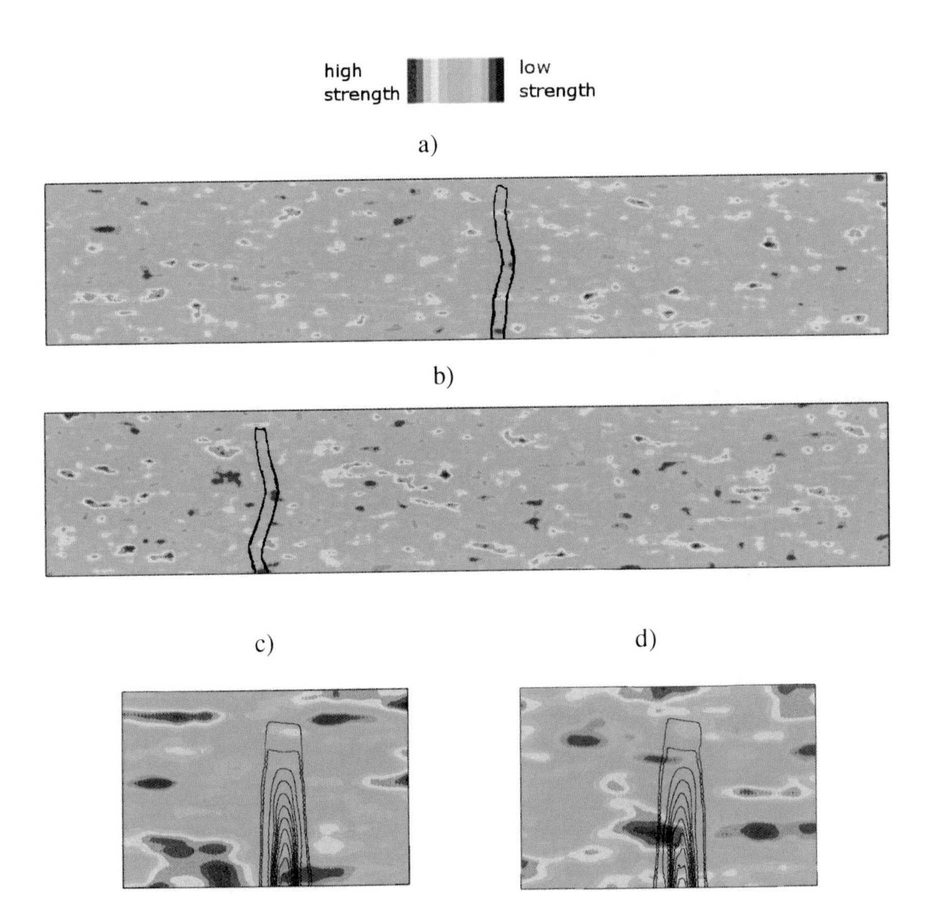

Fig. 8.31 Contours of non-local softening parameter $\overline{\kappa}$ against distribution of tensile strength from stochastic FE calculations in 2 large-size beams D=192 cm and 2 small-size beams D=8 cm (cases 'a' and 'c' correspond to maximum vertical force, cases 'b' and 'd' correspond to minimum vertical force) (tensile strength values are expressed by colour scale) (Syroka et al. 2011)

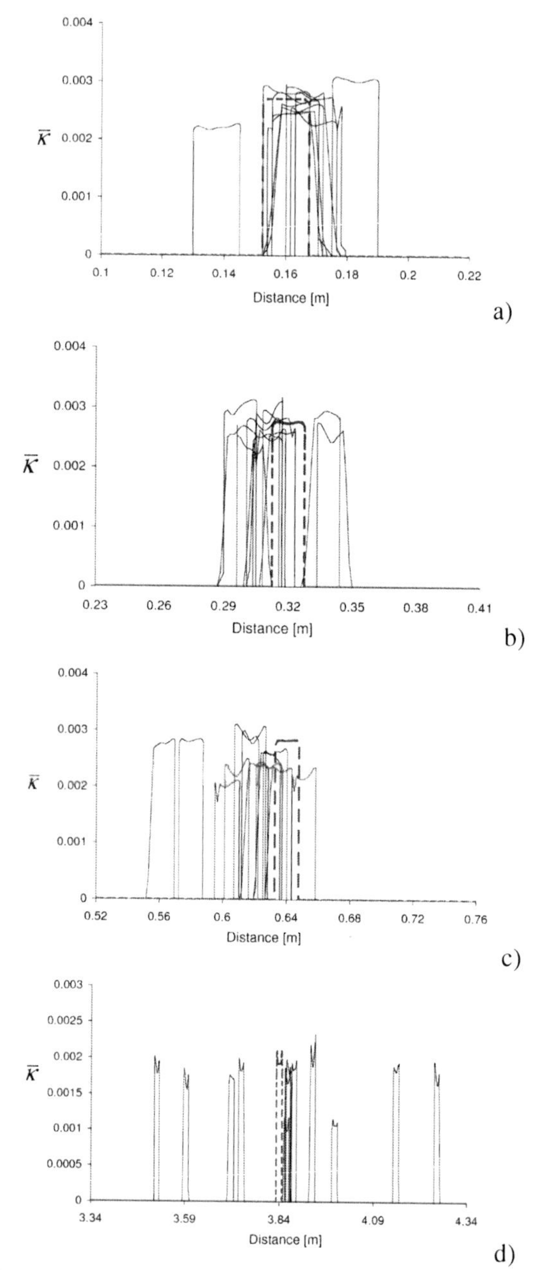

Fig. 8.32 Distribution of non-local softening parameter along beam length for deterministic (dashed lines) and stochastic calculations (solid lines) for 4 beams under three-point bending: a) small D=8 cm, b) medium D=16 cm, c) large D=32 m and d) very large-size beam D=192 cm (Syroka et al. 2011)

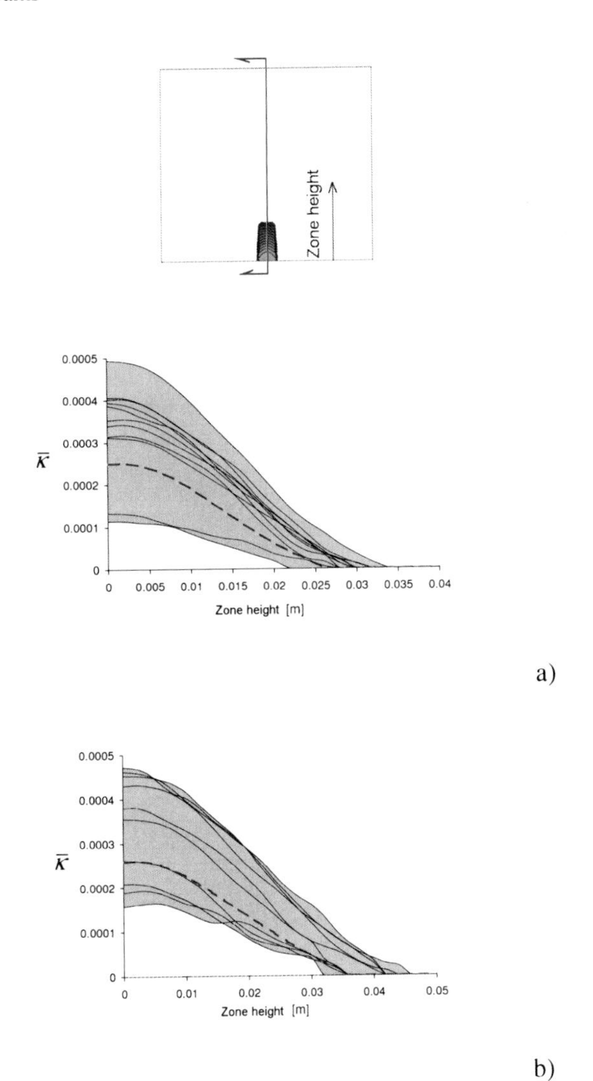

Fig. 8.33 Relationship between non-local softening parameter and localized zone height from deterministic (dashed lines) and stochastic (solid lines) calculations for: a) small D=8 cm, b) medium D=16 cm, c) large D=32 cm and d) very large-size beam D=192 cm (Syroka et al. 2011)

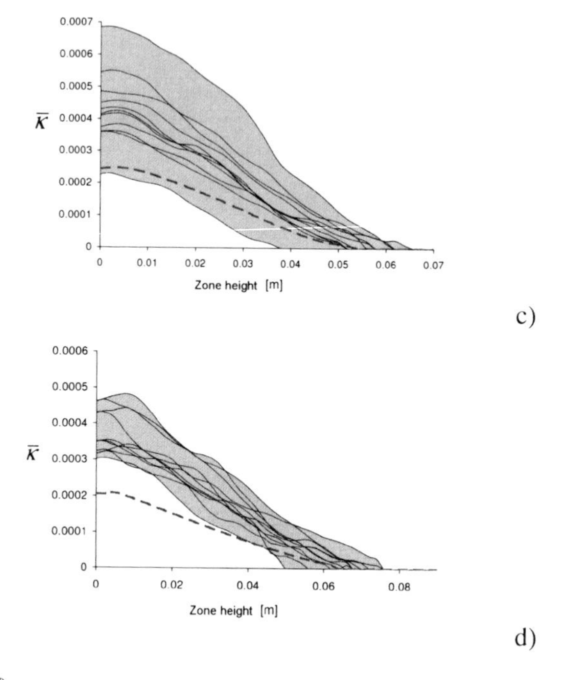

Fig. 8.33 (*continued*)

The maximum vertical force in concrete beams strongly depends on the position of a localized zone. This position is connected with the distribution and magnitude of the tensile strength at the place of a localized zone (within the area $w \times h$) and the magnitude of the horizontal normal stress due to bending σ_{11}. The maximum vertical force increases with increasing ratio $\overline{f}_t(w \times h)/\sigma_{11}$. A localized zone is created, where the mean local tensile strength $\overline{f}_t$ in the localized area $w \times h$ is minimum. In a small-size beam (Figs. 8.31c and 8.31d), the beam mid-region where a localized zone can be created is very limited due to the assumed standard deviation of the tensile strength and correlation range (3 cm in a vertical direction and 8 cm in a horizontal direction). In this limited beam region (with a small number of weak spots, Figs. 8.31c and 8.31d), the tensile strengths are strongly correlated and can be higher or lower than its mean value f_t=3.6 MPa. Therefore, the vertical normal tensile force can be smaller or larger than this in the deterministic study (depending on the spot choice by a localized zone for propagation). With an increase of the beam size, the number of weaker local spots increases with the correlation range assumed (Figs. 8.31a and 8.31b) and the beam mid-region where a localized zone can propagate is significantly larger. In this wide beam region, the tensile strengths are weaker correlated than in a small-size beam. So there exists a very high probability to achieve a smaller vertical force than in a small beam due to the great number of weak spots with the tensile strength smaller than f_t=3.6 MPa, which can be chosen by a localized zone for propagation (Figs. 8.31a and 8.31b).

An extended universal formula for a coupled deterministic-stochastic size effect law involves a deterministic scaling length D_b and a stochastic scaling length L_o (Bažant et al. 2007a, 2007b)

$$\sigma_N(D) = f_r^{\infty}\left(\left(\frac{L_o}{D+L_o}\right)^{\frac{r\cdot n}{m}} + \frac{rD_b}{D+l_p}\right)^{\frac{1}{r}}, \tag{8.10}$$

where m is the Weibull modulus (responsible for the slope of a large-size asymptote) and n is the number of spatial dimensions (n=2 for 2D problems). Thus, the mean size effect is separately divided into a stochastic part and deterministic. The parameter D_b drives the transition from elastic-brittle to quasi-brittle and L_o drives it from constant property to local Weibull via strength random field. The simplest choice for analyses is usually L_o=D_b. Equation 8.10 satisfies 3 asymptotic conditions: a) for small sizes $D\rightarrow0$, it asymptotically reaches the deterministic size effect law (Eq. 8.7), b) for large sizes $D\rightarrow\infty$, it asymptotically reaches the dominating Weibull size effect with the slope equal to $-n/m$ and c) for $m\rightarrow\infty$ and $L_o\rightarrow\infty$, it is equal to the deterministic size effect law. Thus, Eq. 8.10 can be regarded as the asymptotic matching of small-size deterministic and large-size stochastic size effects. With respect to the largest beam, the optimum match for the parameter m is the value of 48 calculated from the coefficient of variation cov (with cov=0.027) - driven by m only

$$cov = \sqrt{\frac{\Gamma\left(1+\frac{2}{m}\right)}{\Gamma^2\left(1+\frac{1}{m}\right)} - 1}\,. \tag{8.11}$$

However, the modulus m in other stochastic FE analyses was equal either to m=24 (Bažant and Novak 2001) or even m=8 (Vorechovsky 2007) with different other parameters (e.g. f_r^{∞}=3.68-3.76 MPa, L_o=D_b=15.53-48.66 mm and r=1.14-1.28, Bažant and Novak 2001). Thus, the stochastic size effect was slightly weaker in our numerical analyses (m=48) being independent of the correlation length. This can be mainly caused by the different loading type (bending versus uniaxial tension), correlation function and sampling type.

Figure 8.34 presents a comparison between our numerical results and size effect law by Bažant (Eqs. 8.7 and 8.10) using the following parameters: L_o= D_b= 30.37 mm, n=2, l_p=0, r =1, m=48 and f_r^{∞}=3.90 MPa with the related asymptotes assuming the Weibull modulus m=12-48. The stochastic outcomes indicate a further decrease of the nominal strength with increasing element size while the deterministic ones reach their lower limit. Our deterministic-statistical results present also a satisfactory agreement with the size effect law by Bažant by assuming the recommended value of m=24 (L_o=D_b=16.95 mm, l_p=0, r=1 and f_r^{∞}=4.753 MPa). However, the Weibull modulus m=48 solely enables a transition from a pure deterministic to a coupled deterministic-statistical size effect. The value m=12 underestimates the calculated deterministic-statistical flexural tensile strength.

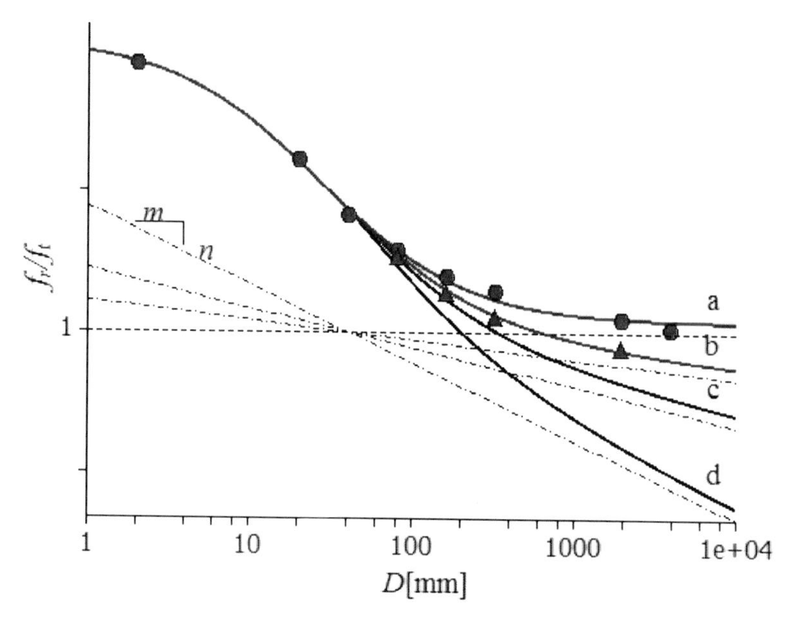

Fig. 8.34 Calculated normalized flexural tensile strength $f_r/f_t=1.5F_{max}L/(f_tD^2t)$ versus beam height D from deterministic (circles) and stochastic (triangles) FE calculations compared with deterministic (line 'a', Eq. 8.7) and deterministic-stochastic size effect law by Bažant (Eq. 8.10) for various Weibull moduli m and constant deterministic parameters (line 'b'- m=48, line 'c'- m=24, line 'd' - m=12) (Syroka et al. 2011)

All size effect results of the normalized nominal (flexural) strength $\sigma_N/f_t=1.5F_{max}L/(D^2tf_t)$ for unnotched and notched concrete beams (Bobiński et al. 2009) are summarized in Fig. 8.35 as compared to the size effect laws by Bažant (Eqs. 8.2, 8.7 and 8.10): Eq. 8.10 with D_b=40 mm, l_p=13.6 mm, r=1.0, f_r^{∞}= 3.78 MPa, n=2, Eq. 8.7 with m=48, L_o=D_b=40 mm, l_p=13.6 mm, r=1, f_r^{∞}=3.78 MPa, n=2 and Eq. 8.2 with B=1.48 and D_o=0.15 m. For notched structures, a random distribution of the tensile strength has obviously no effect on the nominal strength.

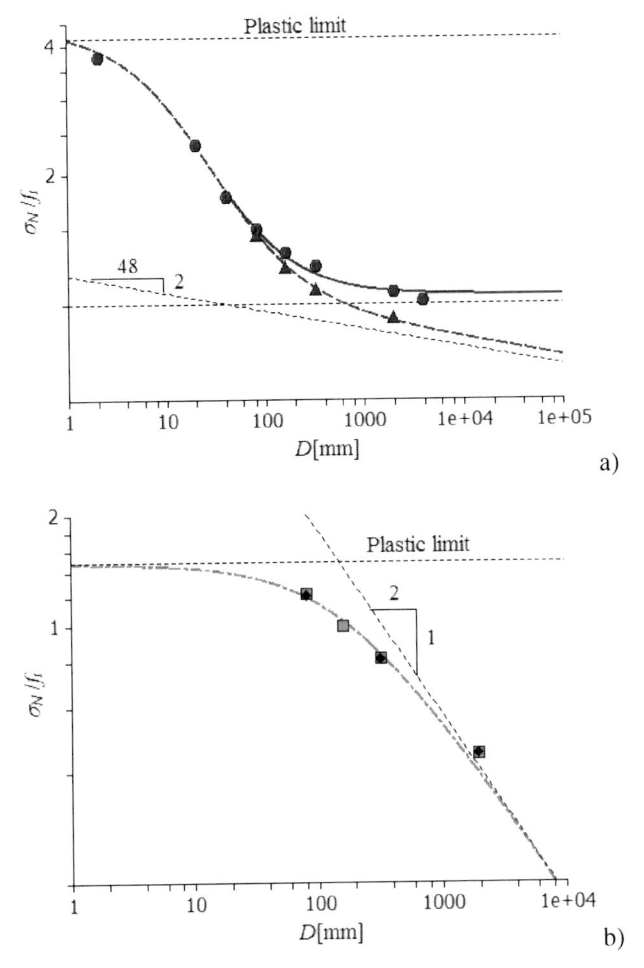

Fig. 8.35 Calculated normalized nominal (flexural) strength σ_N/f_t ($\sigma_N=1.5F_{max}L/(D^2t)$) versus beam height D for: a) unnotched concrete beams from deterministic (red circles) and stochastic (blue triangles); b) for notched concrete beams from deterministic (green squares) and stochastic (green diamonds) FE calculations compared with deterministic size effect law by Bažant (Eq. 8.7) (red solid line), deterministic-stochastic size effect law by Bažant (Eq. 8.10) (blue dashed line) and deterministic size effect law by Bažant (Eq. 8.2) (green dotted-dashed line) (Syroka et al. 2011)

Figures 8.36 and 8.37 compare our FE results on the normalized nominal (flexural) strength with unnotched beams of a coupled deterministic-stochastic size effect with size effect law SEL by Bažant (Eq. 8.10) and MFSL by Carpinteri (Eq. 8.3). In the considered size range of unnotched beams, both size effect laws show almost the same results. In the case of notched beams, there exists, however, a strong discrepancy between the size effect law MFSL (Eq. 8.3 with A_l=2.46 and

A_2=1385.9) and our earlier FE results (Bobiński et al. 2009, Fig. 8.35) for very small (D<0.1 m) and large beams (D>1.0 m) that confirms the conclusions by Bažant and Yavari (2007c) that the size effect law MFSL is not always realistic.

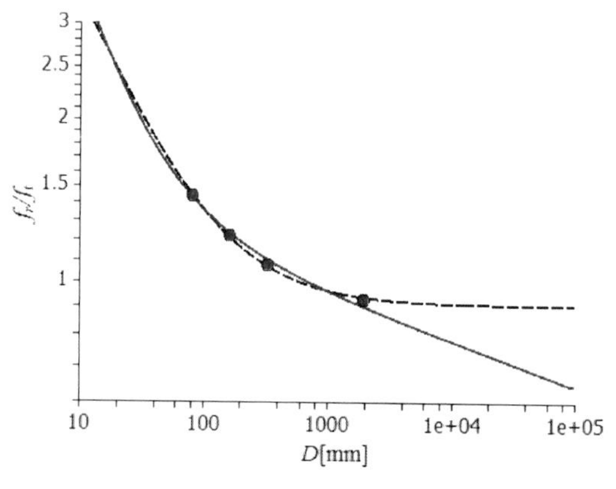

Fig. 8.36 Calculated normalized nominal (flexural) strength f_r/f_t=$1.5F_{max}L/(f_tD^2t)$ versus beam height D from coupled deterministic-stochastic FE calculations (circles) for unnotched beams compared with two size effect laws: SEL by Bažant (Eq. 8.10) – solid line and MFSL by Carpinteri et al. (Eq. 8.3) – dashed line (Syroka et al. 2011)

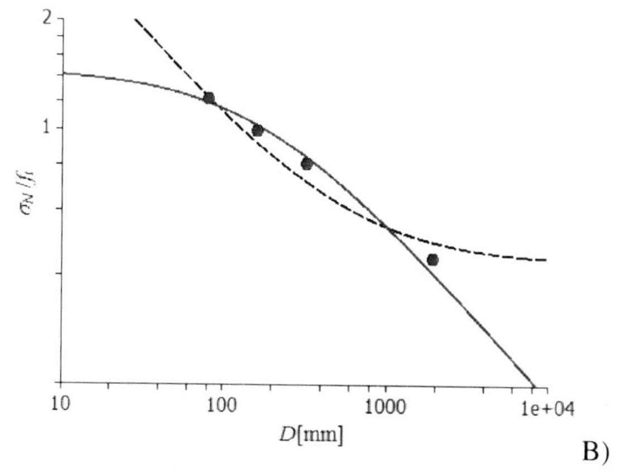

Fig. 8.37 Calculated normalized nominal (flexural) strength σ_N/f_t (σ_N=$1.5F_{max}L/(D^2t)$) versus beam height D from coupled deterministic-stochastic FE calculations (circles) for notched beams (Bobiński et al. 2009) compared with two size effect laws: SEL by Bažant (Eq. 8.2) – solid line and MFSL by Carpinteri et al. (Eq. 8.3) – dashed line (Syroka et al. 2011)

8.3 Elements under Compression

Finally, the effect of an imperfection or notches on a deterministic size effect was investigated during uniaxial compression of a concrete specimen with smooth boundaries (Skuza and Tejchman 2007). Figure 8.38 presents a hardening-softening curve assumed in compression (Drucker-Prager criterion) (Eqs. 3.27-3.30, 3.93 and 3.97). The Young modulus was E=18000 MPa, Poisson's ratio ν=0.2, compressive strength 32 MPa, non-locality parameter m=2, characteristic length l_c=5 mm, hardening/softening parameter $\kappa_{u2}=3\times10^{-3}$, internal friction angle φ=14° and dilatancy angle ψ=8°.

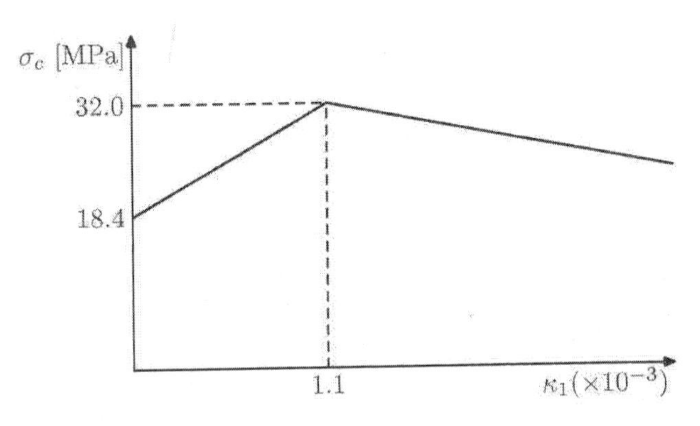

Fig. 8.38 Hardening/softening curve assumed in compression (Skuza and Tejchman 2007)

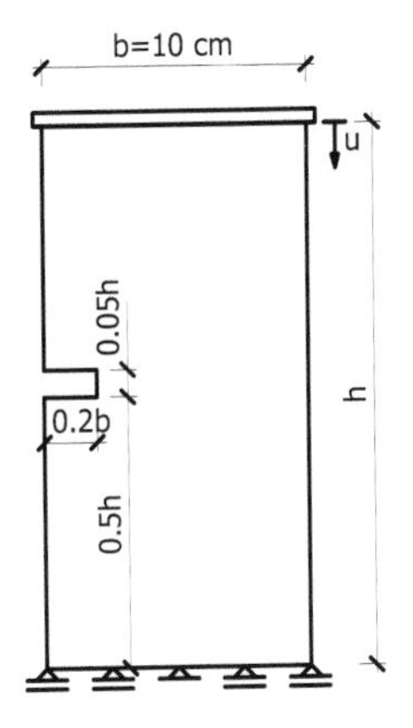

Fig. 8.39 Geometry and boundary conditions of a concrete specimen subjected to uniaxial compression (Skuza and Tejchman 2007)

The FE analyses were performed with 3 different specimens subjected to uniaxial compression: 10×5 cm^2, 10×10 cm^2 and 10×20 cm^2 (Fig. 8.39). The specimens had a weak element or a single notch (mid-point along the left edge). A small deterministic size effect with respect to the strength was obtained in a specimen with a single non-symmetric notch only (Figs. 8.40 and 8.41) due to the fact that damage localization develops faster and is created before the material strength is attained. This outcome is in agreement with calculations by Cusatis and Bažant (2006) using a 3D lattice model for concrete specimens under uniaxial compression. In concrete specimen under compression, shear zones were obtained during a deformation process (Fig. 8.41). In small specimens, they had a tendency to be reflected from rigid boundaries.

The following conclusions can be drawn from our non-linear FE-investigations of a deterministic and statistical size effect under quasi-static conditions:

- The FE results are in agreement with the size effect law by Bažant (notched and unnotched beams) and by Carpinteri (unnotched beams). However, fractality is not needed to induce a size effect, since the stress redistribution and energy release during strain localization cause a size effect (thus, fractality can contribute to a certain refinement of a size effect but not to its replacement). The size effect model by Bažant is universal and has physical foundations and can be introduced into design codes.
- The deterministic size effect (nominal strength decreases with increasing specimen size) is very pronounced in notched and unnotched concrete beams (it is stronger in notched beams). It is caused by occurrence of a straight tensile localized zone with a certain width. The material ductility increases with decreasing specimen size. A pronounced snap-back behaviour occurs for very large-size notched beams ($h/l_c \approx 400$) and for large and very large-size unnotched beams ($h/l_c \approx 8$). The width of the localized zone is similar for all beam sizes.
- The solution of random non-linear problems on the basis of several samples is possible. The statistical size effect is strong in unnotched concrete beams and negligible in notched concrete beams (due to the same position of the localized zone). The larger the beam, the stronger is the influence of a stochastic distribution on the nominal strength due to the presence of a larger number of local weak spots (i.e. the mean stochastic bearing capacity is always smaller than the deterministic one). The stochastic bearing capacity is larger in some realizations with small and medium-large beams than the deterministic value. The randomness of the tensile strength does not change the mean width of the localized zone. The localized zone can be curved and non-symmetric. This position of the localized zone is connected with the distribution and magnitude of the tensile strength in a localized zone at peak and the magnitude of the horizontal normal stress due to bending.
- The calculated stochastic effect is slightly weaker than in works by Bažant, Novak and Vorechovsky (2006, 2007). This can be mainly caused by the loading type, correlation length, correlation function and sampling type assumed in stochastic calculations. The results obtained with the help of Latin hypercube sampling are strongly influenced by the definition of the beam zone where the tensile strength distribution is statistically described. Our FE results match well the combined deterministic-statistical size effect law by Bažant with the Weibull modulus m=24-48. In turn, a prediction of the combined deterministic-statistical size effect based on deterministic results only is possible with the modulus m=48.

• The deterministic size effect can be observed on specimens under uniaxial compression in presence of non-symmetric notches only. In turn, an increase of ductility with decreasing specimen can be observed in all specimens independently of the imperfection type.

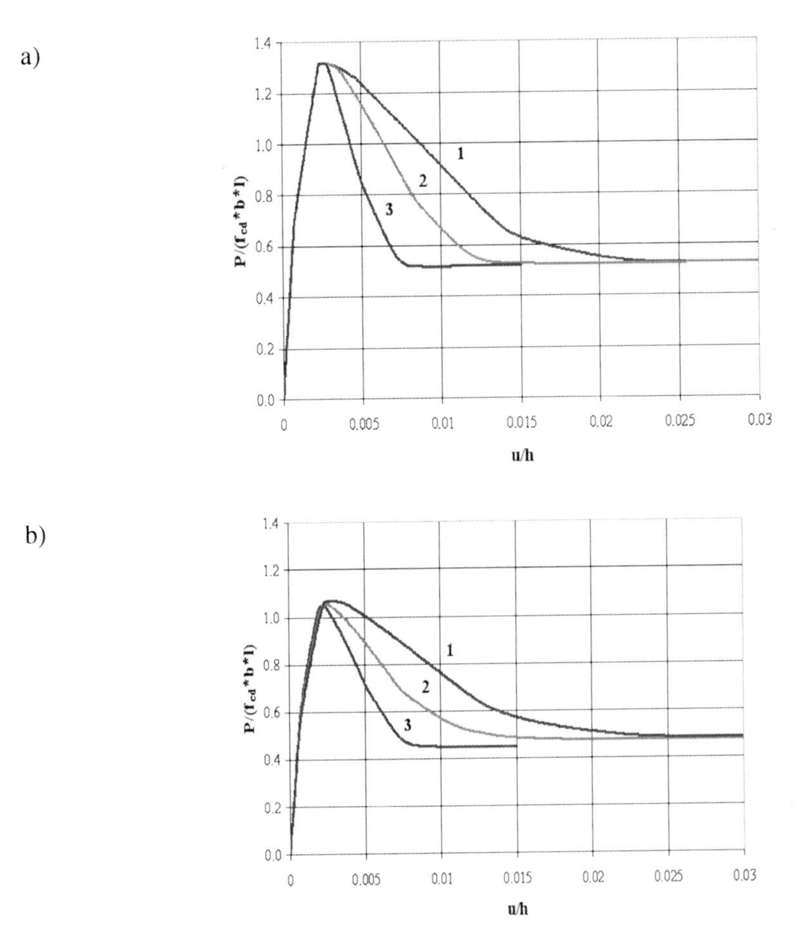

Fig. 8.40 Normalized load-displacement diagrams during uniaxial compression (a) specimen with one weak element, b) specimen with one notch, 1) small-size specimen, 2) medium-size specimen. 3) large-size specimen) (Skuza and Tejchman 2007)

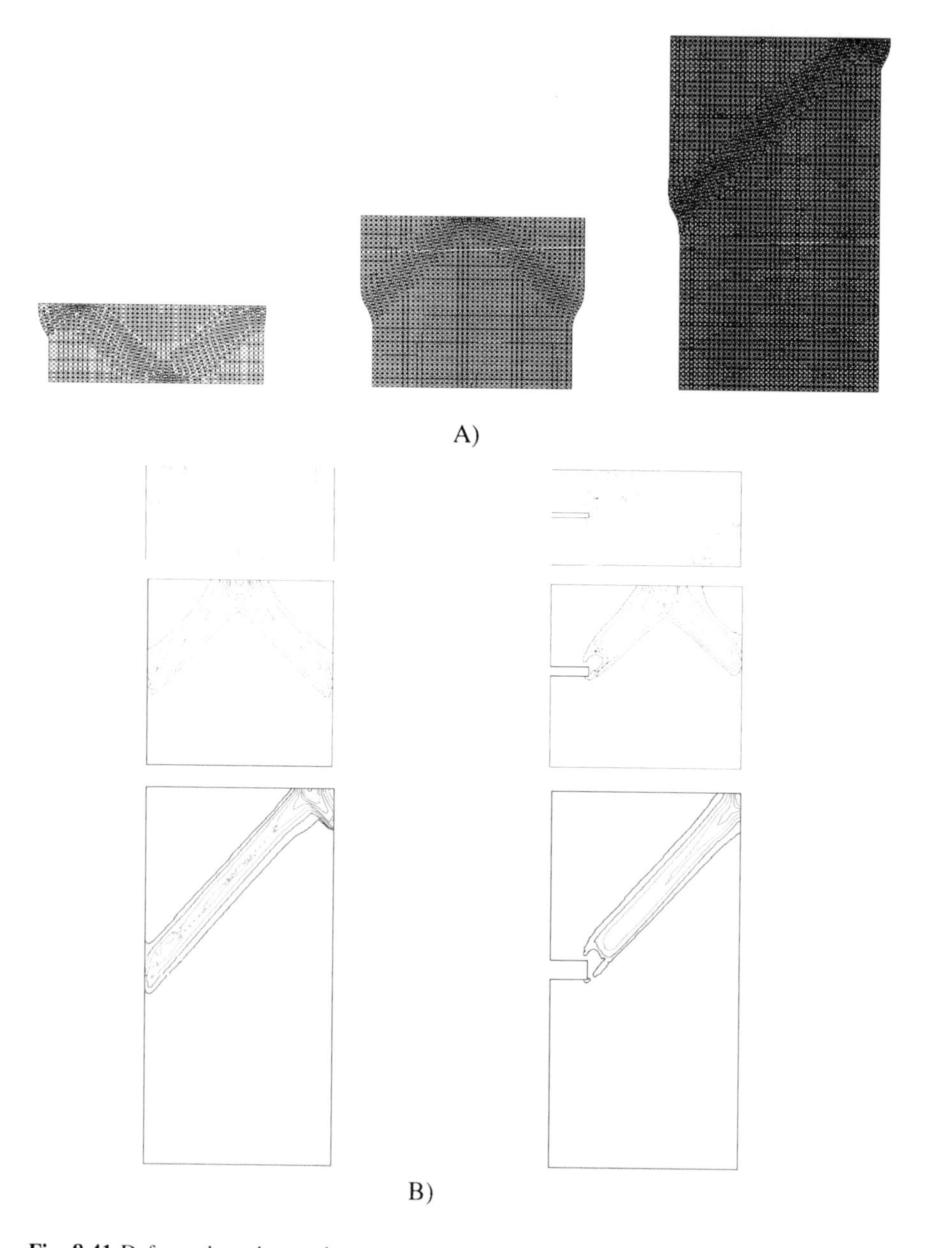

Fig. 8.41 Deformed meshes and contours of non-local softening parameter during uniaxial compression at residual state for small-size, medium-size and large-size concrete specimen: A) unnotched specimens with imperfection, B) specimens with single notch (Skuza and Tejchman 2007)

References

ABAQUS, Theory Manual, Version 5.8, Hibbit. Karlsson & Sorensen Inc. (1998)
Bažant, Z.P.: Size effect in blunt fracture: concrete, rock, metal. Journal of Engineering Mechanics ASCE 110(4), 518–535 (1984)
Bažant, Z.P., Lin, K.L.: Random creep and shrinkage in structures sampling. Journal of Structural Engineering ASCE 111(5), 1113–1134 (1985)
Bažant, Z.P., Chen, E.P.: Scaling of structural failure. Applied Mechanics Reviews 50(10), 593–627 (1997)
Bažant, Z., Planas, J.: Fracture and size effect in concrete and other quasi-brittle materials. CRC Press LLC (1998)
Bažant, Z., Novak, D.: Proposal for standard test of modulus of rupture of concrete with its size dependence. ACI Materials Journal 98(1), 79–87 (2001)
Bažant, Z.: Probability distribution of energetic-statistical size effect in quasibrittle fracture. Probabilistic Engineering Mechanics 19(4), 307–319 (2004)
Bažant, Z.P., Yavari, A.: Is the cause of size effect on structural strength fractal or energetic-statistical? Engineering Fracture Mechanics 72(1), 1–31 (2005)
Bažant, Z., Vorechovsky, M., Novak, D.: Asymptotic prediction of energetic-statistical size effect from deterministic finite-element solutions. Journal of Engineering Mechanics ASCE 133(2), 153–162 (2007a)
Bažant, Z., Pang, S.-D., Vorechovsky, M., Novak, D.: Energetic-statistical size effect simulated by SFEM with stratified sampling and crack band model. International Journal for Numerical Methods in Engineering 71(11), 1297–1320 (2007b)
Bažant, Z., Yavari, A.: Response to A. Carpinteri, B. Chiaia, P. Cornetti and S. Puzzi's comments on "Is the cause of size effect on structural strength fractal or energetic-statistical". Engineering Fracture Mechanics 74(17), 2897–2910 (2007c)
Bielewicz, E., Górski, J.: Shell with random geometric imperfections. Simulation-based approach. International Journal of Non-linear Mechanics 37(4-5), 777–784 (2002)
Bobiński, J., Tejchman, J., Górski, J.: Notched concrete beams under bending – calculations of size effects within stochastic elasto-plasticity with non-local softening. Archives of Mechanics 61(3-4), 1–25 (2009)
Carmeliet, J., Hens, H.: Probabilistic nonlocal damage model for continua with random field properties. Journal of Engineering Mechanics ASCE 120(10), 2013–2027 (1994)
Carpinteri, A.: Decrease of apparent tensile and bending strength with specimen size: two different explanations based on fracture mechanics. International Journal of Solids and Structures 25(4), 407–429 (1989)
Carpinteri, A., Chiaia, B., Ferro, G.: Multifractal scaling law: an extensive application to nominal strength size effect of concrete structures. In: Mihashi, M., Okamura, H., Bažant, Z.P. (eds.) Size Effect of Concrete Structures, vol. 173, p. 185. E&FN Spon (1994)
Carpinteri, A., Chiaia, B., Ferro, G.: Size effects on nominal tensile strength of concrete structures: multifractality of material ligaments and dimensional transition from order to disorder. Materials and Structures (RILEM) 28(180), 311–317 (1995)
Carpinteri, A., Chiaia, B., Cornetti, P., Puzzi, S.: Comments on "Is the cause of size effect on structural strength fractal or energetic-statistical". Engineering Fracture Mechanics 74(14), 2892–2896 (2007)
Chen, J., Yuan, H., Kalkhof, D.: A nonlocal damage model for elastoplastic materials based on gradient plasticity theory. Report Nr.01-13. Paul Scherrer Institute, pp. 1–130 (2001)

Cusatis, G., Bažant, Z.: Size effect on compression fracture of concrete with or without V-notches: a numerical meso-mechanical study. In: Meschke, G., de Borst, R., Mang, H., Bicanic, N. (eds.) Computational Modelling of Concrete Structures, pp. 71–83. Taylor and Francis Group, London (2006)

Elices, M., Guinea, G.V., Planas, J.: Measurement of the fracture energy using three-point bend tests: Part 3—influence of cutting the P -δ tail. Materials and Structures 25(6), 327–334 (1992)

Florian, A.: An efficient sampling scheme: Updated latin hypercube sampling. Probabilistic Engineering Mechanics 7(2), 123–130 (1992)

Frantziskonis, G.N.: Stochastic modeling of hetereogeneous materials – a process for the analysis and evaluation of alternative formulations. Mechanics of Materials 27(3), 165–175 (1998)

Górski, J.: Non-linear models of structures with random geometric and material imperfections simulation-based approach, Habilitation. Gdansk University of Technology (2006)

Gutierrez, M.A., de Borst, R.: Energy dissipation, internal length scale and localization patterning – a probabilistic approach. In: Idelsohn, S., Onate, E., Dvorkin, E. (eds.) Computational Mechanics, pp. 1–9. CIMNE, Barcelona (1998)

Gutierrez, M.A.: Size sensitivity for the reliability index in stochastic finite element analysis of damage. International Journal of Fracture 137(1-4), 109–120 (2006)

Hordijk, D.A.: Local approach to fatigue of concrete. PhD thesis. Delft University of Technology (1991)

Hughes, T.J.R., Winget, J.: Finite Rotation Effects in Numerical Integration of Rate Constitutive Equations Arising in Large Deformation Analysis. International Journal for Numerical Methods in Engineering 15(12), 1862–1867 (1980)

Huntington, D.E., Lyrintzis, C.S.: Improvements to and limitations of Latin hypercube sampling. Probabilistic Engineering Mechanics 13(4), 245–253 (1998)

Hurtado, J.E., Barbat, A.H.: Monte Carlo techniques in computational stochastic mechanics. Archives of Computational Method in Engineering 5(1), 3–30 (1998)

Knabe, W., Przewłócki, J., Różyński, G.: Spatial averages for linear elements for two-parameter random field. Probabilistic Engineering Mechanics 13(3), 147–167 (1998)

Koide, H., Akita, H., Tomon, M.: Size effect on flexural resistance on different length of concrete beams. In: Mihashi, H., Rokugp, K. (eds.) Fracture Mechanics of Concrete, pp. 2121–2130 (1998)

Le Bellego, C., Dube, J.F., Pijaudier-Cabot, G., Gerard, B.: Calibration of nonlocal damage model from size effect tests. European Journal of Mechanics A/Solids 22(1), 33–46 (2003)

McKay, M.D., Conover, W.J., Beckman, R.J.: A comparison of three methods for selecting values of input variables in the analysis of output from a computer code. Technometrics 21(2), 239–245 (1979)

Pijaudier-Cabot, G., Haidar, K., Loukili, A., Omar, M.: Ageing and durability of concrete structures. In: Darve, F., Vardoulakis, I. (eds.) Degradation and Instabilities in Geomaterials. Springer, Heidelberg (2004)

Skarżynski, L., Syroka, E., Tejchman, J.: Measurements and calculations of the width of the fracture process zones on the surface of notched concrete beams. Strain 47(s1), e319–e322 (2011)

Skuza, M., Tejchman, J.: Modeling of a deterministic size effect in concrete elements. Inżynieria i Budownictwo 11, 601–605 (2007) (in Polish)

Syroka, E., Górski, J., Tejchman, J.: Unnotched concrete beams under bending – calculations of size effects within stochastic elasto-plasticity with non-local softening. Internal Report, University of Gdańsk (2011)

Tejchman, J., Górski, J.: Computations of size effects in granular bodies within micro-polar hypoplasticity during plane strain compression. International Journal of Solids and Structures 45(6), 1546–1569 (2007)

Tejchman, J., Górski, J.: Deterministic and statistical size effect during shearing of granular layer within a micro-polar hypoplasticity. International Journal for Numerical and Analytical Methods in Geomechanics 32(1), 81–107 (2008)

Vanmarcke, E.-H.: Random Fields: Analysis and Synthesis. MIT Press, Cambridge (1983)

van Mier, J., van Vliet, M.: Influence of microstructure of concrete on size/scale effects in tensile fracture. Engineering Fracture Mechanics 70(16), 2281–2306 (2003)

van Vliet, M.R.A.: Size effect in tensile fracture of concrete and rock. PhD thesis. University of Delft (2000)

Vorechovsky, M.: Stochastic fracture mechanics and size effect. PhD Thesis. Brno University of Technology (2004)

Vorechovsky, M.: Interplay of size effects in concrete specimens under tension studied via computational stochastic fracture mechanics. International Journal of Solids and Structures 44(9), 2715–2731 (2007)

Walraven, J., Lehwalter, N.: Size effects in short beams loaded in shear. ACI Structural Journal 91(5), 585–593 (1994)

Walukiewicz, H., Bielewicz, E., Górski, J.: Simulation of nonhomogeneous random fields for structural applications. Computers and Structures 64(1-4), 491–498 (1997)

Weibull, W.: A statistical theory of the strength of materials. Journal of Applied Mechanics 18(9), 293–297 (1951)

Wittmann, F.H., Mihashi, H., Nomura, N.: Size effect on fracture energy of concrete. Engineering Fracture Mechanics 33(1-3), 107–115 (1990)

Yang, Z., Xu, X.F.: A heterogeneous cohesive model for quasi-brittle materials considering spatially varying random fracture properties. Computer Methods in Applied Mechanics and Engineering 197(45-48), 4027–4039 (2008)

Yu, Q.: Size effect and design safety in concrete structures under shear. PhD Thesis. Northwestern University (2007)

Chapter 9
Mesoscopic Modelling of Strain Localization in Plain Concrete

Abstract. The Chapter deals with modelling of strain localization in concrete at meso-scale. Concrete was considered as a composite material by distinguishing three phases: cement matrix, aggregate and interfacial transition zones. For FE calculations, an isotropic damage model with non-local softening was used. The simulations were carried out with concrete specimens under uniaxial tension and bending. The effect of aggregate density, aggregate size, aggregate distribution, aggregate shape, aggregate stiffness, aggregate size distribution, characteristic length and specimen size was investigated. The representative volume element was also determined.

A mechanism of strain localization strongly depends upon a heterogeneous structure of materials over many different scales, which changes e.g. in concrete from the few nanometers (hydrated cement) to the millimetres (aggregate particles). Therefore, to take strain localization into account, material composition (micro-structure) has to be taken into account (Nielsen et al. 1995, Bažant and Planas 1998, Sengul et al. 2002, Lilliu and van Mier 2003, Du and Sun 2007, Kozicki and Tejchman 2008, He et al. 2009, Skarżyński and Tejchman 2010). At the meso-scale, concrete can be considered as a composite material by distinguishing three important phases: cement matrix, aggregate and interfacial transition zones ITZs. In particular, the presence of aggregate and ITZs is important since the volume fraction of aggregate can be as high as 70-75% in concrete and ITZs are always the weakest regions in concrete. The concrete behaviour at the meso-scale fully determines the macroscopic non-linear behaviour. The advantage of meso-scale modelling is the fact that it directly simulates micro-structure and can be used to comprehensively study local phenomena at the micro-level such as the mechanism of the initiation, growth and formation of localized zones and cracks (He 2010, Kim and Abu Al-Rub 2011, Shahbeyk et al. 2011). Through that the mesoscopic results allow for a better calibration of continuum models enhanced by micro-structure and an optimization design of concrete with enhanced strength and ductility. The disadvantages are: very high computational cost, inability to model aggregate shape accurately and the difficulty to experimentally measure the properties of ITZs. All FE

J. Tejchman, J. Bobiński: Continuous & Discontinuous Modelling of Fracture, SSGG, pp. 343–405.
springerlink.com

investigations of a heterogeneous three-phase concrete material at the meso-level encompassing cement matrix, aggregates and an interfacial transition zone (ITZ) between the cement matrix and aggregates were performed with an isotropic damage constitutive model with non-local softening using a Rankine failure type criterion to define the equivalent strain measure $\tilde{\varepsilon}$ (Eqs. 3.35-3.40). The inclusions were assumed to be mainly in a circular shape randomly distributed according to a sieve curve (Fig. 9.1) and embedded in a homogeneous cement matrix. There are two widely used methods for the generation of randomly situated aggregate inclusions. The first one allows one to obtain a dense packing of aggregates in two-dimensional body of concrete using a Fuller distribution (van Mier et al. 1995):

$$p = 100\sqrt{\frac{D}{D_{\max}}}, \tag{9.1}$$

where p is the percentage weight of particles passing a sieve with the diameter D and D_{max} is the size of a largest particle. Furthermore by using a cumulative distribution for a two-dimensional cross-section, the circle diameters for a concrete can be generated. The second method of particle generation used by Eckardt and Konke (2006) is more straightforward. First, a grading curve is chosen (based on experimental measurements). Next, the certain amounts of particles with defined diameters d_1, d_2 … d_n are generated according to this curve. In our book, the latter method was used. The circles were randomly placed in the prescribed area starting with the largest ones and preserving a certain mutual distance (van Mier et al. 1995):

$$D > 1.1\frac{D_1 + D_2}{2}, \tag{9.2}$$

where D is the distance between two neighbouring particle centers, and D_1 and D_2 are the diameters of two neighboring aggregate particles. In the next step, the generated particle structure was overlaid with an irregular mesh of triangles. The finite elements belonging to cement matrix, aggregate inclusions and bond zones, respectively, had own different properties. It was assumed that the inclusions and bond zones had the highest and the lowest stiffness, respectively (van Mier et al. 1995).

9.1 Uniaxial Tension

The properties of the cement matrix, aggregate inclusions and bond zones used for FE calculations using an isotropic damage model with non-local softening are shown in Tab. 9.1 (Skarżyński and Tejchman 2009). The size of inclusions varied from a_{min}=2.5 mm up to a_{max}=5 mm. The size of bond zone elements, 0.25 mm (equal to $0.1\times a_{min}$), was smaller than the size of cement matrix elements. The mesoscopic characteristic length of micro-structure was l_c=0.5 mm.

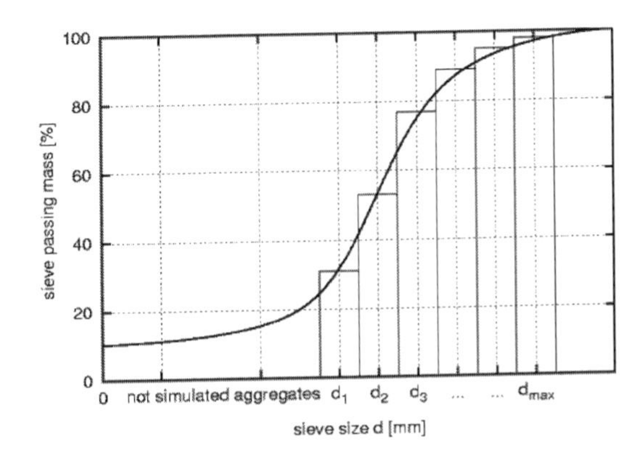

Fig. 9.1 Approximation of non-linear grading curve with discrete numbers of aggregate sizes (Skarżyński and Tejchman 2010)

Table 9.1 Material parameters assumed for uniaxial tension (Skarżyński and Tejchman 2009)

Material parameters	Inclusions	Cement matrix	ITZ
Modulus of elasticity E [GPa]	30	25	20
Poisson's ratio ν [-]	0.2	0.2	0.2
Crack initiation strain parameter κ_0 [-]	0.5	8×10^{-5}	5×10^{-5}
Residual stress level parameter α [-]	0.95	0.95	0.95
Slope of softening parameter β [-]	500	500	500

The calculations were carried out with periodic boundary conditions and material periodicity to avoid the effect of walls (van der Sluis 2001, Gitman 2006, Gitman et al. 2008). In the first case, the positions of nodes along corresponding specimen boundaries were the same before and after deformation. This is illustrated in Fig. 9.2, where an arbitrary periodically deformed unit cell under uniaxial extension conditions is shown. The deformation of each boundary pair is the same and the stresses are opposite in sign for each pair. The displacement boundary conditions are

$$u_{12} - u_4 = u_{11} - u_1 , \tag{9.3}$$

$$u_{22} - u_1 = u_{21} - u_2 , \tag{9.4}$$

$$u_3 - u_2 = u_2 - u_1 , \tag{9.5}$$

where u_{ij} is the displacement for any material point along the boundary Γ_{ij} and u_i is the node displacement. From the periodicity equations (Eqs. 9.3-9.5) can be observed that the independent entities are Γ_{11}, Γ_{21}, u_1, u_2 and u_4, whereas the tied dependent entities are Γ_{22}, Γ_{12} and u_3.

In addition, to eliminate wall effects, the periodicity of the material was assumed (Gitman 2006). Figure 9.3 presents samples different unit cells *A-F* in a concrete specimen. The cells *A, B, D* and *E* are valid in the context of material periodicity. However, the cells *C* and *F* experience wall-effects since some edges are crossed by inclusions. In our calculations, we avoided inclusions penetrating through the unit cell boundaries by letting them re-appear at the opposite edge (Fig. 9.4).

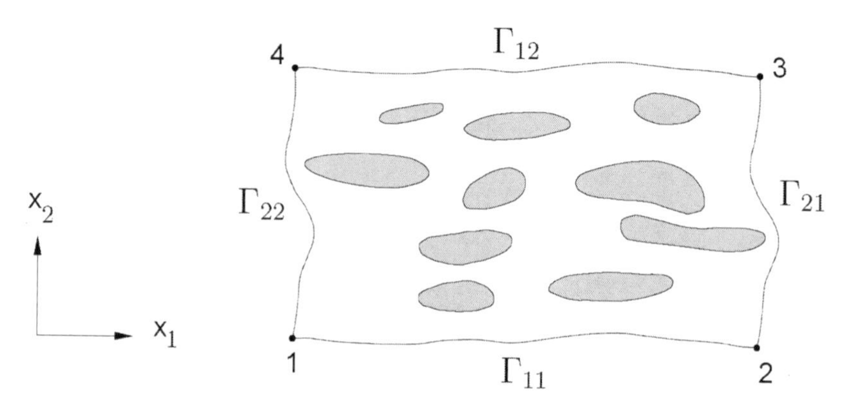

Fig. 9.2 Periodically deformed unit cell with boundaries Γ_R and nodes v_i (van der Sluis 2001)

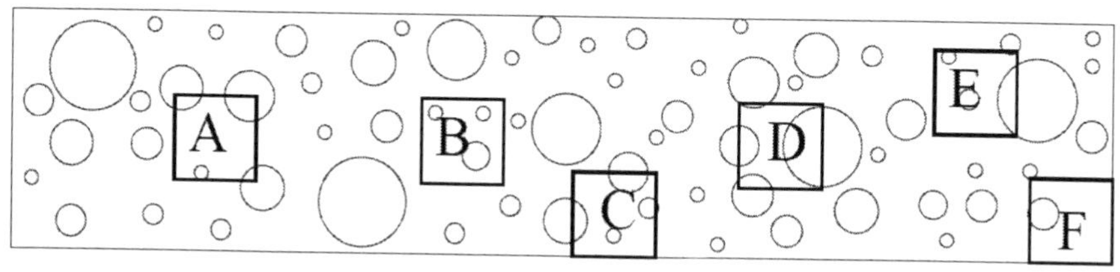

Fig. 9.3 Distribution of different unit cells in a concrete specimen (Gitman 2006)

A two-dimensional uniaxial tension test (Fig. 9.5) was performed with quadratic concrete specimens representing unit cells (Figs. 9.6-9.8) with periodicity of boundary conditions and material. For periodic boundary conditions, the displacements were suppressed in the node '*1*' (Fig. 9.2). Furthermore, in the node '*2*', a non-zero displacement was prescribed in a horizontal direction while the displacement a vertical direction was suppressed. The displacement components of the node '*3*' and '*4*' were free and tied together. The vertical normal stress was obtained from the resultant vertical force along the top boundary divided by the cross-sectional area ($B\times1$ m, where B is the width of the cell) and the strain as the vertical displacement of the top boundary divided by the cell width B.

First, concrete specimens of five different sizes were investigated. The smallest and the largest unit cells were 10×10 mm^2 and 25×25 mm^2, respectively (Fig. 9.6). For each specimen, five different stochastic realizations were performed (Fig. 9.7) with the aggregate density ρ kept constant (ρ=30%, ρ=45% and ρ=60%) (Fig. 9.8). Next, the calculations were carried out with a different characteristic length of micro-structure varying between l_c=0.1 mm-2.0 mm. Later, the effect of an aggregate density (ρ=30%, ρ=45% and ρ=60%) on strain localization was investigated. In the final comparative calculations, non-locality was prescribed to the cement matrix only.

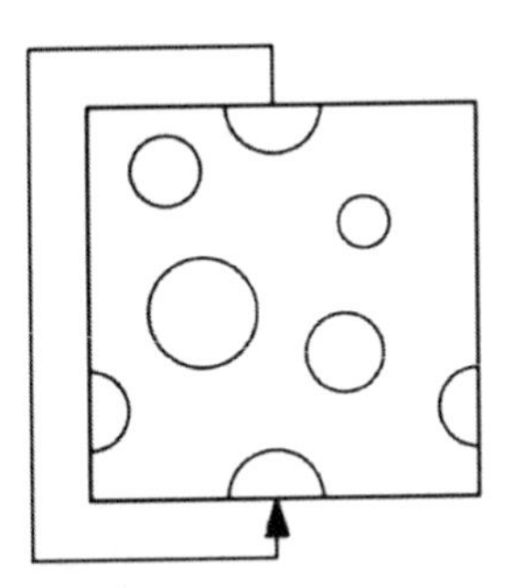

Fig. 9.4 Simulation of material periodicity (Gitman 2006)

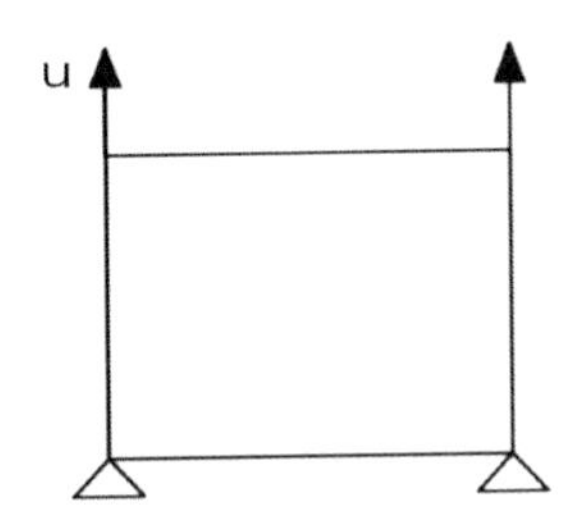

Fig. 9.5 Uniaxial tension test (Skarżyński and Tejchman 2009)

Effect of specimen size and realization

Figures 9.9-9.11 present the resultant mean macroscopic stress-strain relationships for various specimen sizes and random distributions of aggregate with the material constants from Tab. 9.1 (l_c=0.5 mm). The aggregate densities were ρ=30%, ρ=45% or ρ=60%, respectively. In turn, the influence of the specimen size on the evolution of the stress-strain curves for different aggregate densities is demonstrated in Fig. 9.12. The results evidently show that the stress-strain curves are the same independently of the specimen size, aggregate density and distribution of inclusions in an elastic regime only (almost up to the peak).

However, they are completely different in a softening regime after the peak is reached. An increase of the specimen size causes an increase of the material brittleness. The differences in the evolution of stress-strain curves in a softening regime are caused by strain localization contributing to a loss of the material homogeneity (Fig. 9.13). Strain localization in the form of a localized zone propagates between aggregates and can be strongly curved. The width of the calculated zone is about w_c=(4×l_c)=2 mm (with l_c=0.5 mm).

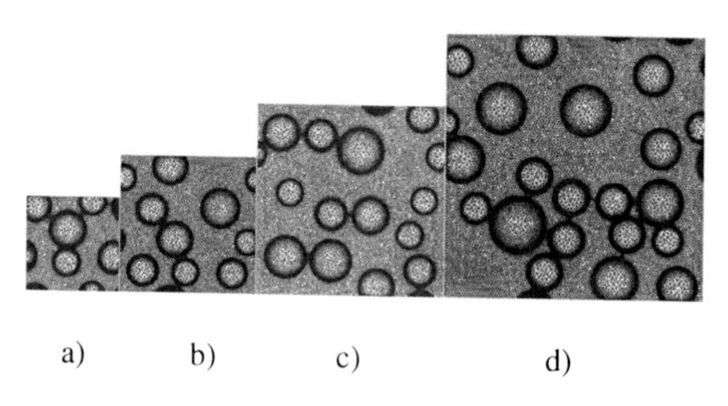

a) b) c) d)

Fig. 9.6 Different size of concrete specimens: a) 10×10 mm^2, b) 15×15 mm^2, c) 20×20 mm^2, d) 25×25 mm^2 (aggregate density ρ=30%) (Skarżyński and Tejchman 2009)

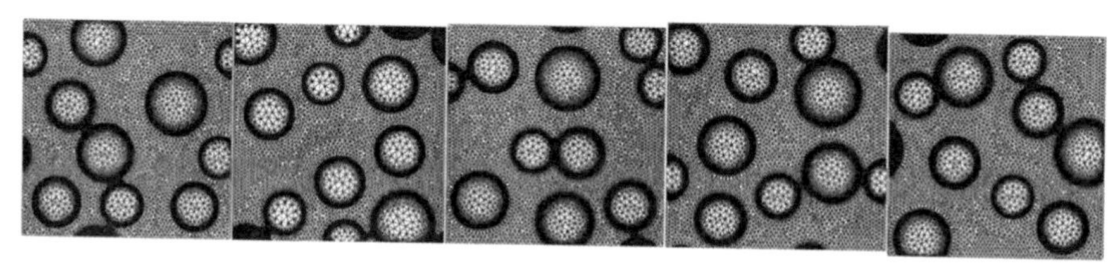

Fig. 9.7 Different stochastic distribution of aggregate for a concrete specimen of 15×15 mm^2 (aggregate density ρ=30%) (Skarżyński and Tejchman 2009)

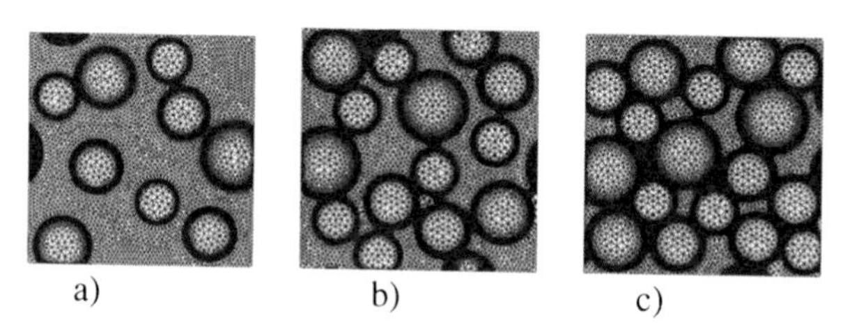

a) b) c)

Fig. 9.8 Different aggregate density ρ in concrete specimens: a) ρ=30%, b) ρ=45%, c) ρ=60% (Skarżyński and Tejchman 2009)

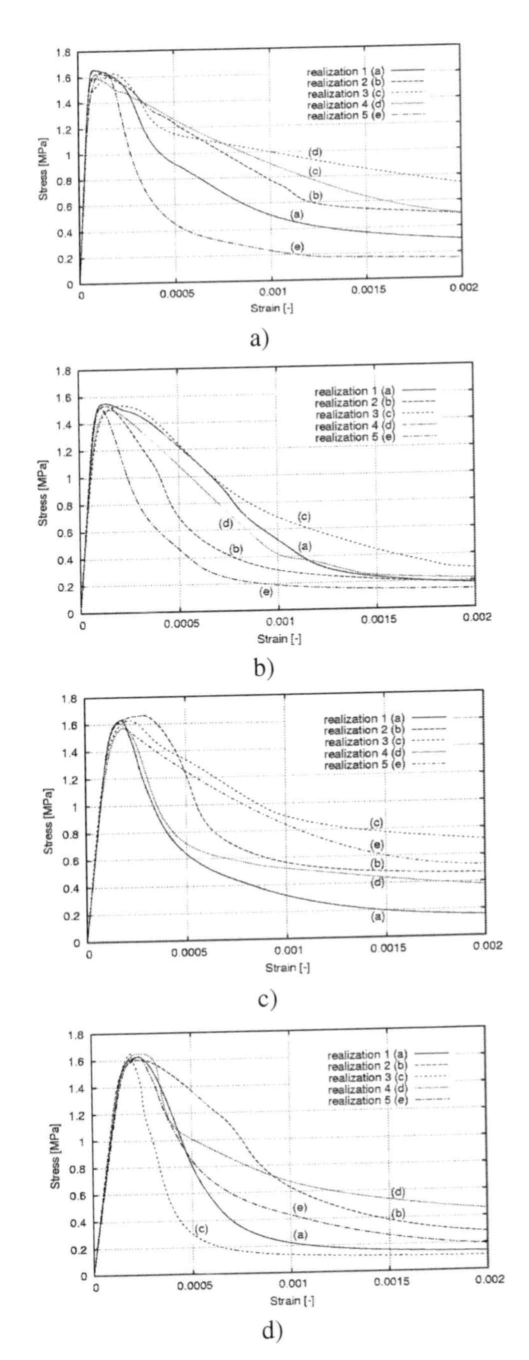

Fig. 9.9 Stress-strain curves with various sizes of concrete specimen and random distributions of aggregates: a) 10×10 mm^2, b) 15×15 mm^2, c) 20×20 mm^2, d) 25×25 mm^2 (characteristic length l_c=0.5 mm, aggregate density ρ=30%) (Skarżyński and Tejchman 2009)

The results indicate that the RVE can be determined in a linear-elastic regime only (due to the lack of differences in the evolution of the stress-strain curves). However, it cannot be determined in a softening regime due to strain localization (Gitman et al. 2008).

Effect of characteristic length of micro-structure

The effect of a characteristic length of micro-structure on the stress-strain curve and strain localization is shown in Figs. 9.14-9.17. Figures 9.14 and 9.16 demonstrate the influence of l_c on the evolution of stress-strain curves with two different specimen sizes: 10×10 mm^2 and 25×25 mm^2, respectively. In turn, Figs. 9.15 and 9.17 present the distribution of a non-local softening strain measure for various l_c changing between 0.1 mm and 2.0 mm.

With increasing characteristic length, both specimen strength and width of a localized zone increase. On the other hand, softening decreases and material behaves more ductile. Thus, a pronounced size effect occurs. The width of a localized zone is about w_c=4×l_c independently of l_c. A localized zone propagating in a cement matrix between aggregates is strongly curved with l_c=0.25 mm-1.0 mm, whereas becomes more straight for l_c>1.0 mm (Fig. 9.17e).

Effect of aggregate density

Figure 9.18 demonstrates the effect of the aggregate density on the stress-strain curves for two specimen sizes: 20×20 mm^2 and 25×25 mm^2, respectively (ρ=30%, ρ=45% or ρ=60% with l_c=0.5 mm).

A localized zone is also influenced by aggregate spacing. With increasing aggregate density, a localized zone becomes slightly narrower (Fig. 9.19). This means that a characteristic length of micro-structure may not be related to the aggregate size only but also to the grain size of the cement matrix.

Effect of non-locality range

Figure 9.20 shows the influence of the range of non-locality on the stress-strain relationship. In contrast to above studies, where non-locality was prescribed to all 3 phases of concrete, here, a cement matrix was solely assumed to be non-local due to fact that strain localization occurred only there. A characteristic length was again 0.5 mm.

The effect of the non-locality range on results turned out to be insignificant since the range of averaging slightly decreased (Figs. 9.20 and 9.21).

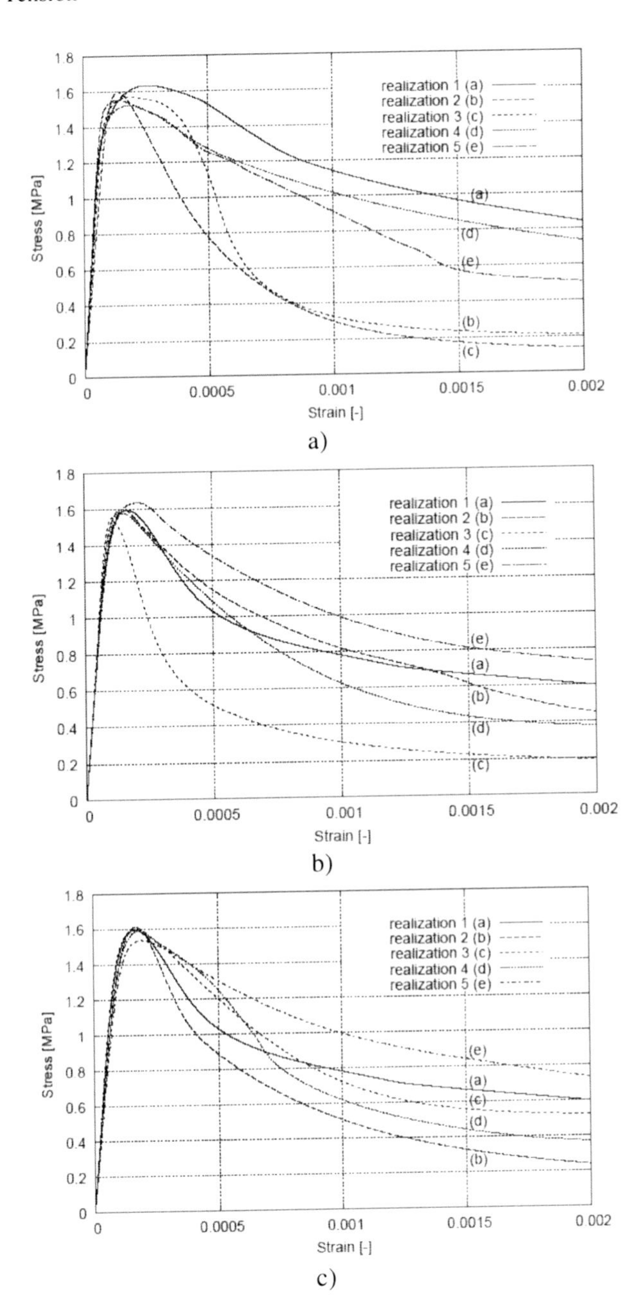

Fig. 9.10 Stress-strain curves for various sizes of concrete specimens and random distributions of aggregates: a) 10×10 mm^2, b) 15×15 mm^2, c) 20×20 mm^2, d) 25×25 mm^2 (characteristic length l_c=0.5 mm, aggregate density ρ=45%) (Skarżyński and Tejchman 2009)

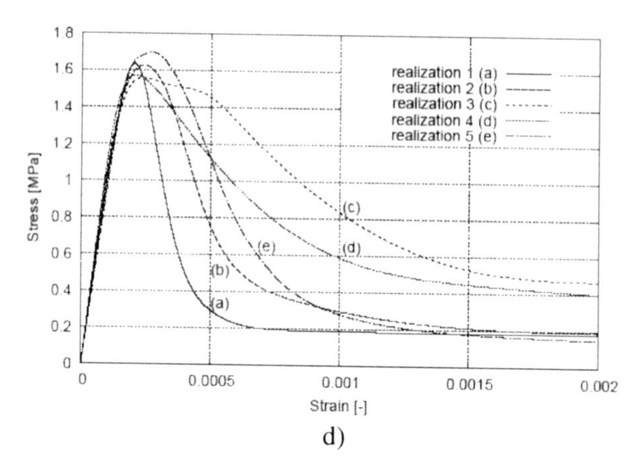

Fig. 9.10 (*continued*)

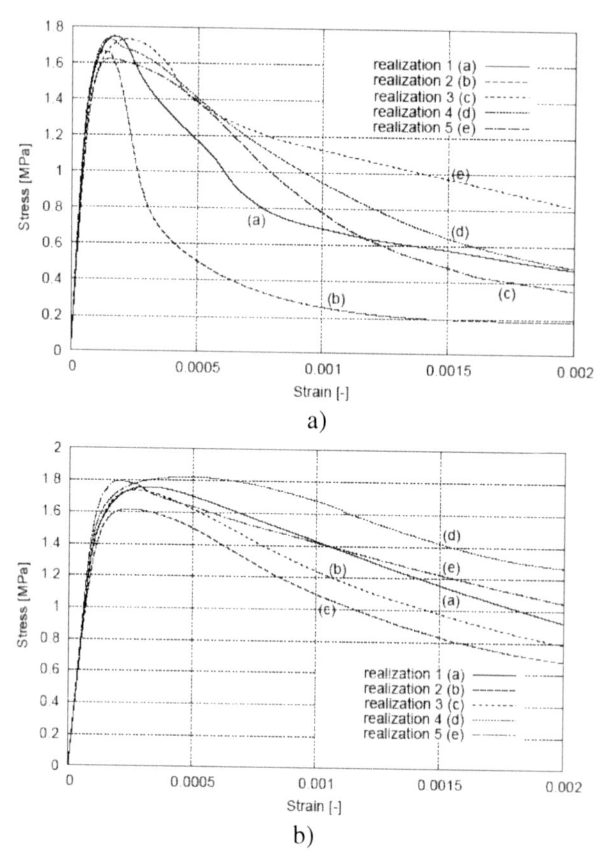

Fig. 9.11 Stress-strain curves for various sizes of concrete specimens and random distributions of aggregates: a) 10×10 mm^2, b) 15×15 mm^2, c) 20×20 mm^2, d) 25×25 mm^2 (characteristic length l_c=0.5 mm, aggregate density ρ=60%) (Skarżyński and Tejchman 2009)

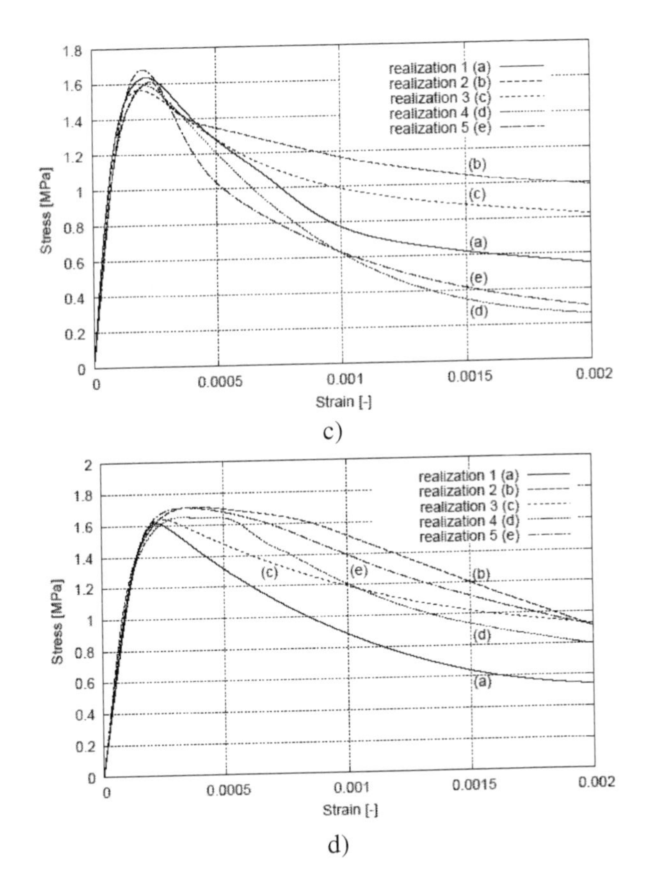

c)

d)

Fig. 9.11 *(continued)*

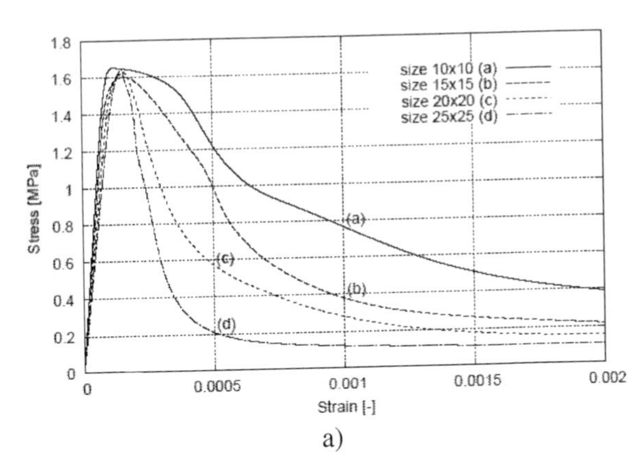

a)

Fig. 9.12 Stress-strain curves for various sizes specimen sizes of concrete specimens and aggregate densities ρ: a) ρ=30%, b) ρ=45%, c) ρ=60% (characteristic length l_c=0.5 mm) (Skarżyński and Tejchman 2009)

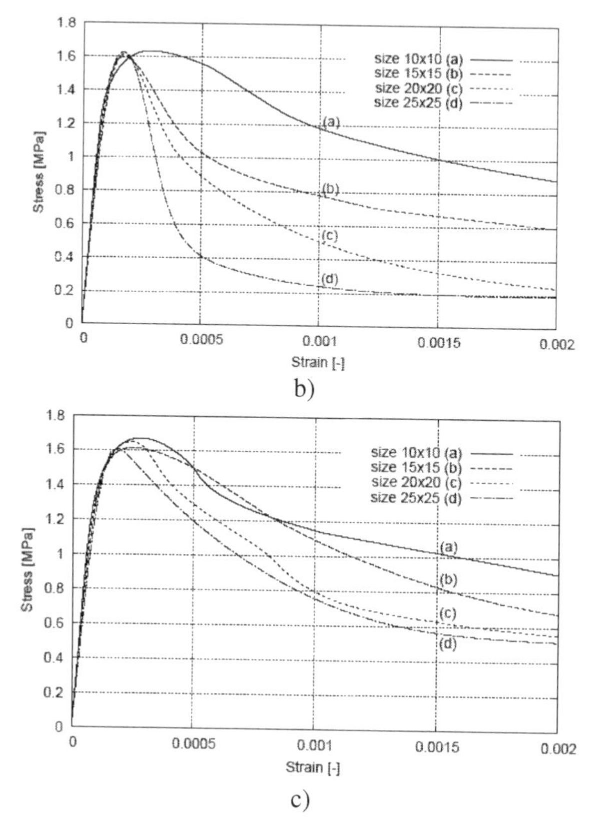

Fig. 9.12 (*continued*)

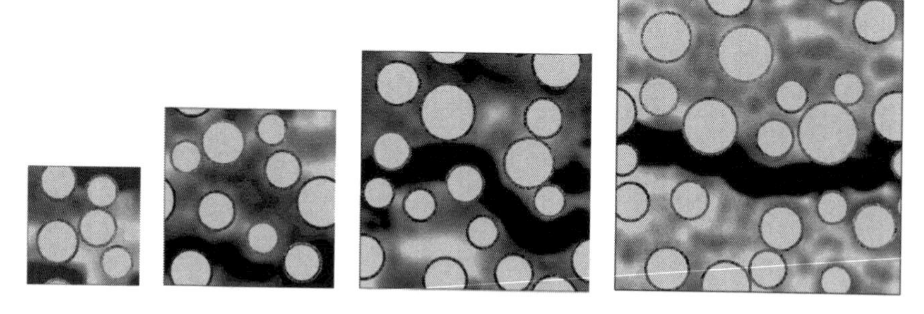

Fig. 9.13 Distribution of non-local strain measure for different specimen sizes with two different stochastic realizations of aggregate density: a) 10×10 mm^2, b) 15×15 mm^2, c) 20×20 mm^2, d) 25×25 mm^2 (characteristic length l_c=0.5 mm, aggregate density ρ=30%) (Skarżyński and Tejchman 2009)

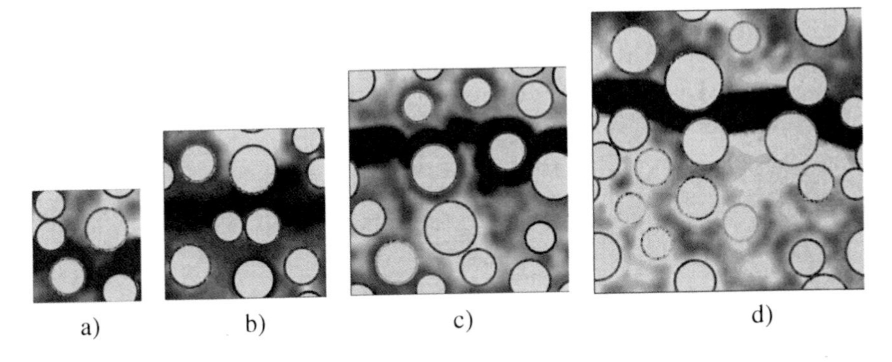

Fig. 9.13 (*continued*)

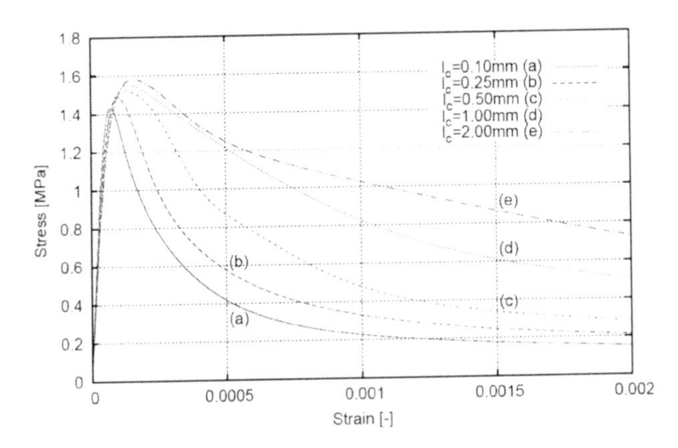

Fig. 9.14 Stress-strain curves for different characteristic lengths: a) l_c=0.1 mm, b) l_c=0.25 mm, c) l_c=0.5 mm, d) l_c=1.0 mm, e) l_c=2.0 mm (specimen size 10×10 mm^2, aggregate density ρ=30%) (Skarżyński and Tejchman 2009)

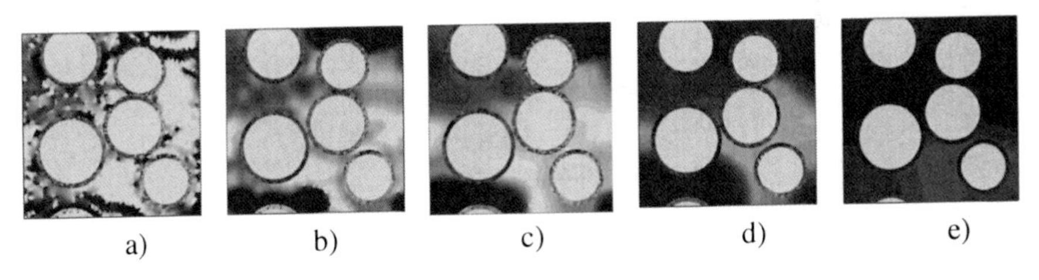

Fig. 9.15 Distribution of non-local softening strain measure for different characteristic lengths l_c: a) l_c=0.1 mm, b) l_c=0.25 mm, c) l_c=0.5 mm, d) l_c=1.0 mm, e) l_c=2.0 mm (specimen size 10×10 mm^2, aggregate density ρ=30%) (Skarżyński and Tejchman 2009)

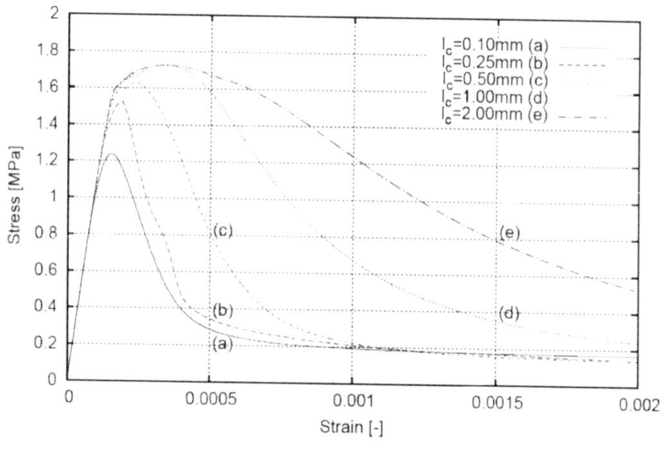

Fig. 9.16 Stress-strain curves for different values of a characteristic length: a) l_c=0.1 mm, b) l_c=0.25 mm, c) l_c=0.5 mm, d) l_c=1.0 mm, e) l_c=2.0 mm (specimen size 25×25 mm^2, aggregate density ρ=30%) (Skarżyński and Tejchman 2009)

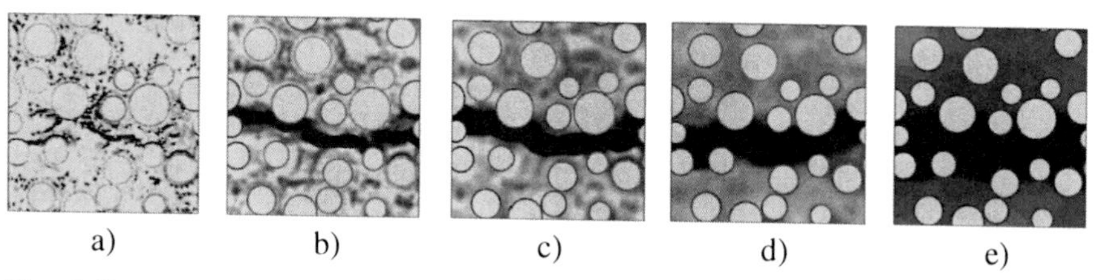

Fig. 9.17 Distribution of non-local softening strain measure for different values of a characteristic length: a) l_c=0.1 mm, b) l_c=0.25 mm, c) l_c=0.5 mm, d) l_c=1.0 mm, e) l_c=2.0 mm (specimen size 25×25 mm^2, aggregate density ρ=30%) (Skarżyński and Tejchman 2009)

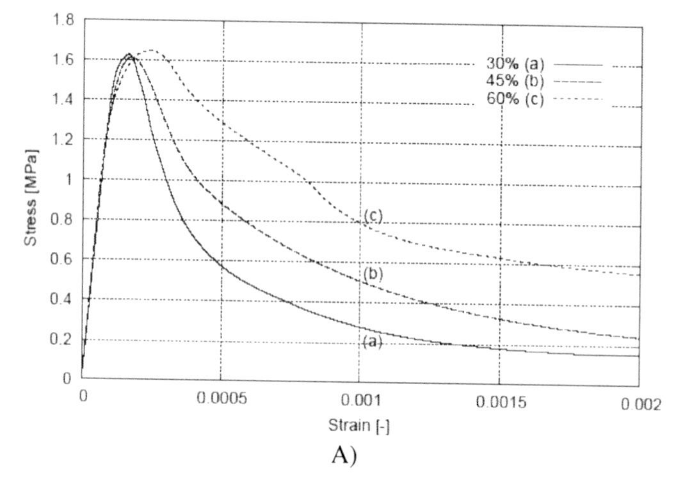

Fig. 9.18 Stress-strain curves for aggregate densities: (a) ρ=30%, (b) ρ=45%, (c) ρ=60% and cell sizes: (A) 20×20 mm^2 (B) 25×25 mm^2 (l_c=0.5 mm) (Skarżyński and Tejchman 2009)

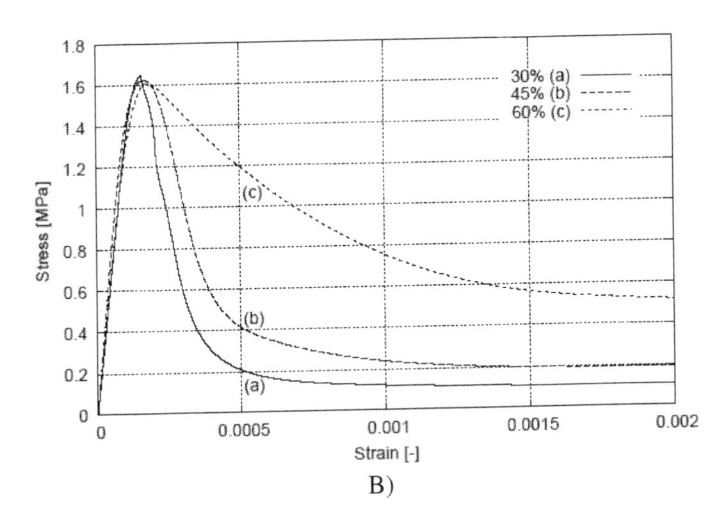

Fig. 9.18 (*continued*)

9.2 Bending

If the meso-structure of concrete is taken into account, such FE modeling is connected with a very large number of finite elements. To solve the problem practically, a macro-meso connection is used. It is done in a direct way, where a region with strain localization is considered at the meso-scale and a remaining region at the macro-level using a constitutive model. Alternatively, a computational homogenization is made using a multi-scale approach (Gitman et al. 2008, Geers et al. 2010, Kaczmarczyk et al. 2010). In this approach, the macro-meso connection is used as a constitutive equation on the macro-level. Thus, instead of an explicit formulation of the stress-strain relation, the data from the meso-level is taken into account. The idea of such technique is as follows: the strain from the macro-level goes in the form of boundary conditions to the meso-level, where a heterogeneous material behaviour is modeled, after which the reaction forces to boundary conditions are transformed by means of a homogenization technique (by changing the macro-level constitutive tangent stiffness) as stresses back to the macro-level. Different models for concrete can be used at meso-scale, e.g. discrete (interface element models (Carol et al. 2001), lattice approaches (Kozicki and Tejchman 2008), discrete element models DEM (Donze at al. 1999)) or continuum models (with cohesive elements (Kaczmarczyk et al. 2010), enhanced by a characteristic length of micro-structure (Gitman et al. 2008) or using displacement discontinuities (Belytschko et al. 2001, 2009).

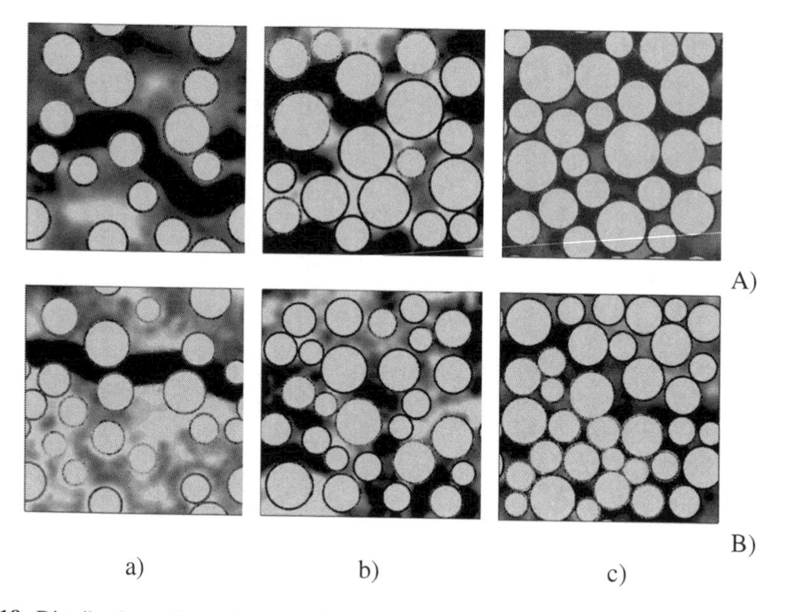

Fig. 9.19 Distribution of non-local softening strain measure for different aggregate densities: a) ρ=30%, b) ρ=45%, c) ρ=60% and specimen sizes: A) 20×20 mm^2, B) 25×25 mm^2 (l_c=0.5 mm) (Skarżyński and Tejchman 2009)

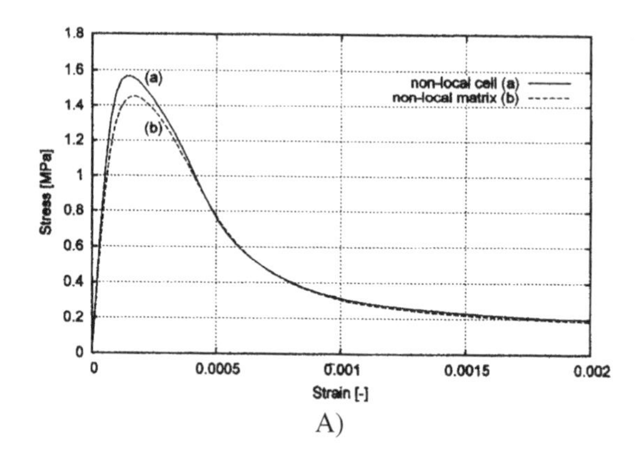

Fig. 9.20 Stress-strain curves for 2 different specimen sizes: A) 15×15 mm^2, B) 25×25 mm^2 with a) non-locality prescribed to three phases and b) non-locality prescribed to cement matrix (aggregate density ρ=30%, characteristic length l_c=0.5 mm) (Skarżyński and Tejchman 2009)

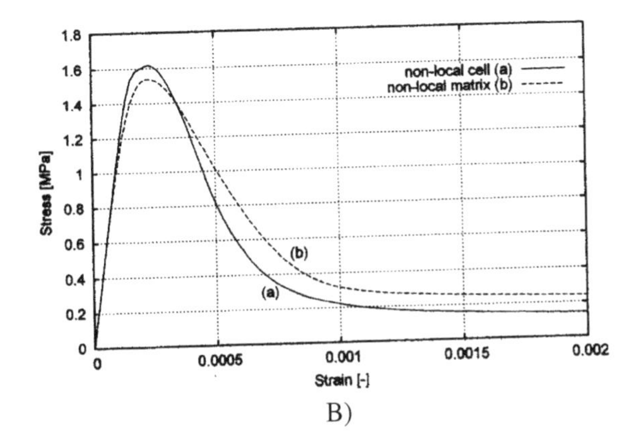

Fig. 9.20 *(continued)*

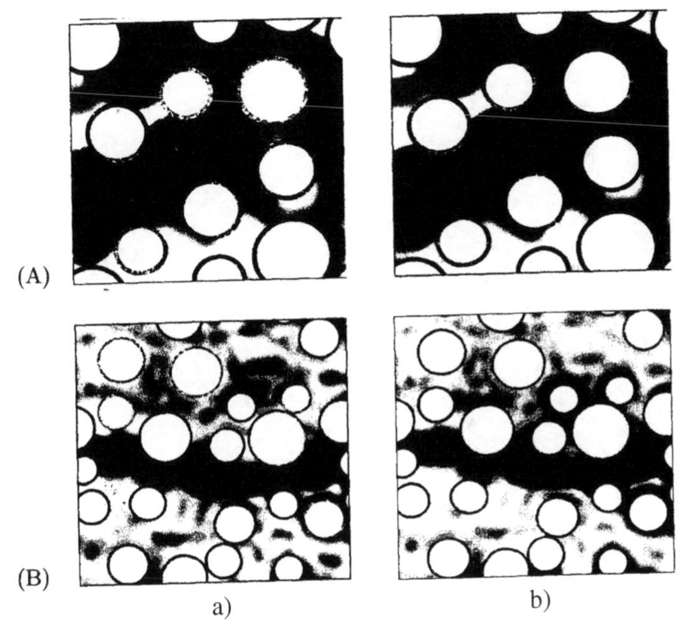

Fig. 9.21 Distribution of non-local softening strain measure for 2 different specimen sizes: A) 15×15 mm^2, B) 25×25 mm^2 with a) non-locality prescribed to three phases and b) non-locality prescribed to cement matrix (aggregate density ρ=30%, characteristic length l_c=0.5 mm) (Skarżyński and Tejchman 2009)

Experiments

The three-point bending laboratory tests were carried out with concrete beams of a different size $D \times L$ (D - beam height, $L=4\times D$ - beam length) with free ends (Skarzyński et al. 2011), Fig. 9.22a. The beams were geometrically similar in two dimensions only for 2 reasons: a) to reduce the number of finite elements and the related computation time in FE calculations at meso-scale (two-dimensional analyses were carried out instead of three-dimensional ones) and b) to avoid differences in the hydration heat effects which are proportional to the thickness of the member (Bažant and Planas, 1998). The following concrete beams were used: a) small-size beams 80×320 mm^2, b) medium-size beams 160×640 mm^2 and c) large-size beams 320×1280 mm^2 (Fig. 9.22a). The thickness of beams was always the same b=40 mm, and the beams' span was equal to 3×D. A notch with a height of D/10 mm was located at the mid-span of the beam bottom. The beams were subjected to a vertical displacement in the mid-point at a very slow rate. Two different fine-grained concrete mixes were composed of ordinary Portland cement, water and fine sand (with a mean aggregate diameter d_{50}=0.5 mm and maximum aggregate diameter d_{max}=3.0 mm) or sand (d_{50}=2.0 mm, d_{max}=8.0 mm) (Fig. 9.22b). The width and shape of a localized zone above the notch on the surface of beams was determined with a Digital Image Correlation (DIC) method which is an optical way to visualize surface displacements by successive post-processing of

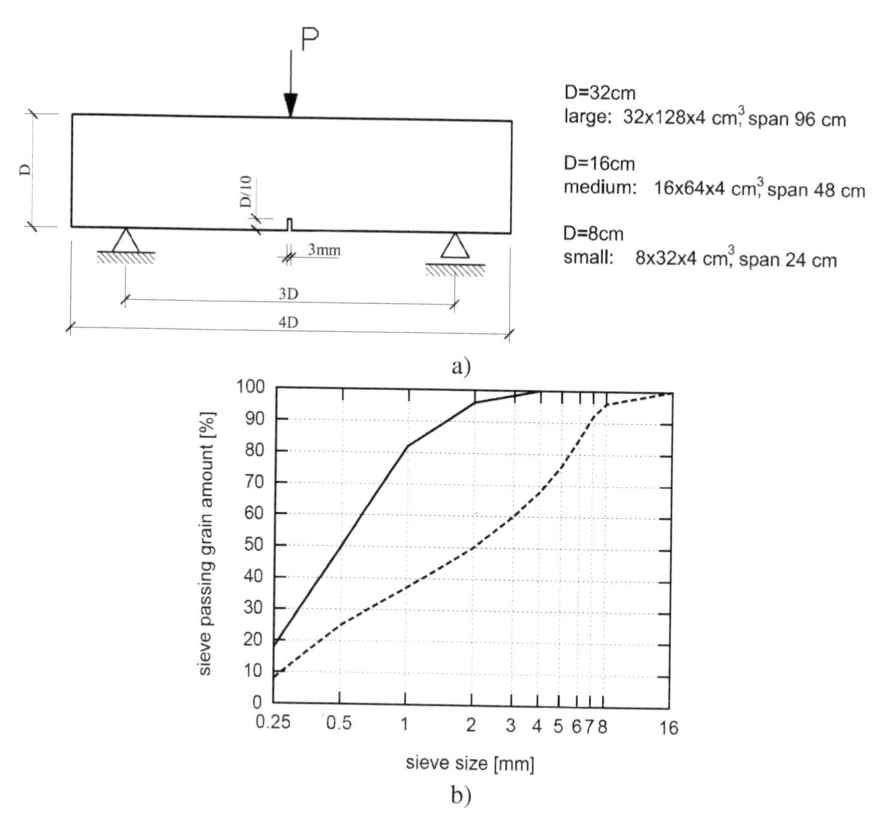

Fig. 9.22 Geometry of experimental concrete beams subjected to three-point bending: a) geometry (Le Bellěgo et al. 2003, Skarżyński et al. 2011), b) grading curve for fine sand (continuous line) and sand (discontinuous line) used for concrete (Skarżyński et al. 2011)

digital images taken at a constant time increment from a professional digital camera (based on displacements, strains can be calculated) (White et al. 2003). The experimental set-up and results were described in detail by Skarżyński et al. (2011). The beams of the same size were also used by Le Bellĕgo et al. (2003).

Figure 9.23 shows the formation of a localized zone on one side of the surface of a fine-grained small-size concrete beam above the notch from laboratory tests

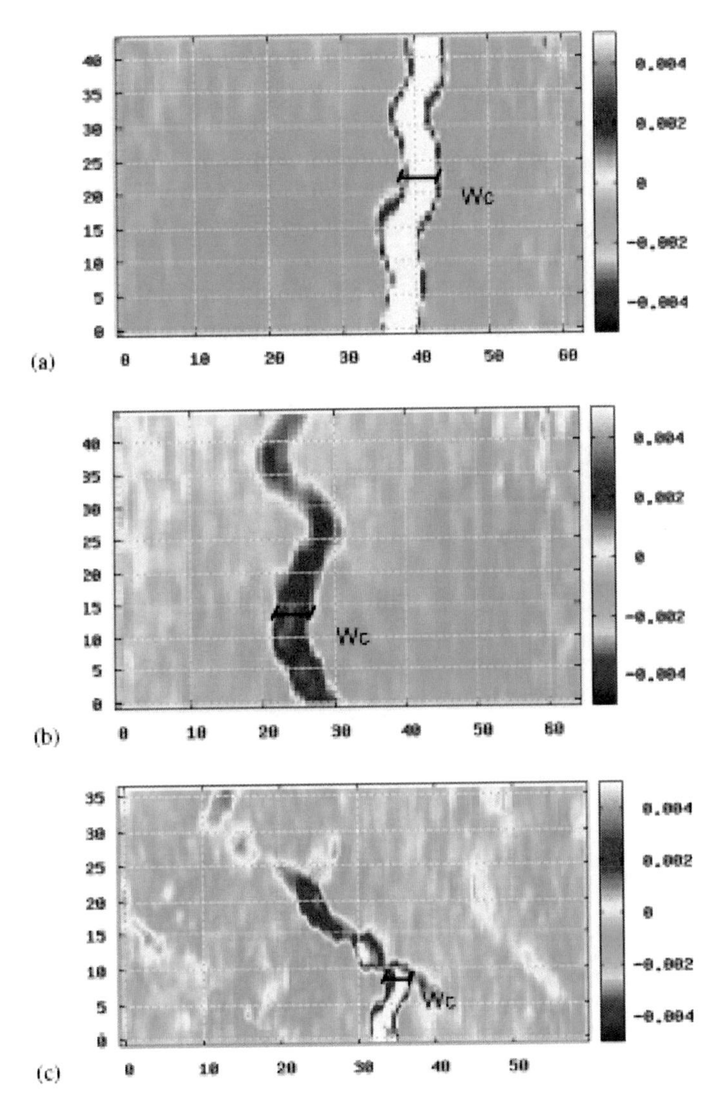

Fig. 9.23 Formation of localized zone with mean width of w_c=3.5-4.0 mm directly above notch in 3 different experiments ('a', 'b' and 'c') with small-size notched fine-grained concrete beam 80×320×40 mm^3 using DIC (vertical and horizontal axes denote coordinates in [mm] and colour scales strain intensity) (Skarżyński et al. 2011)

using a DIC (Skarżyński et al. 2011). A localized zone occurred always before the peak on the force-deflection diagram and was strongly curved. In some cases, it branched. The measured width of a localized zone above the notch increased during deformation due to concrete dilatancy (Fig. 9.24A) up to w_c=3.5-4.0 mm ($\leq d_{max}$) in the range of the deflection u=0.01-0.04 mm until a macro-crack was created. The maximum height of a localized zone above the notch was about h_c=50-55 mm at u=0.04 mm (Fig. 9.24B). The width of a localized zone did not depend upon the concrete mix type and beam size (Skarżyński et al. 2011).

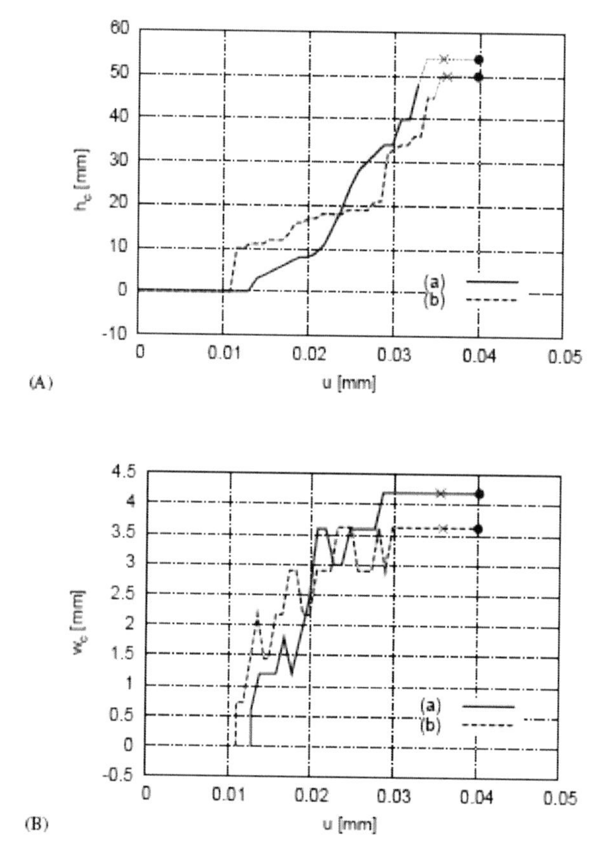

Fig. 9.24 Evolution of width w_c (A) and height h_c (B) of localized zone with deflection u directly above notch in experiments with small-size notched beam 80×320×40 mm^3 of fine-grained concrete using DIC: a) aggregate d_{50}=2 mm and d_{max}=8 mm, b) aggregate d_{50}=0.5 mm and d_{max}=3 mm (× - maximum vertical force, • - formation of macro-crack) (Skarżyński et al. 2011)

FE results

The FE-meshes including 12'000-1'600'000 triangular elements were assumed (Skarżyński and Tejchman 2010). The calculations were carried out with one set of material parameters for usual concrete only which was prescribed to finite elements corresponding to a specified concrete phase (Tab. 9.2) using an isotropic damage model with non-local softening. The interface was again assumed to be the weakest component. In general, the material constants should be determined through laboratory tensile tests for each phase (that is certainly possible for aggregate and cement matrix but not feasible for ITZs). Since the material constants for aggregate and cement matrix were not separately determined with laboratory experiments, other relationships between material constants E and κ_0 were also possible to obtain a satisfactory agreement between experiments and FE analyses.

Table 9.2 Material parameters assumed for three-point bending (Skarżyński and Tejchman 2010)

Material parameters	Inclusions	Cement matrix	Interface
Modulus of elasticity E [GPa]	40	35	30
Poisson's ratio ν [-]	0.2	0.2	0.2
Crack initiation strain parameter κ_o [-]	0.5	1×10^{-4}	7×10^{-5}
Residual stress level parameter α [-]	0.95	0.95	0.95
Slope of softening parameter β [-]	200	200	200

Four different fine-grained concrete mixes were numerically analysed (Fig. 9.25). To reduce the number of aggregate grains in calculations, the size of the smallest inclusions had to be limited (Fig. 9.25). The aggregate size varied between the minimum value d_{min}=2 mm and maximum value d_{max}=8 mm with the mean value of d_{50}=2 mm (aggregate size distribution curve '*a*' of Fig. 9.25 corresponding to the experimental one for sand concrete of Fig. 9.22b), d_{min}=2 mm and d_{max}=10 mm with d_{50}=4 mm (aggregate size distribution curve '*b*' of Fig. 9.25), d_{min}=2 mm and d_{max}=6 mm with d_{50}=4 mm (aggregate size distribution curve '*c*' of Fig. 9.25) and d_{min}=0.5 mm and d_{max}=3 mm with d_{50}=0.5 mm (aggregate size distribution curve '*d*' of Fig. 9.25 corresponding to the experimental one for fine sand concrete of Fig. 9.22b).

The width of ITZs was assumed to be t_b=0-0.75 mm. The size of finite elements was small enough to obtain objective results: s_a=0.5 mm (aggregate), s_{cm}=0.1-0.2 mm (cement matrix) and s_{ITZ}=0.05-0.1 mm (ITZ). The calculation time was about 2-5 days using PC with CPU Q6600 2x2.4 GHz and 4 GB RAM. The aggregate density was ρ=30%, ρ=45% or 60%, respectively.

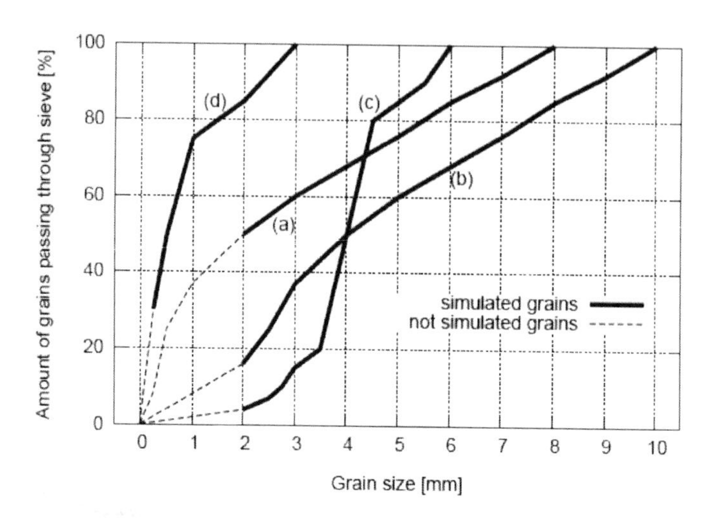

Fig. 9.25 Aggregate size distribution curves assumed for FE calculations (note that small aggregates were cut off to reduce the computation time)

The following numerical calculation program was assumed. First, three beams of a different size of sand concrete were modeled: as a partially homogeneous and partially heterogeneous with a meso-section in the notch neighborhood and as an entirely heterogeneous beam at meso-scale. The width of a heterogeneous meso-scale section b_{ms} varied between $D/2$ (40 mm) and D (80 mm) (D - beam height). These analyses allowed us to determine a representative width of a required heterogeneous region close to the notch. Next, the effect of different parameters was studied in a small-size beam. Finally, calculations were carried out with partially heterogeneous beams of a different size to determine a deterministic size effect. Three-five different stochastic realizations were usually performed for the same case. The width of the fracture process zone above the notch in all beams was determined at the deflection of u=0.15 mm on the basis of a non-local softening strain measure. As the cut-off value, $\bar{\varepsilon}_{\min}$ =0.025 was always assumed at the maximum mid-point value of $\bar{\varepsilon}_{\max}$ =0.08-0.13.

Our combined macro-mesoscopic simulations (Skarżyński and Tejchman 2010) are similar to a multi-scale approach using a Coupled Volume method where the size of a macro-element equals the size of a meso-cell (to avoid the assumption of any size of RVE) (Gitman et al. 2008). However, our simulations are faster because there is no need to continuously move between 2 calculation levels (the effect of an insignificant number of finite elements in a homogeneous beam region on the computation time is practically negligible).

First, the macro-scale calculations were carried out. Concrete was treated as an entirely homogeneous one phase-material with the following material constants: E=38500 MPa, υ=0.2, κ_0=1.3×10^{-4}, α=0.95, β=400 and l_c=2 mm. Totally, 12'000-92'000 triangular elements were assumed. The size of triangular finite elements was s=1.5 mm (in the nearest neighbourhood of the notch). Figure 9.26 presents the FE results of the nominal strength $\sigma_n = 1.5Pl/(bD^2)$ of 3 different concrete beams versus the normalized deflection u/D (P - vertical force, u - beam deflection, D - beam height, b - beam width, $l=3\times D$ - beam span) as compared to laboratory tests by Le Bellĕgo et al. (2003). Figure 9.27 shows the distribution of a non-local softening strain measure in beams.

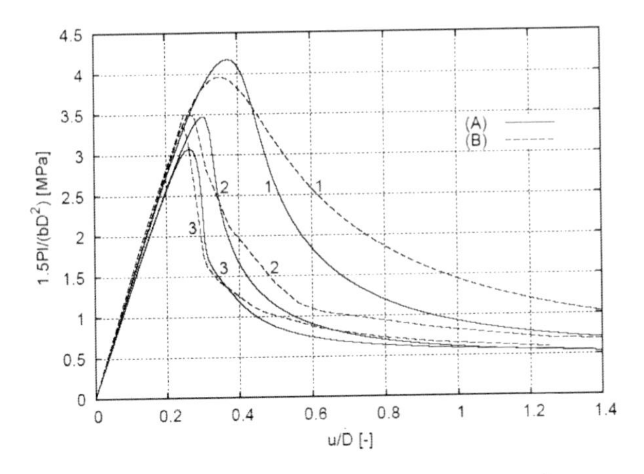

Fig. 9.26 Calculated and experimental nominal strength $1.5Pl/(bD^2)$ versus normalised beam deflection u/D (u - beam deflection, D - beam height): A) FE-results, B) experiments by Le Bellĕgo et al. (2003), 1) small-size beam, 2) medium-size beam, 3) large-size beam (homogeneous one-phase material, l_c=2 mm) (Skarżyński and Tejchman 2010)

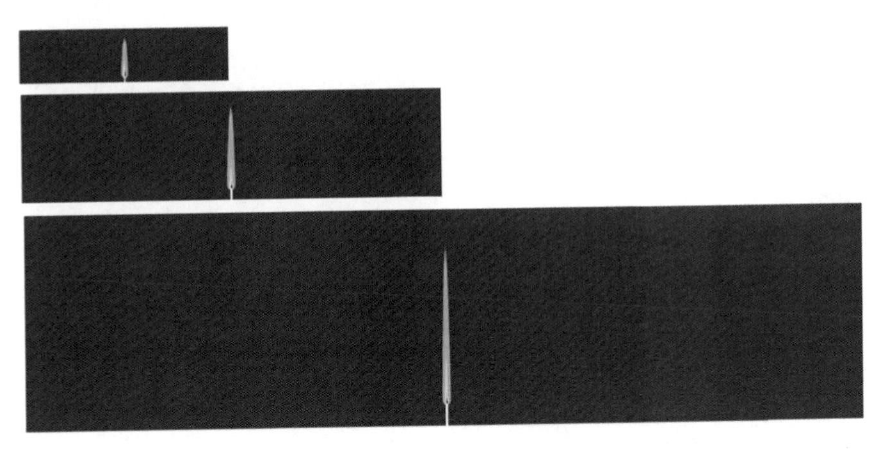

Fig. 9.27 Distribution of non-local strain measure above notch from numerical calculations with homogeneous one-phase material (at u/D=0.5) for small-size (top row), medium-size (medium row) and large-size beam (bottom row) (Skarżyński and Tejchman 2010)

The numerical strength results are in a satisfactory agreement with tests by Le Bellĕgo et al. (2003). The size effect is realistically described (nominal strength and material ductility increase with decreasing beam size). The width of a localized zone above the notch is about w_c=6.0 mm=3×l_c=4×s (at u/D=0.5) and approximately corresponds to the measured maximum value of w_c (5.5 mm) by DIC. However, in contrast to experiments, the calculated localized zones are always straight (an assumption of a stochastic spatially correlated distribution of tensile strength in the beam did not significantly affect their shape, Chapter 8).

Effect of width of meso-scale region

Figure 9.28 demonstrates the load-deflection curves obtained for two different aggregate distributions to determine a realistic width of a meso-scale region close to the notch (to reduce computation time). Concrete was treated in a meso-scale region as a random three-phase heterogeneous material with circularly-shaped aggregate using material constants from Tab. 9.2. In the remaining region, the material was homogeneous one-phase material (E=38500 MPa, υ=0.2, κ_0=1.3× 10^{-4}, α=0.95, β=200). The beam size was 80×320 mm^2. The width of a meso-scale region was b_{ms}=40 mm or b_{ms}=80 mm (Fig. 9.29). Totally 65'000-110'000 finite elements were assumed. The characteristic length was l_c=1.5 mm and the aggregate density 30%. An entirely heterogeneous beam with 365000 elements served as the reference beam. For a comparison, a stochastic distribution of aggregate was always the same in a meso-scale section. Figure 9.30 shows the distribution of a non-local softening parameter above the notch.

The results show that the effect of the width of the meso-scale region on the results can be significant if $b_{ms} \leq D/2$. However, if the width of a meso-scale region close to the notch equals D=80 mm, the results of forces and strains with an entirely and a partially heterogeneous beam are similar. In further calculations to save computational time, a representative meso-scale section was assumed to be always equal to the beam height $b_{ms}=D$ (i.e. 80 mm for a small-size beam, 160 mm for a medium-size beam and 320 mm for a large-size beam).

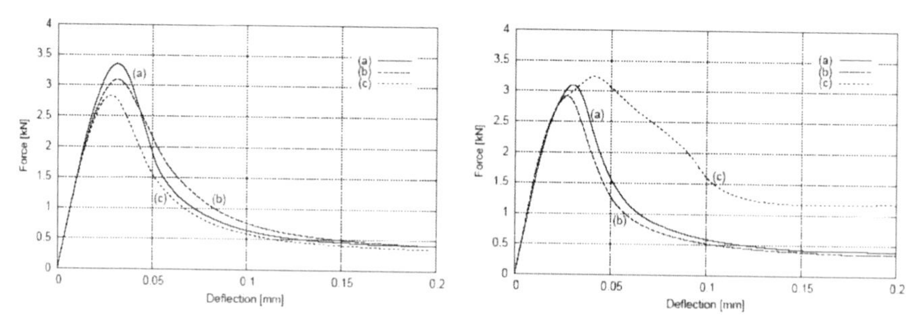

Fig. 9.28 Calculated force-deflection curves for two different random distributions of aggregate in small-size beam 80×320 mm^2 of sand concrete (d_{50}=2 mm, d_{max}=8 mm, l_c=1.5 mm): a) entirely heterogeneous beam, b) partially heterogeneous beam with width of meso-scale section of b_{ms}=80 mm, c) partially heterogeneous beam with width of the meso-scale section of b_{ms}=40 mm (Skarżyński and Tejchman 2010)

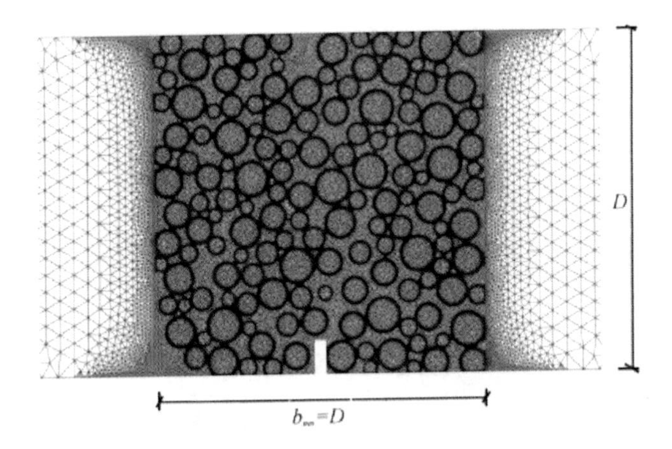

Fig. 9.29 FE mesh: three-phase heterogeneous concrete in notch neighbourhood with round shaped aggregate, cement matrix and interfacial transition zones ITZ and one-phase homogeneous concrete in remaining region (Skarżyński and Tejchman 2010)

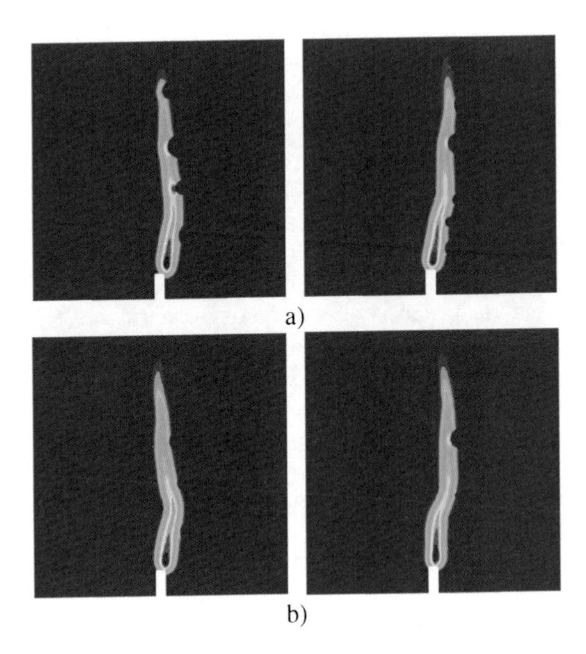

Fig. 9.30 Calculated distribution of non-local strain measure above notch (small size beam 80×320 mm^2, l_c=1.5 mm) for gravel concrete (d_{50}=2 mm, d_{max}=8 mm): a) entirely heterogeneous beam, b) partially heterogeneous beam with width of meso-scale section of b_{ms}=80 mm, c) partially heterogeneous beam with width of the meso-scale section of b_{ms}=40 mm (Skarżyński and Tejchman 2010)

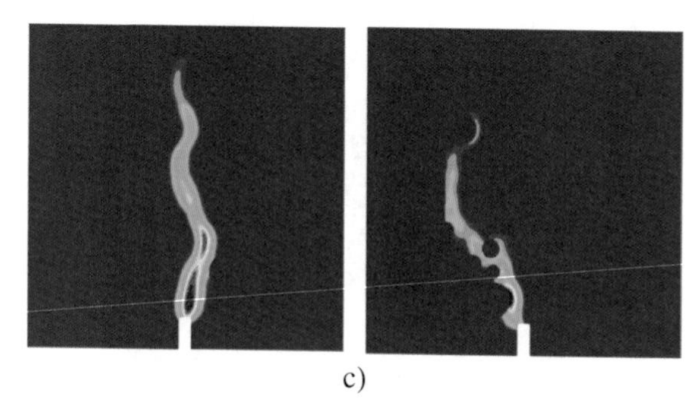

c)

Fig. 9.30 (*continued*)

Next the numerical 2D effect of different parameters such as the aggregate distribution, aggregate volume, aggregate shape, aggregate stiffness, bond thickness, notch size and characteristic length on the material behaviour (load-deflection curve and strain localization) was investigated for the small concrete beam 80×320 mm^2. The parameters were varied independently.

Effect of stochastic aggregate distribution

The effect of a random distribution of round-shaped aggregate particles in the concrete beam on the force-deflection diagram and width of a localized zone is shown in Figs. 9.31 and 9.32. The aggregate volume was ρ=45% using two aggregate size distribution curves '*a*' (d_{50}=2 mm, d_{max}=8 mm) and '*d*' (d_{50}=0.5 mm, d_{max}=3 mm) of Fig. 9.25, respectively. The ITZ thickness was t_b=0.25 mm.

All stochastic force-deflection curves are obviously the same in the almost entire elastic regime. However, they are significantly different at and after the peak (Fig. 9.31) due to a localized zone propagating between aggregate distributed at random, which is always non-symmetric and curved (Fig. 9.32). The difference in the strength is about 10-20%. The calculated width of a localized zone is approximately w_c=4.5 mm=3×l_c=9×s_{cm} independently of d_{max} and d_{50} (as in our tests, Skarżyński et al. 2011). The calculated localized zone is created at about u/D=0.0003 (u=0.024 mm) and its width increases during the deformation process.

A similar strong stochastic effect was also observed in FE calculations by Gitman et al. (2007) and He (2010). Surprisingly, a negligible stochastic effect was found in FE simulations by Kim and Abu Al-Rub (2011).

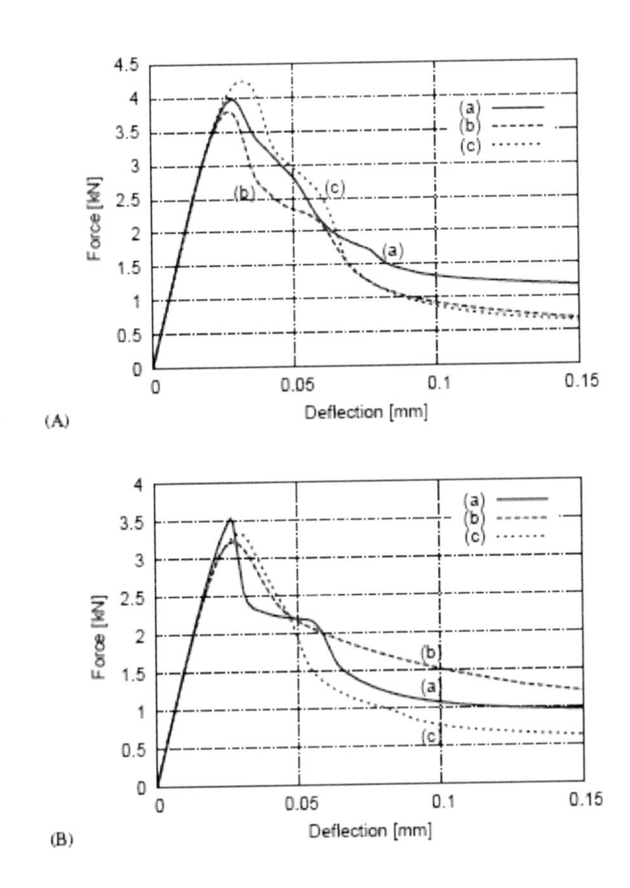

Fig. 9.31 Calculated force-deflection curves for fine-grained concrete beam (l_c=1.5 mm, ρ=45%, t_b=0.25 mm): A) with aggregate size distribution curve '*a*' of Fig. 9.25 (d_{50}=2 mm and d_{max}=8 mm) and B) with aggregate size distribution curve '*d*' of Fig. 9.25 (d_{50}=0.5 mm and d_{max}=3 mm) for three random distributions of circular aggregates (curves '*a*', '*b*' and '*c*')

Effect of aggregate shape and aggregate size distribution

To model the effect of the aggregate shape, four different grain shapes were taken into account, namely: circular, octagonal, irregular (angular) and rhomboidal (Fig. 9.33) keeping always the volume fraction and centres of grains constant (l_c=1.5 mm, ρ=60%, t_b=0.25 mm).

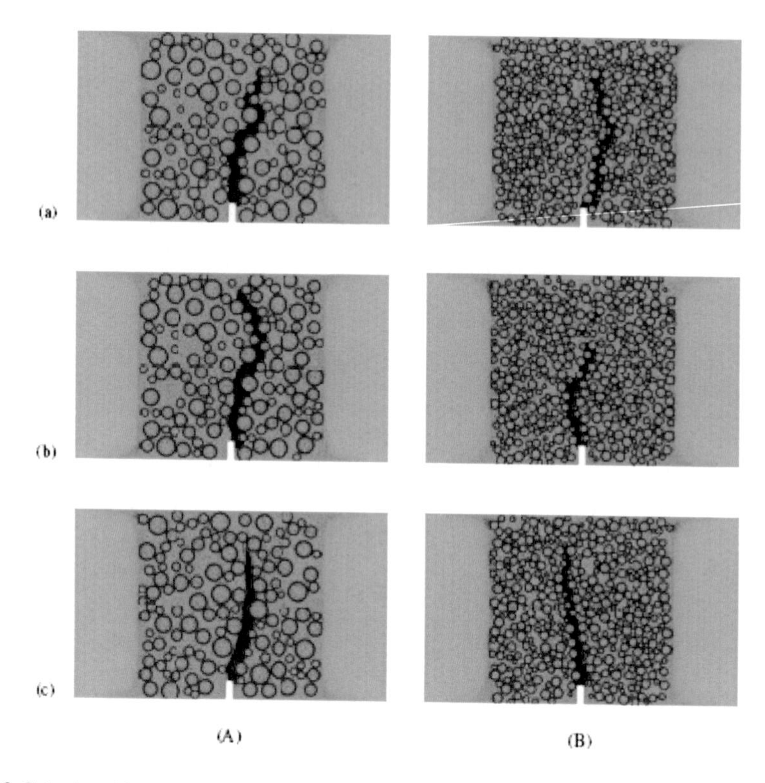

Fig. 9.32 Calculated localized zone in fine-grained concrete beam in notch region based on distribution of non-local strain measure corresponding to load-deflection curves '*a*', '*b*' and '*c*' of Figs. 9.31A and 9.31B (l_c=1.5 mm, p=45%, t_b=0.25 mm)

The aggregate shape can have a different influence on the beam ultimate strength depending upon the aggregate size distribution (Figs. 9.34 and 9.35). For the aggregate size distribution of Fig. 9.25a, the ultimate beam strength is the highest for rhomboidal-shaped particles and the lowest for octagonal-shaped particles (Figs. 9.33a, 9.35B and 9.35D). This difference equals even 30%. In the case of the aggregate size distribution curve of Fig. 9.25b, the ultimate beam strength is similar for all assumed particle shapes (Fig. 9.34B). For the aggregate size distribution of Fig. 9.25c, angular-shaped inclusions have the lower tensile strength than circular grains (Fig. 9.35C). From simulations follows that the mean tensile strength is usually higher with the larger mean grain size and the narrower grain range (Figs. 9.34A, 9.34B, 9.34C and 9.35).

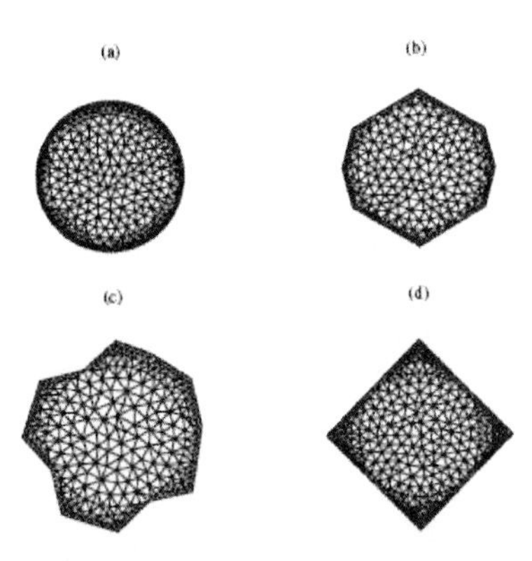

Fig. 9.33 Aggregate shape assumed in calculations: a) circular, b) octagonal, c) irregular (angular), d) rhomboidal

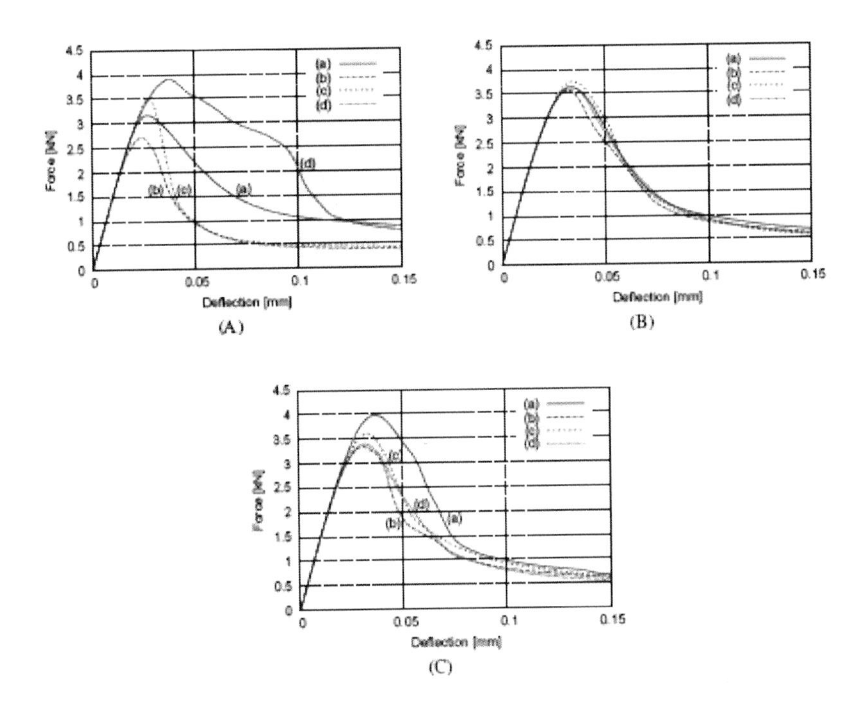

Fig. 9.34 Calculated force-deflection curves for different aggregate shape of Fig. 9.33: a) circular, b) octagonal, c) irregular (angular), d) rhomboidal (fine-grained concrete beam 80×320 mm^2, l_c=1.5 mm, ρ=60%, t_b=0.25 mm) and different aggregate size distributions of Fig. 9.25: A) d_{50}=2 mm and d_{max}=8 mm (curve 'a'), B) d_{50}=4 mm and d_{max}=10 mm (curve 'b'), C) d_{50}=4 mm and d_{max}=6 mm (curve 'c')

The width of a localized zone equals approximately w_c=3 mm for ρ=60% and is not influenced by the aggregate shape, aggregate distribution, mean and maximum grain size (Fig. 9.36). In turn, the form of a localized zone is strongly affected by the aggregate shape contributing thus to the different strength. The calculated width of a localized zone is in good agreement with our experiments with fine-grained concrete (Figs. 9.23 and 9.24A). Our outcome is in contrast to statements by Bažant and Pijauder-Cabot (1989), and Bažant and Oh (1983) wherein the width of a localized zone in usual concrete was estimated to be dependent upon d_{max}. It is also in contrast to experimental results by Mihashi and Nomura (1996) which showed that the width of a localized zone in usual concrete increased with increasing aggregate size. The differences between our and the experimental results (Bažant and Oh 1983, Mihashi and Nomura 1996) lie probably in a different concrete mix, specimen size and loading type. For instance, in our other tests with large reinforced concrete beams 6.0 m long without shear reinforcement under bending, the width of a localized zone in usual concrete was about 15 mm indicating that l_c=5 mm (Syroka and Tejchman 2011). This problem merits further experimental and numerical investigations.

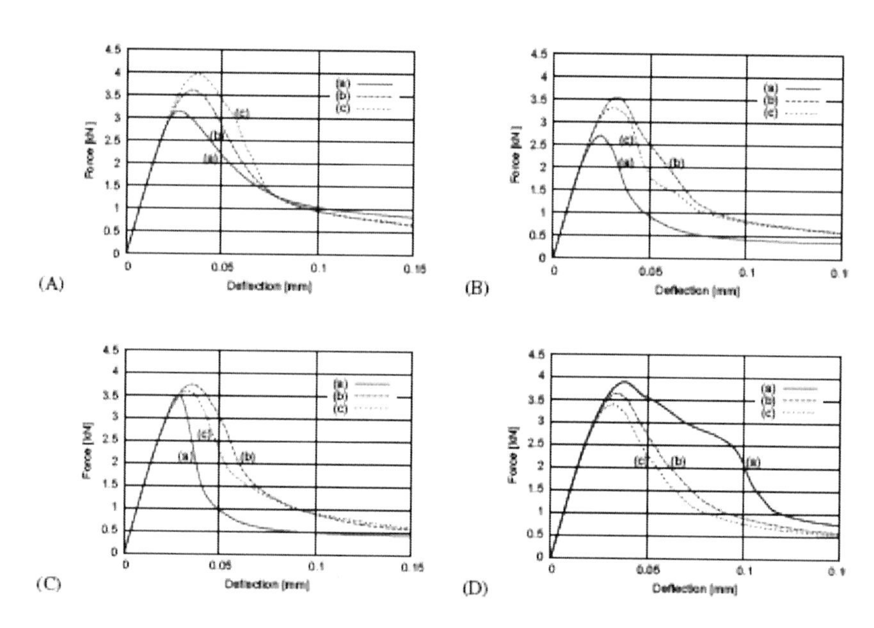

Fig. 9.35 Calculated force-deflection curves for different aggregate shape of Fig. 9.33: A) circular, B) octagonal, C) irregular (angular), D) rhomboidal (fine-grained concrete beam 80×320 mm^2, l_c=1.5 mm, ρ=60%, t_b=0.25 mm) and different aggregate size distribution of Fig. 9.25: a) d_{50}=2 mm and d_{max}=8 mm (curve '*a*'), b) d_{50}=4 mm and d_{max}=10 mm (curve '*b*'), c) d_{50}=4 mm and d_{max}=6 mm (curve '*c*')

According to Kim and Abu Al-Rub (2011) the aggregate shape has a weak effect on the ultimate strength of concrete and on the strain to damage-onset, but significantly affects the crack initiation, propagation and distribution. The stress concentrations at sharp edges of polygonal particles cause that the ultimate tensile strength and strain at the damage onset are the highest for circular grains model. Similar conclusions were derived by He et al. (2009) and He (2010).

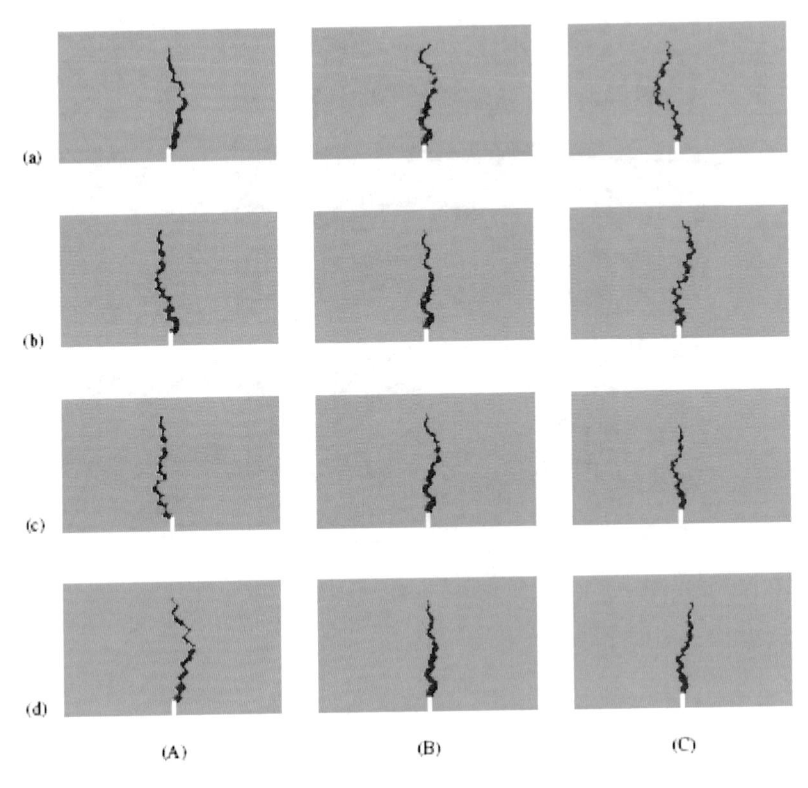

Fig. 9.36 Calculated localized zone based on distribution of non-local strain measure in fine-grained concrete beam in notch region corresponding to load-deflection curves '*a*', '*b*', '*c*' and '*d*' of Figs. 9.34A, 9.34B and 9.34C (l_c=1.5 mm, ρ=60%, t_b=0.25 mm)

Effect of volume fraction of aggregate

Circular grains with the volume of ρ=30%, ρ=45% and ρ=60% were used (l_c=1.5 mm, t_b=0.25 mm), Fig. 9.37. Figures 9.38 and 9.39 demonstrate the effect of the aggregate volume in fine-grained concrete beam with the aggregate size distributions '*a*' of Fig. 9.25 (d_{50}=2 mm, d_{max}=8 mm) and '*d*' of Fig. 9.25 (d_{50}=0.5 mm, d_{max}=3 mm).

In our FE simulations, the Young modulus and ultimate beam strength increase with increasing aggregate density in the range of 30%-60% (Fig. 9.38). This increase certainly depends on material parameters assumed for separated concrete phases, in particular for ITZs being always the weakest parts in concrete.

The width and shape of a localized zone are influenced by the aggregate volume; a localized zone becomes narrower with increasing aggregate volume: w_c=6 mm at ρ=30%, w_c=4.5 mm at ρ=45% and w_c=3 mm at ρ=60% (Fig. 9.39).

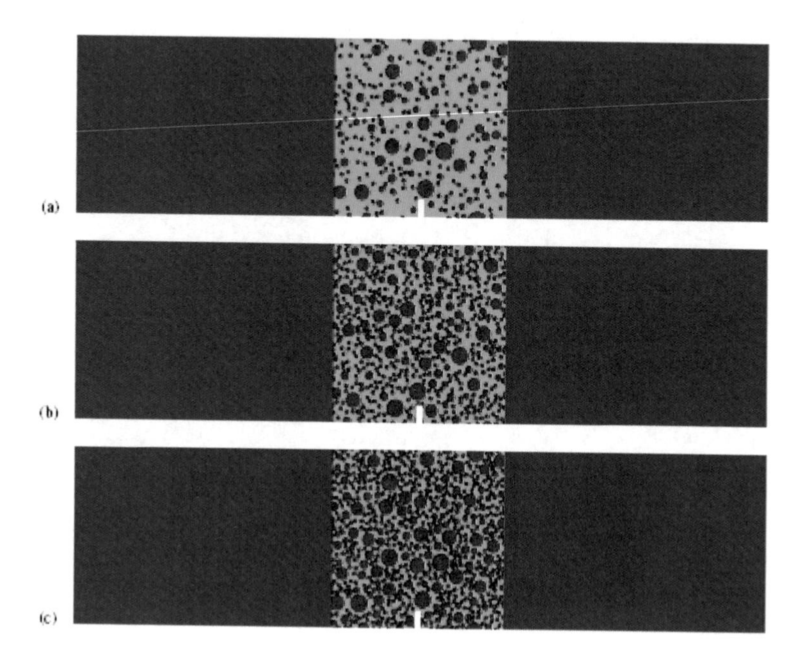

Fig. 9.37 Concrete beams with different volume fraction of aggregate ρ in region close to notch: a) ρ=30%, b) ρ=45% and c) ρ=60% using grain size distribution '*a*' of Fig. 9.25

According to Kim and Abu Al-Rub (2011) the Young modulus linearly increases with increasing aggregate volume, and the tensile strength decreases with increasing aggregate density up to ρ=40% and increases next from ρ=40% up to ρ=60%. The strain at the damage linearly decreases with increasing aggregate volume. He et al. (2009) and He (2010) concluded that concrete with a higher packing density of aggregate up to 50% has a decreasing tensile strength (due to a higher number of very weak interfacial transitional zones around aggregate). It seems that the property of ITZ (stiffness, strength and width) is essential for the global strength versus ρ.

Effect of ITZ thickness

The interfacial transition zone (ITZ) is a special region of the cement paste around particles, which is perturbed by their presence. Its origin lies in the packing of the cement grains against the much larger aggregate which leads to a local increase in porosity (micro-voids) and a presence of smaller cement particles. A paste with lower *w/c* (higher packing density) or made with finer cement particles leads to

ITZ of smaller extent. This layer is highly heterogeneous and damaged and thus critical for the concrete behaviour (Srivener et al. 2004, Mondal et al. 2009). An accurate understanding of the properties and behaviour of ITZ is one of the most important issues in the meso-scale analysis because damage is initiated at the weakest region and ITZ is just this weakest link in concrete. We assumed that ITZs have the reduced stiffness and strength as compared to the cement matrix (Tabl.9.2).

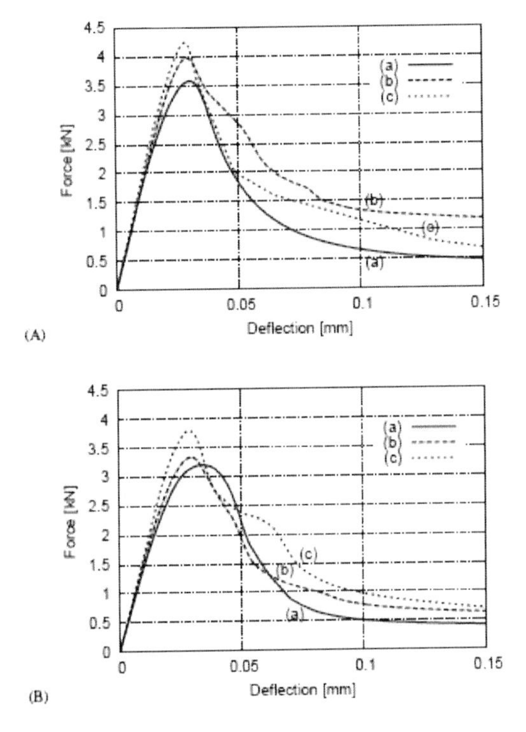

Fig. 9.38 Calculated load-deflection curves for different volume fractions ρ of circular aggregate: a) ρ=30%, b) ρ=45% and c) ρ=60% (concrete beam 80×320 mm^2, l_c=1.5 mm, t_b=0.25 mm, A) aggregate size distribution '*a*' of Fig. 9.25 (d_{50}=2 mm, d_{max}=8 mm), B) aggregate size distribution '*d*' of Fig. 9.25 (d_{50}=0.5 mm, d_{max}=3 mm)

Figures 9.40 and 9.41 demonstrate the effect of the ITZ thickness in a fine-grained concrete beam of circular grains with the aggregate size distribution '*a*' of Fig. 9.25 (d_{50}=2 mm, d_{max}=8 mm) assuming the aggregate volume fraction ρ=45% and ρ=60% (l_c=1.5 mm). Since there is very limited data on the thickness of ITZ, the thickness t_b in our study was assumed to be 0 mm, 0.05mm (He et al. 2011, He 2010), 0.25 mm (Gitman et al. 2007) and 0.75 mm.

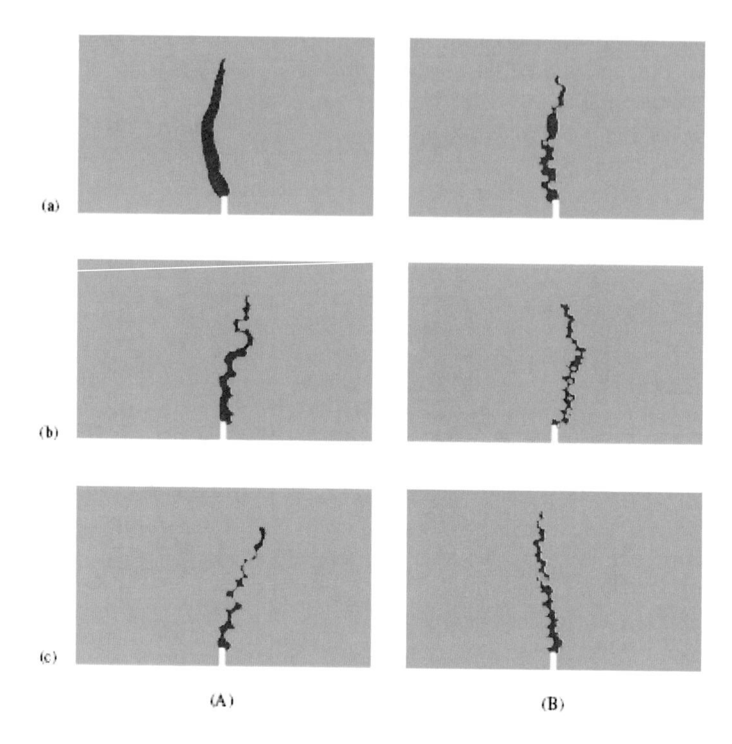

Fig. 9.39 Calculated localized zone based on distribution of non-local strain measure in fine-grained concrete beam 80×320 mm^2 (l_c=1.5 mm, t_b=0.25 mm) corresponding to load-deflection curves '*a*', '*b*' and '*c*' of Figs. 9.38A and 9.38B

The results show that the thickness and strength of ITZs strongly affect both the load-displacement response and shape of localized zone. Since ITZ is the weakest phase, the ultimate beam strength decreases with increasing bond thickness (Fig. 9.40). This result is in agreement with those by He et al. (2009), He (2010) and Kim and Abu Al-Rub (2011). However, the residual strength rather increases with increasing bond thickness as in calculations by Kim and Abu Al-Rub (2011). The width of a localized zone is w_c=4.5 mm (ρ=45%) and w_c=3 mm (ρ=60%) and is not affected by the ITZ size t_b (Fig. 9.41).

Effect of notch size

Figures 9.42 and 9.43 demonstrate the effect of the notch size on the load-deflection diagram and strain localization in a fine-grained concrete beam with a random distribution of aggregate '*a*' of Fig. 9.25 (d_{50}=2 mm to d_{max}=8 mm) using circular aggregate volume ρ=30% and ρ=60% (l_c=1.5 mm, t_b=0.25 mm). The notch size was 0×0 mm^2, 3×3 mm^2 and 6×3 mm^2 (width×height), respectively. The beam without notch was modelled as entirely heterogeneous to be sure that a localized zone occurs in a meso-region.

The ultimate beam strength is higher with decreasing notch size (Fig. 9.42). The notch size has no influence on the width of a localized zone (w_c=6 mm at ρ=30% and w_c=3 mm at ρ=60% (Fig. 9.43).

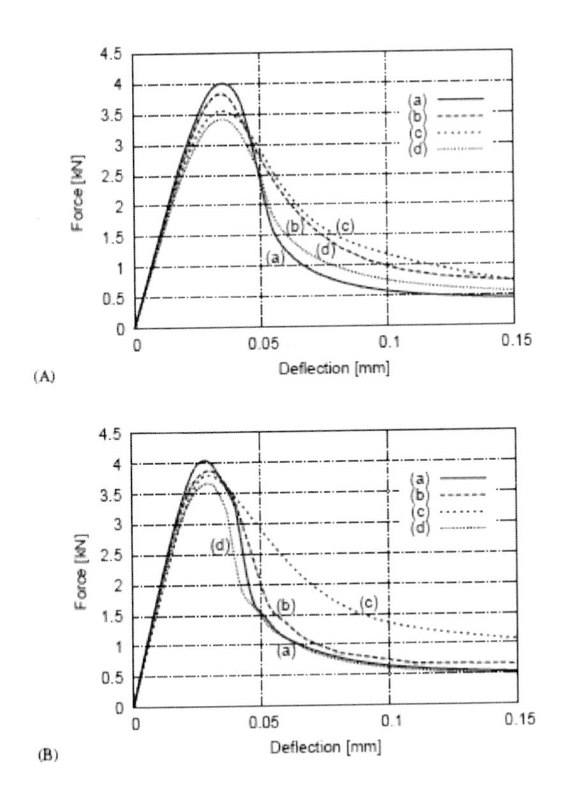

Fig. 9.40 Numerical effect of different ITZ thickness t_b in FE calculations on load-deflection curve: a) t_b=0 mm, b) t_b=0.05 mm, c) t_b=0.25 mm and d) t_b=0.75 mm, A) ρ=45%, B) ρ=60% (fine-grained concrete beam 80×320 mm^2, l_c=1.5 mm, circular grains with size distribution '*a*' of Fig. 9.25 (d_{50}=2 mm, d_{max}=8 mm)

Effect of aggregate stiffness

Figure 9.44 shows the effect of the aggregate stiffness in a small size beam (80×320 mm^2, d_{50}=4 mm and d_{max}=10 mm, ρ=60%, t_b=0.25 mm, l_c=1.5 mm). The calculations were carried out with weak aggregate (which had the same properties as ITZ of Tab. 9.2).

For the weak aggregate, a localized zone can propagate through weak grains. The vertical force is obviously smaller and the width of a localized zone is higher as compared to the results with the strong aggregate (strong aggregate - w_c=3.3 mm, weak aggregate - w_c=5.8 mm).

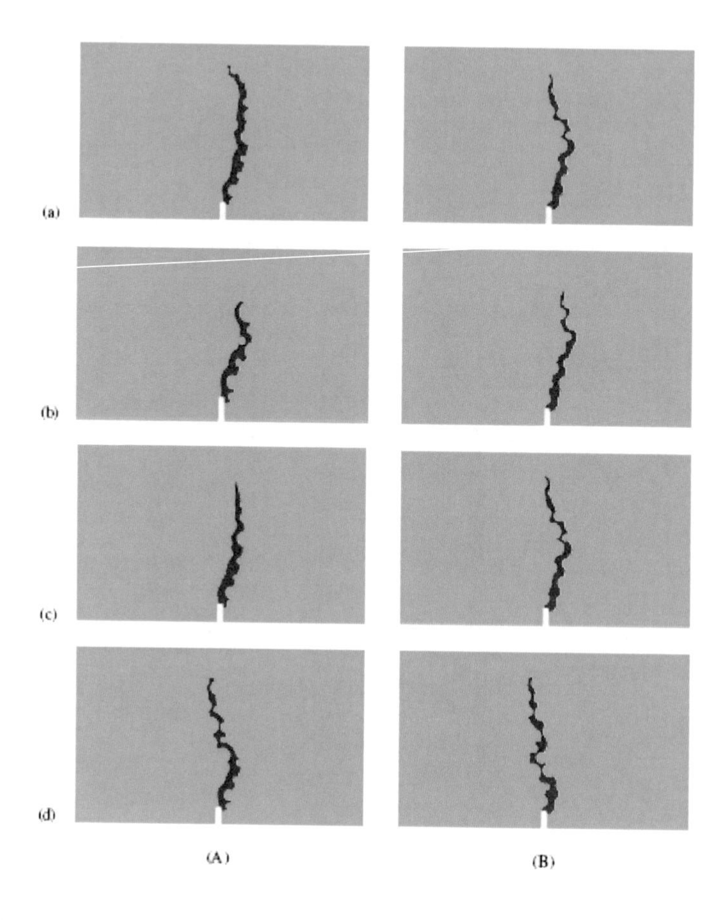

Fig. 9.41 Numerical effect of different bond thickness t_b in FE calculations on distribution of non-local strain measure close to beam notch corresponding to load-deflection curves '*a*', '*b*', '*c*' and '*d*' of Figs. 9.40A and 9.40B

Effect of characteristic length of micro-structure

The effect of a characteristic length of micro-structure on the load-deflection diagram and strain localization is shown in Figs. 9.45 and 9.46 using the same stochastic distribution of circular aggregate (l_c changed between 0.5 mm and 5 mm).

With increasing characteristic length, both beam strength and width of a localized zone strongly increase since the material softening decreases and material becomes more ductile. A pronounced deterministic size effect occurs. A localized zone propagating in a cement matrix between aggregate grains is strongly curved at l_c=0.5-2.5 mm, whereas it becomes more straight at l_c>2.5 mm.

It is about: w_c=2.9-17.6 mm=(3.5-5.9)×l_c=(5.8-35.2)×d_{50} at ρ=30%, w_c=2.5-16.7 mm=(3.0-5.0)×l_c=(1.25-8.35)×d_{50} at ρ=45% and w_c=2.4-13.9 mm=(2.3-4.7)×l_c=(0.6-3.47)×d_{50} at ρ=60% (Tab. 9.3). It always decreases with increasing ρ (Tab. 9.3). A characteristic length of micro-structure is not uniquely connected to the aggregate size.

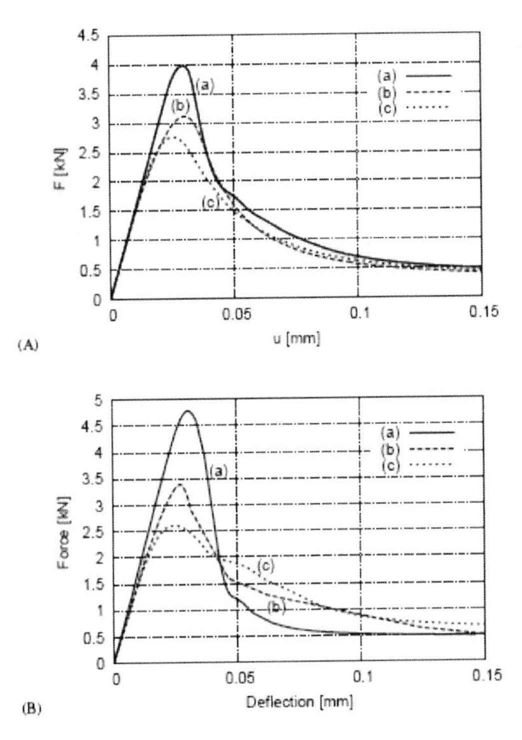

Fig. 9.42 Numerical effect of notch size on force-deflection curve for two different aggregate densities: a) 0×0 mm², b) 3×3 mm² and c) 6×3 mm², A) ρ=30%, B) ρ=60% (fine-grained concrete beam 80×320 mm², l_c=1.5 mm, circular aggregate distribution '*a*' of Fig. 9.25 with d_{50}=2 mm and d_{max}=8 mm)

Figure 9.47 shows the evolution of the width and height of the localized zone from FE calculations. The FE results of Fig. 9.47 are similar as in the experiments (Fig. 9.25). The calculated maximum width is 3.25 mm (3.5-4.0 mm in tests) and height 55 mm (50-55 mm in tests) at u=0.2 mm. The calculated localized zone strongly forms (linearly) before and after the maximum vertical force in the range of u=0.025-0.05 mm (width) and of u=0.025-0.1 mm (length). The mean propagation rate of the calculated localized zone versus the beam deflection is similar as in experiments, although is more uniform (Fig. 9.48). In the experiments, a macro-crack occurred at about u=0.04 mm, which cannot be captured by our model. In order to numerically describe a macro-crack, a discontinuous approach has to be used (e.g. XFEM or cohesive crack model, Chapter 4).

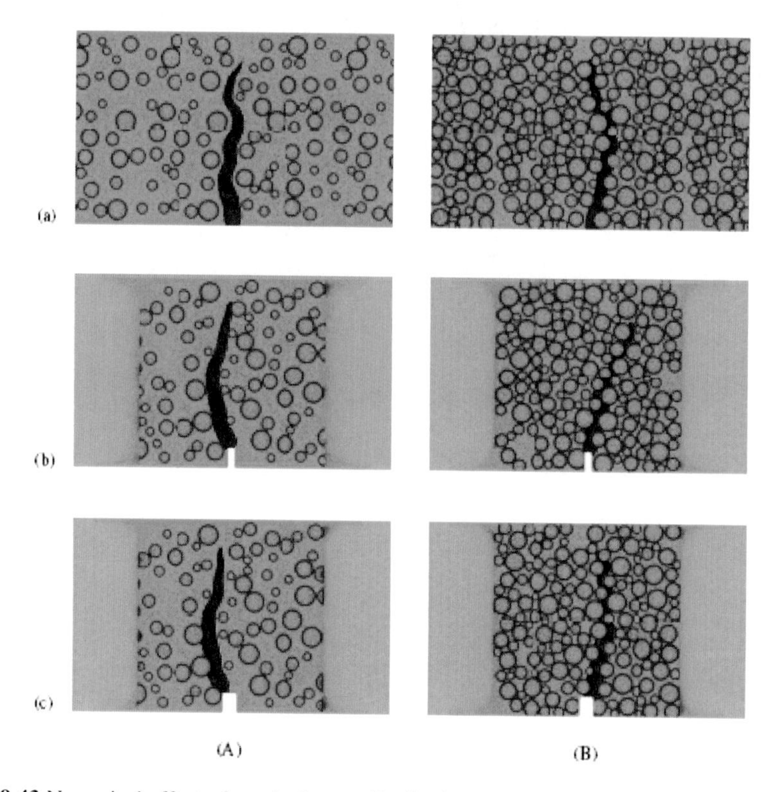

Fig. 9.43 Numerical effect of notch size on distribution of non-local strain measure close to beam notch corresponding to load-deflection curves '*a*', '*b*' and '*c*' of Figs. 9.42A and 9.42B

Effect of beam size

The effect of the beam size is presented in Figs. 9.49 and 9.50. Figure 9.49 shows the numerical results of the nominal strength $\sigma_n = 1.5Pl/(bD^2)$ versus the normalized deflection u/D for three different concrete beams compared to tests by Le Bellĕgo et al. (2003). Concrete was treated as an one-phase material with a heterogeneous three-phase section close to the notch ($b_{ms}=D$) using material constants from Tab. 9.2. The following amount of triangular finite elements was used: 110'000 (small beam), 420'000 medium beam and 1'600'000 (large beam). In turn, Figure 9.50 presents the distribution of a non-local softening strain measure in beams. The calculations were carried out with gravel concrete of d_{max}=8 mm, aggregate density of ρ=30% and a characteristic length of l_c=1.5 mm.

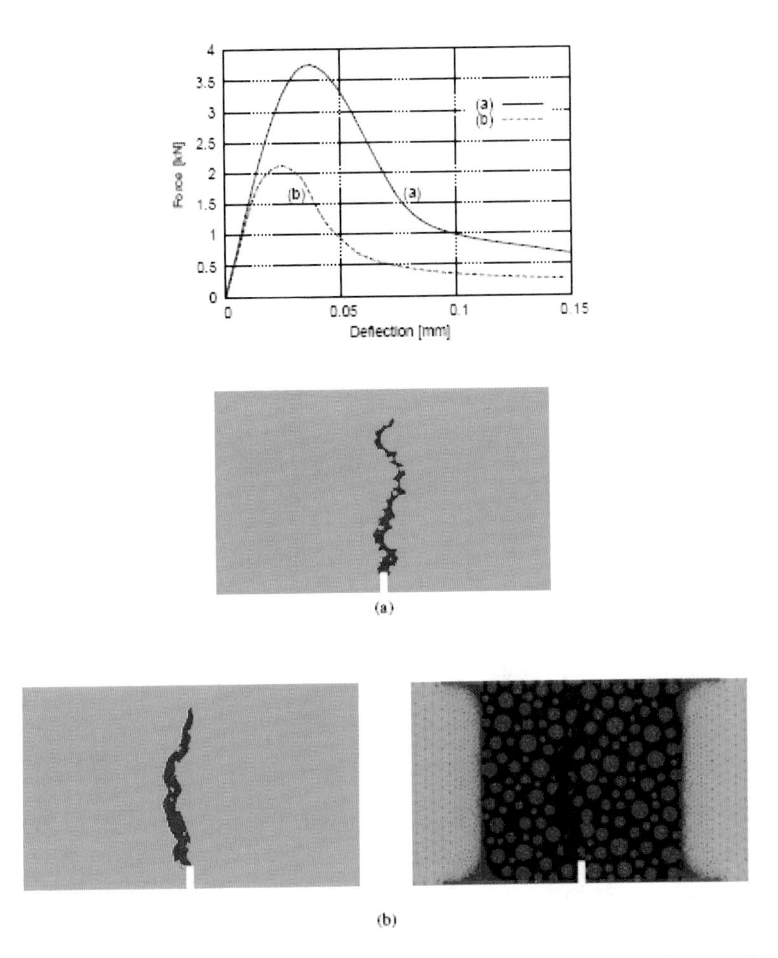

Fig. 9.44 Effect of aggregate stiffness on force-deflection curve and distribution of non-local strain measure close to beam notch: a) strong circular aggregate, b) weak circular aggregate (fine-grained concrete beam 80×320 mm^2, l_c=1.5 mm, circular aggregate distribution '*c*' of Fig. 9.25 with d_{50}=4 mm and d_{max}=10 mm, ρ=60%)

The numerical results are in a satisfactory agreement with tests by Le Bellĕgo et al. (2003). The deterministic size effect is realistically modelled in calculations. The width of the localized zone above the notch at u/D=0.5 is 6 mm (ρ=30%) for all beam sizes. The localized zone propagating between aggregate is always strongly curved, what satisfactorily reflects the experimental results.

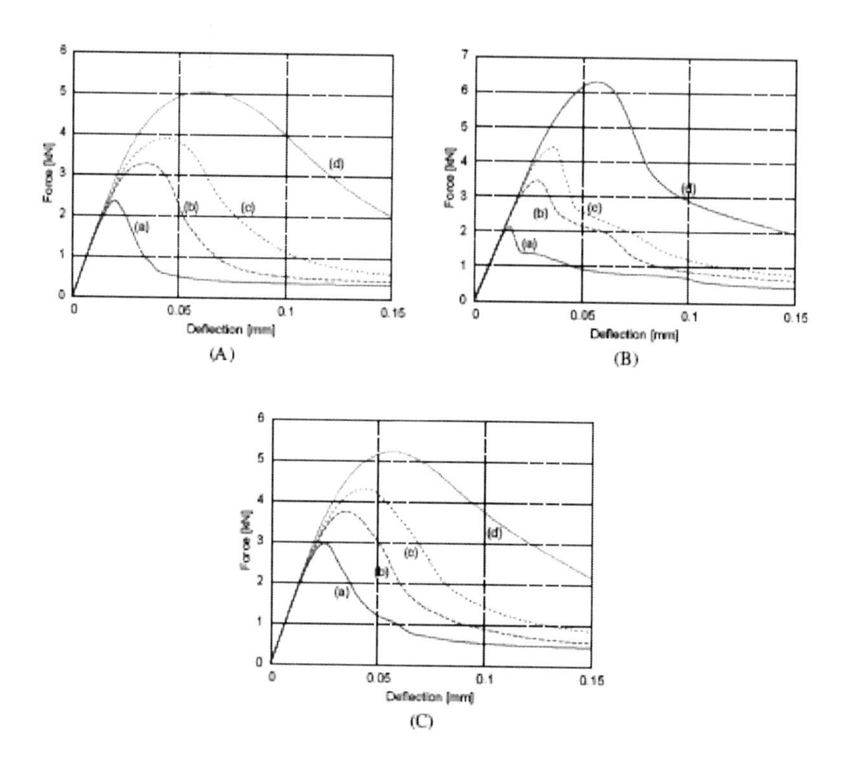

Fig. 9.45 Calculated load-deflection curves for different characteristic lengths l_c: a) l_c=0.5 mm, b) l_c=1.5 mm, c) l_c=2.5 mm and d) l_c=5 mm (concrete beam 80×320 mm^2, ITZ thickness t_b=0.25 mm), A) volume fraction of circular aggregate ρ=30% (concrete mix '*d*' of Fig. 9.25 with d_{50}=0.5 mm and d_{max}=3 mm), B) volume fraction of circular aggregate ρ=45% (concrete mix '*a*' of Fig. 9.25 with d_{50}=2 mm and d_{max}=8 mm), C) volume fraction of angular aggregate ρ=60% (concrete mix '*b*' of Fig. 9.25 with d_{50}=4 mm and d_{max}=10 mm)

Figure 9.51 shows a comparison between the measured and calculated size effect for concrete beams. In addition, the results of a deterministic size effect law by Bažant, Eq. 5.5 (Bažant and Planas 1998, Bažant 2004) are enclosed (which is valid for structures with pre-existing notches, Chapter 8). The experimental and theoretical beam strength shows a strong parabolic size dependence. The experimental and numerical results match quite well the size effect law by Bažant (Bažant and Planas 1998).

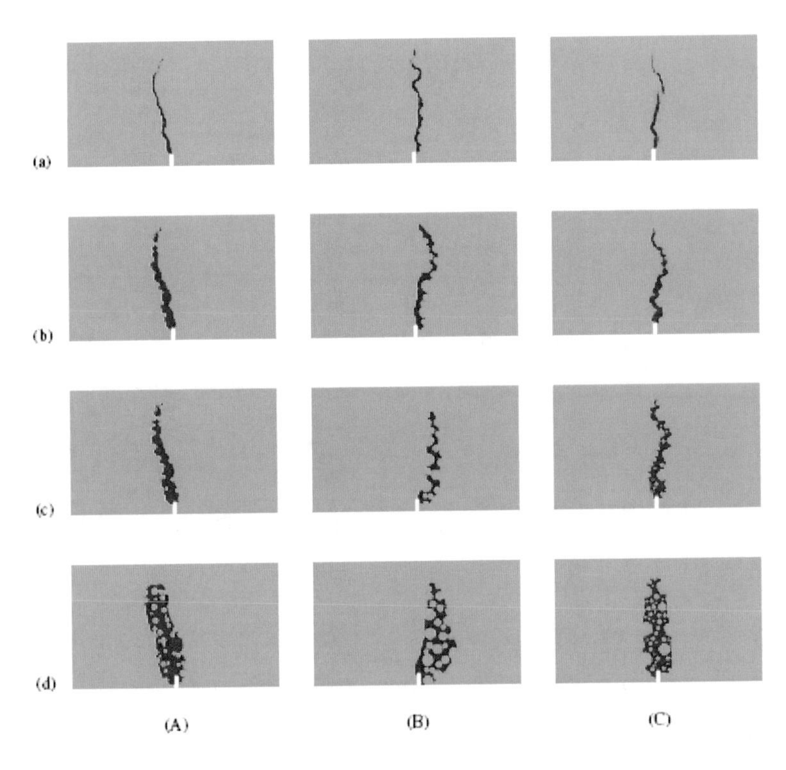

Fig. 9.46 Calculated localized zone based on distribution of non-local strain measure for different characteristic lengths l_c: a) l_c=0.5 mm, b) l_c=1.5 mm, c) l_c=2.5 mm and d) l_c=5 mm (concrete beam 80×320 mm^2, bond thickness t_b=0.25 mm), A) volume fraction of circular aggregate ρ=30% (concrete mix '*d*' of Fig. 9.25 with d_{50}=0.5 mm and d_{max}=3 mm), B) volume fraction of circular aggregate ρ=45% (concrete mix '*a*' of Fig. 9.25 with d_{50}=2 mm and d_{max}=8 mm), C) volume fraction of angular aggregate ρ=60% (concrete mix '*b*' of Fig. 9.25 with d_{50}=4 mm and d_{max}=10 mm)

Table 9.3 Calculated width of localized zone with different characteristic length l_c and volume fraction ρ

Characteristic length l_c	Width of localized zone w_c [mm] for different volume fraction of aggregate ρ					
	ρ = 30%		ρ = 45%		ρ = 60%	
0.5 mm	2.9	5.9 × l_c	2.4	5.0 × l_c	2.4	4.7 × l_c
1.5 mm	6.2	4.1 × l_c	4.5	3.0 × l_c	3.5	2.3 × l_c
2.5 mm	9.3	3.7 × l_c	8.7	3.5 × l_c	6.9	2.7 × l_c
5 mm	17.6	3.5 × l_c	16.7	3.4 × l_c	13.9	2.8 × l_c

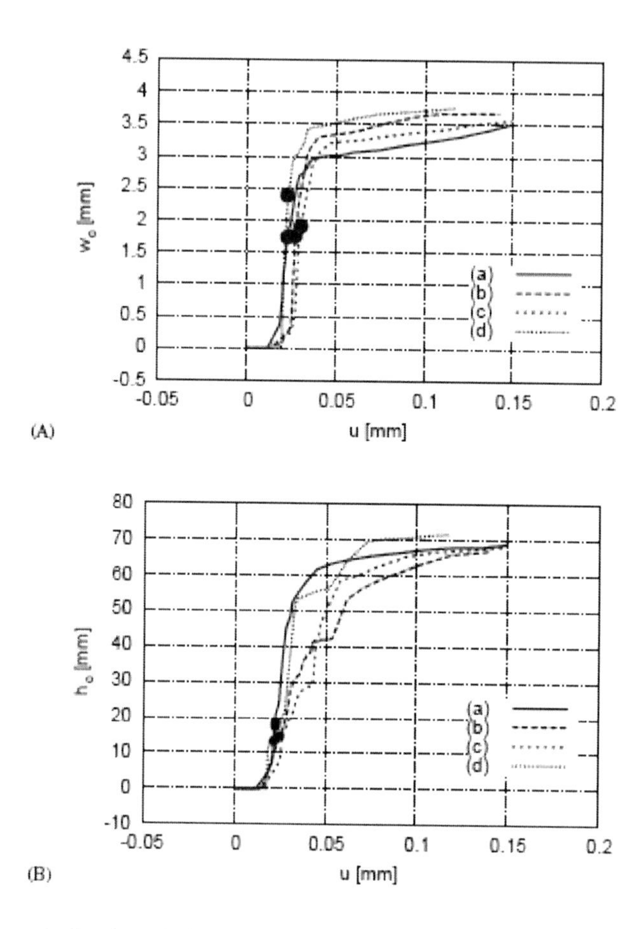

Fig. 9.47 The calculated evolution of width (A) w_c and height h_c (B) of localized zone versus beam deflection u: a) concrete mix 'a' of Fig. 9.25 with d_{50}=2 mm and d_{max}=8 mm, irregular aggregate, ρ=60%, l_c=1.5 mm, b) concrete mix 'b' of Fig. 9.25 with d_{50}=4 mm and d_{max}=10 mm, octagonal aggregate, ρ=60%, l_c=1.5 mm, c) concrete mix 'c' of Fig. 9.25 with d_{50}=4 mm and d_{max}=6 mm, circular aggregate, ρ=60%, l_c=1.5 mm, d) concrete mix 'a' of Fig. 9.25 with d_{50}=2 mm and d_{max}=8 mm, circular aggregate, ρ=60%, beam without notch, l_c=1.5 mm (• - maximum vertical force)

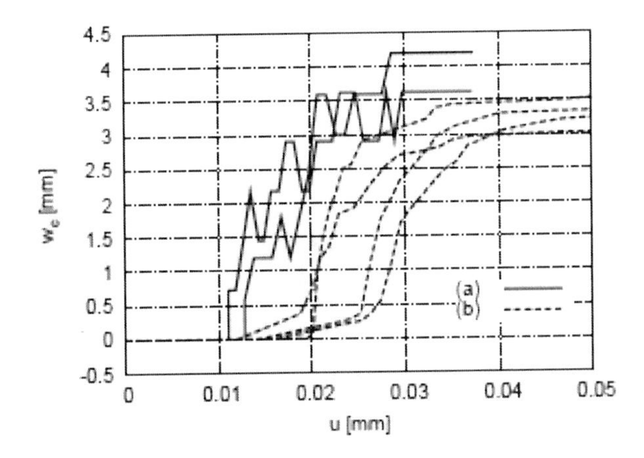

Fig. 9.48 Comparison between measured (a) and calculated (b) evolution of width of localized zone w_c versus beam deflection u (maximum vertical force occurs at deflection u=0.035 mm)

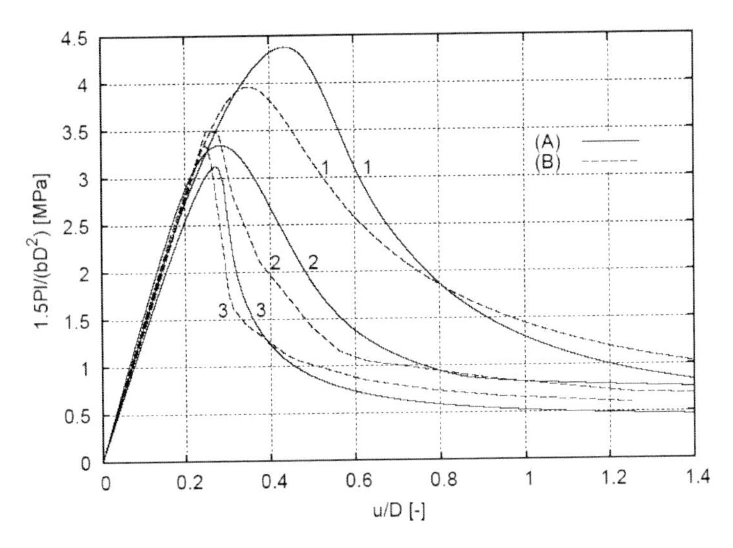

Fig. 9.49 Calculated nominal strength $1.5Pl/(bD^2)$ versus normalised beam deflection u/D (u - beam deflection, D - beam height): A) FE-results, B) experiments by Le Bellěgo et al. (2003): 1) small-size beam, (2) medium-size beam, (3) large-size beam (three-phase random heterogeneous material close to notch, b_{ms}=D) (Skarżyński and Tejchman 2010)

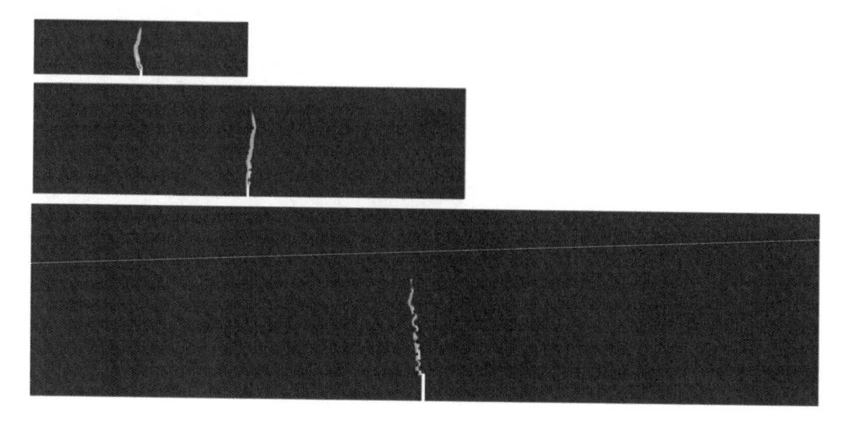

Fig. 9.50 Calculated distribution of non-local strain measure above notch from numerical calculations (at u/D=0.5) in small-size, medium-size and large-size beam (random heterogeneous three-phase material close to notch, $b_{ms}=D$) (Skarżyński and Tejchman 2010)

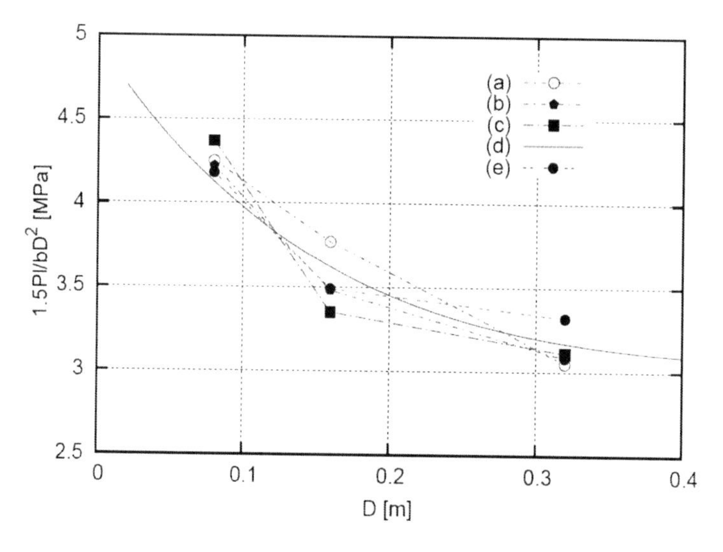

Fig. 9.51 Calculated and measured size effect in nominal strength $1.5Pl/(bD^2)$ versus beam height D for concrete beams of a similar geometry (small-, medium- and large-size beam): a) our laboratory experiments, b) our FE-calculations (homogeneous one-phase material), c) our FE-calculations (heterogeneous material close to notch, $b_{ms}=D$), d) size effect law by Bažant (2004), Eq. 5.5, e) experiments by Le Bellěgo et al. (2003) (Skarżyński and Tejchman 2010)

9.3 Representative Volume Element

Thus, the most important issue in multi-scale analyses is determination of an appropriate size for a micro-structural model, so-called representative volume element RVE. The size of RVE should be chosen such that homogenized properties become independent of micro-structural variations and a micro-structural domain is small enough such that separation of scales is guaranteed. Many researchers attempted to define the size of RVE in heterogeneous materials with a softening response in a post-peak regime (Hill 1963, Bažant and Pijauder-Cabot 1989, Drugan and Willis 1996, Evesque 2000, van Mier 2000, Bažant and Novak 2003, Kanit et al. 2003, Kouznetsova et al. 2004, Gitman et al. 2007, Skarżyński and Tejchman 2009). The last outcomes in this topic show, however, that RVE cannot be defined in softening quasi-brittle materials due to strain localization since the material loses then its statistical homogeneity, Chapter 9.1 (Gitman et al. 2007, Skarżyński and Tejchman 2009, 2010). Thus, each multi-scale approach always suffers from non-objectivity of results with respect to a cell size. RVE solely exists for linear and hardening regimes.

The intention of our FE investigations is to determine RVE in concrete under tension using two alternative strategies (one of them was proposed by Nguyen et al. 2010) (Skarżyński and Tejchman al. 2012). Concrete was assumed at meso-scale as a random heterogeneous material composed of three phases. The FE calculations of strain localization were carried out again with a scalar isotropic damage with non-local softening (Tab. 9.4). The interface was assumed to be the weakest component (Lilliu and van Mier 2003) and its width was 0.25 mm (Gitman et al. 2007). For the sake of simplicity, the aggregate was assumed in the form of circles. The number of triangular finite elements changed between 4'000 (the smallest specimen) and 100'000 (the largest specimen). The size of triangular elements was: s_a=0.5 mm (aggregate), s_{cm}=0.25 mm (cement matrix) and s_{itz}=0.1 mm (interface). To analyze the existence of RVE under tension, a plane strain uniaxial tension test was performed with a quadratic concrete specimen representing a unit cell with the periodicity of boundary conditions and material periodicity (Chapter 9.1), Fig. 9.52.

The unit cells of six different sizes were investigated $b{\times}h$: 5×5 mm^2, 10×10 mm^2, 15×15 mm^2, 20×20 mm^2, 25×25 mm^2 and 30×30 mm^2, respectively. For each specimen, three different stochastic realizations were performed with the aggregate density of ρ=30% (the results for ρ=45% and ρ=60% showed the same trend). A characteristic length of micro-structure was assumed to be l_c=1.5 mm based on DIC and numerical studies with an isotropic damage model (Chapter 9.2). Thus, the maximum finite element size in 3 different concrete phases was not greater than $3{\times}l_c$ to obtain mesh-objective results (Bobiński and Tejchman 2004, Marzec et al. 2007).

Table 9.4 Material properties assumed for FE calculations of 2D random heterogeneous three-phase concrete material (Skarżyński and Tejchman 2012)

Parameters	Aggregate	Cement matrix	ITZ
Modulus of elasticity E [GPa]	30	25	20
Poisson's ratio υ [-]	0.2	0.2	0.2
Crack initiation strain κ_0 [-]	0.5	8×10^{-5}	5×10^{-5}
Residual stress level α [-]	0.95	0.95	0.95
Slope of softening β [-]	200	200	200

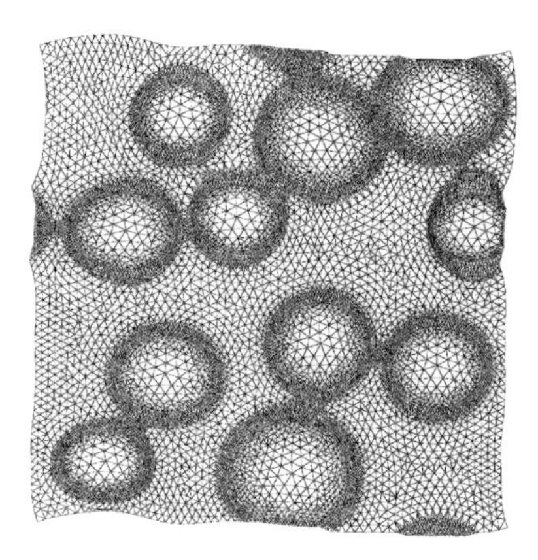

Fig. 9.52 Deformed three-phase concrete specimen with periodicity of boundary conditions and material periodicity (Skarżyński and Tejchman 2012)

Standard averaging approach

The standard averaging is performed in the entire specimen domain (Chapter 9.1). The homogenized stress and strain are defined in two dimensions as

$$<\sigma>=\frac{f_y^{int}}{b} \quad \text{and} \quad <\varepsilon>=\frac{u}{h}, \tag{9.4}$$

where f_y^{int} denotes the sum of all vertical nodal forces in the 'y' direction along the top edge of the specimen (Fig. 9.5), u is the prescribed vertical displacement in the 'y' direction and b and h are the width and height of the specimen.

Figure 9.53 presents the stress-strain relationships for various cell sizes and two random aggregate distributions with the material constants of Tab. 9.5 (l_c= 1.5 mm). In the first case, the aggregate distribution was similar and in the second case it was at random in different unit cells. The results show that the stress-strain curves are the same solely in an elastic regime independently of the specimen size, aggregate density and aggregate distribution. However, they are completely different at the peak and in a softening regime. An increase of the specimen size causes a strength decrease and an increase of material brittleness (softening rate). The differences in the evolution of stress-strain curves in a softening regime are caused by strain localization (in the form of a curved localized zone propagating between aggregates, Figs. 9.54 and 9.55) contributing to a loss of material homogeneity (due to the fact that strain localization is not scaled with increasing specimen size). The width of a calculated localized zone is approximately w_c=3 mm=2×l_c=12×s_{cm} (unit cell 5×5 mm^2), w_c=5 mm=3.33×l_c=20×s_{cm} (unit cell 10×10 mm^2) and w_c=6 mm=4×l_c=24×s_{cm} (unit cells larger than 10×10 mm^2).

Figure 9.56 presents the expectation value and standard deviation of the tensile fracture energy G_f versus the specimen height h for 3 different realizations. The fracture energy G_f was calculated as the area under the strain-stress curves g_f multiplied by the width of a localized zone w_c

$$G_f = g_f \times w_c = \left(\int_{a_1}^{a_2} <\sigma> \mathrm{d} <\varepsilon> \right) \times w_c \,. \tag{9.4}$$

The integration limits 'a_1' and 'a_2' are 0 and 0.001, respectively. The fracture energy decreases with increasing specimen size without reaching an asymptote, i.e. the size dependence of RVE exists (since a localized zone does not scale with the specimen size). Thus, RVE cannot be found for softening materials and a standard averaging approach cannot be used in homogenization-based multi-scale models.

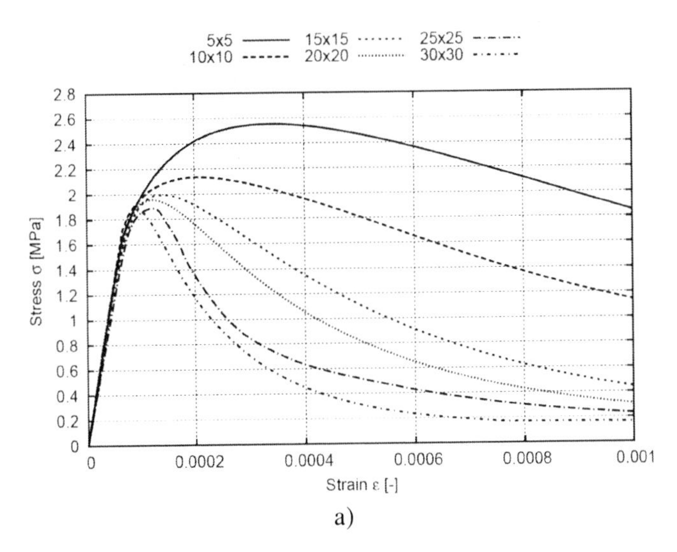

a)

Fig. 9.53 Stress-strain curves for various sizes of concrete specimens and two different random distributions of aggregate (a) and (b) using standard averaging procedure (characteristic length l_c=1.5 mm, aggregate density ρ=30%) (Skarżyński and Tejchman 2012)

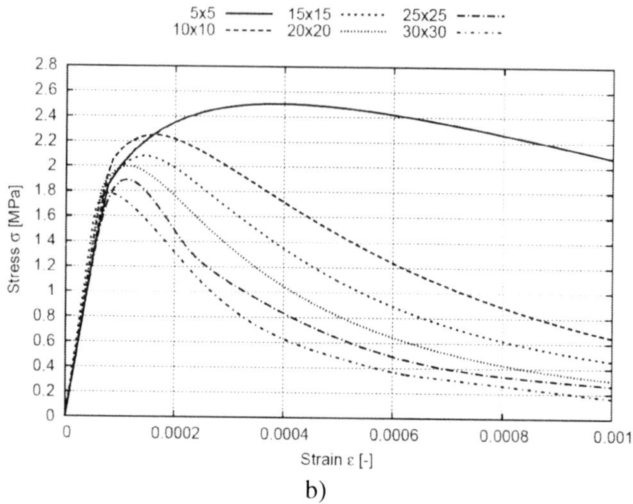

Fig. 9.53 (*continued*)

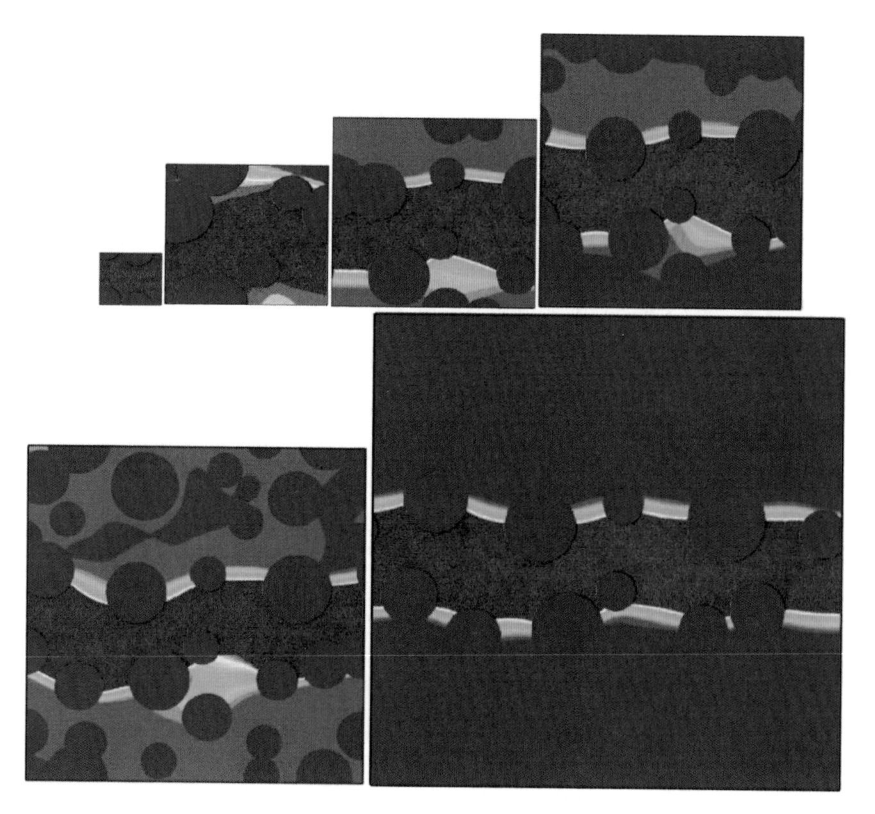

Fig. 9.54 Distribution of non-local softening strain measure for various specimen sizes and stress-strain curves of Fig. 9.53a using standard averaging procedure (characteristic length l_c=1.5 mm, aggregate density ρ=30%) (Skarżyński and Tejchman 2012)

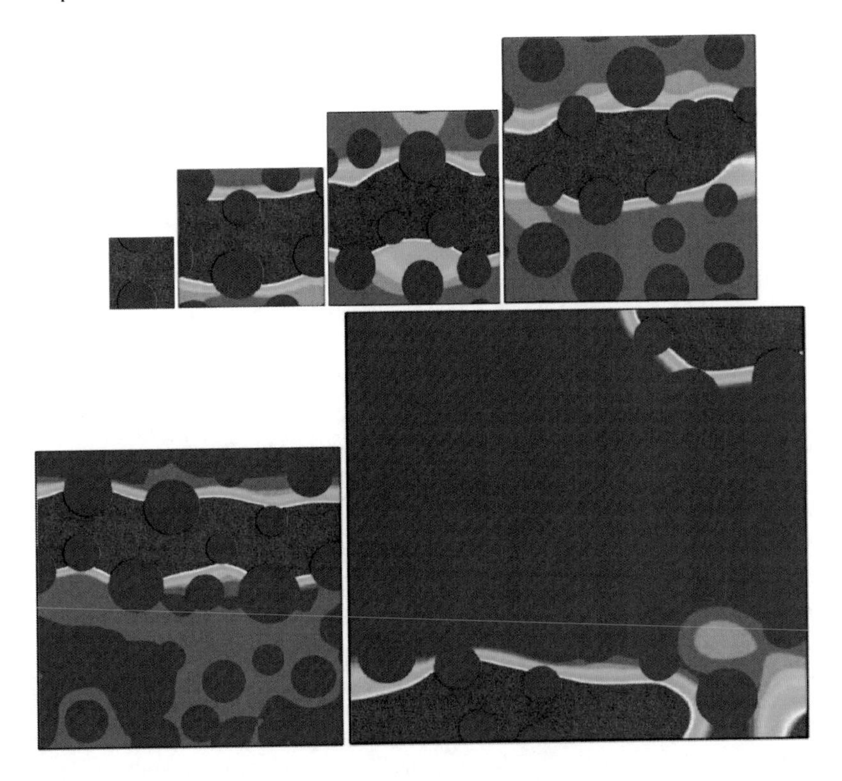

Fig. 9.55 Distribution of non-local softening strain measure for various specimen sizes and stress-strain curves of Fig. 9.53b using standard averaging procedure (characteristic length l_c=1.5 mm, aggregate density ρ=30%) (Skarżyński and Tejchman 2012)

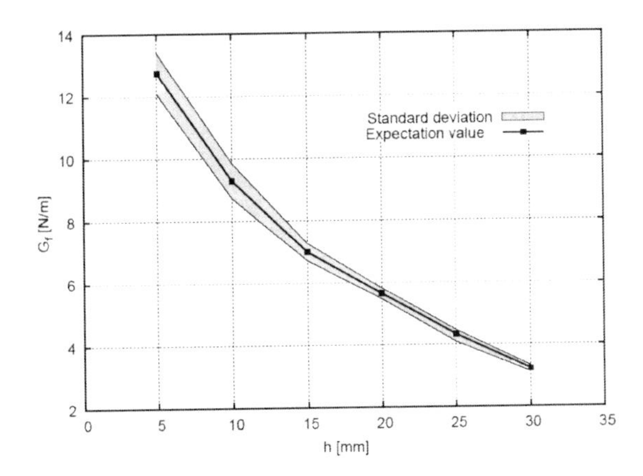

Fig. 9.56 Expected value and standard deviation of tensile fracture energy G_f versus specimen height h using standard averaging (aggregate density ρ=30%) (Skarżyński and Tejchman 2012)

Localized zone averaging approach

Recently, the existence of RVE for softening materials was proved (based on Hill's averaging principle) for cohesive and adhesive failure by deriving a traction-separation law (for a macro crack) instead of a stress-strain relation from microscopic stresses and strains (Verhoosel et al. 2010a, 2010b). This was indicated by the uniqueness (regardless of a micro sample size) of a macro traction-separation law which was obtained by averaging responses along propagating micro discrete cracks. Prompted by this approach and the fact that a localized zone does not scale with the micro specimen size, Nguyen et al. (2010) proposed an approach where homogenized stress and strain were averaged over a localized strain domain in softening materials rather (which is small compared with the specimen size) than over the entire specimen. We used this method in this paper. In this approach, the homogenized stress and strain are

$$<\sigma>=\frac{1}{A_z}\int_{Az}\sigma_m dA_z \qquad \text{and} \qquad <\varepsilon>=\frac{1}{A_z}\int_{A_z}\varepsilon_m dA_z\,, \tag{9.5}$$

where A_z is the localized zone area and σ_m and ε_m are the meso-stress and meso-strain, respectively. The localized zone area A_z is determined on the basis of a distribution of the non-local equivalent strain measure $\bar{\varepsilon}$. As the cut-off value $\bar{\varepsilon}_{\min}=0.005$ is always assumed at the maximum mid-point value usually equal to $\bar{\varepsilon}_{\max}=0.007-0.011$. Thus, a linear material behaviour is simply swept out (which causes the standard stress-strain diagrams to be specimen size dependent), and an active material plastic response is solely taken into account.

Figure 9.57 presents the stress-strain relationships for various specimen sizes and two random aggregate distributions with the material constants from Tab. 9.5 (l_c=1.5 mm) for the calculated localized zones of Figs. 9.54 and 9.55. These stress-strain curves in a softening regime (for the unit cells larger than 10×10 mm^2) are in very good accordance with respect to their shape. In this case, the statistically representative volume element exists and is equal to 15×15 mm^2.

Figure 9.58 presents the expectation value and standard deviation of the tensile fracture energy G_f versus the specimen height h for 3 different realizations. The integration limits were a_1=0 and a_2=0.004. The fracture energy decreases with increasing specimen size approaching an asymptote when the cell size is 15×15 mm^2. Thus, the homogenized stress-strain relationships obtained are objective with respect to the micro sample size. RVE does not represent the entire material in its classical meaning, but the material in a localized zone.

Varying characteristic length approach

With increasing characteristic length, both specimen strength and width of a localized zone increase. On the other hand, softening decreases and material behaves more ductile. Taking these two facts into account, a varying characteristic length related to the reference specimen size (assumed as 15×15 mm^2 or 30×30 mm^2) is introduced (to scale the width of a localized zone with varying specimen height) according to the formula

$$l^v_c = l_c^{15\times 15} \times \frac{h\ [\mathrm{mm}]}{15\ [\mathrm{mm}]} \tag{9.6}$$

or

$$l^v_c = l_c^{30\times 30} \times \frac{h\ [\mathrm{mm}]}{30\ [\mathrm{mm}]}, \tag{9.7}$$

where $l_c^{15\times 15}=l_c^{30\times 30}=1.5$ mm is a characteristic length for the reference unit cell 15×15 mm^2 or 30×30 mm^2 and h is the unit cell height. A larger unit cell than 30×30 mm^2 can be also used (the width of a localized zone in the reference unit cell cannot be too strongly influenced by boundary conditions, as e.g.. the cell size smaller than 10×10 mm^2). The characteristic length l^v_c is no longer a physical parameter related to non-local interactions in the damaging material, but an artificial parameter adjusted to the specimen size.

The stress-strain relationships for various specimen sizes and various characteristic lengths are shown in Figs. 9.59 and 9.60. A characteristic length varies between l_c=0.5 mm for the unit cell 5x5 mm^2 and l_c=3.0 mm for the unit cell 30×30 mm^2 according to Eq. 9.6, and between l_c=0.25 mm for the unit cell 5x5 mm^2 and l_c=1.5 mm for the unit cell 30×30 mm^2 according to Eq. 9.7. The width of a calculated localized zone (for the reference unit cell 15×15 mm^2) is approximately w_c=2 mm=4×l_c=8×s_{cm} (cell 5×5 mm^2), w_c=4 mm=4×l_c=16×s_{cm} (cell 10×10 mm^2), w_c=6 mm=4×l_c=24×s_{cm} (cell 15×15 mm^2), w_c=8 mm=4×l_c=32×s_{cm} (cell 20×20 mm^2), w_c=10 mm=4×l_c=40×s_{cm} (cell 25×25 mm^2) and w_c=12 mm=4×l_c=48×s_{cm} (cell 30×30 mm^2) (Figs. 9.61 and 9.62). The width of a calculated localized zone (for the reference unit cell 30×30 mm^2) is approximately w_c=1 mm=4×l_c=4×s_{cm} (cell 5×5 mm^2), w_c=2 mm=4×l_c=8×s_{cm} (cell 10×10 mm^2), w_c=3 mm=4×l_c=12×s_{cm} (cell 15×15 mm^2), w_c=4 mm=4×l_c=16×s_{cm} (cell 20×20 mm^2), w_c=5 mm=4×l_c=20×s_{cm} (cell 25×25 mm^2) and w_c=6 mm=4×l_c=24×s_{cm} (cell 30×30 mm^2) (Figs. 9.63 and 9.64). A localized zone is scaled with the specimen size. Owing to that the material does not lose its homogeneity and its response during softening is similar for the cell 15×15 mm^2 and larger ones. Thus, the size of the representative volume element is again equal to 15×15 mm^2.

The expected value and standard deviation of the unit fracture energy g_f=G_f/w_c versus the specimen height h are demonstrated in Fig. 9.65. With increasing cell size, the value of g_f stabilizes for the unit cell of 15×15 mm^2.

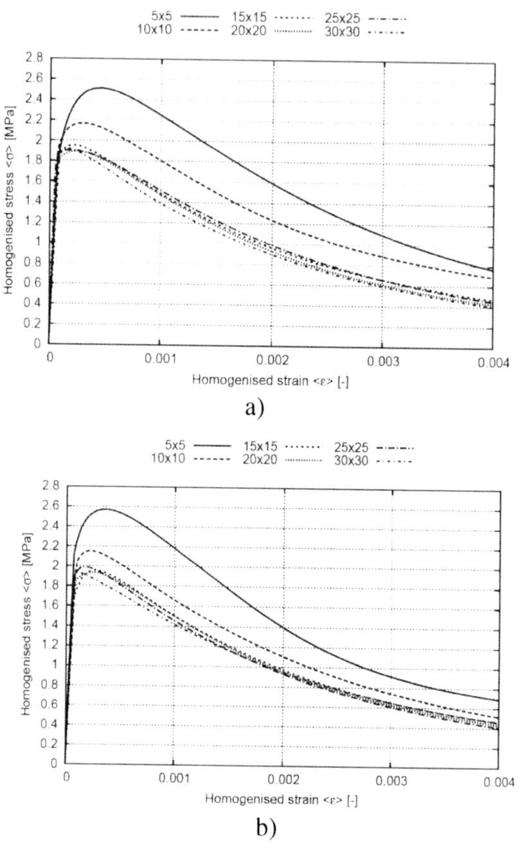

Fig. 9.57 Stress-strain curves for various sizes of concrete specimens and two different random distributions of aggregate (a) and (b) using localized zone averaging procedure (characteristic length l_c=1.5 mm, aggregate density ρ=30%) (Skarżyński and Tejchman 2012)

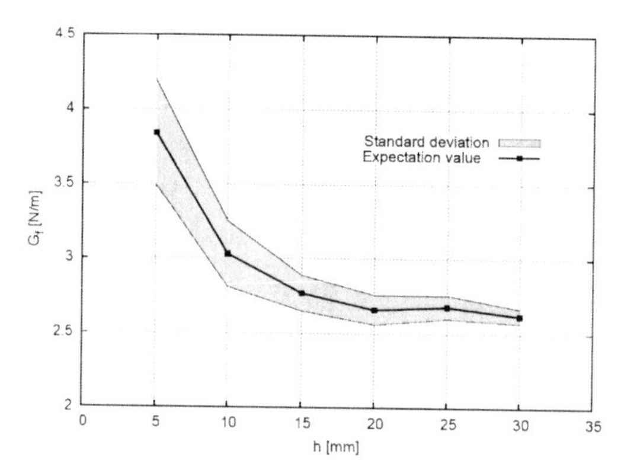

Fig. 9.58 Expected value and standard deviation of tensile fracture energy G_f versus specimen height h using localized zone averaging (aggregate density ρ=30%) (Skarżyński and Tejchman 2012)

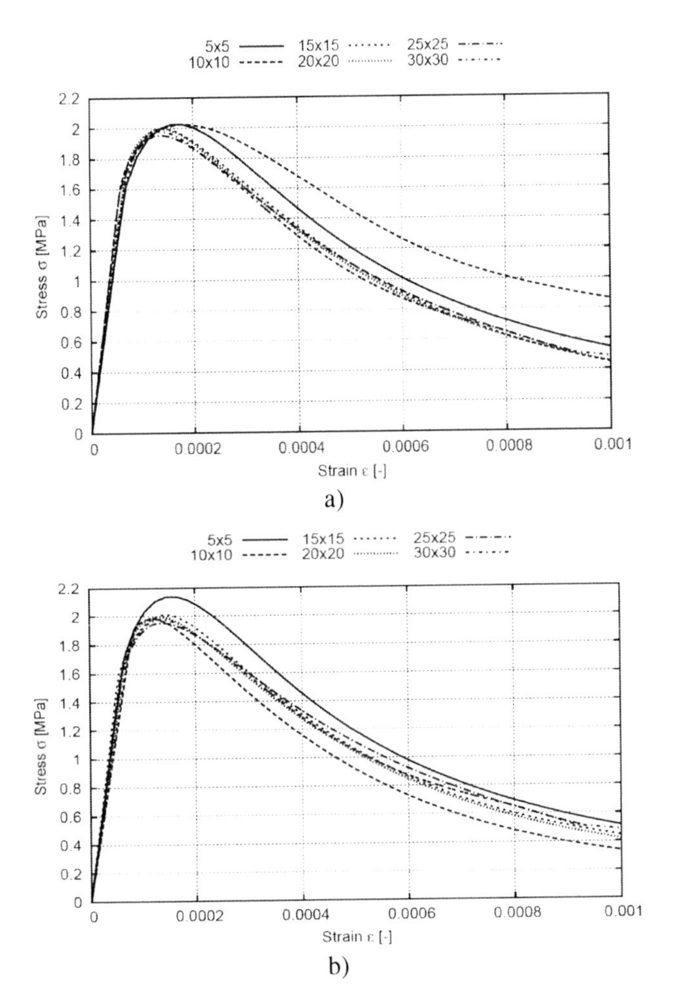

Fig. 9.59 Stress-strain curves for various sizes of concrete specimens and two different random distributions of aggregate (a) and (b) using varying characteristic length approach (reference unit size 15×15 mm^2, characteristic length according to Eq. 9.6, aggregate density ρ=30%) (Skarżyński and Tejchman 2012)

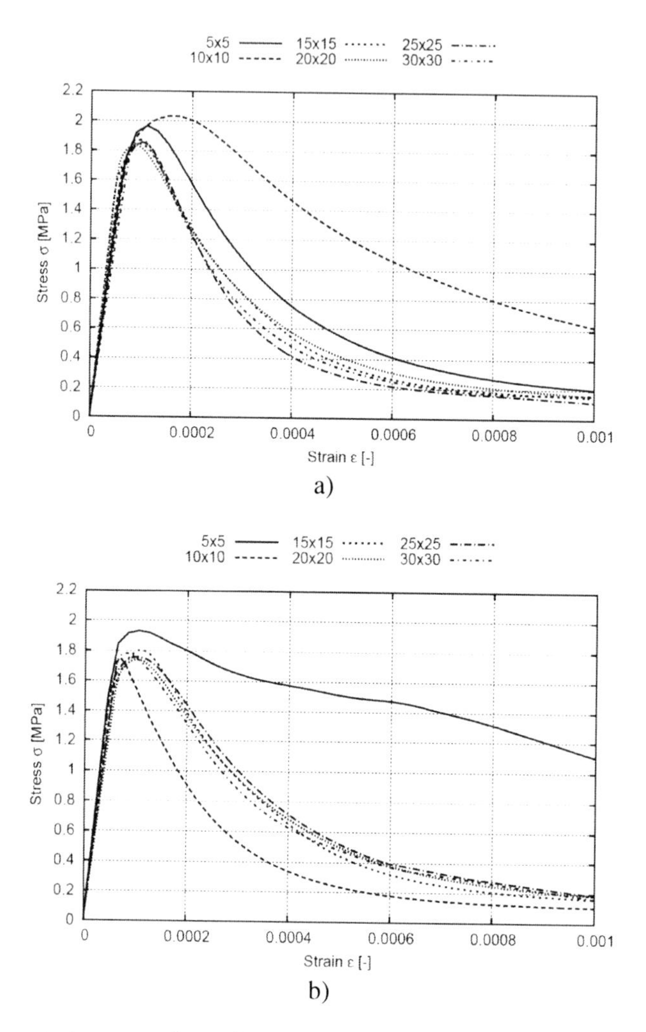

Fig. 9.60 Stress-strain curves for various sizes of concrete specimens and two different random distributions of aggregate (a) and (b) using varying characteristic length approach (reference unit size 30×30 mm^2, characteristic length according to Eq. 9.7, aggregate density ρ=30%) (Skarżyński and Tejchman 2012)

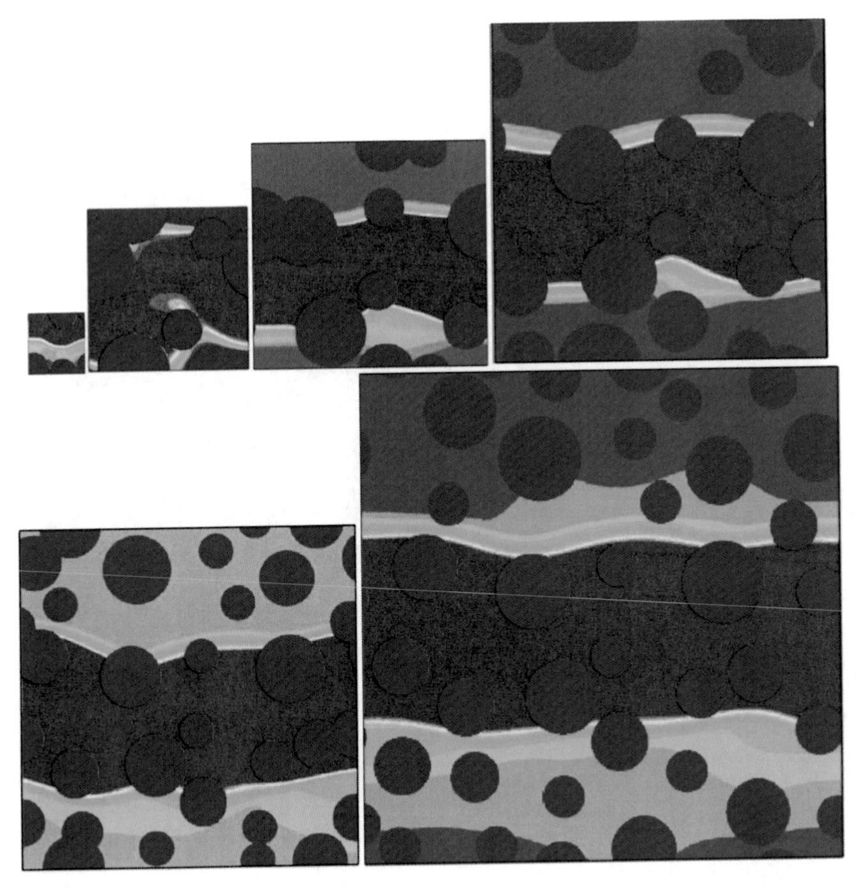

Fig. 9.61 Distribution of non-local softening strain measure for various specimen sizes and stress-strain curves from Fig. 9.59a using varying characteristic length approach (reference unit size 15×15 mm^2, characteristic length according to Eq. 9.6, aggregate density ρ=30%) (Skarżyński and Tejchman 2012)

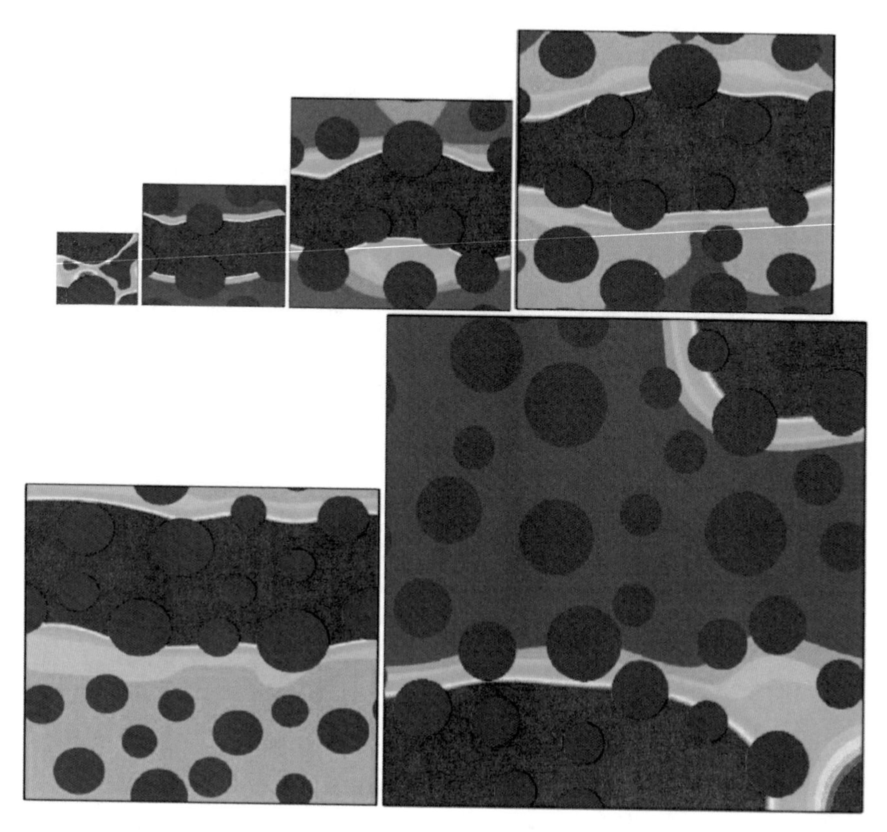

Fig. 9.62 Distribution of non-local softening strain measure for various specimen sizes and stress-strain curves from Fig. 9.59b using varying characteristic length approach (reference unit size 15×15 mm^2, characteristic length according to Eq. 9.6, aggregate density ρ=30%) (Skarżyński and Tejchman 2012)

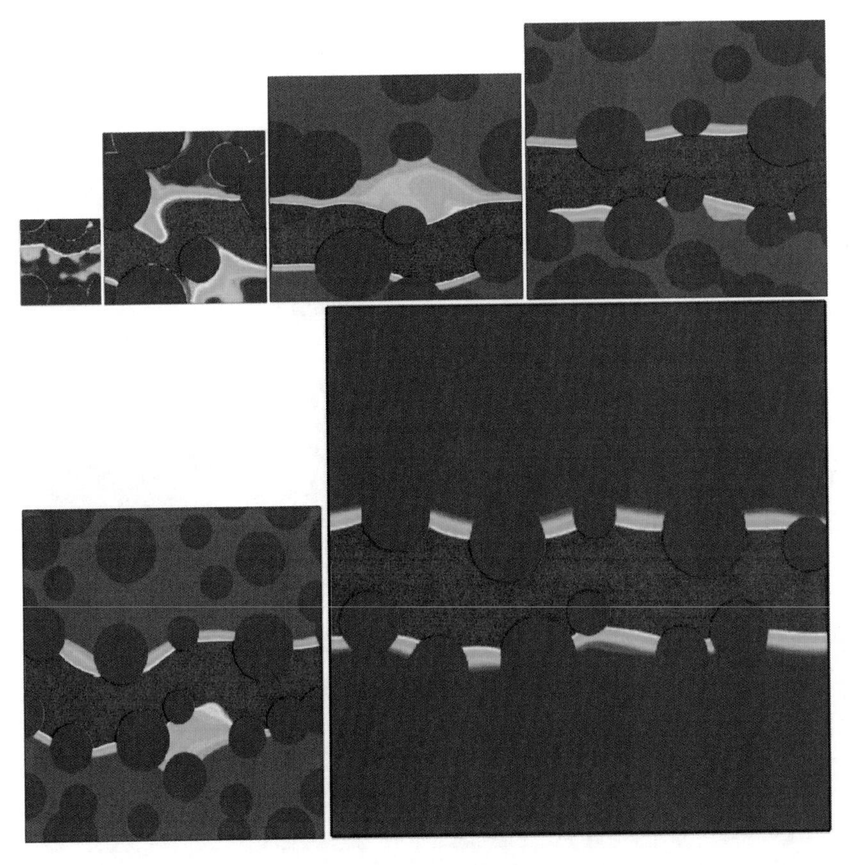

Fig. 9.63 Distribution of non-local softening strain measure for various specimen sizes and stress-strain curves from Fig. 9.60a using varying characteristic length approach (reference unit size 30×30 mm^2, characteristic length according to Eq. 9.7, aggregate density ρ=30%) (Skarżyński and Tejchman 2012)

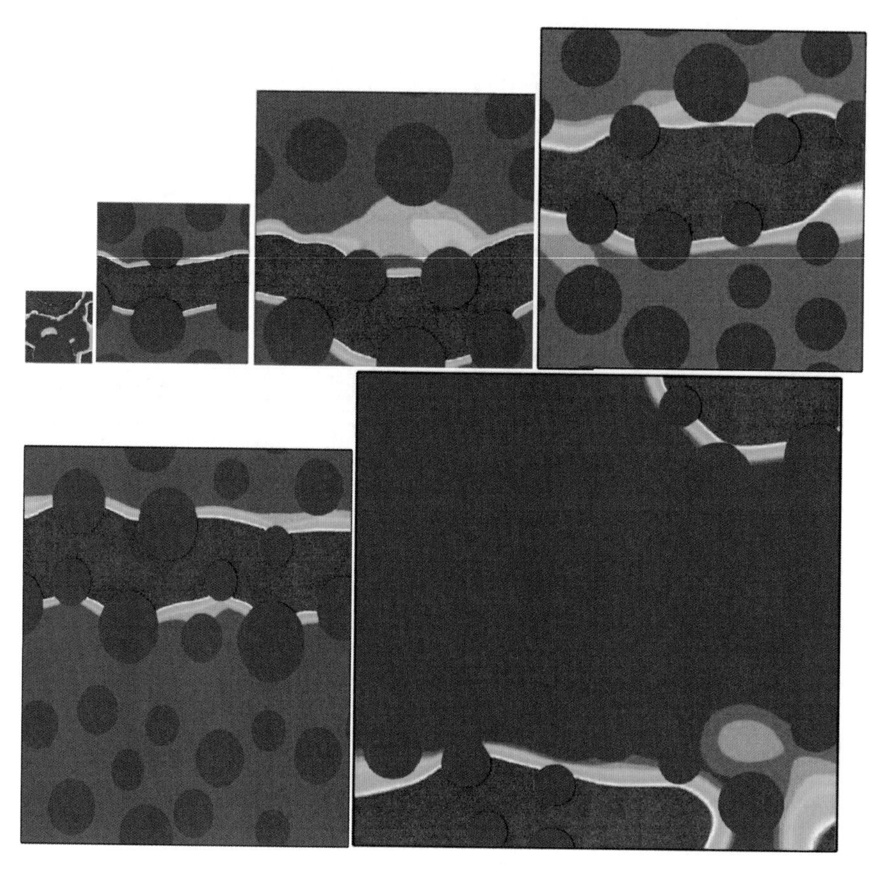

Fig. 9.64 Distribution of non-local softening strain measure for various specimen sizes and stress-strain curves from Fig. 9.60b using varying characteristic length approach (reference unit size 30×30 mm^2, characteristic length according to Eq. 9.7, aggregate density ρ=30%) (Skarżyński and Tejchman 2012)

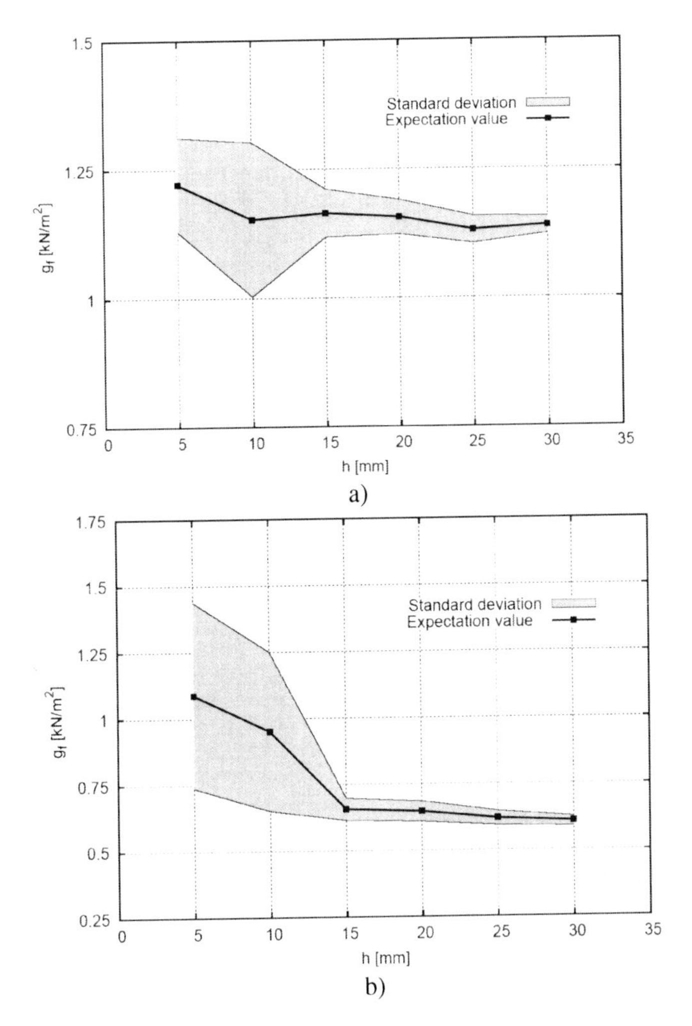

Fig. 9.65 Expected value and standard deviation of unit fracture energy g_f versus specimen height h using varying characteristic length approach: a) reference cell size 15×15 mm^2, b) reference cell size 30×30 mm^2 (aggregate density ρ=30%) (Skarżyński and Tejchman 2012)

The following conclusions can be drawn from our mesoscopic non-linear FE-investigations of strain localization in concrete under tensile loading:

- The 2D representative volume element (RVE) can be determined in quasi-brittle materials using both a localized zone averaging approach and a varying characteristic length approach. In the first case, the averaging is performed over the localized domain rather than over the entire domain, by which the material contribution is swept out. In the second case, the averaging is performed over the entire domain with a characteristic length of micro-structure being scaled with the specimen size. In both cases, convergence of the stress-strain diagrams for different RVE sizes of a softening material is obtained for tensile loading. The size of a two-dimensional statistically representative volume element is approximately equal to 15×15 mm^2.

- An isotropic continuum damage model with non-local softening is able to capture the mechanism of evolution of strain localization in concrete under tensile loading treated at the meso-scale as a heterogeneous three-phase material.
- Material micro-structure on the meso-scale has to be taken into account in calculations of strain localization to obtain a proper shape of a localized zone.
- The representative volume element (RVE) cannot be defined in quasi-brittle materials with a standard averaging approach (over the entire material domain) due to occurrence of a localized zone whose width is not scaled with the specimen size. The shape of the stress-strain curve depends on the specimen size beyond the elastic region. RVE can be found in homogeneous materials only.
- The 2D representative volume element (RVE) can be determined in quasi-brittle materials using both a localized zone averaging approach and a varying characteristic length approach. In the first case, the averaging is performed over the localized domain rather than over the entire domain, by which the material contribution is swept out. In the second case, the averaging is performed over the entire domain with a characteristic length of micro-structure being scaled with the specimen size. In both cases, convergence of the stress-strain diagrams for different RVE sizes of a softening material is obtained for tensile loading. The size of a two-dimensional statistically representative volume element is approximately equal to 15×15 mm^2.
- The calculated strength, width and geometry of the localized zone are in a satisfactory agreement with experimental measurements when a characteristic length is about 1.5 mm.
- The load-displacement evolutions strongly depend on material parameters assumed for separated concrete phases and a statistical distribution of aggregate. The ultimate beam strength certainly increases with increasing characteristic length, aggregate stiffness, mean aggregate size and decreasing ITZ thickness. It may increase with increasing volume fraction of aggregate. It is also dependent upon aggregate shape.
- Tensile damage is initiated first in the ITZ region. This region is found to have a significant impact on the fracture behaviour and strength of concrete.
- The width of a localized zone increases with increasing characteristic length and decreasing aggregate volume. It may increase if it propagates through weak grains. It is not affected by the aggregate size, aggregate shape, stochastic distribution, ITZ thickness and notch size. The width of a calculated localized zone above the notch changes from about $2\times l_c$ (ρ=60%) up to $4\times l_c$ (ρ=30%) at l_c=1.5 mm. If l_c=5 mm, the width of a calculated localized zone above the notch changes from $2.8\times l_c$ (ρ=60%) up to $3.5\times l_c$ (ρ=30%).
- The calculated increment rate of the width of a localized zone is similar as in experiments.

- Concrete softening is strongly influenced by the statistical distribution of aggregate, characteristic length, volume fraction of aggregate, aggregate shape, aggregate stiffness and ITZ thickness.
- Beams strength increases with increasing characteristic length, aggregate density and aggregate roughness and decreasing beam height. It depends also on the aggregate distribution.
- The localized zone above the notch is strongly curved with l_c=1.0-2.5 mm.

The mesoscopic modelling allows for a better understanding of the mechanism of strain localization. However, it cannot be used for engineering problems due to a long computation time and too small knowledge on both properties of meso-phases in concrete and a stochastic distribution of aggregate which are of a major importance. A direct link between a characteristic length and material micro-structure remains still open. To realistically describe the entire fracture process in concrete, a combined continuous-discontinuous numerical approach has to be used.

References

Bažant, Z.P., Novak, D.: Stochastic models for deformation and failure of quasibrittle structures: recent advances and new directions. In: Bicanić, N., de Borst, R., Mang, H., Meschke, G. (eds.) Computational Modelling of Concrete Structures, pp. 583–598 (2003)

Bažant, Z.P., Oh, B.H.: Crack band theory for fracture of concrete. Materials and Structures RILEM 16(93), 155–177 (1983)

Bažant, Z.P., Pijauder-Cabot, G.: Measurement of characteristic length of non-local continuum. ASCE Journal of Engineering Mechanics 115(4), 755–767 (1989)

Bažant, Z., Planas, J.: Fracture and size effect in concrete and other quasi-brittle materials. CRC Press LLC (1998)

Belytschko, T., Moes, N., Usui, S., Parimi, C.: Arbitrary discontinuities in finite elements. International Journal for Numerical Methods in Engineering 50(4), 993–1013 (2001)

Belytschko, T., Gracie, R., Ventura, G.: A review of extended/generalized finite element methods for material modeling. Modelling and Simulation in Material Science and Engineering 17(4), 1–24 (2009)

Bobiński, J., Tejchman, J.: Numerical simulations of localization of deformation in quasi-brittle materials with non-local softening plasticity. Computers and Concrete 1(4), 433–455 (2004)

Carol, I., López, C.M., Roa, O.: Micromechanical analysis of quasi-brittle materials using fracture-based interface elements. International Journal for Numerical Methods in Engineering 52(12), 193–215 (2001)

Donze, F.V., Magnier, S.A., Daudeville, L., Mariotti, C.: Numerical study of compressive behaviour of concrete at high strain rates. Journal for Engineering Mechanics 125(10), 1154–1163 (1999)

Drugan, W.J., Willis, J.R.: A micromechanics-based nonlocal constitutive equations and estimates of representative volume element size for elastic composites. Journal of the Mechanics and Physics of Solids 44(4), 497–524 (1996)

Du, C.B., Sun, L.G.: Numerical simulation of aggregate shapes of two dimensional concrete and its application. Journal of Aerospace Engineering 20(3), 172–178 (2007)

Eckardt, S., Konke, C.: Simulation of damage in concrete structures using multi-scale models. In: Bicanić, N., de Borst, R., Mang, H., Meschke, G. (eds.) Computational Modelling of Concrete Structures, EURO-C, pp. 77–89. Taylor and Francis (2006)

Evesque, P.: Fluctuations, correlations and representative elementary volume (REV) in granular materials. Poudres et Grains 11(1), 6–17 (2000)

Geers, M.G.D., Kouznetsova, V.G., Brekelmans, W.A.M.: Multi-scale computational homogenization: trends and challenges. Journal of Computational and Applied Mathematics 234(7), 2175–2182 (2010)

Gitman, I.M.: Representative Volumes and multi-scale modelling of quasi-brittle materials, PhD Thesis. Delft University of Technology (2006)

Gitman, I.M., Askes, H., Sluys, L.J.: Representative volume: existence and size determination. Engineering Fracture Mechanics 74(16), 2518–2534 (2007)

Gitman, I.M., Askes, H., Sluys, L.J.: Coupled-volume multi-scale modelling of quasi-brittle material. European Journal of Mechanics A/Solids 27(3), 302–327 (2008)

He, H., Guo, Z., Stroeven, P., Stroeven, M., Sluys, L.J.: Influence of particle packing on elastic properties of concrete. In: Proc. First International Conference on Computational Technologies in Concrete Structures (CTCS 2009), Jeju, Korea, pp. 1177–1197 (2009)

He, H.: Computational modeling of particle packing in concrete. PhD Thesis. Delft University of Technology (2010)

Hill, R.: Elastic properties of reinforced solids: some theoretical principles. Journal of the Mechanics and Physics of Solids 11(5), 357–372 (1963)

Kaczmarczyk, L., Pearce, C.J., Bicanic, N., de Souza Neto, E.: Numerical multiscale solution strategy for fracturing heterogeneous materials. Computer Merhods in Applied Mechanics and Engineering 199(17-20), 1100–1113 (2010)

Kanit, T., Forest, S., Galliet, I., Mounoury, V., Jeulin, D.: Determination of the size of the representative volume element for random composites: statistical and numerical approach. International Journal of Solids and Structures 40(13-14), 3647–3679 (2003)

Kim, S.M., Abu Al-Rub, R.K.: Meso-scale computational modelling of the plastic-damage response of cementitious composites. Cement and Concrete Research 41(3), 339–358 (2011)

Kouznetsova, V.G., Geers, M.G.D., Brekelmans, W.A.M.: Size of Representative Volume Element in a second-order computational homogenization framework. International Journal for Multiscale Computational Engineering 2(4), 575–598 (2004)

Kozicki, J., Tejchman, J.: Modeling of fracture processes in concrete using a novel lattice model. Granular Matter 10(5), 377–388 (2008)

Le Bellĕgo, C., Dube, J.F., Pijaudier-Cabot, G., Gerard, B.: Calibration of nonlocal damage model from size effect tests. European Journal of Mechanics A/Solids 22(1), 33–46 (2003)

Lilliu, G., van Mier, J.G.M.: 3D lattice type fracture model for concrete. Engineering Fracture Mechanics 70(7-8), 927–941 (2003)

Marzec, I., Bobiński, J., Tejchman, J.: Simulations of crack spacing in reinforced concrete beams using elastic-plastic and damage with non-local softening. Computers and Concrete 4(5), 377–403 (2007)

Mihashi, H., Nomura, N.: Correlation between characteristics of fracture process zone and tension softening properties of concrete. Nuclear Engineering and Design 165(3), 359–376 (1996)

Nguyen, V.P., Lloberas Valls, O., Stroeven, M., Sluys, L.J.: On the existence of representative volumes for softening quasi-brittle materials. Computer Methods in Applied Mechanics and Engineering 199(45-48), 3028–3038 (2010)

Nielsen, A.U., Montiero, P.J.M., Gjorv, O.E.: Estimation of the elastic moduli of lightweight aggregate. Cement and Concrete Research 25(2), 276–280 (1995)

Pijaudier-Cabot, G., Bažant, Z.P.: Nonlocal damage theory. Journal of Engineering Mechanics ASCE 113(10), 1512–1533 (1987)

Sengul, O., Tasdemir, C., Tasdemir, M.A.: Influence of aggregate type on mechanical behaviour of normal- and high-strength concretes. ACI Materials Journal 99(6), 528–533 (2002)

Shahbeyk, S., Hosseini, M., Yaghoobi, M.: Mesoscale finite element prediction of concrete failure. Computational Materials Science 50(7), 1973–1990 (2011)

Skarżyński, L., Tejchman, J.: Mescopic modeling of strain localization in concrete. Archives of Civil Engineering LV(4), 521–540 (2009)

Skarżyński, L., Tejchman, J.: Calculations of fracture process zones on meso-scale in notched. concrete beams subjected to three-point bending. European Journal of Mechanics A/Solids 29(4), 746–760 (2010)

Skarżynski, L., Syroka, E., Tejchman, J.: Measurements and calculations of the width of the fracture process zones on the surface of notched concrete beams. Strain 47(s1), 319–322 (2011)

Skarżyński, L., Tejchman, J.: Determination of representative volume element in concrete under tensile deformation. Computers and Concrete 1(9), 35–50 (2012)

Syroka, E., Tejchman, J.: Experimental investigations of size effect in reinforced concrete beams without shear reinforcement. Internal Report. Gdańsk University of Technology (2011)

van der Sluis, O.: Homogenisation of structured elastoviscoplastic solids. PhD Thesis. Technical University Eindhoven (2001)

van Mier, J.G.M., Schlangen, E., Vervuurt, A.: Latice type fracture models for concrete. In: Mühlhaus, H.-B. (ed.) Continuum Models for Material and Microstructure, pp. 341–377. John Wiley & Sons (1995)

van Mier, J.G.M.: Microstructural effects on fracture scaling in concrete, rock and ice. In: Dempsey, J.P., Shen, H.H. (eds.) IUTAM Symposium on Scaling Laws in Ice Mechanics and Ice Dynamics, pp. 171–182. Kluwer Academic Publishers (2000)

Verhoosel, C.V., Remmers, J.J.C., Gutierrez, M.A.: A partition of unity-based multiscale approach for modelling fracture in piezoelectric ceramics. International Journal for Numerical Methods in Engineering 82(8), 966–994 (2010a)

Verhoosel, C.V., Remmers, J.J.C., Gutieerrez, M.A., de Borst, R.: Computational homogenization for adhesive and cohesive failure in quasi-brittle solids. International Journal for Numerical Methods in Engineering 83(8-9), 1155–1179 (2010b)

White, D.J., Take, W.A., Bolton, M.D.: Soil deformation measurement using particle image velocimetry (PIV) and photogrammetry. Geotechnique 53(7), 619–631 (2003)

Chapter 10
Final Conclusions and Future Research Directions

The book analyzes quasi-static fracture in plain concrete and reinforced concrete by means of constitutive models formulated within continuum mechanics. A continuous and discontinuous modelling approach was used. Using a continuous approach, analyses were performed using a finite element method and four different continuum concrete models: enhanced isotropic elasto-plastic, enhanced isotropic damage, enhanced anisotropic smeared crack and enhanced coupled elasto-plastic-damage model. The models were equipped with a characteristic length of micro-structure by means of a non-local and a second-gradient theory, so they could describe the formation of localized zones with a certain thickness and spacing and a related deterministic size effect. FE results converged to a finite size of localized zones via mesh refinement. In addition, numerical results of cracks in plain concrete using a discontinuous approach including cohesive (interface) elements and XFEM were presented which were also properly regularized. Numerical results were compared with corresponding laboratory tests from the scientific literature and own tests.

The following the most important conclusions can be derived on the basis of our quasi-static FE simulations of plain concrete specimens under monotonic and cyclic loading and of reinforced concrete specimens under monotonic loading:

Plain Concrete

- The calculations on strain localization in concrete demonstrate that conventional constitutive continuum models suffer from a mesh-dependency when material softening is included. The thickness and inclination of localized zones inside specimens, and load-displacement diagram in a post-peak regime depend strongly upon the mesh discretization. In turn, the continuum models enhanced by a characteristic length of micro-structure cause a full regularisation of the boundary value problem. Numerical results converge to a finite size of the strain localization upon mesh refinement. The load-displacement curves are similar. The effect of the mesh alignment on the inclination of localized zones is negligible.

J. Tejchman, J. Bobiński: Continuous & Discontinuous Modelling of Fracture, SSGG, pp. 407–412.
springerlink.com

• A choice of a suitable local state variable for non-local averaging strongly depends on the model used. It should be carefully checked to avoid problems with non-sufficient regularization.
• The representative volume element (RVE) cannot be defined in quasi-brittle materials with a standard averaging approach (over the entire material domain) due to occurrence of a localized zone whose width is not scaled with the specimen size. The shape of the stress-strain curve depends on the specimen size beyond the elastic region.
• The 2D representative volume element (RVE) can be determined in quasi-brittle materials using both a localized zone averaging approach and a varying characteristic length approach. In the first case, the averaging is performed over the localized domain rather than over the entire domain, by which the material contribution is swept out. In the second case, the averaging is performed over the entire domain with a characteristic length of micro-structure being scaled with the specimen size. In both cases, convergence of the stress-strain diagrams for different RVE sizes of a softening material is obtained for tensile loading. The size of a two-dimensional statistically representative volume element under tensile loading is approximately equal to 15×15 mm^2.
• The material micro-structure on the meso-scale has to be taken into account in calculations of strain localization to obtain a proper shape of localized zones. The load-displacement evolutions strongly depend on material parameters assumed for separated concrete phases and a statistical distribution of aggregate. The ultimate material strength certainly increases with increasing characteristic length, aggregate stiffness, mean aggregate size and decreasing ITZ thickness. It may increase with increasing volume fraction of aggregate. It is also dependent upon aggregate shape. The calculated strength, width and geometry of the localized zone are in a satisfactory agreement with experimental measurements when a characteristic length is about 1.5 mm. Tensile damage is initiated first in the ITZ region. This region is found to have a significant impact on the fracture behaviour and strength of concrete. The width of a localized zone increases with increasing characteristic length and decreasing aggregate volume. It may increase if it propagates through weak grains. It is not affected by the aggregate size, aggregate shape, stochastic distribution, ITZ thickness and notch size. The width of a calculated localized zone above the notch in fine-grained concrete changes from about $2\times l_c$ (ρ=60%) up to $4\times l_c$ (ρ=30%) at l_c=1.5 mm. Concrete softening is strongly influenced by the statistical distribution of aggregate, characteristic length, volume fraction of aggregate, aggregate shape, aggregate stiffness and ITZ thickness. The material strength increases with increasing characteristic length, aggregate density and aggregate roughness and decreasing specimen height. It depends also on the aggregate distribution.
• The enhanced continuous models show a different capability to capture localized zones during complex shear-extension tests. In general, the elasto-plastic model with the Rankine's failure criterion is the most effective among continuous models. The usefulness of an isotropic damage model strongly depended on the

definition of the equivalent strain measure. The influence of the material description in a tensile-compression regime has to be taken into account to improve results. A smeared crack model is not able to reproduce curved cracks (the worst results are obtained with a rotating smeared crack model). An approach with cohesive elements provides the best approximation of experiments under a mixed shear-tension mode. However, the crack propagates along the mesh edges only. The calculations using XFEM are also realistic if the crack evolution is not blocked when using an conventional criterion based on a direction of the maximum principal stress).

• The width of the localized zone is larger in FE-analyses with a damage model than with an elasto-plastic model using a similar characteristic length of micro-structure during extension and bending. The width of localized zones grows during the entire deformation process within damage mechanics, whereas it is almost constant within elasto-plasticity. It increases with increasing characteristic length.

• The normalized strength of concrete specimens increases with increasing characteristic length and non-locality parameter and decreasing specimen size. The larger the ratio between a characteristic length of micro-structure and the specimen size, the higher usually both the specimen strength and the ductility of the specimen during extension and bending.

• The size and number of imperfections, and the distance between them do not influence the thickness and inclination of localized zones.

• An enhanced coupled elasto-plastic-damage describing plastic strains and stiffness degradation in tension and compression and stiffness recovery is able to reproduce the concrete behaviour during quasi-static cyclic loading. Its drawback is no clear distinction between elastic, plastic and damage strain rates, and a relatively large number of material constants to be calibrated. Most of material constants may be calibrated independently with a monotonic uniaxial compression and tension (bending) test. Cyclic simple tests are needed to calibrate damage scale factors.

• The deterministic size effect (nominal strength decreases with increasing specimen size) is very pronounced in notched and unnotched concrete under bending (it is stronger in notched beams). It is caused by occurrence of a straight tensile localized zone with a certain width. Therein the material ductility increases with decreasing specimen size. A pronounced snap-back behaviour occurs for large-size beams. The deterministic size effect can be observed on specimens under uniaxial compression in presence of non-symmetric notches only. A deterministic size effect can be also captured by the cohesive crack model and XFEM.

• The statistical size effect is strong in unnotched concrete beams and negligible in notched concrete beams. The larger the beam, the stronger is the influence of a stochastic distribution on the nominal strength due to the presence of a larger number of local weak spots (i.e. the mean stochastic bearing capacity is always smaller than the deterministic one). The stochastic bearing capacity is larger in some realizations with small and medium-large beams than the deterministic value. This position of the localized zone is connected with the distribution and magnitude of the tensile strength in a localized zone at peak and the magnitude of the horizontal normal stress due to bending.

• The calculated combined deterministic-statistical size effect is in agreement with the size effect law by Bažant and by Carpintieri for unnotched beams and the size effect law by Bažant for notched beams (in the considered size range). Our numerical results with beams match well the combined deterministic-stochastic size effect law by Bažant with the Weibull modulus equal to 24-48. The fractality is not needed to induce a size effect (it can contribute to its certain refinement but not to its replacement). The size effect model by Bažant has physical foundations and can be introduced into design codes.

Reinforced concrete

• The normalized strength of reinforced concrete specimens increases with increasing reinforcement ratio, characteristic length, non-locality parameter, initial stiffness of the bond-slip, tensile and compressive fracture energy and confining compressive pressure and decreasing specimen size. The width of localized zones increases with increasing characteristic length and non-locality parameter. It increases insignificantly with initial bond stiffness. It does not depend on the reinforcement ratio, shape of the softening curve, distribution of tensile strength, cross-section size and compressive confining pressure. The spacing of localized zones increases with increasing characteristic length, non-locality parameter and softening modulus, and decreasing reinforcement ratio, fracture energy and initial bond stiffness. It does not depend on the distribution of the tensile strength and stirrups. It is also not affected by the type of the bond-slip.

• For reinforced concrete elements, the enhanced continuum models provide a satisfactory agreement with experiments when the tensile failure mode is considered. In the case of a mixed shear-tension mode, the isotropic damage model cannot always reproduce the experimental geometry of localized zones (e.g. in short reinforced concrete beams).

• The results for reinforced concrete within a non-local and second-gradient approach are similar. The computational time of FE calculations within second-gradient mechanics is shorter by ca. 30%.

• The calculated and experimental spacing of localized zones in reinforced concrete elements is significantly smaller than this from analytical formulae.

• Linear elastic analysis is not always suitable for a proper engineering dimensioning of reinforced concrete structures (e.g. in the case of wall corners loaded by positive bending moments). In contrast, a non-linear elasto-plastic analysis is capable to capture strain localization. An internal diagonal crack in the corner region under positive bending moments has to be covered by reinforcement.

• The enhanced elasto-plastic model is the most realistic with respect to strain localization in reinforced concrete described at macro-level.

We will continue the FE calculations. Continuous models will be connected with a discontinuous one by means of XFEM (Moonen et al. 2008, Moonen 2009) to properly describe the entire fracture process within a single macroscopic framework. To obtain a better match with experiments, more refined continuum models should be used at macro-level to take into account strain localization. A more advanced concrete model in compression can be implemented in elasto-plasticity (e.g. model proposed by Menétrey and Willam (1995)). In addition, the evolution of internal friction and dilatancy against plastic deformation should be taken into account. In the case of damage mechanics, anisotropy will be considered. Within a smeared crack approach, plastic strains can be added (de Borst 1986). The calculations will be also carried out at meso-scale. The effect of aggregate shape, aggregate density, aggregate roughness, aggregate size, bond, cement particle size and reinforcement will be studied. The representative volume element (RVE) will be determined for shear and mixed mode loading. Two-scale approach will be used to link a meso-level with macro-level by means of a Coupled–Volume Approach (Gitman et al. 2008, Chapter 9) or a novel mixed computational homogenization technique, where at small-scale level concrete micro-structure will be considered (simulated using the discrete element method), and at macroscopic level, the finite element method will be used (Nitka et al. 2011, Nitka and Tejchman 2011). The up-scaling technique will take into account a discrete model at each Gauss point of the FEM mesh to derive numerically an overall constitutive response. The dynamic behaviour of plain concrete and reinforced concrete will be also simulated by taking into account inertial forces and elastic and plastic viscosity (Ožbolt et al. 2006, Pedersen et al, 2008, Pedersen 2010). The numerical results will be checked by own laboratory experiments.

Acknowledgement. Scientific research has been carried out as a part of the Project: "Innovative resources and effective methods of safety improvement and durability of buildings and transport infrastructure in the sustainable development" financed by the European Union.

References

de Borst, R.: Non-linear analysis of frictional materials. Phd Thesis. University of Delft (1986)

Gitman, I.M., Askes, H., Sluys, L.J.: Coupled-volume multi-scale modelling of quasi-brittle material. European Journal of Mechanics A/Solids 27(3), 302–327 (2008)

Menétrey, P., Willam, K.J.: Triaxial failure criterion for concrete and its generalization. ACI Structural Journal 92(3), 311–318 (1995)

Moonen, P., Carmeliet, J., Sluys, L.J.: A continuous-discontinuous approach to simulate fracture processes. Philosophical Magazine 88(28-29), 3281–3298 (2008)

Moonen, P.: Continuous-discontinuous modeling of hygrothermal damage processes in porous media. PhD Thesis. Univcersity of Delft (2009)

Nitka, M., Combe, G., Dascalu, C., Desrues, J.: Two-scale modeling of granular materials: a DEM-FEM approach. Granular Matter 13(3) (2011)

Nitka, M., Tejchman, J.: A two-scale numerical approach to granular systems. Archives of Civil Engineering LVII(3), 313–330 (2011)

Ožbolt, J., Rah, K.K., Mestrovia, D.: Influence of loading rate on concrete cone failure. International Journal of Fracture 139(2), 239–252 (2006)

Pedersen, R.R., Simone, A., Sluys, L.J.: An analysis of dynamic fracture in concrete with a continuum visco-elastic-plastic damage model. Engineering Fracture Mechanics 75(13), 3782–3805 (2008)

Pedersen, R.R.: Computational modelling of dynamic failure of cementitious materials. PhD Thesis. University of Delft (2010)

Symbols and Abbreviations

Symbol	Description
a_{min}	minimum aggregate diameter
a_{max}	maximum aggregate diameter
b	specimen width
c	concrete cover
C^e	elastic stiffness matrix
C^{ep}	elasto-plastic stiffness matrix
C^s_{ijkl}	secant stiffness matrix
C^{cr}_{ijkl}	secant cracked stiffness matrix
d	effective height
d_a^{max}	maximum aggregate size
D	damage parameter, specimen size
e	strain deviator, eccentricity
E	modulus of elasticity
E_c	modulus of elasticity of concrete
E_s	modulus of elasticity of steel
E_0	penalty (dummy) stiffness
f	failure function
f_t	tensile strength
f_c	compressive strength ,
f_{bc}	biaxial compressive strength
g	potential function
G	shear modulus
G_f	tensile fracture energy
G_c	compressive fracture energy
H_t, H_c	softening modulus in tension and compression
h	height
I_1	first stress tensor invariant
J_2	second deviatoric stress tensor invariant
J_3	third deviatoric stress tensor invariant

I_1^{ε}	first strain tensor invariant
J_2^{ε}	second deviatoric strain tensor invariant
J_3^{ε}	third deviatoric strain tensor invariant
k	ratio between compressive and tensile strength
l	length
l_c	characteristic length of micro-structure
m	non-locality parameter
r_{bc}^{σ}	ratio between uniaxial and biaxial compression strength
s	standard deviation
s_{ij}	deviatoric stress tensor
s_c, s	crack spacing, spacing of localized zones
p	mean stress
q	von Mises equivalent deviatoric stress
t_{eff}	effective traction
t	out-plane thickness
u	slip
w	localized zone width
V	volume
α	damage evolution law parameter
β	damage evolution law parameter, shear retention parameter
γ	volumetric weight
δ_{ij}	Kronecker delta
δ_{eff}	effective crack opening displacement
δ_n	normal crack opening displacement
δ_s	sliding crack opening displacement
ε^e, ε^e_{ij}	elastic strain tensor
ε^p, ε^{pl}_{ij}	plastic strain tensor
ε^{cr}, ε^{cr}_{ij}	crack strain
$\overline{\varepsilon}_{kl}$	non-local strain tensor
$\tilde{\varepsilon}$	equivalent strain measure
ε_{nu}	ultimate cracked strain in tension
ε_{su}	ultimate cracked strain in shear
ε_r	radial strain
ϕ_s	reinforcing bar diameter
φ	internal friction angle
η	cohesive zone approach parameter
κ_i	local hardening/softening parameters
$\overline{\kappa}_i$	non-local softening parameters
κ_0	initial threshold value of parameter κ

λ_{xi}, λ	decay coefficients, factor of proportionality, Lame's constant, slenderness
μ	Lame's constant
ν	Poisson's ratio
θ	mesh inclination
ρ	deviatoric axis, reinforcement ratio, weight
ρ_t	deviatoric length
σ_i	principal stresses
σ, σ_{ij}	stress tensor
σ_i^{eff}	effective stress tensor
σ_c	compressive yield stress
σ_t	tensile yield stress
σ_y	yield stress
τ_b	bond stress
ξ	hydrostatic axis
ω	weighting function
ψ	dilatancy angle
CMOD	crack mouth opening displacements
DIC	digital image correlation
FPZ	fracture process zone

Summary

Fracture process is a fundamental phenomenon in concretes. It is a major reason of damage under mechanical loading contributing to a significant degradation of the material strength leading to a total loss of load-bearing capacity. During fracture process first micro-cracks arise which change gradually into macro-cracks. Thus, the entire fracture process includes two main stages: narrow zones of intense deformation (where micro-cracks are created) and discrete macro-cracks. Strain localization can be captured by a continuous approach and macro-cracks by a discontinuous one. Usually, to describe the behavior of concrete, one type approach is used.

The book analyzes a quasi-static fracture process in concrete and reinforced concrete by means of constitutive models formulated within continuum mechanics. A continuous and discontinuous modeling approach was used. Using a continuous approach, numerical analyses were performed using a finite element method and four different enhanced continuum models: isotropic elasto-plastic, isotropic damage, anisotropic smeared crack and isotropic coupled elasto-plastic-damage one. The models were equipped with a characteristic length of micro-structure by means of a non-local and a second-gradient theory. So they could properly describe the formation of localized zones with a certain thickness and spacing and a related deterministic size effect. FE results converged to a finite size of localized zones via mesh refinement and boundary value problems became mathematically well-posed at the onset of strain localization. Using a discontinuous FE approach, numerical results of cracks using a cohesive crack model and XFEM were presented which were also properly regularized. Finite element analyses were performed with concrete elements under monotonic uniaxial compression, uniaxial tension, bending and shear-extension. Concrete beams under cyclic loading were simulated using a coupled elasto-plastic-damage approach as well. Numerical simulations were performed at macro- and meso-level. In the case of reinforced concrete specimens, FE calculations were carried out with bars, slender and short beams, columns, corbels and tanks. Tensile and shear failure mechanisms were studied. A stochastic and deterministic size effect was carefully investigated in plain concrete. Numerical results were compared with results from corresponding own and known in the scientific literature laboratory and full-scale tests.